OUTPUT, EMPLOYMENT, AND PRODUCTIVITY IN THE UNITED STATES AFTER 1800

This thirtieth volume in the National Bureau's Studies in Income and Wealth shows the interest of economists not only in improving existing statistical estimates but also in extending their observations of economic activities back into earlier historical periods. As the Introduction states, "with skill and ingenuity, with imagination and plain hard work, the authors have extended the measurement of related economic magnitudes for the entire economy back to the 1840's and for particular industries and even for individual firms back almost to their beginnings." This book continues the work of Volume 24 in the same series, *Trends in the American Economy in the Nineteenth Century*.

The discussion of sources of output growth is divided into five sec-

(continued on back flap)

*Output, Employment,
and Productivity
in the United States
After 1800*

National Bureau of Economic Research
Conference on Research in Income and Wealth

Output, Employment, and Productivity in the United States After 1800

Studies in Income and Wealth
Volume Thirty
by the Conference on Research
in Income and Wealth

Published by
NATIONAL BUREAU OF ECONOMIC RESEARCH
NEW YORK
Distributed by COLUMBIA UNIVERSITY PRESS
NEW YORK AND LONDON
1966

Relation of
National Bureau Directors to
Publications Reporting Conference Proceedings

Since the present volume is a record of conference proceedings, it has been exempted from the rules governing submission of manuscripts to, and critical review by, the Board of Directors of the National Bureau. It has, however, been reviewed and accepted for publication by the Director of Research.

Resolution adopted July 6, 1948,
as revised
November 21, 1949

Copyright © 1966 by the National Bureau of Economic Research
All Rights Reserved
L. C. Card: 65-15964

Printed in the United States of America

NATIONAL BUREAU OF ECONOMIC RESEARCH
1965

OFFICERS

Frank W. Fetter, *Chairman*
Arthur F. Burns, *President*
Theodore O. Yntema, *Vice-President*
Donald B. Woodward, *Treasurer*
William J. Carson, *Secretary*

Geoffrey H. Moore, *Director of Research*
Douglas H. Eldridge, *Executive Director*
Hal B. Lary, *Associate Director of Research*
Victor R. Fuchs, *Associate Director of Research*

DIRECTORS AT LARGE

Robert B. Anderson, *New York City*
Wallace J. Campbell, *Foundation for Cooperative Housing*
Erwin D. Canham, *Christian Science Monitor*
Solomon Fabricant, *New York University*
Marion B. Folsom, *Eastman Kodak Company*
Crawford H. Greenewalt, *E. I. du Pont de Nemours Company*
Gabriel Hauge, *Manufacturers Hanover Trust Company*
A. J. Hayes, *International Association of Machinists*
Walter W. Heller, *University of Minnesota*
Albert J. Hettinger, Jr., *Lazard Frères and Company*

H. W. Laidler, *League for Industrial Democracy*
Geoffrey H. Moore, *National Bureau of Economic Research*
Charles G. Mortimer, *General Foods Corporation*
J. Wilson Newman, *Dun & Bradstreet, Inc.*
George B. Roberts, *Larchmont, New York*
Harry Scherman, *Book-of-the-Month Club*
Boris Shishkin, *American Federation of Labor and Congress of Industrial Organizations*
George Soule, *South Kent, Connecticut*
Gus Tyler, *International Ladies' Garment Workers' Union*
Joseph H. Willits, *Langhorne, Pennsylvania*
Donald B. Woodward, *A. W. Jones and Company*

DIRECTORS BY UNIVERSITY APPOINTMENT

V. W. Bladen, *Toronto*
Francis M. Boddy, *Minnesota*
Arthur F. Burns, *Columbia*
Lester V. Chandler, *Princeton*
Melvin G. de Chazeau, *Cornell*
Frank W. Fetter, *Northwestern*
R. A. Gordon, *California*

Harold M. Groves, *Wisconsin*
Gottfried Haberler, *Harvard*
Maurice W. Lee, *North Carolina*
Lloyd G. Reynolds, *Yale*
Paul A. Samuelson, *Massachusetts Institute of Technology*
Theodore W. Schultz, *Chicago*

Willis J. Winn, *Pennsylvania*

DIRECTORS BY APPOINTMENT OF OTHER ORGANIZATIONS

Percival F. Brundage, *American Institute of Certified Public Accountants*
Nathaniel Goldfinger, *American Federation of Labor and Congress of Industrial Organizations*
Harold G. Halcrow, *American Farm Economic Association*
Murray Shields, *American Management Association*

Willard L. Thorp, *American Economic Association*
W. Allen Wallis, *American Statistical Association*
Harold F. Williamson, *Economic History Association*
Theodore O. Yntema, *Committee for Economic Development*

DIRECTORS EMERITI

Shepard Morgan, *Norfolk, Connecticut*
Jacob Viner, *Princeton, New Jersey*
N. I. Stone, *New York City*

RESEARCH STAFF

Moses Abramovitz
Gary S. Becker
William H. Brown, Jr.
Gerhard Bry
Arthur F. Burns
Phillip Cagan
Frank G. Dickinson
James S. Earley
Richard A. Easterlin
Solomon Fabricant

Albert Fishlow
Milton Friedman
Victor R. Fuchs
H. G. Georgiadis
Raymond W. Goldsmith
Jack M. Guttentag
Challis A. Hall, Jr.
Daniel M. Holland
Thor Hultgren
F. Thomas Juster

C. Harry Kahn
John W. Kendrick
Irving B. Kravis
Hal B. Lary
Robert E. Lipsey
Ruth P. Mack
Jacob Mincer
Ilse Mintz
Geoffrey H. Moore
Roger F. Murray

Ralph L. Nelson
G. Warren Nutter
Richard T. Selden
Lawrence H. Seltzer
Robert P. Shay
George J. Stigler
Norman B. Ture
Herbert B. Woolley
Victor Zarnowitz

This volume of Studies in Income and Wealth contains most of the papers presented at the joint sessions of the Economic History Association and the Conference on Research in Income and Wealth, held at Chapel Hill, North Carolina, in September 1963. We are indebted to the University of North Carolina for making its facilities available to us; to the Planning Committee, which consisted of William N. Parker (chairman), Richard A. Easterlin, and Raymond W. Goldsmith; to Dorothy S. Brady, who served as Conference Editor with the assistance of Richard A. Easterlin, William N. Parker, and Albert Fishlow; to Marie-Christine Culbert and Joan R. Tron, who prepared the volume for press; and to H. Irving Forman, who drew the charts.

Executive Committee, 1965-66

Robert J. Lampman, *Chairman*
Jack Alterman
Donald J. Daly
Evsey D. Domar
Graeme S. Dorrance
Richard A. Easterlin
Robert Eisner
F. Thomas Juster
Milton Moss
Charles L. Schultze
Mildred E. Courtney, *Secretary*

Contents

Introduction *Dorothy S. Brady* ix

CONSUMPTION, INVESTMENT, AND EMPLOYMENT

Gross National Product in the United States, 1834–1909
    Robert E. Gallman 3
Comment *Richard A. Easterlin* 76

Price Deflators for Final Product Estimates *Dorothy S. Brady* 91

Labor Force and Employment, 1800–1960 *Stanley Lebergott* 117

General Comment *Brinley Thomas* 205

OUTPUT OF FINAL PRODUCTS

The New England Textile Industry, 1825–60: Trends and Fluctuations *Lance E. Davis and H. Louis Stettler, III* 213
Comment *Paul F. McGouldrick* 239
Reply by *Davis and Stettler* 240

Building in Ohio Between 1837 and 1914 *Manuel Gottlieb* 243
Comment *Paul A. David* 281
Reply by *Gottlieb* 288

MINERALS AND FUELS

Development of the Major Metal Mining Industries in the United States from 1839 to 1909 *Orris C. Herfindahl* 293
Comment *Paul W. McGann* 347

The American Petroleum Industry *Harold F. Williamson, Ralph L. Andreano, and Carmen Menezes* 349
Comment *Arthur M. Johnson* 403

Some Aspects of Development in the Coal Mining Industry,
    1839–1918 *Vera F. Eliasberg*   405
Comment *Paul W. McGann*   436
Harold J. Barnett   437

POWER AND MACHINES

Growth and Diffusion of Power in Manufacturing, 1838–1919
    Allen H. Fenichel   443

Changing Production of Metalworking Machinery, 1860–1920
    Ross M. Robertson   479

Machine Tool Output, 1861–1910 *Duncan McDougall*   497

Comment on Robertson and McDougall *W. Paul Strassman*   518

SOURCES OF PRODUCTIVITY CHANGE

Productivity Growth in Grain Production in the United States,
    1840–60 and 1900–10 *William N. Parker and*
    Judith L. V. Klein   523
Comment *Glen T. Barton*   580

Productivity and Technological Change in the Railroad Sector,
    1840–1910 *Albert Fishlow*   583

Index   647

Introduction

DOROTHY S. BRADY

UNIVERSITY OF PENNSYLVANIA

This thirtieth volume of Studies in Income and Wealth presents papers on output, employment, and productivity in the United States after 1800. It manifests the continued interest of a group of scholars in extending coherent systems of observations on the magnitudes of economic activities back into ever earlier historical periods. With Volume 24 of these studies,[1] the 1840's became an outpost. Observations on some activities were extended as far back as 1790 and 1800, and some estimates for periods after 1869 were significantly improved in concept and in coverage. Like that earlier volume, the present one stems from a joint meeting of the Conference on Income and Wealth and the Economic History Association,[2] and it presents the results of more extensive and more intensive explorations. With skill and ingenuity, with imagination and plain hard work, the authors have extended the measurement of related economic magnitudes for the entire economy back to the 1840's and for particular industries and even for individual firms back almost to their beginnings.

The Significance of Measurement

The trouble with measurement is the deceptive simplicity, clarity, and finality of the results. The columns of numbers presenting observations of phenomena are all too neat, too ready for use in the discovery or testing of hypotheses. Despite the pages of footnotes, annotations, and evaluations, the work of measurement, to the uncritical reader, seems to be an operation something like the totaling of entries in an accountant's journal. The "quantifier" with his routines contributes to the advance of knowledge by producing the empirical estimates of magnitudes for

[1] *Trends in the American Economy in the Nineteenth Century*, Studies in Income and Wealth 24, Princeton for NBER, 1960.
[2] The meeting was held at the University of North Carolina, Chapel Hill, September 4 and 5, 1963. Four of the papers presented at the meeting were published in the *Journal of Economic History*.

the purposes of the analyst. Upon publication, the estimates are all too readily accepted as reliable observations.

If measurement were simply a matter of counting and totaling, there would be no way to determine which of two observations was the more credible. In reality, the process of measurement is an essential part of a theory and specifies how its propositions are to be related to experience. Inevitably, theories have two connections with the world of events, both of which are established by measurements. One is that the observations produced by measurement fill in the detail of a theory; the other is that the logical extensions of the theory must then be confronted with new sets of observations. The instruments of measurement determine how the nature of the observations is prescribed by the theory and how the logic of the theory is supported by the observations. With the progress of quantification, theories became bodies of propositions about observed magnitudes, and this development has made up much of the history of aggregative economic theory.

Observations that are the result of measurement relate the abstractions of theoretical concepts to documentary evidence of economic activity. The crude material that measurement converts into observations is found in the many kinds of records which preserve information about persons, events, activities, and transactions. Even with censuses and surveys, observers of the economic scene can have little influence on the nature of the information, because typically the primary data for their compilations have been recorded according to the customary practices or legal requirements of the time. Our knowledge of the world of economic activities comes through the kind of evidence contemporaries put on paper, and forgetting this medium leads to greatly oversimplified conceptions of the processes of measurement and the meaning of observations.

Knowledge of the source material necessarily becomes more important as the observer attempts to construct measurements of economic activities in remote periods. Methods and procedures pertaining to customs and usages of the present time must be adapted to the historical material in order to maintain consistency over time in the meaning of the measured magnitudes. Extending the province of measurement involves a greater understanding of the nature of the source materials and also the collection of information from sources not hitherto utilized. The papers in this volume show both expanded use of the comprehensive sources (the Censuses and other government collections of data) and considerable advance in the fusion of materials from diverse sources into estimates for sectors and segments of the economy.

The contributions of additional information gathered from "old" and "new" sources appear on several levels. The quality of the existing estimates for Census years or other benchmarks may be improved through the association of data coming from different sources. New annual series on elements of the aggregates may lead to substantial improvements in the estimates of year-to-year changes. Information on the rates and relationships between associated magnitudes that could be utilized in the estimation of aggregates or in their evaluation may expand the possibilities of interpolation or extrapolation. Thus measurements are accumulated, improved, and extended. The importance of this work is to be gauged by the questions raised about interrelationships of particular developments in the course of economic growth and change and the answers that are suggested.

Explanation in Measurement and Observation

The work of measurement in different investigations is carried on within the same framework of accounting, i.e., the same general scheme of identification and classification, and all studies use modern statistical concepts and techniques to weigh the historical evidence found in the various sources relating to the same time period. As studies accumulate, two or more based on different procedures and source materials may lead to estimates of the same magnitude, say, employment in industry i during year t. Such replications of observations are a sign of real progress in measurement, for the comparison of two estimates, particularly if they are discordant, provides a testing ground for the blend of explanatory relationships and source materials used in their derivation. Examination of the divergence between two or more estimates of the same magnitude, an operation known as reconciliation, may ultimately result in sharper tools of measurement and, consequently, in increased validity of the observations. As the scope of measurement is extended, reconciliation becomes an integral part of the work of estimation, for investigators must build on interpretations of the results of earlier work, sometimes their own.

The explanation of a difference between two estimates may be traced through careful accounting to identification of a missing element in the sources. If the size of the difference, at various points of time, accords with historical facts about the changing importance of the activity, the explanation provides a basis for estimation of its magnitude. Much of the work described in Gallman's paper on his measures of the gross national product for the Census years between 1834 and 1909 and in Lebergott's paper on his measures of employment depended on tracing differences between estimates to their explanation.

The methods for estimating missing observations are many, but the investigator's choice may be severely limited by available information. When observations made of related magnitudes in particular years form a complete set, missing observations for other years may be estimated through a projection of ratios or regression coefficients that quantify the association between two or more magnitudes. Clearly the form of the explanatory relationship has a determining effect on the estimates but generally only the comparison with estimates from other studies can give an indication of the range of possibilities. Studies of particular industries, like those in this volume by Mrs. Eliasberg on coal, by Herfindahl on metal mining, by Williamson, Andreano, and Menezes on oil, and by Fishlow on railroads, may lead to industrywide estimates of magnitudes—value of output, employment, investment—which, in procedures and sources, are independent of the estimates of the same magnitudes obtained for the studies of the entire economy. The interpretation of the differences between estimates will, very likely, focus on the nature of the relationships projected. Some references to divergent estimates and attempts at reconciliation will be found in the papers in this volume, but, since much of the work represents additions to the stock of measurements, thorough reconciliation and absorption into a general framework will require more time for study of the results.

Missing observations may be estimated from information on a segment of a total, as when estimates of the year-to-year changes in a total value for a group of commodities are based on the changes in the value of a subgroup or on changes in the total in one geographic area. Studies like those of Davis and Stettler on the New England textile industry, McDougall and Robertson on machine tools, and Gottlieb on construction in Ohio offer explanatory detail about changes from year to year that will ultimately lead to a better understanding of the variations in annual movements from region to region, among the commodities in a group, and among the firms in an industry. The use of indexes as interpolators and extrapolators can then be given a firmer empirical base than can be proffered at the present time.

When the general sources do not permit disaggregation below a certain level, those ratios which are based on primary aggregates, various input-output ratios, may be estimated from existing evidence on their magnitude in individual situations. In his study of productivity in cereal production, Parker assembled enough evidence on employment per acre from agricultural publications, manuscripts, and other sources to permit averages to be drawn and statistical analysis to be undertaken. Although the exploration of the sources requires much time and effort, this kind of

INTRODUCTION xiii

evidence will be sought increasingly as the work of measurement is focused on more and more detailed explanations of the course of economic change.

The Directions of Research

Do these empirical studies provide any guideposts that indicate where further research might make the greatest contributions? The general interest in national accounts has given some unity to work in progress and work in view. The order of studies undertaken is of little consequence as long as they fill in the records of the national wealth and income and add to related accounts. Even where investigations seem to have exhausted the general source material, there is still much that can and will be done as individuals pursue their own interests in the work of measurement. The results of their studies are certainly additions to knowledge and of great general value.

This second meeting of the Conference with the Economic History Association pointed to the need for other general frames of reference, in particular some kinds of input-output tables and geographic distributions. Investigations of the reasons for productivity changes and descriptions of diffusion processes can lead to a mass of isolated bits of information unless their integration is assured through some common perspective. Although the completion of an input-output table for even one year in the nineteenth century might take a long time, the design for the matrix could exert a salutary influence on the conduct of research in the immediate future. It could stimulate research in areas where information is wanting, but, of more significance, it could make investigators aware of the kinds of observations needed to relate the results of different studies. The plan for the matrix with a detailed examination of the problems of measurement would make a large contribution to the advance of knowledge by demonstrating one way to combine the results of different studies. Geographic distributions provide another frame that could be developed into a general scheme for tracing the interrelationships in the diffusion of new commodities and techniques of production.

The papers presented at the meeting, including those published in the *Journal of Economic History*, invite some synthesis in the form of proposals for analytic summaries that could enhance the value of new research on specific topics.[3] Changes in outputs and inputs between and within

[3] Paul H. Cootner, "The Role of Railroads in U.S. Economic Growth," Nathan Rosenberg, "Technological Change in the Machine Tool Industry, 1840–1910," and Peter Temin, "The Composition of Iron and Steel Products, 1869–1909," in the *Journal of Economic History*, December 1963, and Dorothy S. Brady, "Relative Prices in the Nineteenth Century," in *ibid.*, June 1964.

geographic areas are found in the Parker study of cereal production, in the Davis and Stettler study of cotton textiles, in the studies of machine tools by McDougall and Robertson, and in the study of power in manufacturing by Fenichel. These papers and others from the Conference including Rosenberg's on machine tools and mine on prices suggest that, in the design of an input-output frame, it would be necessary to use narrowly defined product classes and to make explicit provisions for the production of equipment and materials for "own use" in manufacturing and other activities. Fishlow's and Parker's analyses illustrate the potential value of the input columns for at least two years that mark off a period of migration and diffusion of new techniques. The problems encountered in drawing up the general plans could give the impetus toward work on historical measurements that would have the greatest likelihood of deepening our understanding of the processes of growth and change.

Consumption, Investment, and Employment

Gross National Product in the United States, 1834–1909

ROBERT E. GALLMAN

UNIVERSITY OF NORTH CAROLINA

This paper is a short summary of the main findings of a study of American national product since 1834. Concepts, estimating procedures, and a few tests of the estimates are described in the appendix. We have attempted to keep these matters out of the body of the paper. The reader is duly warned to look into the appendix before he makes use of the series in his own work.

Levels and Rates of Growth of National Product and National Product Per Capita

The findings described in this section are similar to the findings of two earlier papers. Therefore, they are discussed very quickly and the reader is referred to the earlier papers.[1]

NOTE: My obligations are many: to the Ford Foundation for the support of a Faculty Research Fellowship and research assistance; to the Development Fund of Ohio State University for support of a research leave of one academic quarter and for research assistance; to the following scholars who let me see and use their unpublished work and who were willing to discuss my work with me—Simon Kuznets, Dorothy Brady, Manuel Gottlieb, Robert Fogel, Paul David, Albert Fishlow, Martin Primack, Moses Abramovitz, William N. Parker; to Larry Shotwell, Stephen Hu, and Marz Garcia, who helped in the gathering and processing of data; to my wife, Jane, who read and summarized several monographs on early American industries for me; to William P. McGreevey, who helped to design and carry out the distribution estimates; to Barry Poulson, who assisted me ably for more than a year and who has provided me with data arising out of his study of U.S. industrial growth, 1809–39; and, finally, to the discussant, Richard A. Easterlin, who made important suggestions for the improvement of the paper. None of these people or institutions should be held liable for the results of my work.

[1] "Commodity Output, 1839–1899," *Trends in the American Economy in the Nineteenth Century*, Studies in Income and Wealth 24, Princeton for NBER, 1960; "Estimates of American National Product Made Before the Civil War," *Economic Development and Cultural Change*, April 1961.

LEVELS OF NATIONAL PRODUCT AND NATIONAL PRODUCT PER CAPITA CIRCA 1840 AND 1950

In 1839 the United States was a very young country, not long on the path of industrialization. One tends to think of it as still small and relatively weak; but, in fact, this is not correct. American aggregate product was probably smaller, but not much smaller, than the aggregate product of each of the two major world powers of the time, Great Britain and France (see Table 1). International comparisons are difficult to make today and the difficulties are multiplied when we attempt comparisons for a date well over a hundred years in the past. But there are several different ways of going about it and the results obtained are similar enough to yield useful conclusions.

First, Deane and Cole have recently estimated British national income for 1841.[2] Converting their figure into dollars by use of the par of exchange yields a result about one-quarter above the level of American GNP (less changes in inventories) for 1839. The American figure probably exceeds American national income by about 5 per cent. Consequently, British national income probably exceeded American national income by about 30 per cent.

Ideally, international comparisons should be made by valuing components of the national products to be compared by a common price system. Gilbert and Kravis have shown that this procedure may give results quite different from those of a comparison conducted through the rate of exchange. Seaman made estimates of this kind for 1839, which show that British national income exceeded the American by 50 per cent; French national income exceeded the American by nearly 60 per cent. Three points should be made about the Seaman estimates. First, Seaman's American estimate is very much lower than ours; the margin is too great to reflect only conceptual differences. Second, we derived the British estimate by multiplying Seaman's per capita figure for England and Wales by the 1841 population of Great Britain. But per capita product in England and Wales may very well have exceeded per capita product in Great Britain. Third, Seaman used American prices in all of his computations. When a comparison is made between two countries, one of which is more highly developed (industrialized) than the other, the use of the prices of the less-developed country to make the comparison will lead to results relatively favorable to the more highly developed country, and vice

---

[2] The next few paragraphs refer to Table 1. Citations should be sought in the sources of that table.

GROSS NATIONAL PRODUCT, 1834–1909

TABLE 1

COMPARISONS OF BRITISH, FRENCH, AND AMERICAN NATIONAL PRODUCTS AND NATIONAL PRODUCTS PER CAPITA, CIRCA 1840 AND 1950

	In Prices Circa 1840 (as per cent of U.S.)		In Prices of 1950 (as per cent of U.S.)	
	Great Britain	France	United Kingdom	France
1. National income, circa 1840				
A. Deane-Cole-Gallman	130			
B. Seaman	150	157		
2. GNP, circa 1840			85–112	123–156
3. GNP, 1950			18–22	12–15
4. National income per capita, circa 1840				
A. Deane-Cole-Gallman	120			
B. Seaman	140[a]	78		
5. GNP per capita, circa 1840			78–103	60–76
6. GNP per capita, 1950			53–63	42–53

Source

Line 1A: British national income, 1841: Phyllis Deane and W. A. Cole, *British Economic Growth, 1688-1959*, Cambridge, Eng., 1962, p. 166.
  American national income, 1839: GNP, Table A-1, minus 5 per cent (see text).
  Par of exchange: *Hunt's Merchants' Magazine*, XXXX, Volume III, p. 345 ($4.40).
Line 1B: Line 4B adjusted for population differences, the latter taken from sources for lines 4A and 2 (France).
Line 2: U.S., U.K., and French GNP, 1950, extrapolated to 1840 (circa) on constant price national product series (see text).
  1950 estimates: Milton Gilbert and Irving B. Kravis, *An International Comparison of National Products and the Purchasing Power of Currencies*, OEEC, Paris, 1954, p. 37.
  Extrapolators: U.S., 1899-1908 to 1944-53, Simon Kuznets, *Capital in the American Economy*, Princeton for NBER, 1961, p. 521, GNP, 1929 prices, Variant I; U.S., 1834-43 to 1899-1908, Table A-1, GNP, 1860 prices; U.K., 1870-79 to 1949-53, Simon Kuznets, "Quantitative Aspects of the Economic Growth of Nations," *Economic Development and Cultural Change*, October 1956, p. 53, national income, 1912-13 prices; Great Britain, 1841-51 to 1871-81 (carried back to 1831-41 at the same rate of change), Deane and Cole, *British Economic Growth*, p. 170, national income, mean prices of 1865 and 1885; France, 1831-50 to 1949-53, Kuznets, in *Economic Development and Cultural Change*, October 1956, p. 53, net national product, 1938 prices.
Line 3: Computed from data in Gilbert and Kravis, *International Comparison*, p. 37.
Line 4A: Data underlying line 1A divided by population estimates in Deane and Cole, *British Economic Growth*, p. 8, and *Historical Statistics of the United States, Colonial Times to 1957*, Washington, 1960, Series A-2.
Line 4B: Ezra C. Seaman, *Essays on the Progress of Nations*, 2nd ed., New York, 1852, pp. 445, 462.
Line 5: Data underlying line 2 divided by population estimates from sources of lines 4A and 2 (France).
Line 6: Gilbert and Kravis, *International Comparison*, p. 39.

[a] England and Wales.

versa.³ Since Britain was more highly developed around 1840 than was the United States, Seaman's procedure tends to maximize the difference between the national products of the two countries.

Gilbert and Kravis have produced comparisons for 1950 in both American and foreign prices. That is, they have estimated American GNP in dollars, francs, and pounds; British GNP in dollars and pounds; French GNP in dollars and francs. The estimates can be extrapolated back to 1840 on a constant price national product series so that we can compare, e.g., British and American GNP in 1840 in terms of both dollars and pounds. There is a second advantage to this procedure. The constant price extrapolating series consist of ten- to twenty-year averages. Therefore, the effects of cyclical phenomena on the comparisons are limited.

The procedure has one very important disadvantage: the results are difficult to interpret. As a first approximation, we are comparing 1840 (circa) national products valued in 1950 prices, since the extrapolated estimates are in 1950 prices. But the extrapolators are not based on 1950. The price base of a constant price national product series affects the rate of growth of the series; in general, the earlier the price base, the higher the rate of growth.⁴ Since the base years of the extrapolators are earlier than 1950, the extrapolated 1840 values are really smaller than 1840 national products in 1950 prices. If the extent of "bias" in the three series were identical, the comparisons would be unaffected, of course; but there is no good reason for supposing that they are. The date of the price base differs from series to series. In addition, the extent to which an early price base raises the rate of growth of a national product series, compared with a late price base, depends on the extent of changes in the price structure over time. There is no good reason for believing that the price structures of the three economies changed at the same pace.⁵

In view of the above remarks, it is a little surprising to find that the results of the third procedure are not very far from the results of the first two. French national product is shown to be between 23 and 56 per cent above American product. The upper limit is almost identical

---

³ See Milton Gilbert and Irving B. Kravis, *An International Comparison of National Products and the Purchasing Power of Currencies*, OEEC, Paris, 1954, p. 39. Simon Kuznets has provided the theoretical explanation for these results. See his *Economic Change*, New York, 1953, p. 171.

⁴ For the same reason discussed by Kuznets (*ibid.*).

⁵ Additionally, the French and the early segment of the British extrapolators are national income series, not GNP series; the early segment of the British extrapolator refers to Great Britain, while the 1950 comparison and the later segment of the extrapolator refer to the United Kingdom; the early segment of the American extrapolator excludes changes in inventories.

with the result obtained from Seaman's work. The American-British comparison worked out through the third procedure is less favorable to the British than are the comparisons resting on Seaman's work and on the exchange rate conversion of the Deane and Cole figure.[6] But even in the American-British case, the range of results of the three procedures is not very great, when put in the context of the differences among national products of developed countries in more recent years. That is, we find that American product may have been slightly higher (unlikely) or as much as a third lower than British product in 1840. In 1950, according to Gilbert and Kravis, American product was four or five times the size of British product; six or eight times the size of French product; roughly twice the size of the combined products of Italy, Germany, France, and the United Kingdom. According to Bornstein, American GNP was roughly two to four times the magnitude of Soviet GNP in 1955.[7] The position has changed somewhat in the last several years, but not enough to invalidate the main point, which is that, relative to the variations observed since, the margins between American product and both British and French product around 1840 were slight. Very early in her history the United States was one of the great economic powers.

In 1840, American population was somewhat smaller than British and very much smaller than French population. Consequently, the per capita comparisons are more favorable to the United States than are the national product ones. French product per capita ran between 24 and 40 per cent below the American; British product per capita, between 22 per cent below and 40 per cent above the American. Again, the American situation was closer to the British and French in 1840 than in recent years (see Table 1). According to Gilbert and Kravis, American product per capita was roughly double French product per capita in 1950, and about 60 to 100 per cent larger than British.

LONG-TERM RATES OF GROWTH OF REAL GNP
AND REAL GNP PER CAPITA

The data on the relative positions of the three countries in 1840 and 1950 imply widely disparate rates of growth between these dates and dramatize the rapidity with which major changes in relative economic strength can take place during the process of growth. This is especially

---

[6] Note, also, that the third procedure involves comparisons with the U.K., whereas the first two involve comparisons with Great Britain.

[7] Morris Bornstein, "Comparison of Soviet and United States National Product," *Comparisons of the United States and Soviet Economies*, Part II, Joint Economic Committee, 86th Cong., 1st Sess., Washington, 1959, p. 385.

clear if growth is measured in terms of the rate of change of national product, rather than product per capita.

Comparisons with other countries are harder to come by, since the records of few go back as far as the American. However, for eleven countries, including Britain, France, and most of the other developed economies of the world, there are records running back as far as the 1860's or 1870's. The average decade rates of change of real national product of these countries (to the 1950's) range from 13 to 42 per cent.[8] Only two of these nations show rates in excess of 36 per cent. The American rate of growth over the longer period, 1834–43 through 1944–53, is 42 per cent, or an increase of very nearly forty-seven-fold.[9] The growth of American national product, then, was exceptionally rapid, compared with growth in the rest of the developing world.

The growth of American national product per capita, however, was not exceptionally rapid. Between 1834–43 and 1944–53, there was a fivefold increase; the average decade rate of growth was just under 16 per cent. Of the eleven countries mentioned above, four exhibited rates of growth substantially higher (19–28 per cent) and three substantially lower (10–14 per cent).

LONG-TERM CHANGES IN RATES OF GROWTH OF REAL GNP AND REAL GNP PER CAPITA

The rate of growth of American GNP was higher in the nineteenth century than it has been in the twentieth (see Table 2). The average decade rate of growth between 1834–43 and 1894–1903 was 48 per cent; between 1894–1903 and 1944–53, only 34 per cent.[10] The magnitudes of the computed rates depend, to some extent, on the locations of the terminal dates within the long-swing chronology. But shifting the terminal dates by a decade does not alter the results appreciably. The finding does refer to a long-term change in the rate of growth of the series.

There is no reason to suppose that the series misrepresents the broad course of change or even that it overstates the degree of retardation. An inspection of the appendix will show that there is a possibility that we

[8] Simon Kuznets, "Quantitative Aspects of the Economic Growth of Nations," *Economic Development and Cultural Change*, October 1956, p. 13.

[9] The rate of growth was calculated from a linked series composed of the series described in this paper and Kuznets' Variant I, in 1929 prices, 1899–1908 through 1944–53. See Simon Kuznets, *Capital in the American Economy: Its Formation and Financing*, Princeton for NBER, 1961, p. 521.

[10] This rate would have been slightly higher had the series been constructed along the lines of the Department of Commerce concept instead of the Kuznets concept. See John W. Kendrick, *Productivity Trends in the United States*, Princeton for NBER, 1961, p. 62.

TABLE 2

DECENNIAL RATES OF GROWTH OF GNP AND GNP PER CAPITA,
CONSTANT PRICES, 1834-43 THROUGH 1944-53
(per cent)

Decades	GNP I^a (1)	GNP II^b (2)	GNP I^a ÷ Population (3)	GNP II^b ÷ Population (4)
1834-43 to 1844-53	63	51	20	11
1839-48 to 1849-58	70	65	25	22
1844-53 to 1854-63				
1849-58 to 1859-68				
1854-63 to 1864-73				
1859-68 to 1869-78				
1864-73 to 1874-83				
1869-78 to 1879-88	65	60	31	27
1874-83 to 1884-93	50	49	19	18
1879-88 to 1889-98	36	34	9	8
1884-93 to 1894-1903	36	35	13	12
1889-98 to 1899-1908	51		25	
1894-1903 to 1904-13	49		23	
1899-1908 to 1909-18	35		12	
1904-13 to 1914-23	28		11	
1909-18 to 1919-28	38		20	
1914-23 to 1924-33	29		11	
1919-28 to 1929-38	4		-5	
1924-33 to 1934-43	17		9	
1929-38 to 1939-48	50		44	
1934-43 to 1944-53	52		33	
Average rates^c				
1834-43 to 1894-1903	48	45	16	13
1894-1903 to 1944-53	34		16	

Source

Col. 1: 1834-44 through 1899-1908, derived from data of Table A-1, 1860 prices; 1894-1903 through 1944-53, derived from data in Kuznets, *Capital in the American Economy*, p. 521, Variant I, 1929 prices.
Col. 2: Derived from data in Tables A-1, A-4, and A-5. It was assumed that the ratio of value added by home manufacturing to GNP was the same, in current and constant prices. Benchmark ratios were interpolated and extrapolated on GNP.
Col. 3: Derived from data underlying col. 1 divided by Series A-2 of *Historical Statistics* (1840, 1845, 1850, etc.).
Col. 4: Derived from data underlying col. 3 divided by Series A-2 of *Historical Statistics* (1840, 1845, 1850, etc.).

^{a}Excludes changes in inventories, 1834-43 to 1899-1908.

^{b}Excludes changes in inventories; includes value added by home manufacturing and the value of improvements to farm land (Variant I, since this is the more relevant measure—see appendix). See text.

^{c}Computed from terminal values.

have overestimated the size of GNP in the pre-Civil War decades and that, therefore, the computed rate of change for the nineteenth century is somewhat too low, rather than too high. The price bases for the nineteenth- and twentieth-century segments of the series are located toward the middle of each segment. Therefore there is no reason to suppose that the deflation procedure imparts bias to the rate of growth of one segment relative to the rate of growth of the other. The finding of retardation appears to be secure.

In some measure, the rapidity of nineteenth century growth reflected the transfer of economic activities out of the home into the market place, where the consequences of these activities could be measured. It is a little surprising to see how limited the impact of these transfers really was. Our measure of GNP, which includes value added by home manufacturing and the value of improvements made to farm land with farm materials (GNP II), increases at a rate of 45 per cent per decade during the nineteenth century—a rate only 3 points below the rate of growth of GNP less the value of these activities (GNP I, see Table 2).

The rate of growth of population was also subject to sharp retardation, so that the ratio of GNP to population—GNP per capita—increased at the same rate in the twentieth century as in the nineteenth, 16 per cent per decade. There is an alluring quality to the constancy of this rate of increase—a constancy which masks any number of secular decisions on immigration, child labor, the length of the workweek, etc.—which asks for simple explanations from the unwary. Warily, we move on and note that nineteenth century GNP per capita, including value added by home manufacturing and the value of improvements to farm land (GNP II), increased at a somewhat lower rate, 13 per cent. The more inclusive measure, then, gives us a slight acceleration in the rate of increase over the long run.

Composition of GNP and GNP*, 1834–43 Through 1899–1908

SHARE OF CAPITAL FORMATION IN GNP AND GNP*

We have computed four variants of the share of capital formation in product (see Table 3), two referring to the share of gross national capital formation (GNCF) in GNP, two to gross domestic capital formation (GDCF) in gross national product less changes in claims against foreigners (GNP*), the closest approximation to gross domestic product that we have. All are in prices of 1860. GDCF refers to domestic investment, however financed; GNCF to investment financed by Americans, whether carried out at home or abroad. The share of GNCF in GNP is the share of domestic savings in product; the share of GDCF in GNP* is the share of domestic investment in product.

The shares of GNCF I in GNP I and GDCF I in GNP* I are calculated from capital formation and product data which *exclude* value added by home manufacturing and the value of farm improvements made with farm construction materials; i.e., the data conform to the capital formation and product definitions which are in common use today. The shares of

TABLE 3

SHARES OF CAPITAL FORMATION IN GROSS NATIONAL PRODUCT, 1860 PRICES, 1823-44 THROUGH 1899-1908
(per cent)

	Shares of Capital Formation in			
Decades	GNP I (1)	GNP II (2)	GNP* (3)	GNP* II (4)
1834-43	9	16	10	16
1839-48	11	14	11	14
1844-53	13	14	13	15
1849-58	14	16	15	17
1854-63				
1859-68				
1864-73				
1869-78	22	24	23	24
1874-83	21	22	21	22
1879-88	22	23	23	24
1884-93	26	26	27	27
1889-98	28	28	28	28
1894-1903	27	27	26	26
1899-1908	28	28	27	28

Source: Cols. 1-2, see source to Table 2, cols. 1-2, and Table A-3; cols. 3-4, calculated from data underlying cols. 1-2 less changes in claims against foreigners. See text.

GNCF II in GNP II and GDCF II in GNP* II are calculated from capital formation and product data which *include* value added by home manufacturing and the value of farm improvements made with farm materials (see the notes to Table 2 and the appendix). We have computed these "II" series because economic activities conducted beyond the market[11] were relatively more important in the early years of the period under review than in the later years and, therefore, changes in the composition, as well as the level (see Table 2), of GNP II and GNP* II might very

[11] It is convenient to make the distinction in terms of the market. However, one should bear in mind that modern measures of GNP do cover important outputs which do not flow through markets, such as the imputed rents of owner-occupied houses and agricultural products consumed on the farms where they are produced. Our GNP I includes these products, of course.

well be different from changes in GNP I and GNP* I. Therefore, before the fact, there was reason to suppose that the additional series might provide important information on the nature of nineteenth century American growth.

All four series show that the share of capital formation in product increased markedly. In the series in which the most moderate increase occurs, the share doubles; in the series in which the most pronounced increase occurs, the share triples. All of the postwar shares are higher, by wide margins, than all of the prewar shares. There is no question that the rise is a long-term phenomenon.

The duration of the increase varies among the four series. The two variant I series rise from 1834–43 through 1889–98, a period of five and a half decades. There is no clear evidence of increase in the variant II series in the prewar period, nor is the evidence for an increase after the Civil War completely convincing. The entire long-term increase may have taken place in the twenty-year period 1849–58 through 1869–78, or in the forty-year period 1849–58 through 1889–98. But we cannot really say much about the timing or duration of the long-term increase in any of the four series, since the data are surely affected by long swings and by the results of the Civil War. The break in the series over the period 1849–58 through 1869–78 makes it very difficult to appraise the effects of these phenomena on the series. The increase was a long-term phenomenon, but we cannot presently establish precisely when it began and when it ended.

The capital formation and product data from which the series in Table 3 were calculated are lacking an important component of capital formation, changes in inventories. The volume of inventories depends upon the structure of the economy and the level of economic activity; changes in inventories, on changes in the structure of the economy and the level of activity. Some of the structural changes taking place between 1834–43 and 1899–1908 no doubt tended to reduce inventories relative to output (e.g., the decline in the relative importance of agriculture), whereas others worked in the opposite direction (e.g., the increase of economic specialization and interdependence). We do not know what the net effects were. Assuming that the net effects were zero and that the ratio of inventories to output remained constant, then the ratio of inventory change to output at any given time would depend upon the rate of increase of output; the higher the rate of increase, the larger the ratio of inventory increase to output, and vice versa.

The data in Table 2 (and Table 6) show that the rate of increase of output during the first decade of the postwar period was roughly the same as during the prewar period, which suggests that the share of inventory increase in product may have been roughly the same in the two periods. Assume that this share was 4 per cent (and ignore the differences among the various product concepts). This means that the share of capital formation (including inventory change) in product rose from between 18 and 21 per cent in the decade 1849–58 to between 26 and 28 per cent in the decade 1869–78. The absolute magnitudes of the increases are, of course, the same as the absolute magnitudes of the increases of the shares of capital formation, less inventory changes, in product. In relative terms, they are somewhat smaller, but still very large.

The assumption of an average inventory increase of 4 per cent of product in 1869–78 may be roughly correct. Kuznets' estimate of average inventory change in that decade (current prices)[12] amounts to something less than 5 per cent of our GNP estimate. For reasons given in the appendix, we think that Kuznets' estimate is somewhat too high.

The data in Table 2 (and Table 6) show that the rate of growth of product fell quite sharply after the interval 1869–78 through 1879–88 and then rose again during the interval 1889–98 through 1899–1908. Presumably, then, the share of inventory change in product also fell and then rose, while the share of the remainder of capital formation in product was rising and falling (see Table 3). The share of total capital formation in product during the postwar period, then, probably was very much more stable than the share of capital formation, less inventory change, in product. As noted above, the short-term influences on the share of capital formation in product in the first decade of the postwar period are difficult to appraise, so that the interpretation which should be placed on this finding is uncertain. However, it is important to notice that, in the more comprehensive measures we have used, the period during which the share of capital formation in product rose tends to be much shorter than in the less comprehensive measures, and that it appears to be centered on the two decades around the Civil War.

We have been dealing only with the share of gross capital formation in gross product, since we have no capital consumption estimates for the prewar decades. However, net capital formation estimates can be derived for the postwar decades from the gross estimates and Kuznets' data on capital consumption,[13] deflated by our implicit price index for capital formation (Table A-3). The share of net national capital formation (NNCF I) in net national product (NNP I) thus derived runs between 14 per cent (1869–78 through 1879–88) and 17 per cent (1889–98) over

[12] Kuznets, *Capital in the American Economy*, p. 524.
[13] *Ibid.*, p. 528.

the postwar decades. That is, the share of NNCF I in NNP I, after the Civil War, was roughly as high as the share of GNCF I in GNP I, before the war (Table 3). Then the share of NNCF I in NNP I must have been smaller before the Civil War than after it. That is, there must have been a long-term increase in the share of NNCF I in NNP I over the period under study. While the example used here refers only to the NNCF I and NNP I concepts, the results would have been similar had we used any of the other three net capital formation and net product concepts.

However measured, the share of capital formation in product (1860 prices) rose. It is important to notice that the increase was from a high level to an exceptionally high level. For example, if we assume (as above) that the ratio of inventory increase to GNP was 4 per cent in all of the prewar decades, then the share of gross capital formation (including inventory changes) in gross product ran between 13 and 21 per cent in these decades. Assuming that capital consumption accounted for one-third of gross capital formation—an estimate which is probably too high, since it is based on the ratio of capital consumption to gross national capital formation (including inventory changes) in 1869–78—the share of net capital formation (including inventory changes) in net product ran between 9 and 14 per cent before the Civil War. These rates are by no means low.[14]

THE COMPOSITION OF CAPITAL FORMATION

The shares of GNCF (I and II) in GNP (I and II) increased slightly more rapidly than did the shares of GDCF (I and II) in GNP* (I and II) (see Table 3).[15] That is, the domestic savings rates increased more rapidly than did the domestic investment rates. In the early decades, Americans financed part of domestic investment by net borrowing abroad; in the later decades, they reduced their net foreign indebtedness. But these international transactions were quantitatively relatively unimportant, and this is why the behavior of the domestic savings rates over time was so similar to the behavior of the domestic investment rates. Changes in claims against foreigners amounted to only 7 per cent of GNCF in the decade in which this component of capital formation was most prominent (Table 4, column 1).

Much more striking, especially in the prewar period, are the differences between the variants I and II of domestic savings and investment rates.[16]

[14] See Kuznets in *Economic Development and Cultural Change*, July 1961, Part II, pp. 10, 11.
[15] We are comparing here the share of GNCF I in GNP I with the share of GDCF I in GNP* I; the share of GNCF II in GNP II with the share of GDCF II in GNP* II.
[16] We are comparing here the share of GNCF I in GNP I with the share of GNCF II in GNP II; the share of GDCF I in GNP* I with the share of GDCF II in GNP* II.

The variant II series are much higher, before the Civil War, than the variant I series. Also, the increases over time are less pronounced in the variant II series, especially in the prewar period (Table 3). The explanation for this, of course, is that the variant II series include farm improvements in capital formation, while the variant I series do not. Farm improvements constituted a very important component of capital formation in the early decades. Over time, the relative importance of this component declined precipitately (Table 4, column 2).

TABLE 4

SHARES OF COMPONENTS OF CAPITAL FORMATION IN VARIOUS AGGREGATES OF CAPITAL FORMATION, 1860 PRICES, 1834-43 THROUGH 1899-1908
(per cent)

Decades	Share in GNCF I of Changes in Claims Against Foreigners (1)	Share in GDCF II of Farm Improvements (2)	Shares in GDCF I of		
			Manufact. Producer Durables (3)	New Gross Construc. (4)	Col. 3 Plus Col. 4 (5)
1834-43	-7	47	21	79	100
1839-48	5	28	22	78	100
1844-53	-3	18	23	77	100
1849-58	-4	20	23	77	100
1854-63					
1859-68					
1864-73					
1869-78	-5	9	31	69	100
1874-83	a	7	39	61	100
1879-88	-2	4	45	55	100
1884-93	-3	3	43	57	100
1889-98	1	2	45	55	100
1894-1903	4	2	51	49	100
1899-1908	3	1	57	43	100

Source: Derived from data in Tables A-1, A-3, A-4, and A-5. See source to Table 2.

[a] Less than .5 per cent.

There is some question of whether the data in Table 4 overstate the relative importance of farm improvements in the early decades and understate it in the later decades. The figures on most of the other components of capital formation were derived by deflating current price estimates of value of output or by valuing physical outputs in 1860 prices (see the appendix). In the case of farm improvements, however, inputs were valued in 1860 prices. We estimated inputs required to carry out improvements with 1860 techniques, so that the effects of technical changes on the constant price farm improvements series should be the same as the effects of technical changes on the other constant price components of capital formation. However, we estimated the input requirements for the clearing of forest and nonforest land separately. Forest land took many more inputs per acre to clear than nonforest land. As the frontier moved westward, out of the forest and into the prairies and the plains, input requirements per acre cleared fell and this is reflected in our series by a decline in the constant price value of new farm improvements per new acre cleared.

If the output involved was a single, homogeneous one—acres cleared, fenced, etc.—then this procedure overestimates the constant price value of improvements in the early years, and underestimates it in the later years. In a way, the economies in clearing realized by the westward movement are analogous to economies in the production of, e.g., plows arising out of the discovery of new and better sources of iron ore. The evaluation of a homogeneous plow series in constant prices involves applying a single price, determined in a particular resource context, to the output series. One could argue that the same procedure should be followed in the case of farm improvements. To follow the procedure we did is to assert that improvements to forest land were different products from improvements to nonforest land and that they commanded premium prices. Whether this, in fact, was true and whether, if it was, the premium equaled the differential input cost, we cannot say, but it seems reasonable to suppose that these things were broadly true. That being the case, the data in Table 4 give an accurate representation of the changing relative importance of farm improvements in capital formation. However, if total acres improved were treated as a single, homogeneous product, then the share of farm improvements in GDCF II would be somewhat smaller (but still large) in the decades before 1860, and somewhat larger (but still relatively small) in the decades after 1860 than the shares shown in Table 4. The main points of interest would remain, however: a major share of American investment before the Civil War went into the improvement of land; this investment involved the direct commitment of agricultural manpower—mainly the manpower of land owners—in the creation of capital; the market, savings in money form, financial intermediaries, etc., played virtually no direct role in the process.

The main drift of the composition of GDCF I is very clear and very familiar. The share of construction fell from about 80 per cent to less than 50 per cent, while the share of manufactured producer durables rose from about 20 per cent to over 50 per cent (Table 4, columns 3 and 4). However, there are long periods during which the shares change little (1834–43 through 1849–58, 1874–83 through 1889–98) and relatively short periods when the movements are striking (during the Civil War, 1869–78 through 1879–88, 1889–98 through 1899–1908). The explanation may lie in the long swing. During periods of rising or peak long-swing rates of advance for construction, the share of construction in GDCF I remained roughly constant; during periods of declining or trough rates of advance, the share of construction fell sharply (compare Tables 4 and 6). We return to this point below.

The long-term rise in the share of manufactured producer durables in GDCF I may be overstated somewhat, since the estimates miss production in the home and, probably, production by some of the hand trades, sources of output relatively more important in the early decades than in the later. However, we do not believe that bias from this source is important.

The supply of manufactured producer durables came mainly from domestic sources. Before the Civil War, both imports and exports were limited. Interestingly, the balance of trade in durables, including ships, was apparently an export balance during this period. Omitting ships, there was a small import balance before 1850 and a small export balance thereafter. Of the imports, most important were saddlery and harnesses. Apparently Americans produced their own machines in the process of industrialization.

THE COMPOSITION OF GOODS FLOWING TO CONSUMERS

In the period following the Civil War, there are two major changes in the distribution, among major groups, of goods flowing to consumers in current prices (GFC I). The share of perishables declines by 4 percentage points, while the share of services rises by 5 points. The shares of the other two major groups remain roughly constant (Table 5, panel A).

The pattern displayed by the deflated series is quite different. The share of durables increases and the share of services declines slightly, whereas the shares of the other two groups remain constant (Table 5, panel A). Apparently the prices of durables fell and the prices of services rose relative to the price index of goods flowing to consumers. The finding is not surprising. However, the price index of services is not fully representative (see the appendix). Also, the changes in the composition of the constant price aggregate involve only a few percentage points. Consequently, it would be safest to regard the distribution of flows (1860 prices) as roughly unchanging over the entire period 1869–78 through 1899–1908.

The current price estimates for the prewar years refer only to single years, not to decade averages. Therefore, it is not easy to identify trends. The

TABLE 5

COMPOSITION OF GOODS FLOWING TO CONSUMERS, CURRENT PRICES
AND 1860 PRICES, 1834-43 THROUGH 1899-1908
(per cent)

Years or Decades	Current Prices					1860 Prices				
	Perishables (1)	Semi-durables (2)	Durables (3)	Services (4)	Total (5)	Perishables (6)	Semi-durables (7)	Durables (8)	Services (9)	Total (10)
	PANEL A: SHARES OF MAJOR GROUPS IN GFC I[a]									
1839	57	12	3	27	99	58	7	3	32	100
1844	50	16	4	29	99	57	12	3	28	100
1849	50	18	5	27	100	53	15	4	27	99
1854	54	16	6	25	101	52	15	6	27	100
1859	52	17	6	25	100	51	17	6	26	100
1834-43						57	9	2	32	100
1839-48						57	11	3	29	100
1844-53						53	16	4	27	100
1849-58						51	17	6	26	100
1854-63										
1859-68										
1864-73										
1869-78	51	17	7	25	100	51	17	8	24	100
1874-83	51	16	7	26	100	51	17	8	24	100
1879-88	50	17	7	26	100	51	17	10	22	100
1884-93	48	17	8	27	100	50	18	11	20	99
1889-98	48	17	8	28	101	51	18	11	20	100
1894-1903	48	16	7	29	100	52	18	10	20	100
1899-1908	47	16	8	30	101	50	18	10	22	100
	PANEL B: SHARES OF MAJOR GROUPS IN GFC II[b]									
1839	56	16	3	24	99	58	10	2	30	100
1849	50	20	5	26	101	52	18	4	26	100
1859	52	18	5	24	99	51	19	5	25	100
1869-78	52	17	7	24	100	51	18	8	23	100
1874-83	51	17	7	25	100	51	17	8	24	100
1884-93	48	18	7	27	100	50	18	11	20	99

(continued)

TABLE 5 (concluded)

PANEL C: SHARES OF COMPONENTS OF MAJOR GROUPS IN GFC II[b] (current prices)

Years	Unmanu-factured Perishables (1)	Factory Manufact. Perishables (2)	Home Manufact. Perishables (3)	Factory Manufact. Semi-durables (4)	Home Manufact. Semi-durables (5)	Durables (6)	Rents (7)	Other Services (8)	Total (9)
1839	36	16	4	11	5	3	11	13	99
1849	30	17	3	17	3	5	11	14	100
1859	29	20	3	16	2	5	11	13	99

Source: Computed from data in or underlying Tables A-2 and A-5. See source to Table 2.

[a] Excludes value added by home manufacturing.
[b] Includes value added by home manufacturing.

shares of perishables and services fluctuate fairly widely from year to year. The trend of each series appears to be downward. The share of semidurables rises to 1844 and then fluctuates around a level maintained into the postwar period. The share of durables rises steadily.

The same trends appear somewhat more clearly and in more pronounced form in the constant price series (both single-year and decade average estimates). In particular, the year-to-year fluctuations of the shares of perishables and services disappear and the downward movements of these series are strong and persistent. The share of semidurables in the early years is smaller and it is not until 1859 (or 1849–58) that it reaches the postwar level.

In both the current and constant price series, changes in the composition of goods flowing to consumers are more pronounced in the prewar than in the postwar period. However, when value added by home manufacturing is included in final flows (Table 5, panel B), the prewar composition of final flows becomes somewhat more stable. The decline over time in the share of perishables and the rise in the share of durables are very nearly as strong as before. But, in the constant price aggregates, the increase over time of the share of semidurables and the decrease of the share of services are less marked, while in the current price aggregates the shares of these components remain almost constant over time.

In panel C of Table 5 the shares of several components of the major groups in GFC II (current prices) are distinguished. Perhaps the most interesting finding is the sharp decline in the share of unmanufactured perishables (column 1), which fully accounts for the decline in the share of perishables (panel B, column 1). The share of factory production of perishables actually increased (panel C, column 2), more than offsetting a small decline in the share of home production of perishables (column 3). The relative importance of home manufacturing of semidurables declined, of course (column 5).

Fluctuations of the Rates of Growth of GNP and Components

The rates of change of GNP and components fluctuate fairly widely, presumably reflecting long swings. The series are not well designed for analysis of long swings, since they are overlapping decade averages rather than cycle averages and since the record is seriously broken by the Civil War. We are concerned here only with whether the movements of the series appear reasonable in the light of research on long swings.

All of the series (1860 prices), with the exception of perishables and services, show a clear prewar peak rate of growth centered on 1846 (Table 6), matching the long-swing peak rate of growth of output found by Abramovitz.[17] As for the exceptions, perishables shows no marked variations in rates of growth, which is not surprising. The series is heavily weighted with components which either did not enter markets at all or entered only local markets (firewood, components of animal products). The effects of long swings on this series ought to have been limited.

The rate of change of the services series rises over the entire period, possibly peaking over 1851, although we cannot be sure of this. Again, the finding is not unreasonable. A major component of services is rents, which depends upon the stock of dwellings (see the appendix). The rate of increase of the stock of dwellings varies with the absolute level of new construction, while the peak in the level of construction lags well behind the peak *rate of increase* of construction. Therefore, the rate of increase of the rents component of services ought to follow a long-swing path lagged well behind the long swing of construction.

Abramovitz identifies three postwar trough rates of change, centered on 1874, 1886, and 1892, in the measure which he prefers. A second measure yields a trough in 1891.[18] Our series gives us quinquennial rates of change centered on years five years apart—1876, 1881, 1886, 1891, etc. Consequently, it is impossible for the series to show trough rates over both 1886 and 1891 (or 1892, of course). As it turns out, four of the series exhibit trough rates over 1891 and two more over 1886. One rate, centered on 1876, is so low that it may very well be a trough rate, nearly matching Abramovitz' trough rate over 1874.

Abramovitz identifies postwar peak rates of change over 1881, the end of 1889, and 1899 in the preferred measure, and over 1876 (tentative) and 1901 in the second measure.[19] Of course, none of our series which trough over 1886 or 1891 (all but one) could possibly show the Abramovitz peak of 1889. However, all of the series do show what may be peaks over 1901 or 1896 (one series), as close to the Abramovitz dates as these series could come.

Only two series, consumer and producer durables, pick out the Abramovitz 1881 peak. Of the remaining series, all but one show high rates over 1876, which may be peak rates, matching the 1876 peak in

---

[17] Moses Abramovitz, "Long Swings in U.S. Economic Growth," in *The Study of Economic Growth*, 39th Annual Report of the NBER, New York, 1959, p. 25.
[18] *Ibid.*
[19] *Ibid.*

TABLE 6

QUINQUENNIAL RATES OF GROWTH, GNP AND COMPONENTS,
1860 PRICES, 1834–43 THROUGH 1899-1908
(per cent)

Quinquennia		GNP (1)	Gross New Construct. (2)	Producer Durables (3)	Perishables (4)	Semi-durables (5)	Consumer Durables (6)	Services (7)
1834-43	to 1839-48	24	34	43	22	49	61	11
1839-48	to 1844-53	31	58	65	21	87	91	18
1844-53	to 1849-58	30	49	52	22	38	54	25
1849-58	to 1854-63							
1854-63	to 1859-68							
1859-68	to 1864-73							
1864-73	to 1869-78							
1869-78	to 1874-83	31	6	55	33	33	35	31
1874-83	to 1879-88	26	24	58	24	25	46	16
1879-88	to 1884-93	20	46	35	11	15	31	4
1884-93	to 1889-98	13	15	22	13	12	10	8
1889-98	to 1894-1903	20	0	28	22	20	12	25
1894-1903	to 1899-1908	26	15	48	21	25	22	32

Source: Computed from data in Tables A-1, A-2, and A-3.

Abramovitz' second measure. However, the 1876 rates of change are not much higher than the rates centered on 1881. Also, our smoothing device is not well designed to handle the very long business cycle of 1873–82. We tested to see whether the peaks at the early date were the results of the smoothing procedure used. We computed GNP cycle averages and calculated average annual rates of change between cycle averages. The cycle averages running from peak to peak show a clear peak rate of change centered on 1881; the trough-to-trough averages, a high rate, perhaps a peak, centered on the two years 1877 and 1878.

There is a peak rate of change in the construction series centered on 1886, a trough date, in the Abramovitz chronology. However, Abramovitz also gives evidence of a peak rate of change in urban building over 1884 and a peak rate for transportation investment over 1891.[20] The differential movements of these two components of construction could easily result in a peak for the aggregate over 1886. The last trough for construction is also late (1896). But, again, it is not far from the troughs in urban building and transportation (1892 and 1893).[21]

In the main, then, the series conform well to the long-swing chronology.

Summary

At the beginning of the period of this study, American national product was somewhat smaller than national product in Britain or in France. American product per capita was also probably smaller than the British, but was considerably larger than the French. Between 1834–43 and 1944–55, American GNP increased at an exceptionally high rate of 42 per cent per decade, a rate perhaps never equaled elsewhere for such an extended period. GNP per capita also increased at a high rate, compared with British and French growth. However, several developed countries have experienced higher rates of growth, at least since the 1860's and 1870's.

The rate of growth of American GNP has been subject to quite sharp retardation. But GNP per capita has increased at a roughly constant rate. If value added by home manufacturing and farm improvements made with farm materials are included in GNP, then the rate of growth of GNP per capita has probably undergone a slight acceleration.

The share of capital formation (gross and net) in real national product was relatively high before the Civil War. Sometime between 1834–43 and

---

[20] *Employment, Growth, and Price Levels*, Joint Economic Committee, 86th Congress, 1st Session, 1959, Part 2, p. 434.
[21] *Ibid.*

1899–1908, there was a marked long-term increase in the investment proportion, although the precise timing and duration of the increase cannot be established.

The share of changes in claims against foreigners in capital formation was small throughout the period. In the early decades it was negative; in the later decades, positive. Farm improvements made with farm materials accounted for nearly half of real gross domestic investment (less changes in inventories) in the decade 1834–43. Thereafter, the share fell sharply, reaching a level of about 2 per cent of real investment by 1899–1908. At the beginning of the period, gross new construction was roughly four times as important as gross new investment in manufactured producer durables, but by 1899–1908 the two forms of investment were of roughly equal importance.

Following the Civil War, the share of perishables in flows of goods to consumers declined, while the share of services increased and the share of semidurables and durables remained constant, if measurements are made in current prices. In 1860 prices, the shares of all the major groups in flows to consumers are roughly constant over time.

Prior to the Civil War, both the current and constant price series show that the share of perishables in flows to consumers decreased, while the share of durables increased. The share of semidurables increased, reaching the postwar level before the Civil War. The movement is most pronounced in the constant price measure which excludes value added by home manufacturing from flows to consumers; it is barely discernible in the current price measure which includes value added by home manufacturing in flows to consumers. The share of services in current price flows to consumers, including value added by home manufacturing, is roughly constant over the prewar period. However, in the constant price variant which excludes value added by home manufacturing from flows to consumers, the share of services declines quite markedly.

The movements of the real GNP series and the main components conform well to the Abramovitz long-swing chronology.

Appendix

This appendix describes the derivation of the various series produced in the course of the study. Limitations of space prevent the reproduction of detailed estimating procedures. The general descriptions given, however, should allow a careful reader to make a preliminary appraisal of the series. Hopefully, a subsequent, longer publication will provide greater detail.

GROSS NATIONAL PRODUCT, FINAL PRODUCT FLOWS,
1869–1909, CURRENT PRICES

The estimates for 1869–1909 are revisions of Kuznets' figures, which rest in part on the work of Shaw.[22] Kuznets offers three statistical variants which differ in the magnitudes of the various flows to consumers. For present purposes, there is very little to gain in working with three variants and there is little basis for choice among the three. All three embody the same concepts and differ only in estimating procedures. For the period 1869–1909, the components of Variants II and III are extrapolated on the components of Variant I, which are based on the series contained in *National Product since 1869*.[23] The trends displayed by the three series over this period ought to be, and are, essentially the same. We chose to work only with Variant I, which reflects the basic Kuznets estimates for the period.

Commodities Flowing to Consumers, in Producer Prices

The nineteenth century trends of the Kuznets series are determined, in the main, by comprehensive commodity flow estimates at ten-year intervals (1869, 1879, 1889, 1899) based largely on manufacturing Census data. The components of these benchmark figures were interpolated for the intercensal years on change indexes drawn up from less comprehensive data.[24] Kuznets believes that the 1869 benchmark figure is short because of deficiencies of the Census. For this reason, his 1869 GNP estimate is too low, but probably less than 10 per cent too low. The effect of this, in turn, is to make the decade average GNP estimate for 1869–78 short by something under 5 per cent and to give the rate of change of the series a slight upward bias.[25]

There are several reasons for believing that Kuznets overestimated the effect of deficiencies of the 1869 Census on his series. He gave some weight to Census Commissioner Walker's estimate that returns were short by 13 per cent; but Walker attributed the shortage to poor returns of the

---

[22] Kuznets, *Capital in the American Economy;* William Howard Shaw, *Value of Commodity Output since 1869*, New York, NBER, 1947. Kuznets kindly allowed us to use the unpublished annual estimates for the period 1869–89.

[23] Kuznets, *Capital in the American Economy*, pp. 471, 472, 517; Kuznets, *National Product since 1869*, New York, NBER, 1946.

[24] Shaw, *Commodity Output*, pp. 79, 92–107; Kuznets, *Capital in the American Economy*, pp. 545–546. The procedure is somewhat more involved than our description of it.

[25] Kuznets, *National Product*, p. 60. See also Kuznets, "Long-term Changes in the National Income of the United States of America since 1870," *Income and Wealth of the United States*, Income and Wealth Series II, Cambridge, Eng., 1952, p. 37.

hand trades, especially the construction hand trades.[26] Shaw and Kuznets made no use of these data.

Kuznets also took into account Shaw's estimate that the 1869 returns were low by about 5 per cent, chiefly because several minor industries were omitted from the canvass. Study of Shaw's tables shows that the industries

TABLE A-1

GROSS NATIONAL PRODUCT, VARIANT I[a], SINGLE-YEAR ESTIMATES AND OVERLAPPING DECADE AVERAGES, CURRENT PRICES, 1860 PRICES, AND IMPLICIT PRICE INDEX, 1834-1908
(billion dollars)

Year or Decade[b]	GNP, Current Prices (1)	GNP, 1860 Prices (2)	Implicit Price Index (3)
1839	1.54	1.62	94
1844	1.80	1.97	90
1849	2.32	2.43	96
1854	3.53	3.37	105
1859	4.17	4.10	102
1834-43	--	1.56	--
1839-48	--	1.94	--
1844-53	--	2.54	--
1849-58	--	3.30	--
1869-78	7.87	6.40	123
1874-83	9.54	8.40	115
1879-88	11.2	10.6	106
1884-93	12.3	12.7	97
1889-98	13.2	14.4	92
1894-1903	16.2	17.3	94
1899-1908	22.4	21.8	103

Source: Tables A-2 and A-3.

[a]Excludes the value of improvements made to farm land with farm construction materials (Table A-4), value added by home manufacturing (Table A-5), and changes in inventories.

[b]1834-59 are Census years; 1869-1908 are calendar years.

covered in 1879 but apparently left out in 1869 accounted for only about 2.7 per cent of final product in 1879. Almost half of the total is due to the mixed textiles industry, unenumerated in 1869, according to Shaw. It is likely that the product of this industry *was* counted in 1869, but was included with the product of the cotton and woolen industries. Apparently this is what happened in 1879. The separate identification of the industry in Census tables of that year was accomplished at the Census office and

[26] *Ninth Census of the United States: 1870*, Washington, 1872, Vol. III, p. 376.

TABLE A-2

FLOWS OF GOODS TO CONSUMERS[a], SINGLE-YEAR ESTIMATES AND OVERLAPPING DECADE AVERAGES, CURRENT PRICES, 1860 PRICES, AND IMPLICIT PRICE INDEX, 1834-1908
(billion dollars)

Year or Decade[b]	Perishables (1)	Semi-durables (2)	Durables (3)	Services (4)	Total (5)	Implicit Price Index[c] (6)
PANEL A: CURRENT PRICES						
1839	.775	.160	.044	.366	1.35	
1844	.804	.260	.069	.474	1.61	
1849	1.03	.364	.113	.564	2.07	
1854	1.60	.466	.180	.744	2.99	
1859	1.87	.623	.207	.924	3.63	
1869-78	3.34	1.08	.449	1.64	6.51	
1874-83	4.04	1.28	.518	2.03	7.88	
1879-88	4.52	1.51	.646	2.40	9.08	
1884-93	4.64	1.64	.762	2.59	9.63	
1889-98	4.96	1.70	.786	2.88	10.3	
1894-1903	6.18	2.00	.936	3.76	12.9	
1899-1908	8.24	2.79	1.35	5.36	17.7	
PANEL B: PRICES OF 1860						
1839	.826	.108	.031	.457	1.42	94
1844	1.01	.217	.052	.502	1.78	90
1849	1.15	.335	.097	.594	2.17	95
1854	1.46	.446	.162	.758	2.82	106
1859	1.83	.623	.200	.919	3.57	102
1834-43	.809	.125	.033	.454	1.42	
1839-48	.985	.186	.053	.504	1.72	
1844-53	1.19	.348	.101	.593	2.23	
1849-58	1.45	.480	.156	.739	2.83	
1869-78	2.56	.858	.412	1.19	5.02	130
1874-83	3.40	1.14	.555	1.56	6.65	119
1879-88	4.21	1.43	.810	1.81	8.27	110
1884-93	4.68	1.65	1.06	1.88	9.26	104
1889-98	5.30	1.85	1.17	2.04	10.4	100
1894-1903	6.48	2.22	1.31	2.56	12.6	102
1899-1908	7.83	2.77	1.60	3.39	15.6	114

Source: See text.

[a] Excludes value added by home manufacturing (Table A-5).
[b] 1834-59 are Census years; 1869-1908 are calendar years.
[c] Col. 5 of panel A divided by col. 5 of panel B multiplied by 100.

involved double-counting of at least part of the product of mixed textiles.[27]

In connection with another study, the results of which were published in Volume 24 of Studies in Income and Wealth, we evaluated the Ninth and Tenth Manufacturing Censuses, deducted the returns of nonmanufacturing industries, added estimates for manufacturing industries which had been omitted by the Censuses, and corrected the returns of industries which we judged to be in error.[28] There is no reason to repeat here the justifications for these adjustments. However, the data published in Volume 24 are value-added data, whereas we are concerned here with value of output. Therefore, we will review the adjustments described in Volume 24, indicating how they affect value of output and comparing them with the adjustments made to the same data by Shaw. We begin with the subtraction of nonmanufacturing returns from the Census totals:

	Shaw (p. 200)		Data Underlying Volume 24 Estimates (pp. 56–58)	
Value of Output of:	1869	1879	1869	1879
		(million dollars)		
1. Census totals	4,232	5,370	4,232	5,370
2. Nonmanufact. industries				
a. Mech. and hand trades	249	262	228	215
b. Agricultural industries		6		3
c. Roofing work		3		
d. Mining			5	5
e. Personal and health services			6	10
f. Kindling wood			1	2
g. Total	249	271	240	235
3. Line 1 minus line 2g	3,983	5,099	3,992	5,135

The Shaw and Volume 24 totals of value of output of nonmanufacturing industries are fairly close in both 1869 and 1879. Shaw's figures are somewhat larger, probably mainly because Shaw estimated part of the value of output of mechanical and hand trades, whereas the data listed under the Volume 24 heading were taken directly from the Censuses. Differences between the detailed deductions probably reflect, in some measure, differences between the classification systems. For example, Shaw may have included personal and health services in his total for mechanical and hand trades.

[27] *Eleventh Census of the United States: 1890*, Washington, 1897, Vol. VI, Part I, p. 4.
[28] Gallman in *Trends in the American Economy*, pp. 56–60.

As noted above, the Tenth Census double-counted part of the value of output of the textile industries:

4. Double-counting in textiles   66
5. Line 3 minus line 4   3,983   5,099   3,992   5,069

According to Volume 24 (p. 58), various manufacturing industries were either omitted or badly returned in the Ninth and Tenth Censuses. The appropriate adjustments to value of output are:

6. a. Fish curing   6
   b. Coopering   16   23
   c. Wheelwrighting   12
   d. Periodical press   16   89
   e. Manufactured gas   30
   f. Petroleum refining   44
   g. Hydraulic lime and cement   2
   h. Smelting   18
   i. Products of steam R. R.
      shops   38
   j. Total   44   250
7. Line 5 plus line 6j   3,983   5,099   4,036   5,319

By the standard of the estimates underlying the Volume 24 series, then, it appears that the aggregates Shaw worked with were about 1 per cent too low in 1869, and about 4 per cent too low in 1879. However, we are only interested in Shaw's estimates of final product. Therefore, only the entries in line 6 which refer to industries that contributed to final product are relevant here. Eliminating the rest (entries b, c, e, f, and h) brings the totals listed under the Volume 24 heading to within 1 per cent and well within 3 per cent of the Shaw totals in 1869 and 1879, respectively. That is, the deficiencies of the Ninth and Tenth Manufacturing Censuses had insignificant effects on the aggregate data with which Shaw worked. We conclude that there is no clear evidence that the shortcomings of the first two manufacturing Censuses bias Shaw's or Kuznets' series importantly.[29]

In addition to the manufacturing Censuses, Shaw's main sources were various Department of Agriculture publications, which he used in estimating most of his final flows of unmanufactured commodities. With enough

[29] Milton Friedman believes that Kuznets' 1869 GNP estimate is low (and the 1879 estimate high), but the source of the problem is not necessarily the manufacturing Census (see below). See Milton Friedman, "Monetary Data and National Income Estimates," *Economic Development and Cultural Change*, April 1961, pp. 281–282. One step in Shaw's procedure tends to make the estimates for the early years too large relative to the estimates for the later years. He attempts to eliminate custom production from his series, but believes that his procedure includes relatively more custom production in the early years than in the later ones. See Shaw, *Commodity Output*, p. 80. Shaw's estimate of the value of custom production included in the 1869 manufacturing Census is about 2.5 per cent of the value of manufacturing output (*ibid.*, p. 200).

data, one ought to be able to determine whether agricultural outputs are consistent with Shaw's manufacturing data and his estimates of final flows of unmanufactured commodities. To be more precise, the drafts on farm output are as follows: (1) intermediate use on the farm (e.g., as animal feed); (2) consumption by final consumers (in unmanufactured form); (3) net exports; (4) consumption in manufacturing; and (5) inventory accumulation.

The Department of Agriculture data are net of the first draft and Shaw was obliged to estimate the second and part of the third. The fourth can be derived from Census data, sometimes in value terms only, but often in physical terms, and the elements of the third not supplied by Shaw (e.g., net exports of raw cotton) can be taken from Treasury publications. This leaves only the fifth draft, which must be derived as a residual. Presumably it should be small, relative to total output, and there should be no marked trend in this relationship over time. If the residual is large and if the relationship of the residual to total output shows a marked trend over time, then we may question the consistency of the agricultural data with Shaw's manufacturing data and his estimates of final flows of unmanufactured goods.

We attempted calculations of this kind for textile fibers, grains, and animal products. Outputs of textile fibers are very close to the totals of the first four drafts in every year. In the case of grains, there is always a positive residual and it is not small (e.g., in value terms, 10–20 per cent of the value of grain production). However, this is probably due in no small measure to the fact that we were unable to take into account the consumption of oats by nonfarm horses. Additionally, there seems to be no marked trend in the residual and the residual is small, compared with the value of commodities flowing to consumers (1–2 per cent). Therefore, if there are inconsistencies among the agricultural output, manufacturing, and final flow data they are not of much significance for present purposes.

In the case of animal products, however, the residual is very large in 1869 and falls sharply thereafter. As a percentage of gross farm income derived from animal products, the residual, in value terms, runs as follows:

1869	1879	1889	1899	1904	1909
86	53	29	29	24	24

As a percentage of Shaw's estimates of the value of finished consumer commodities, it runs:

1869	1879	1889	1899	1904	1909
28	9	4	4	3	3

GROSS NATIONAL PRODUCT, 1834–1909

These results are probably due to Shaw's failure to estimate the value of animal products flowing from retail slaughterers to final consumers. He measured only farm consumption of farm production and production by the manufacturing sector (i.e., production reported in the manufacturing Census, which excludes retail slaughtering). With the development and adoption of the refrigerator car in the 1870's and 1880's, large slaughtering and packing firms displaced small establishments. Presumably this is why the relative magnitude of the residual falls so markedly after 1869. Clearly the omission is a serious one which biases the rate of growth of commodity production and national product. Even though accurate data on production in retail establishments are not available, we were obliged to approximate the magnitude of the missing component.

We wanted as smooth a link at 1909 with the Kuznets series as possible. Compared with the total value of finished consumer commodities, the missing component is relatively unimportant in that year. Therefore, we adopted Shaw's 1909 estimate of the value of output of animal products destined for consumption (manufactured and unmanufactured) and extrapolated it backward to earlier benchmark years on a series which consists of the Shaw estimates plus the value of animal products (farm prices) in excess of the first four drafts. We substituted these estimates for Shaw's figures. The valuation of the second component of the extrapolator takes no account of the cost of transportation from the farm to the retail slaughtering firm (presumably a minor cost) or of value added by retail slaughter. Therefore, the estimates are low and the rate of change of the series has an upward bias, probably slight.

For the interbenchmark years, we interpolated the benchmark adjustments (the new estimates minus Shaw's original figures) on the Department of Agriculture aggregate series.[30]

Shaw intentionally omitted a second important flow, the value of firewood entering consumption, for lack of an adequate series. Since the publication of Shaw's book, Barger has derived estimates and incorporated them into his work on distribution.[31] Kuznets considered the estimates for inclusion in his series and rejected them on the ground that they appeared of dubious quality.[32] He compared Barger's firewood estimates with Shaw's food series and found that the ratio of the first to the second in 1869 was so large as to cast serious doubt on the firewood

---

[30] U.S. Department of Agriculture, Technical Bulletin 703, December 1940, gross income from animal products, less exports of live cattle and changes in inventory values, pp. 23 and 111.

[31] Harold Barger, *Distribution's Place in the American Economy since 1869*, Princeton for NBER, 1955, p. 128.

[32] Kuznets, *Capital in the American Economy*, pp. 516–517, footnote.

estimate. However, as just noted, Shaw's food estimate in 1869 is very much too low. Additionally, Barger valued his 1869 firewood consumption estimate by use of an 1879 price extrapolated to 1869 on what amounts to an index of coal prices. If one substitutes Dorothy Brady's firewood price index (prepared for this Conference) for the price extrapolator, the value of firewood consumed in 1869 drops by over 40 per cent and the relationship between the value of firewood consumed and the value of food flowing to consumers becomes more reasonable, compared with subsequent experience. For example, the ratio of the value of firewood consumed to the value of food flowing to consumers falls fairly gradually from .19 in 1869 to .04 in 1909; the ratio of the value of all fuel and lighting products flowing to consumers, less gas and electricity, to the value of food flowing to consumers falls from .23 in 1869 to .09 in 1909.

Barger's estimates rest ultimately on physical consumption estimates published by the U.S. Department of Agriculture, which are derived from regional per capita consumption data assembled by the Forest Service (1907, 1918, 1925–29, 1933) and the Tenth Census (1879).[33] The Department of Agriculture estimators associated the regional consumption data with climate, type of timber available, type of population in the region, housing conditions, alternative fuels available, and heating appliances in use (fireplaces, stoves). They then interpolated between benchmark years and extrapolated to years before 1879, regionally, on the characteristics listed above. No details of the estimating procedure are given in the publication, apart from those described above. However, the estimates appear to be carefully made and rest on appropriate considerations.

While the estimates are given as decade aggregates, they depend ultimately on Census year estimates, which can be quite simply derived from the decade data. The Census year estimates imply the following per capita consumption (in cords):

1819	1839	1859	1879	1899	1907
4.2	4.4	4.1	2.8	1.3	.9

The general downward drift of these figures, over time, seems reasonable and requires no extended comment. Additionally, the 1907 figure, as noted above, is based on Forest Service records and seems firm. A question remains as to whether the *rate* at which per capita consumption declines is appropriate. The answer depends in no small measure on the faith that can be put in the 1879 figure, an estimate (by states) produced by a Census forestry expert. The only piece of data from the Census

---

[33] U.S. Department of Agriculture, Circular 641.

enumeration which bears on this estimate is cordwood production on farms. Aggregate farm production amounted to about one cord per member of the total U.S. population.[34] Of course, farm production does not exhaust the universe. Additionally, there is reason for believing that cordwood might be one of the more poorly returned farm products, since the production of cordwood was not the major occupation of the farmer. However, the farm data do suggest that the U.S.D.A. figure for 1879 is more likely to be too high than too low.

Three scraps of data from the prewar period suggest that the U.S.D.A. figures for this period may be too high. The 1839 agricultural Census returned cordwood sold off farms. Assuming that this cordwood was sold to the nonfarm population and that the population was divided between farm and nonfarm as the labor force was, the per capita consumption implied by the Census return is roughly one cord per head. Stanley Lebergott has sent us an 1817 estimate of one cord per head from *Niles' Register*.[35] Finally, Seaman believed that consumption ran about one and a half cords for each free person and domestic servant (15 million out of a total population of 17 million, according to Seaman) in 1839.[36] But Seaman no doubt was influenced by the Census return.

All these prewar figures are surely much too low. We have a reliable figure of .9 cords per head for 1907, as noted above. Clearly, in an earlier era when wood was more plentiful and more easily acquired, when heating appliances were far less efficient, and when coal and other fuels were not used much for heating, wood consumption by final users must have been much higher than one to one and a half cords per head. On the other hand, it is hard to imagine that Seaman and Niles could have been as much in error as the U.S.D.A. figures suggest.

The U.S.D.A. figures, then, may be somewhat too high in the early years. Our use of them to derive estimates of the value of firewood consumption may make our early GNP estimates too large and the rate of growth of GNP too small. However, the broad trends described by the series are surely correct and firewood was too important an item of consumption for us to neglect it. Finally, one can generally assume that estimates of national product for early years are more likely to be too low than too high. Seen in this light, the probable deficiencies of the firewood series are not quite so serious. Therefore, we used the U.S.D.A. data in the estimation of the value of firewood consumed.

[34] Gallman in *Trends in the American Economy*, pp. 49–50.
[35] June 28, 1817, p. 278.
[36] Ezra C. Seaman, *Essay on the Progress of Nations*, 2nd ed., New York, 1852, p. 281.

TABLE A-3

GROSS CAPITAL FORMATION[a] (LESS CHANGES IN INVENTORIES), SINGLE-YEAR ESTIMATES AND OVERLAPPING DECADE AVERAGES, CURRENT PRICES, 1860 PRICES, AND IMPLICIT PRICE INDEX, 1834-1908
(billion dollars)

Year or Decade[b]	Manufact. Durables (1)	New Construction (2)	Changes in Claims Against Foreigners (3)	Total (4)	Implicit Price Index[c] (5)
PANEL A: CURRENT PRICES					
1839	.032	.137	.031	.200	
1844	.048	.135	.004	.187	
1849	.076	.206	-.025	.257	
1854	.131	.423	-.013	.541	
1859	.140	.385	.007	.532	
1869-78	.389	1.07	-.095	1.37	
1874-83	.490	1.17	-.002	1.65	
1879-88	.591	1.57	-.040	2.12	
1884-93	.616	2.10	-.076	2.64	
1889-98	.637	2.21	.018	2.86	
1894-1903	.860	2.31	.190	3.36	
1899-1908	1.30	3.19	.221	4.70	
PANEL B: PRICES OF 1860					
1839	.027	.140	.033	.200	100
1844	.044	.146	.004	.194	96
1849	.067	.217	-.026	.258	99
1854	.124	.431	-.012	.542	99
1859	.133	.392	.007	.532	100
1834-43	.032	.120	-.010	.141	
1839-48	.045	.161	.011	.216	
1844-53	.074	.254	-.009	.319	
1849-58	.113	.379	-.018	.474	
1869-78	.441	1.00	-.070	1.37	99
1874-83	.685	1.07	-.001	1.75	94
1879-88	1.08	1.33	-.040	2.36	90
1884-93	1.47	1.93	-.077	3.32	80
1889-98	1.80	2.22	.023	4.04	71
1894-1903	2.30	2.22	.196	4.72	71
1899-1908	3.40	2.55	.214	6.17	76

Source: See text.

[a] Excludes the value of improvements to farm land made with farm construction materials (Table A-4).

[b] 1834-59 are Census years; 1869-1908 are calendar years.

[c] Col. 4 of panel A divided by col. 4 of panel B multiplied by 100.

TABLE A-4

VALUE OF IMPROVEMENTS TO FARM LAND MADE WITH FARM CONSTRUCTION MATERIALS, VARIANT I AND VARIANT II, OVERLAPPING DECADE AVERAGES, 1860 PRICES, 1834-1908
(million dollars)

Decade[a]	Variant I	Variant II
1834-43	133	92
1839-48	80	55
1844-53	72	50
1849-58	118	81
1869-78	147	105
1874-83	125	89
1879-88	102	72
1884-93	92	66
1889-98	96	68
1894-1903	95	67
1899-1908	89	63

Source: See text.

[a]1834-59 are Census years; 1869-1908 are calendar years.

TABLE A-5

VALUE ADDED BY HOME MANUFACTURING,[a] CURRENT PRICES, 1839-89
(million dollars)

	1839	1849	1859	1869	1879	1889
Baked goods	10	11	24	43	32	19
Animal products	51	58	105	132	26	3
Textiles	29	28	25	23	--	--
Clothing	47	48	64	77	43	67
Total	137	145	218	275	101	89

Source: See text.

[a]More precisely, value added (by home manufacturing) to materials which would have been processed in factories had the structure of the economy been the same as in 1899. See text.

Barger also added to Shaw's figures the value of federal alcoholic beverage taxes, which he believed were left out of the Census value of output.[37] The adjustment is minor, affects trend little, and we accepted it.

Trade Markups

Kuznets applied a constant trade markup to each class of commodities flowing to consumers. He reviewed Barger's finding that the ratio of value added by trade to the value of commodities flowing to consumers rose somewhat over time and expressed the view that the rise reflects a tendency for retail prices to rise relative to wholesale prices. Since his price deflators really measure the movement of wholesale prices rather than retail prices, he was faced with a choice. If he accepted Barger's results and adjusted his series, his current price series would be improved, but his constant price series would be biased. His prime interest was in the constant price series, so he chose to leave his estimates as they were and simply allow for the fact that the rate of change of the current price series may be biased in a downward direction. Since he believed that other factors tend to bias the rate in the other direction (see above), the possible bias arising out of the trade markup was not a source of great concern.[38]

Our position was different from Kuznets' in three ways: (1) We had Dorothy Brady's retail price indexes for deflators and therefore did not have to depend on wholesale price indexes. (2) We think we have located and eliminated the main causes of upward bias in the rate of change of the commodity flow series, as noted above. (3) Our series goes back three and a half decades farther than Kuznets', which makes the extrapolation of a constant markup a much more doubtful expedient in our case than in his. Therefore, we used Barger's data. In general, we attempted to follow benchmark flows of commodities through distribution and to compute value added by distribution for each class of commodities flowing to consumers.[39] We used these series to extrapolate Kuznets' figures for 1909, again, in order to maintain as close a link with the Kuznets series as possible.[40] We interpolated the benchmark estimates of flows in final prices on Kuznets' series to get interbenchmark estimates.

---

[37] Barger, *Distribution's Place*, p. 128.
[38] Kuznets, *Capital in the American Economy*, pp. 513–516.
[39] Barger, *Distribution's Place*, Tables 24, 25, B-3, B-4, B-5, pp. 81, 84, 130–140. Barger's margin data refer to types of outlets, not to types of commodities. We had to assume that all commodities handled by an outlet carried the same margin.
[40] Since Kuznets' trade markup incorporates an adjustment for changes in inventories of final goods, our trade estimates also incorporate this adjustment.

Services Flowing to Consumers

Kuznets estimated services flowing to consumers by deriving from budget studies the ratio of consumer expenditures on services to consumer expenditures on commodities in each benchmark year and multiplying his commodity flow estimate by this ratio.[41] Since we changed the commodity flow estimates, we had to change the service estimates proportionately.

Manufactured Producer Durables

We followed the procedure described above in connection with flows of commodities to consumers.[42]

Gross New Construction

The Kuznets estimates are extrapolations on constant price materials flows, marked up for distribution in the manner described above.[43] We used Barger's margin data, for the reasons given above. Additionally, we made our extrapolations in current prices for reasons given in the Volume 24 study.[44] Finally, at the suggestion of Albert Fishlow, we estimated railroad and nonrailroad construction separately. We derived estimates of new railroad construction from Ulmer.[45] We then extrapolated Kuznets' 1909 estimate of new construction, less the figure for new railroad construction, on materials flowing into construction (current final prices), less railroad construction materials.

Our gross new construction estimate differs markedly from Kuznets' figure in 1869 and less markedly in later benchmark years. The principal explanation for this is that Kuznets assumed that the value of construction is a constant ratio of the value of materials flowing into construction. But in heavy construction the value of construction is typically a larger ratio of the value of materials used than in building. In effect, Kuznets

---

[41] Kuznets, *National Product*, pp. 77, 104–105. See also Kuznets, *Capital in the American Economy*, p. 523. Again, the procedure is a little more complicated than the description indicates.

[42] The figures should be adjusted for sales of ships to foreigners, an adjustment Shaw failed to make. But the change called for is slight and we did not bother to make it.

[43] Kuznets, *National Product*, p. 100, and *Capital in the American Economy*, pp. 512–513.

[44] Gallman in *Trends in the American Economy*, pp. 60–61.

[45] Melville J. Ulmer, *Capital in Transportation, Communications, and Public Utilities*, Princeton for NBER, 1960, p. 256, "Gross Capital Expenditure, Excluding Land" minus the value of equipment (derived from Shaw, *Commodity Output*, pp. 56–57; *Historical Statistics*, 1960, Series E-214 and E-215; and Ulmer, *Capital*, p. 274, col. 2).

assumed that the construction mix changed little over time. In 1869 railroad construction was relatively much more important than in any benchmark year thereafter. Our procedure yields an estimate for 1869 much larger than Kuznets' estimate.

The estimates based on Ulmer are available annually for 1869–1909 and we used them. We interpolated the benchmark new nonrailroad construction estimates on Shaw's construction materials series, less railroad materials, for 1889–1909, and on Kuznets' construction series, for 1869–89.

The procedure proposed by Fishlow and followed here is also different from the procedure used in the Volume 24 study to derive the value of total construction.[46] The Volume 24 study contains two variants of the value of total construction. Both rest on the materials flows and markup data used in the present study. But Variant A is based on the assumption that value added by construction in current prices is equal to the value of materials flowing into construction; Variant B is based on the assumption of equality between value added and value of materials in *constant prices* (essentially the Kuznets assumption). Neither variant takes into account the changing composition of construction and, therefore, both are subject to the Fishlow criticism.

The following tabulation compares the Volume 24 estimates of the value of total construction with our new estimates of gross new construction (in million dollars, current prices, for major benchmark years):

	Value of Construction		Value of Gross New Construction	Ratio of Col. 3 to Col. 1	Ratio of Col. 3 to Col. 2
	Variant A	Variant B			
	(1)	(2)	(3)	(4)	(5)
1839	150	132	137	.91	1.06
1849	220	206	206	.94	1.00
1859	456	436	385	.84	.88
1869	1,072	1,075	1,064	.99	.99
1879	1,180	1,180	953	.81	.81
1889	2,192	2,315	1,830	.84	.79
1899	2,576	2,813	2,090	.81	.74

The Volume 24 estimates are more comprehensive than the gross new construction estimates and therefore should exceed them. However, we are interested here in the movements of the series, not the levels of the estimates, or, more accurately, in the relative movements of the series, which are described by the ratios in columns 4 and 5. The effect of the Fishlow adjustment can be seen clearly in the ratios of column 4. The

[46] Gallman in *Trends in the American Economy*, pp. 60–64.

ratios in column 5 show the combined effects of the Fishlow adjustment and the extrapolation on current price materials flows. The new estimates of gross new construction imply a lower rate of growth for construction than do the Volume 24 estimates. Additionally, prior to 1879, the movements of the new series from one benchmark year to the next are quite different from the movements of the old series.

Changes in Claims Against Foreigners

Improvements of Kuznets' estimates of changes in claims against foreigners could probably be worked out from Matthew Simon's series on the balance of payments.[47] But in the context of national product, the improvements would probably be slight. We were unable to carry out the necessary calculations and therefore accepted the Kuznets estimates.

Changes in Inventories

Kuznets' estimates of changes in inventories are, in considerable measure, extrapolations on rates of change of output. Since we have altered these rates of change, the inventory figures should be adjusted. But Kuznets himself has limited confidence in the procedures he used.[48] Application of these procedures to pre-Civil War data would appear to be even more dubious, but no other method is presently available. Consequently, we decided to leave this component out of both the pre- and post-Civil War series.

GROSS NATIONAL PRODUCT, FINAL PRODUCT FLOWS, 1869–1909, 1860 PRICES

The current price series on firewood and animal products rest on output estimates. Consequently, we produced constant price series by applying 1860 prices to the output estimates.[49]

Dorothy Brady has produced final price indexes on an 1860 base for the benchmark years 1869–99. The commodity price indexes refer to the narrowest classification provided by Shaw. We derived indexes for the broad classes of perishables, semidurables, durables, and manufactured producer durables in the following way. We deflated Shaw's detailed estimates of the value of final output; then we divided the aggregated

---

[47] Matthew Simon, "The United States Balance of Payments, 1861–1900," *Trends in the American Economy*.

[48] Kuznets, *Capital in the American Economy*, pp. 159–160. See also Kuznets, *National Product*, pp. 108–109.

[49] The 1879 firewood price was carried to 1860 on the Brady firewood index (see above). The Census year 1859 animal products prices underlying Table A-2 in Gallman (in *Trends in the American Economy*, pp. 46–48) were used in place of 1860 prices.

current price estimate for each broad class in each benchmark year by the appropriate aggregated constant price estimate.[50] The weighting scheme is not quite appropriate since the Shaw estimates used to derive weights are gross of exports and net of imports and are expressed in producer prices, while the indexes are applied to flows into domestic consumption expressed in final prices. However, this is not a serious shortcoming. The Brady indexes are, without question, the most comprehensive and best-designed nineteenth century indexes national income estimators have had to work with. The deflation of commodity flows in benchmark years is relatively strong.

TABLE A-6

REVISED ESTIMATES EXPRESSED AS RATIOS OF KUZNETS' VARIANT I ESTIMATES, OVERLAPPING DECADE AVERAGES, CURRENT PRICES, 1869-1903

Decade	GNP Less Inventory Changes (1)	Perishables (2)	Semi-durables (3)	Durables (4)	Services (5)	Gross New Construction (6)	Gross Manufactured Producer Durables (7)
1869-78	1.16	1.30	.92	.96	1.12	1.34	1.08
1874-83	1.10	1.20	.93	.95	1.06	1.21	1.07
1879-88	1.08	1.14	.93	.96	1.03	1.18	1.07
1884-93	1.04	1.10	.96	.95	1.00	1.11	1.06
1889-98	1.03	1.09	.96	.96	1.01	1.03	1.03
1894-1903	1.03	1.09	.98	.96	1.02	.97	1.02

Source: Revised series, see text; Kuznets series, *Capital in the American Economy*, pp. 522, 524.

The only Brady index referring to services is an index of rents. Ethel Hoover has published some price indexes of medical services (1869-80) and we used these also.[51] But the weights that could legitimately be given them were slight. The rent index dominates the services price index. Rents apparently accounted for about 45 per cent of the value of services during this period.[52]

With some hesitancy we used Ulmer's cost index to deflate the value of new railroad construction.[53] Building wage rates figure in the index, whereas the wage rates of common labor would be more appropriate. No doubt the index could be improved, but we have not been able to carry out the necessary research and calculations.

[50] The computations were actually carried out first at the minor group level.
[51] Ethel Hoover, "Retail Prices after 1850," *Trends in the American Economy*, p. 176.
[52] Kuznets, *National Product*, p. 144.
[53] Ulmer, *Capital*, p. 275, col. 8, shifted to the base 1860.

Dorothy Brady has two building price indexes (true price indexes), one referring to houses and churches, the other to factories and stores. We weighted them equally and used them to deflate nonrailroad construction.

Benchmark constant price estimates (except for firewood and animal products) were interpolated to interbenchmark years (and extrapolated from 1899 to 1909) on Kuznets' constant price estimates.

Changes in claims against foreigners were deflated by the implicit price index of GNP, excluding changes in claims against foreigners.

Table A-6 compares the revised estimates with the Kuznets Variant I estimates.

GROSS NATIONAL PRODUCT, FINAL PRODUCT FLOWS, 1834–59, CURRENT PRICES

Mrs. Brady has produced price index numbers for major benchmark years 1839, 1849, and 1859 and for minor benchmark years 1834, 1836, 1844, and 1854. Therefore, we had the means to build up both current and constant price (1860) national product estimates for all of these years.[54] We made the interbenchmark estimates in constant prices, since this was the easiest and most secure procedure. We did not have adequate interbenchmark price indexes and were unable to construct them in the time available. Therefore, we could not construct current price estimates for these years.

Manufactured Commodities Flowing to Consumers, in Producer Prices

Major Benchmark Years.[55] Our principal source of data on the value of manufacturing output was the federal Census. In connection with the study described in Volume 24 of this series, we had tested the federal decennial Censuses of 1839 through 1879, had made corrections where these seemed called for, and had classified the data according to the Standard Industrial Classification Manual (classification of 1945, two-digit groups).[56] Census data, especially after 1839, are available in great detail (631 industries are distinguished in the general tables for 1859). Once the data had been distributed among the two-digit groups, it was not an unduly difficult matter to construct industry series, at the lowest level of aggregation employed by Shaw, covering the Census years 1839 through 1869.

---

[54] But we have not yet constructed complete current price GNP estimates for the first two minor benchmark years.

[55] The years are Census years (see Gallman in *Trends in the American Economy*, p. 15), as opposed to the calendar years of the postwar series.

[56] *Ibid.*, pp. 56–60.

The relationships between Census industry data and Shaw's commodity output estimates are described by the following schema:

1. Value of output of industry 1 (Census data)
2. *minus* value of intermediate output (unfinished goods) of industry 1
3. *minus* value of output of final commodity non-A by industry 1
4. *plus* value of output of final commodity A by industries 2, 3, etc.,
5. *equals* value of output of final commodity A.

Shaw's work is so detailed and lucid that it is possible to reconcile his estimates with Census data; i.e., one can work from Census data through entries (2) through (4) to the value of output of each final commodity. We attempted reconciliations for 1869 and, while we could not make a perfect reconciliation in every case, the disparities between the reconstructed data and Shaw's estimates were minor.[57]

In summary, then, the data available to us to construct prewar benchmark final commodity output estimates consisted of industry value of output series for 1839–69 and reconciliations between the 1869 members of these series and Shaw's estimates of final commodity output in 1869. Therefore, we were in a position to extrapolate the 1869 detailed final commodity output estimates to the prewar years on industry data. The form of the extrapolation depended, in each instance, on the values of entries (1) through (5) above and on the complexity of the relationships implicit in entries (2) through (4).

In a surprising number of instances, entries (1) and (5) were identical in 1869 and all other entries were zero; i.e., industry data were, in fact, final commodity output data. In these instances, we simply assumed that prewar data could be treated as final commodity output data, too. We will speak of these estimates as Class A estimates.

In a second important group of cases, entry (2) was large, the commodity flows measured by entry (2) could be traced to a limited number of user industries, and the estimation of entries (3) and (4) posed no very serious problems. In these instances, we estimated entries (2) through (4) and took entry (5) as a residual. We will speak of these estimates as Class B estimates.

We derived entries (3) and (4) from the limited product data reported in the Censuses[58] or, more often, by extrapolation from 1869 on entry (1)

---

[57] The problem was complicated by Shaw's attempt to eliminate custom production, an attempt which we did not make for the prewar years. See footnote 29.

[58] All the Census volumes contain some product data, generally in tables in the introduction. In the case of the Census of 1839, many of the general tables carry product data, rather than industry data. Of course, where product data were available, the estimating procedure was less complicated than that described in the text.

or by a similar procedure. No very important errors could arise out of the estimation of entries (3) and (4), since these entries were usually very small in 1869 and, in any case, the estimating procedure often involved simply distinguishing among commodities belonging to the same minor commodity group (e.g., commodity non-A belonged to the same minor commodity group as commodity A). Since we do not propose to use levels of aggregation below the minor commodity group for general analytical purposes, errors affecting the classification of goods within minor commodity groups have no significance.

Entry (2) was estimated by extrapolation from 1869 on the value of materials consumed (value of output in 1839) by using industries. For example, Shaw gives the value of flows from the woolen and worsted goods industries into the manufacturing of furniture, hats and caps, and clothing. We extrapolated the value of each flow on the value of all materials consumed by the using industry and, in each prewar year, summed the resulting three intermediate flows to get the total intermediate flow of woolens and worsteds. The extrapolation assumes that the input mix of the using industry remained the same, in value terms, over the period of the extrapolation. In fact, there were surely changes. However, errors arising from this source frequently tend to cancel out within minor groups. Suppose that between 1849 and 1869 the input mix of the clothing industry shifted in such a way that relatively more cotton was used at the later date. Then our procedure understates woolen intermediate flows in 1849 and overstates cotton intermediate flows. But these errors tend to cancel out within minor group 6, dry goods and notions.

One other feature of this estimating procedure should be noticed. In general, it is easier to test Census returns of the early stages of manufacturing than of the later stages. The Census sometimes provides physical input data for industries in the early stages (e.g., textile industries) and these can be checked against supplies of raw materials implicit in agricultural output and Treasury foreign trade data in the manner described above for the postwar estimates. As a result of the testing we conducted, we have greater confidence in the data on the earlier stages of manufacturing than in those referring to the later stages. The procedure described above incorporates a partial hedge against errors in the returns of the later stages of production. For example, suppose that the returns of the clothing industries were short in 1849. Then our procedure results in an overstatement of the value of final output of the textile industries, which partially compensates for the underreturn of the clothing industries.[59] In

---

[59] Any remaining underreturn of, e.g., the clothing industries should be compensated for by an overestimate of value added by home manufacturing.

this case, errors tend to cancel out within the major group (semidurables), but not within the relevant minor groups (6 and 7).

All of the remaining estimates of the value of final commodity output were made by extrapolation from 1869 on entry (1). We will speak of these as Class C estimates. We followed the Class C procedure when the case did not warrant the more involved operations of the Class B procedure (entry 2 or 5 was very small) or, less often, when these operations could not be carried out effectively (there was no very secure way of extrapolating entry 2 to the prewar years). The Class C procedure was used most often in estimating the value of final output of durables, for reasons which should be evident. Some durable goods industries produce both consumer and producer durables (e.g., carriages and wagons). Since we estimated value of final output by extrapolation on industry value of output from 1869, the prewar relationships between these consumer and producer goods are determined by the relationships of 1869, a point discussed in the Volume 24 paper.[60] This represents the chief weakness of the Class C estimates. Of course, it is a weakness affecting only the Class C estimates of certain durables.

Seaman's work on final commodity flows served as an extremely valuable check on our results for 1839.[61] Since our results were achieved principally by extrapolation, it was very important to have an independent set of prewar estimates for this purpose. Seaman was an exceptionally talented national income estimator. It is reassuring to find that, in the main, his work supports ours.

The following tabulation of the value of manufacturing output in 1849 (in million dollars) suggests the importance of the three classes of estimates described above.

Estimating Procedure	Intermediate Commodities (1)	Final Commodities				Total (6)	Total (col. 1 + col. 6) (7)
		Perishables (2)	Semi-durables (3)	Consumer Durables (4)	Producer Durables (5)		
Class A	0	54	73	10	1	138	138
Class B	82	153	51	1	0	204	286
Class C	104	22	58	51	63	194	298
Total	186	229	182	62	64	536	722

The entry under intermediate commodities for Class C is only roughly accurate, since the Class C procedure did not generate estimates of intermediate production directly. We derived the data by subtracting

---
[60] Gallman in *Trends in the American Economy*, pp. 65–67.
[61] Seaman, *Essay*, 2nd ed., pp. 274–284.

value of final product estimates from value of industrial output data and ignoring entry 3 (see above) unless it was important enough to figure explicitly in an estimate of final product (e.g., a distribution of a durable commodity between output destined for consumers and output destined for producers). Presumably errors arising from this source (minor, in any case) tend to cancel out in the aggregate. Additionally, in a few Class C cases it was necessary to use an industry series more or less comprehensive than the industry data with which Shaw worked in order to achieve a comparable extrapolating series from 1869 to 1849. Consequently, the value of industry output minus the value of final output, in these instances, is larger or smaller than the true value of intermediate product. Again, however, in the aggregate these discrepancies tend to cancel out.

Tables A-7 through A-9, which cover 1839, 1849, and 1859, reconcile Census data with the data underlying the manufacturing estimates in the Volume 24 paper and the latter with the data used in the present study. The value of intermediate commodities for 1849 in Table A-7 exceeds the corresponding figure in the text tabulation by the value of output of industries which produced only intermediate commodities and therefore did not figure in the estimation of the value of final product.

We have now described how the value of manufactured final domestic production was derived for the major prewar benchmark years. To move from final production to final flows, it is necessary to subtract the value of final goods exported (at producer prices at the plant), add the value of final goods imported (at port prices, including duties) and subtract net increases (add net decreases) in the value of inventories of final commodities. The inventory adjustment is comprehended in the estimate of the distribution markup discussed below.

In principle, the international trade adjustment should have been made at the industry level before final product was distinguished. However, it is possible to reduce the number of calculations and problems of classification significantly by reversing the procedure and making the international trade adjustments at a higher level of aggregation, the minor commodity group level. The losses occasioned by the reordering of the procedure are trivial. We lose the detail of final flows at the commodity level, but it would be impossible to put much trust in this detail in any case.

A slightly more serious problem is introduced by imports of "mixed commodities," i.e., commodities which flow in part to producers as intermediate goods and in part to consumers as final goods (such as textiles). In the main, Shaw's estimates of the value of domestic intermediate production are based on total consumption (or purchases) by

TABLE A-7

RECONCILIATION OF VALUE OF PRODUCT DATA UNDERLYING VOLUME 24 ESTIMATES OF VALUE ADDED BY MANUFACTURING AND VOLUME 30 ESTIMATES OF COMMODITY FLOWS, CLASSIFIED BY TWO-DIGIT INDUSTRY GROUPS AND MAJOR COMMODITY GROUPS, 1839[a]
(million current dollars)

S.I.C. No.	S.I.C. Title	Vol. 24 Total[b] (1)	Vol. 30 Corrections (2)	Commodity Flow Total[c] (3)	Perishables (4)	Semi-durables (5)	Consumer Durables (6)	Producer Durables (7)	Construction Materials (8)	Intermediate Commodities[d] (9)	Not Elsewhere Classified (10)
	Manufacturing Industries										
19	Ordnance and accessories	1.0		1.0		1.0					
20	Food and kindred products	119.0	+47.4[e]	166.4	135.0					31.4	
21	Tobacco manufactures	5.8		5.8	5.8						
22	Textile mill products	108.7		108.7		70.8	2.2	.7		35.0	
23	Apparel, etc.	43.7		43.7			.8			33.8	
24	Lumber and wood products	7.6		7.6			6.5	1.0	9.1	.1	
25	Furniture and fixtures										
26	Pulp, paper, and allied products	6.2		6.2	.9				.1	5.2	
27	Printing, publishing, and allied indus.	23.4		23.4	5.0		.8			17.6	
28	Chemicals and allied products	36.8		36.8	13.2				1.8	21.8	
29	Products of petroleum and coal	1.1		1.1	f					.4	.7[g]
30	Rubber products										
31	Leather and leather products	62.0		62.0		22.5		3.7		35.8	
32	Stone, clay, and glass products	18.2		18.2			2.1		13.7	2.4	
33	Primary metal industries	52.3		52.3			2.6	2.0	12.6[h]	35.1	
34	Fabricated metal products										
35	Machinery (incl. agric. implements)	11.0		11.0			6.1	8.7		2.3	
36	Transport. equipment	32.3		32.3			.6	8.4		17.8	
37	Prof. instruments	1.3		1.3			.6	.4		.3	

(continued)

TABLE A-7 (concluded)

S.I.C. No.	S.I.C. Title	Vol. 24 Total[b] (1)	Vol. 30 Corrections (2)	Commodity Flow Total[c] (3)	Perishables (4)	Semi-durables (5)	Consumer Durables (6)	Producer Durables (7)	Construction Materials (8)	Intermediate Commodities[d] (9)	Not Elsewhere Classified (10)
	Manufacturing Industries										
38	Miscellaneous	18.0		18.0		2.0	3.6	.7	.2[h]	11.5[i]	
	Total	548.2	+47.4[e]	595.6	159.9	96.3	25.2	25.5	37.5	250.4	.7[g]
	Nonmanuf., Nonagric. Indus.										
	Fisheries			4.5	4.5						
	Forest products			121.3	115.3[j]				6.0[k]		
	Grand total			721.4	279.7	96.3	25.2	25.5	43.5[h,k]	250.4	.7[g]

Source: Col. 1: Worksheets underlying Gallman in *Trends in the American Economy*, pp. 56-60; col. 2-10: Worksheets underlying Tables A-2 and A-3.

[a] Detail may not add to total due to rounding.

[b] The worksheets underlying the Volume 24 estimates do not contain complete estimates of value of product. Therefore, several of the entries in this colum were derived by dividing value added in 1839 by the ratio of value added to value of product in 1849.

[c] Total of cols.1-2 and also total of cols. 4-10.

[d] Partly estimated as residuals. See text.

[e] Underestimate of flour milling in Volume 24. The Volume 24 estimate of value added is about $10.4 million too low, which amounts to about 4.3 per cent of total value added by manufacturing in 1839.

[f] Less than .05.

[g] Manufactured gas.

[h] Excludes railroad rails. See text.

[i] Includes a large Census return of "miscellaneous" industries, some of which probably belong in other classifications.

[j] Consumption of firewood.

[k] Excludes railroad ties. See note h.

TABLE A-8

RECONCILIATION OF MANUFACTURING CENSUS, VALUE OF PRODUCT DATA UNDERLYING VOLUME 24 ESTIMATES OF VALUE ADDED BY MANUFACTURING, AND VOLUME 30 ESTIMATES OF COMMODITY FLOWS, CLASSIFIED BY TWO-DIGIT INDUSTRY GROUPS AND MAJOR COMMODITY GROUPS, 1849[a]
(million current dollars)

S.I.C. No.	S.I.C. Title	Census Industry Total (1)	Vol. 24 Corrections (2)	Vol. 24 Total (cols.1+2) (3)	Vol. 30 Corrections (4)	Commodity Flow Total[b] (5)	Perishables (6)	Semi-durables (7)	Consumer Durables (8)	Producer Durables (9)	Construction Materials (10)	Intermediate Commodities[c] (11)	Not Elsewhere Classified (12)
	Manufacturing Industries												
19	Ordnance and accessories	2.5		2.5		2.5		1.2			d	1.0	.3[e]
20	Food and kindred products	203.3	+.6[f]	203.9	+1.6[f]	205.5	177.0					28.4	
21	Tobacco manufactures	13.5		13.5		13.5	13.5						
22	Textile mill products	131.7		131.7		131.7		52.1	6.3	.3		73.1	
23	Apparel, etc.	84.5		84.5		84.5		65.6		1.3		17.7	
24	Lumber and wood products	75.3	+16.0[g]	91.3		91.3			2.0	.2	12.9	76.2	
25	Furniture and fixtures	19.5		19.5		19.5			16.6	2.5		.4	
26	Pulp, paper, and allied products	12.5		12.5		12.5	.9				.1	11.5	
27	Printing, publishing, and allied indus.	15.8	+27.6[h]	43.4		43.4	15.0		2.3			26.1	
28	Chemicals and allied products	46.5	−1.9[i]	44.6	+1.9[i]	46.5	21.9				2.5	22.0	
29	Products of petroleum and coal	1.9		1.9		1.9	.1				d		1.8[j]
30	Rubber products	3.0		3.0		3.0		1.1				1.9	
31	Leather and leather products	108.4		108.4		108.4		54.6	1.5	6.0		46.4	
32	Stone, clay, and glass products	16.5	+5.7[k]	22.1	+2.6[k]	24.7			3.8	.2	19.3	1.3	
33	Primary metal industries	52.3		52.3		52.3					2.0[ℓ]	50.3	
34	Fabricated metal products	46.8		46.8		46.8			8.4	6.3	12.9	19.3	
35	Machinery (incl. agric. implements)	36.7		36.7		36.7				28.6		8.1	
36	Transport. equipment	38.9	+10.8[m]	49.7		49.7			8.1	16.8		24.9	
37	Prof. instruments	2.4		2.4		2.4	d		1.2	1.0		.2	

(continued)

TABLE A-8 (concluded)

S.I.C. No.	S.I.C. Title	Census Industry Total (1)	Vol. 24 Corrections (2)	Vol. 24 Total (cols.1+2) (3)	Vol. 30 Corrections (4)	Commodity Flow Total[b] (5)	Perishables (6)	Semi-durables (7)	Consumer Durables (8)	Producer Durables (9)	Construction Materials (10)	Intermediate Commodities[c] (11)	Not Elsewhere Classified (12)
	Manufacturing Industries												
38	Miscellaneous	34.4		34.4		34.4	.5	7.7	12.1	.7	3.2[n]	10.2[n]	2.1
	Total	946.4	+58.9	1,005.3	+6.1	1,011.4	228.9	182.2	62.3	64.0	52.9[ℓ]	419.0	2.1
	Nonmanuf., Nonagric., Indus.												
	Fisheries	10.1	-.6[f]	9.5	-1.6[f]	7.9	2.8						
	Forest products	.1				173.7	161.3[o]					5.1	
	All other	62.6									12.4[p]		
	Grand total	1,019.2				1,193.0	393.0	182.2	62.3	64.0	65.3[ℓ,p]	424.1	2.1

Source: Cols. 1-3: Worksheets underlying Gallman in *Trends in the American Economy*, pp. 56-60; cols. 4-12: Worksheets underlying Tables A-2 and A-3.

[a]Detail may not add to total due to rounding.
[b]Total of cols. 3-4 and also total of cols. 6-12.
[c]Partly estimated as residuals. See text.
[d]Less than .05.
[e]U.S. armories.
[f]Fish preserving and packing.
[g]Coopers.
[h]Periodical press.
[i]Salt. The Census combines mining and manufacturing components of the industry. The two were separated for the Volume 24 study and recombined for the present study.

[j]Manufactured gas.
[k]Marble and stonework. The Census combines the quarrying and manufacturing components of the industry. The two were separated for the Volume 24 study and recombined for the present study.
[ℓ]Excludes railroad rails, not separately identified in the census. See text.
[m]Wheelweights.
[n]Includes a Census return of "miscellaneous" industries, some of which probably belong in other classifications.
[o]Consumption of firewood.
[p]Excludes railroad ties. See note ℓ.

TABLE A-9

RECONCILIATION OF MANUFACTURING CENSUS, VALUE OF PRODUCT DATA UNDERLYING VOLUME 24 ESTIMATES OF VALUE ADDED BY MANUFACTURING, AND VOLUME 30 ESTIMATES OF COMMODITY FLOWS, CLASSIFIED BY TWO-DIGIT INDUSTRY GROUPS AND MAJOR COMMODITY GROUPS, 1859[a]
(million current dollars)

S.I.C. No.	S.I.C. Title	Census Industry Total (1)	Vol. 24 Corrections (2)	Vol. 24 Total (cols.1+2) (3)	Vol. 30 Corrections (4)	Commodity Flow Total[b] (5)	Perishables (6)	Semi-durables (7)	Consumer Durables (8)	Producer Durables (9)	Construction Materials (10)	Intermediate Commodities[c] (11)	Not Elsewhere Classified (12)
	Manufacturing Industries												
19	Ordnance and accessories	2.9		2.9		2.9		2.3			.3	.1	.2[d]
20	Food and kindred products	421.0		421.0	+2.6[e]	423.6	371.2					52.4	
21	Tobacco manufactures	30.9		30.9		30.9	30.9						
22	Textile mill products	226.1		226.1		226.1		88.1	11.0	.2		126.7	
23	Apparel, etc.	127.0		127.0		127.0		105.0		1.3		20.7	
24	Lumber and wood products	146.9	+20.2[f]	167.1		167.1			4.1	.5	30.5	131.9	
25	Furniture and fixtures	29.3		29.3		29.3		g	22.4	3.8		3.1	
26	Pulp, paper, and allied products	26.7		26.7		26.7	1.8	g			2.1	22.8	
27	Printing, publishing, and allied indus.	37.1	+46.9[h]	84.0		84.0	28.0		4.4			51.6	
28	Chemicals and allied products	83.3		83.3	+.3[i]	83.6	32.8				7.9	42.9	
29	Products of petroleum and coal	18.6		18.6		18.6	6.9	2.0				3.6	
30	Rubber products	5.6		5.6		5.6							11.7[j]
31	Leather and leather products	186.7		186.7		186.7		93.1	2.7	8.6		82.3	
32	Stone, clay, and glass products	46.8	−4.8[k]	41.9	+4.8[k]	46.8			7.2	.6	35.4[l]	3.6	
33	Primary metal industries	116.9		116.9		116.9			.2	.9	6.0[m]	109.7	
34	Fabricated metal products	63.0		63.0		63.0			21.0	7.4	24.8	9.8	
35	Machinery (incl. agric. implements)	82.8		82.8		82.8			3.0	66.1		13.7	

(continued)

TABLE A-9 (concluded)

S.I.C No.	S.I.C. Title	Census Industry Total (1)	Vol. 24 Cor- rections (2)	Vol. 24 Total (cols.1+2) (3)	Vol. 30 Cor- rections (4)	Com- modity Flow Total[b] (5)	Perish- ables (6)	Semi- durables (7)	Consumer Durables (8)	Producer Durables (9)	Con- struction Materials (10)	Inter- med- iate Com- modi- ties[c] (11)	Not Else- where Clas- si- fied (12)
	Manufacturing Industries												
36	Transport. equipment	63.1	+12.4[n]	75.5		75.5			19.5	26.5		29.6	
37	Prof. instruments	5.5	−.4[o]	5.1		5.1	.1		2.2	1.9		.8	
38	Miscellaneous	38.6		38.6		38.6	.8	10.8[g]	18.1	.9		7.7	
	Total	1,758.6	+74.3	1,833.0	+7.7	1,840.7	472.4	301.4	115.8	118.7	107.4[ℓ,m]	713.1	11.9
	Nonmanuf., Nonagric.Indus.												
	Fisheries	14.4			−2.6[e]	11.8	5.9					5.9	
	Forest products	1.3				231.5	208.2[p]				23.3[q]		
	All other	111.5											
	Grand total	1,885.8				2,084.0	686.5	301.4	115.8	118.7	130.7[ℓ,m]	719.0	11.9

Source: Cols. 1–3: Worksheets underlying Gallman in *Trends in the American Economy*, pp. 56–60; cols. 4–12: Worksheets underlying Tables A-2 and A-3.

[a] Detail may not add to total due to rounding.
[b] Total of cols. 3–4 and also total of cols. 6–12.
[c] Partly estimated as residuals. See text.
[d] Ammunition.
[e] Fish preserving and packing (+2.7); bottled liquors (−.1).
[f] Coopers.
[g] Less than .05.
[h] Periodical press.
[i] Estimate based on salt mining.
[j] Manufactured gas.
[k] Marble and stonework. The Census combines the quarrying and manufacturing components of the industry. The two were separated for the Volume 24 study and recombined for the present study.
[ℓ] This represents a reduction of 5.7 from the Volume 24 total, due to the discovery of an error in the Volume 24 computations.
[m] Excludes railroad rails, not separately identified in the Census. See text.
[n] Wheelwrights.
[o] Watch repairing.
[p] Consumption of firewood.
[q] Excludes railroad ties. See note m.

domestic users of intermediate production.[62] Therefore, intermediate flows of "mixed" commodities are fully accounted for by domestic production and hence imports of "mixed" goods must be treated as final goods. Since some imported mixed goods certainly entered domestic production, the procedure understates final domestic output and overstates imports of final goods by a like amount. But we have no adequate data to apportion domestic consumption of intermediate mixed goods between domestic production and imports of these goods. Therefore, there is no way in which the error can be corrected.

We had understood that Shaw had followed the procedure described above in every case and we had made our prewar estimates on that assumption. But John Dales of the University of Toronto has pointed out in correspondence that, in fact, some of Shaw's estimates of the value of intermediate production were based on the shipping records of domestic producers of mixed commodities; i.e., they cover only domestic production of intermediate mixed goods, not domestic consumption of them. Shaw was, therefore, obliged to estimate in some cases the imports of mixed goods flowing into domestic intermediate uses and to subtract these estimates from total imports of mixed goods. Apparently only one minor commodity group (6) was involved, however.[63] Shaw divided group 6 mixed imports between final and intermediate classes in the same proportions as he had divided domestic product (group 6, mixed) between these classes.

The procedure is quite weak and produces nonsensical results if applied in the prewar years. An alternative would be to estimate the value of minor group 6 mixed intermediate imports by extrapolation from 1869 (in detail) on the value of materials consumed by using industries. Since Shaw gives little detail on imports, the method would require a substantial amount of work to reconstruct the data underlying Shaw's import estimates. Rough tests suggest that the results to be expected would not be worth the work. The minor group 6 estimates in major benchmark years would be changed by between 1 and 2.5 per cent at most. Additionally, as will become clear, other factors tend to compensate for whatever error is introduced by our failure to rework the estimates at this point. Consequently, we did not carry out these corrections. Of course, before the tests were made, it was by no means obvious what their results would be. Therefore, we are grateful to Dales for taking the trouble to apprise us of our misinterpretation of Shaw.

The data on exports and imports were taken from Treasury reports,

[62] Shaw, *Commodity Output*, pp. 186–199.
[63] *Ibid.*, p. 276.

which contain fiscal year data. Through 1842, the fiscal year ended on September 30; thereafter, on June 30. We made no effort to adjust the data to fit the Census year more closely.

Treasury data are classified by commodities. Where details are provided, we encountered no serious problems of classification. However, there are various N.E.S. (not elsewhere specified) categories which we were unable to distribute and which, therefore, we could not use. Further work in this area might lead to improvement of the estimates. For the major benchmark years, the N.E.S. export categories included values equal to between 4 and 14 per cent of the value of exports of final goods (including producer durables); the N.E.S. import categories included between 8 and 15 per cent of the value of imports of final goods. The problem on the export side is somewhat less serious than on the import side. A substantial fraction (about one-half in 1839, one-seventh in 1849, and one-third in 1859) of N.E.S. exports are classified as "raw" and therefore presumably are not importantly involved in final flows. Additionally, the value of N.E.S. exports is only equal to between one-third and one-tenth of the value of N.E.S. imports and therefore affects flows much less markedly than does the value of N.E.S. imports.

Three major N.E.S. import categories probably contained few items of final flow. One, which appeared only in 1859, covered imports from British provinces under the reciprocity treaty, probably mainly unfinished goods. The other two are "N.S. manufactures of iron and steel" and "N.S. articles paying duties of 1–5 per cent," which North classifies with raw materials and foods.[64] Eliminating these items, the remaining N.E.S. categories cover values equal to about 5 per cent of imports of final goods in 1839, 9 per cent in 1849, and 4 per cent in 1859. In the main, these are goods paying over 5 per cent in duties, presumably manufactured goods which might figure in final flows.[65]

Exports were valued at the port. Shaw estimated the differences between port and producer prices, by minor groups, and adjusted export valuations to the producer price level. The adjustments ran between 5 and 20 per cent of export values at port prices, but produced exceedingly limited modifications in minor group final flows.[66] We had no bases for carrying the adjustments into the prewar years and therefore did not make prewar adjustments. In the aggregate, the resulting overstatement of the value of exports compensates for the understatement arising from

---

[64] Douglass C. North, *The Economic Growth of the United States, 1790–1860*, Englewood Cliffs, 1961, p. 289.
[65] *Ibid.*
[66] Shaw, *Commodity Output*, p. 271.

our failure to use the N.E.S. categories and from reporting deficiencies (slight) noted by North.[67] There is no reason to believe, however, that there is appropriate compensation at the minor group, or even the major group, level.

Import valuations were net of shipping costs and duties. Shaw left out the former on the ground that the required adjustment is slight.[68] But North's data suggest that shipping costs ran between 6 and 9 per cent of the value of total imports.[69] Presumably shipping costs of final goods were relatively less important, although probably not much less. In any case, we had no basis for distributing shipping costs among minor groups. A case might be made for distributing them on the basis of import values; but we chose to disregard them altogether.

We computed duties from tariff schedules. This laborious procedure should perhaps be justified. Given the character of output and trade data and the manipulations to which they were subjected, a shortcut to the calculation of duties, which implied more limited claims to accuracy, might have been preferable to the procedure followed. However, we could devise no acceptable shortcut which reduced the number of computations significantly. Furthermore, historical reconstructions are never complete. They are always subject to revision. That being the case, it is reasonable to take as much care as possible at each step, even though this means ultimately combining the results of detailed and accurate procedures with the results of the crudest extrapolation. The crude extrapolation may some day be replaced by a better series.

The export estimates, in the aggregate, may measure fairly accurately the flows we attempted to measure. Factors making for underestimates are probably balanced by factors making for overestimates. In the case of imports, however, there is no such balancing. The estimates are clearly short. We failed to take into account the N.E.S. categories and shipping costs. Additionally, North thinks that imports were undervalued by about 2 per cent until 1846, and by about 4 per cent thereafter.[70] In the aggregate, the import estimates are probably no more than 20 per cent short, and perhaps much less. Suppose that the import estimates were, in fact, 20 per cent short, in the aggregate and for each of the major groups. Then our estimate of the final flow of consumer commodities (in producer prices) would be a little less than 2 per cent short in 1839,

---

[67] Douglass C. North, "The United States Balance of Payments, 1790–1860," *Trends in the American Economy*, p. 602.

[68] Shaw, *Commodity Output*, p. 272.

[69] North in *Trends in the American Economy*, pp. 607–608. The range in nonbenchmark years is from a little over 5 to a little under 12 per cent.

[70] North in *Trends in the American Economy*, p. 604.

and a little more than 2 per cent short in 1849 and 1859. For the major groups, the shortages would be roughly as follows (in per cent):

	1839	1849	1859
Perishables	1	1	1–2
Semidurables	4	6	7
Durables	3	3	3

The shortages may bias the trend of the series, but only very slightly.

Minor Benchmark Years. Domestic final flows were interpolated to the minor benchmark years 1844 and 1854 on the returns of state censuses, described in another publication.[71] Most of the flows were also extrapolated to 1836 on the returns of the Massachusetts Census and some of the important ones to 1834 on the returns of the New York Census.[72] (As noted above, we did not construct full current price GNP estimates for these two years.) The foreign trade adjustments were made in precisely the way in which they were made for the major benchmark years. However, the N.E.S. import categories in 1844 and 1854 were relatively more important than in the major benchmark years. After "N.S. manufactures of iron and steel" and "N.S. articles paying duties of 1–5 per cent" were deducted they were equal in value to 14 and 11 per cent, respectively, of imports of final commodities.

Unmanufactured Commodities Flowing to Consumers in Producer Prices

Estimates in current prices were made for the major benchmark years and for the minor benchmark years 1844 and 1854. Estimates of domestic final flows (except firewood) were extrapolated from 1869 on series appearing in the Volume 24 study.[73] The procedures followed to make the estimates were similar to those described for the estimates of manufactured commodities flowing to consumers. We used the Class B procedure principally.

The Shaw and Volume 24 series on the value of farm products rest, in the main, on the same sources and are largely consistent. However, Shaw has a series on the value of output of small fruits that is missing from the Volume 24 study. We extrapolated Shaw's 1869 estimate to the

---

[71] Gallman in *Trends in the American Economy*, pp. 56, 60. The states are Massachusetts, New York, and Connecticut (1844 only).

[72] *Statistical Tables: Exhibiting the Condition and Products of Certain Branches of Industry in Massachusetts for the Year ending April 1, 1837*, Boston, 1838; *Census of the State of New York for 1835*, Albany, 1836.

[73] Gallman in *Trends in the American Economy*, pp. 46–47.

prewar years on the value of output of orchard fruits. Additionally, Shaw's estimates of the value of farm and market garden produce are larger than the Volume 24 estimates. Therefore, we adjusted the Volume 24 prewar figures upward. Finally, our postwar estimates of nonmanufactured animal products rest on series consistent with the Volume 24 animal products series. However, as noted above, we adjusted the basic series downward to make a smooth link with Kuznets' estimates at 1909. Consequently, it was necessary to adjust the prewar Volume 24 animal products estimates downward also, before these estimates were used to derive final flows. The following tabulation (in million dollars) compares the adjustments we have made in the prewar Volume 24 series for the major benchmark years with the Volume 24 estimates of aggregate value of output of agriculture (excluding firewood, improvements to farm lands, and home manufactures, all treated separately below):

	1839	1849	1859
1. Value of output of agriculture, excl. value of firewood, farm improvements, and home manufact.	631	738	1,377
2. Adjustments to component series:			
a. Small fruits	+2.5	+2.7	+7.0
b. Farm and market garden produce	+6.5	+7.2	+31.5
c. Animal products	−32.6	−38.3	−68.4
d. Lines a + b + c	−23.6	−28.4	−29.9
3. Ratio of line 2d to line 1	.037	.038	.022

The adjustments, in the aggregate, are of quite limited significance.

The value of firewood consumed was estimated directly from consumption (physical quantity) and price (see the earlier discussion of the postwar estimates). The 1879 price was carried to the prewar years on Dorothy Brady's firewood price index. We made no prewar estimates for other nonmanufactured fuels flowing to ultimate consumers. The omission is quite unimportant, but serves to offset, at least to a limited extent, the probable overestimate of firewood consumption.

Trade Markups

Postwar estimates of the shares of output flowing into distribution and distributive spread,[74] at the minor commodity group level, were

[74] Defined here as the ratio of value added by distribution (including transportation of finished goods) to the value of goods entering distribution, in producer prices.

extrapolated to the prewar years and checked against Seaman's 1839 estimates. In the main, Seaman's work confirms our results:[75]

	Minor Groups	Distributive Spread (per cent)	
		Seaman	Extrapolation
1a, 1b	Food and kindred products	50	62
2	Cigars, cigarettes, tobacco	50	53
6–9, 14	Dry goods and notions; clothing and personal furnishings; shoes and other footwear; semidurable and durable housefurnishings	39	39

Marburg estimated income originating in trade in 1839 from figures on employment, capital stock, wage rates, and rates of return on capital.[76] He also estimated the relationship between spread and income originating, so that one can derive value added by distribution (including the transportation of finished goods) from his data.[77] According to Marburg's figures, value added by trade was $203 million in 1839. Our estimate is well within 10 per cent of this figure.

Services Flowing to Consumers

Perhaps there are, somewhere, usable budget studies for the prewar period which might be used to work out service estimates along the lines of Kuznets' procedure, but we have not come across them. There are various series related to services which could enter into an extrapolator: the stock of residential units (available back to 1850, but easily carried back to 1840), state and local government tax receipts (back to 1849), the value of the stock of churches (back to 1860), estimates of the labor force attached to the service sector. Flow estimates depending on labor force or capital stock figures have less analytical value than estimates independently made. But we could not dispense with both the labor force and capital stock series. We chose to depend exclusively on the latter, partly because this seemed the more secure procedure and partly because we

---

[75] Seaman, *Essays*, 2nd ed., pp. 278, 280, 283. The spread for food and kindred products refers only to domestically produced goods. Apparently Seaman believed that the spread on imported goods was much greater (*ibid.*, pp. 279–280). In another place (*ibid.*, pp. 458–459), Seaman provided different estimates, but apparently he intended that they measure only income originating in trade, not distributive spread. See Gallman in *Economic Development and Cultural Change*, April 1961, pp. 403–406, which also discusses Seaman's estimates of flows into distribution. These also tend to confirm our results.

[76] Theodore F. Marburg, "Income Originating in Trade, 1799–1869," *Trends in the American Economy*, p. 322.

[77] *Ibid.*, p. 321.

believed that the immediate analytical losses would be less serious than if we used labor force figures. However, the choice was not much more than a matter of taste, as will become clear.

The estimates were made in 1860 prices and were then adjusted to current prices, since the form of the data made this the simplest and most reasonable method. As in the case of the postwar period, we had price indexes representing rents and medical services, the latter only back to 1851. The two indexes are almost identical; therefore we did not bother to combine them and used only the rent index. The construction of the constant price estimates is described below. Here we consider only the tests of the estimates.

Seaman's estimate of the value of rents to the free population (including imputed rents of owner-occupied houses) in 1839 comes to $90 million.[78] Assuming that the value of per capita consumption of shelter by slaves was half that of free persons, the value of rents for the entire population would be $96 million, substantially below our estimate of $166 million.

Seaman's procedure involves the estimation of average family rent and average family size, the latter presumably standing for the average number of persons per dwelling. Average family size is taken to be seven, which may be too large. The ratio of population to dwellings, according to Gottlieb's data, is under 6 to 1 in 1850 and about 5.6 to 1 in 1860 (including slaves),[79] while our estimate of the stock of dwellings in 1840 (derived from Gottlieb's data) implies a ratio of 5.3 to 1 in that year. Were Seaman's estimate of average family size the same as our estimate of average number of persons per dwelling, his rental estimate would be $128 million, closer to, but still below, ours.

Seaman's rent estimate was made in the context of a national product estimate. The share of rents in Seaman's estimate of the flow of goods to consumers (excluding nonrent services) is very close to the share implied by our work (in per cent):

Seaman, Unadjusted	Seaman, Adjusted for Slaves and Family Size	Gallman
10	14	12

According to E. W. Martin, rents accounted for perhaps 17 per cent of the expenditures of families of urban working men in the 1850's.[80] Presumably

---

[78] Seaman, *Essay on the Progress of Nations*, 1st ed., New York and Detroit, 1846, p. 305.

[79] Manuel Gottlieb, *Estimates of Residential Building, 1840–1939*, NBER Technical Paper 17, New York, 1964, p. 44; *Historical Statistics*, 1960, Series A-1.

[80] Edgar W. Martin, *The Standard of Living in Chicago in 1860*, Chicago, 1942, pp. 396–397. However, the evidence on which the judgment rests is not impressive.

this share would be well above the average for all families and, consequently, Martin's judgment provides a modicum of support for the distribution of final flows shown by Seaman's work and ours.

In his second edition, Seaman estimated the value of residences (including land and associated outbuildings) in 1840 at "... over a thousand million dollars...."[81] Assuming that Seaman intended to exclude slave dwellings from his estimate of residences, then annual rents amounted to about 9 per cent of the value of residences; assuming that he intended to include slave dwellings (unlikely), then rents probably amounted to under 10 per cent of the value of residences. Our estimate of rents comes to about 8 per cent of our estimate of the value of residences. Goldsmith's work implies a ratio of between 7 and 8 per cent in 1850.[82]

It is fairly clear, then, that the relationships among rents, totals flows to consumers, and the value of residences are roughly the same in Seaman's work and ours. Additionally, other scraps of evidence suggest that these relationships are appropriate. The difference between us has to do with the levels of the various aggregates. Seaman puts forward smaller values than we do.

In his first edition, Seaman estimated the value of nonrent services, "... ordinary domestic labor, medical and professional services, education, religious instruction, amusements, and ... the expenses of government and the administration of justice ... ," at $310 million.[83] In his second edition, the list of items covered is slightly different and perhaps more inclusive: "... housekeeping, the labor of domestic servants, all professional business, teaching of all kinds, all matters of pleasure and amusement, official labor, military services, and the administration of justice."[84] The estimate is very much lower than in the first edition—$150–$200 million—and is fairly close to our figure of $200 million. But our estimate excludes the value of the services of housewives, whereas Seaman's may include it. If that is the case, then Seaman's estimate less the value of the services of housewives would be lower than our figure, and perhaps substantially lower. It is possible that, as in the case of rents, Seaman's estimate may bear roughly the same relationship to other components of final flow, within his framework, as our estimate bears to components of final flow in our framework. The first edition estimate, however, seems far out of line. It implies that nonrent services accounted for over three-quarters of

[81] Seaman, *Essay*, 2nd ed., p. 282.
[82] Raymond W. Goldsmith, "The Growth of Reproducible Wealth of the United States of America from 1805 to 1950," *Income and Wealth of the United States*, p. 319.
[83] Seaman, *Essay*, 1st ed., p. 305.
[84] Seaman, *Essay*, 2nd ed., p. 284. See also Gallman in *Economic Development and Cultural Change*, April 1961, pp. 409–410.

the flow of services to consumers, compared with something over 50 per cent in the postwar years. Surely the share of nonrents in services must have risen, not fallen, between 1839 and the postwar years. (This is the pattern of change which our estimates show, incidentally.)

A second test of the nonrent component can be best conducted in terms of the constant price estimates. The following tabulation compares rates of change of our constant price series with rates of change of the labor force attached to the service sectors (in per cent):

	1839–59	1859–79	1879–99	1839–99	1859–99
Nonrent services	111	105	52	555	210
Number of teachers and free domestics[85]	180	93	34	626	159
Number of workers attached to the service sectors (less trans., pub ut., trade, and fin.)[86]	126	60	98	617	218

The movements of the final flow series are broadly similar to the movements of the labor force series and the degree of correspondence in the prewar period is at least as close as in the postwar period. There is a suggestion that the rate of growth of the final flow series may be too small—surely there were some improvements in the productivity of the service industries over the period. However, the service sector is heterogeneous and shifts in the composition of the sector need not always promote higher output per worker for the sector as a whole. For example, Lebergott's data show that the number of free domestics increased faster than the number of teachers between 1839 and 1859. Presumably this development tended to reduce output per worker in the service sector.

Manufactured Producer Durables

The estimating procedures are described above on pages 37 and 43.

Gross New Construction

We followed the procedure and used the materials flow series described for the postwar estimates. Fishlow made available to us his new railroad construction series.[87] We also estimated canal construction separately, using Cranmer's figures for this purpose.[88] We had no figures on the value

---

[85] Based on preliminary estimates by Stanley Lebergott, transmitted by letter dated July 18, 1960.
[86] *Historical Statistics*, 1960, Series D, 66–70.
[87] Typescript supplied by Albert Fishlow, June 22, 1964.
[88] H. Jerome Cranmer, "Canal Investment, 1815–1860," *Trends in the American Economy*, pp. 555–556.

GROSS NATIONAL PRODUCT, 1834–1909   61

of materials flowing into canal construction and simply assumed, arbitrarily, that the ratio of materials used to value of construction was .30, or slightly higher than the ratio of materials used in railroad construction to the value of railroad construction.

Our estimate of new, nonrailroad, noncanal construction in 1839 is $108 million. The Census return of the value of houses constructed, which Seaman and Gottlieb regard as roughly accurate,[89] is $42 million, leaving $66 million for all other construction. Seaman has a series of estimates (in million dollars) which bear on the residual:[90]

Draining and other improvements made on agricultural land, except land newly cleared	6
Increase of manufacturing, milling, mechanical, and mining capital ...	16
Increase in capital employed in commerce, retail trade, navigation, transportation and fisheries ...	15
Increase in other public property, such as roads, bridges, churches, national, state, and county buildings, forts, harbours etc. ...	10
Total	47

Seaman's estimates are apparently of net investment, whereas our figures refer to gross new construction. Additionally, Seaman has no estimate for farm, nonresidential building. On the other hand, Seaman apparently includes changes in the value of nonagricultural inventories and net investment in nonagricultural manufactured producer durables, items missing from our figure, of course. The impression conveyed by these data is that our construction estimate may be a little high.

Changes in Claims Against Foreigners

We used North's estimates of annual net balance.[91]

Summary Appraisal

The data underlying the estimates of the value of output of final consumer and producer commodities have been tested in a variety of ways and we believe them to be strong. The data on highly fabricated manufactures are more difficult to test than those on the early stages of manufacturing, but the procedure for estimating final flows incorporates hedges against errors in these data. Unfortunately, we cannot give a quantitative

---

[89] Seaman, *Essay*, 2nd ed., pp. 282, 456; Gottlieb, *Estimates*, pp. 51–57. Gottlieb is more interested in, and more impressed by, the Census count of houses built than the returns of the value of houses built.

[90] Seaman, *Essay*, 2nd ed., p. 284.

[91] North in *Trends in the American Economy*, p. 581.

allowance for error in the aggregates or the major components. Census procedures improved over time and one would expect that each Census was more complete than the one preceding it. But we have worked with the Census materials in detail, adjusting where it could be shown that the data were deficient. We doubt that the final series is significantly biased.

The foreign trade estimates are weaker than the value of output estimates. Without question, they understate the value of the net flows of foreign final goods into domestic consumption, conceivably by as much as 20 per cent, or something less than 2 per cent of GNP.

The estimates of value added by distribution are weaker than the value of output estimates, but probably at least as strong as the foreign trade estimates.

Of all the estimates, the poorest are those of the value of services flowing to consumers. We do not know what margin for error to assign to these figures. If they are in error, the chances are that they are too high. Services account for roughly one-quarter of GNP in the prewar years. Consequently, an error as large as 20 per cent in the service component would throw GNP off by only 5 per cent.

The construction estimates are probably at least as strong as the estimates of value added by distribution. The tests suggest that the 1839 estimate may be high—at a guess, perhaps as much as 10 per cent high, that is, less than 1 per cent of GNP.

The estimates of changes in claims against foreigners are strong. In any case, the component is so small that errors in it would have an exceedingly limited effect on GNP.

GROSS NATIONAL PRODUCT, FINAL PRODUCT FLOWS, 1834–59, PRICES OF 1860, BENCHMARK YEARS

We used Dorothy Brady's price index numbers to deflate all components which had been originally estimated in current prices, except for firewood, some components of unmanufactured food, and railroad and canal construction. The index numbers were briefly discussed in connection with the postwar estimates. Two areas of weakness should be mentioned. First, there are no nonrent service deflators, as noted above. Second, there are gaps in the coverage of manufactured producer durables which had to be filled by interpolation and extrapolation on the few continuous series. However, all of the interpolations and extrapolations were made within the prewar period; i.e., no price index had to be extrapolated from the postwar period.

A few estimates, such as those for firewood, were based on physical output series. Where we had an appropriate 1860 price, we produced the

GROSS NATIONAL PRODUCT, 1834–1909

constant price series directly, rather than by deflation. We used Fishlow's price index to deflate railroad construction and his index of wage rates of common labor to deflate canal construction.

As noted above, the service estimates were made in constant prices and then inflated. The constant price estimates are extrapolations from the postwar period. The principal extrapolator was a series on the value of residences in 1860 prices, weighted .1 to approximate the value of rents.[92] The remaining extrapolators were a series on tax receipts of state and local governments[93] (running back to 1849), deflated by use of the rent price index, and a series on the value of churches (running back to 1860), deflated by use of the price index of houses and churches and weighted .1.[94]

We derived the series on the value of residences by valuing the estimates of the stock of dwellings. We used Gottlieb's stock data for 1850 and 1860, associating each with a rental estimate for the Census year.[95]

The stock of nonfarm dwellings in 1840 was calculated by subtracting Gottlieb's estimate of housing production in the 1840's from his figure for the nonfarm stock in 1850.[96] We estimated the farm increment of the 1840's by extrapolation from the 1850's on the nonfarm increment and derived the 1840 farm stock by subtracting the farm increment of the 1840's from the 1850 farm stock. We checked our results by comparing the ratio of farm stock to agricultural labor force with the ratio of nonfarm stock to nonagricultural labor force in 1840, 1850, and 1860. The test suggests that our estimate for 1840 may be slightly high.

We used Goldsmith's 1850 unit values,[97] divided by Brady's price index of houses and churches, to value the stock estimates. We then raised the values by 20 per cent to take into account the value of land.[98]

The three extrapolators were then combined and the value of services in 1860 prices was extrapolated from the postwar years to 1859. The 1859 estimate was then extrapolated to 1849 on the sum of the two extrapolating series available, and the 1849 estimate was extrapolated to 1839 on the one remaining extrapolator. The tests of the estimates have been discussed in a preceding section.

Several components of the 1834 and 1836 benchmark GNP estimates were produced in constant prices only. The procedures used are described below with the nonbenchmark year estimates.

[92] See, e.g., Goldsmith in *Income and Wealth*, p. 319.
[93] *Census of 1890*, Vol. 15, Part II, 1892, p. 61. Compendium of the Census of 1850.
[94] *Historical Statistics*, 1960, Series H-530.
[95] Gottlieb, *Estimates*, p. 44.
[96] *Ibid.*, pp. 44 and 61.
[97] Goldsmith in *Income and Wealth*, p. 319.
[98] *Ibid.*

GROSS NATIONAL PRODUCT, FINAL PRODUCT FLOWS, 1834–59, PRICES OF 1860, INTERBENCHMARK YEARS

The interbenchmark estimates of changes in claims against foreigners are North's figures (see above, benchmark estimates), deflated by the implicit GNP price index, interpolated or extrapolated from benchmark years on the Warren and Pearson all commodities wholesale price index.[99]

The benchmark construction estimates were interpolated on the Fishlow, Gottlieb, and Cranmer series (deflated), described above (see the benchmark estimates), and extrapolated to the pre-1839 years on the Fishlow and Cranmer series, together with the lumber series described in Table A-10.

Gottlieb's stock and flow data, together with our 1840 stock estimate, were used to derive annual housing stock estimates, 1840–60. Gottlieb's flow series was extrapolated to 1836 on the lumber series of Table A-10, and annual stock estimates were than derived for the years 1835–39. The benchmark service estimates were interpolated and extrapolated to non-benchmark years on the housing stock series.

Table A-10 describes the derivations of all other interbenchmark estimates. The table is intended to show the exact weighting schemes of the interpolating series and is, therefore, a little complicated. For example, benchmark estimates of perishables were interpolated and extrapolated on annual estimates for various minor groups, among them Minor Group 1a. The Minor Group 1a annual estimates, in turn, were interpolated and extrapolated from benchmark years on annual estimates of various components of Minor Group 1a, among them domestic production of flour and bread flowing to domestic consumers and net imports of manufactured and unmanufactured food (see Table A-10, item 1a). The latter series was assembled from Treasury reports, but the former was produced by interpolation of benchmark estimates on a flour trade series or a series measuring wheat or wheat and corn flowing into domestic production (see Table A-10, item 1a, 1).

The following general points concerning the interpolators and extrapolators should be noticed:

1. Net imports receive relatively too much weight, since the annual series relating to net imports exhaust the universe (except for the N.E.S. categories, see above), whereas the series referring to domestic flows do not. This is especially important in the case of semidurables and accounts for our willingness to use the relatively weak leather series in the 1834–42 interpolations. Over this period net imports oscillate fairly widely. We

[99] *Historical Statistics*, 1949, Series L-2.

TABLE A-10

INTERPOLATING AND EXTRAPOLATING SERIES, COMMODITY FLOWS TO
CONSUMERS AND MANUFACTURED PRODUCER DURABLES,
1860 PRICES, 1834-59

Component Interpolated or Extrapolated	Interpolating and Extrapolating Series	Years[a]	Sources
1. Perishables, less firewood & unmanufactured animal products	*Minor Group 1a* Manufactured food & kindred products		
	Minor Group 1b: Unmanufactured food & kindred products		
	Minor Group 5a: Manufactured fuel & lighting products		
1a. Minor Groups 1a & 1b	Domestic production flowing to domestic consumers: Bread & other bakery products Flour Clean rice Meat products Rum, whiskey, & other distilled products Liquors, malt Canned fish & oysters, smoked, salted, & pickled fish Confectionery, Sugar, granulated, refined, & brown Salt		
	Imports of coffee	1834-38 1840-43 1845-48 1850-53 1855-58	*Historical Statistics* (1960) Series U-95
	Net imports of manufact. &[b] unmanufact. food (less coffee)	1835-37 1837-38 1840-43 1845-48 1850-53 1855-58	Treasury Reports
1a.(1) Bread & other bakery products Flour	Bbls. of flour rec. at tidewater (N.Y.)	1835 1837-38 1840-46	*Hunt's Merchants' Magazine* (Vol. 28, p.481)
	(Wheat prod.[c] plus imports less exports) times .85 times 1859 price of wheat plus (corn prod. plus imports less exports) times .20 times 1859 price of corn	1841-45 1847-48	Prod.: Annual Reports of the Patent Office[d] Exports & imports: Treasury Reports
	Wheat prod. plus imports less exports	1850-53 1855-58	e

(continued)

TABLE A-10 (continued)

Component Interpolated or Extrapolated	Interpolating and Extrapolating Series	Years[a]	Sources
1a.(2) Clean rice	Rice prod. less exports	1834-38 1840-43 1845-48 1850-53 1855-58	U.S.D.A. Circular 33 (1912)
1a.(3) Meat products	Pork prod. at Cincinnati	1834-38 1840-41	T.S. Berry[f] Western Prices before 1861 (1943, p. 223)
	Pork prod. in the West	1842-43 1845-48 1850	
	Pork prod.[g] in the West times 1859 price of pork plus beef prod. in Chicago times 1859 price of beef	1851-53 1855-58	Hunt's Merchants' Magazine[h] (Vol. 40, p.230)
1a.(4) Rum, whiskey, & other distilled products	Corn prod. plus imports less exports	1841-43 1845,47 1848	See item 1a.(1)
1a.(5) Liquors, malt	Barley prod.	1841-43 1845,47 1848	Annual Reports of the Patent Office[d]
1a.(6) Canned fish & oysters, smoked, salted, & pickled fish	Mackerel catch	1834-35 1837,38 1840-43 1845-48 1850-53 1855-58	Historical Statistics (1949) Series F-164
1a.(7) Confectionery, sugar, granulated, refined, & brown	Louisiana sugar prod. less exports	1834-35 1837-38 1840-43 1845-48 1850-53 1855-58	Prod.: L.C. Gray, History of Agriculture in the Southern U.S. to 1860 (1933, Vol. II, p. 1033)
1a.(8) Salt	Salt made at the Onondaga Springs	1834-38 1840-43 1845-48 1850-53 1855-58	Transactions of the N.Y. State Agricultural Society (Vol. XIII, 1853, pp. 172-173)
1b. Minor Group 5a	Sperm oil prod. x 2 plus whale oil prod.	1834-35 1837-38 1840-43 1845-48 1850-53 1855-58	G.B. Goode, The Fisheries & the Fishing Industry of the U.S. (1884-87, Vol. 2, p. 168)
2. Firewood	Straight-line interpolation	1834-38 1840-43 1845-48 1850-53 1855-58	

(continued)

TABLE A-10 (continued)

Component Interpolated or Extrapolated	Interpolating and Extrapolating Series	Years[a]	Sources
3. Unmanufactured animal products	Domestic prod. flowing to domestic consumers: Beef Pork	i	i
4. Semidurables	Minor Group 6: Dry goods & notions Minor Group 8: Boots & shoes Net imports of nonfood consumer goods	1835,37,38 1840-43 1845-48 1850-53 1855-58	Treasury Reports[j]
4a. Minor Group 6	Domestic production flowing to domestic consumers: All other cotton woven goods Woolen & worsted woven goods, except shawls, blankets & carriage equipment		
4a.(1) All other cotton woven goods	Consumption of raw cotton by manu.	1835,37 1838 1840-43 1845-48 1850-53 1855-58	U.S.D.A. Office of the Experiment Stations *Bulletin 33* (pp. 41-42)
4a.(2) Woolen & worsted woven goods	Raw wool prod. plus imports	1835,37 1838 1840-43 1845-48 1950-53 1855-58	k
4b. Minor Group 8	Leather inspections in New York	1835,37 1838 1840-42	*Hunt's Merchants' Magazine* (Vol. 30)
5. Consumer durables	Minor Group 12: Household furniture Minor Group 20c: Passenger vehicles, horse-drawn Minor Group 13a: Heating & cooking apparatus Minor Group 14a: Floor coverings Minor Group 14b: Miscellaneous Housefurnishings		
5a. Minor Groups 12 & 20c	Lumber trade: Bangor	1834,35 1837,38 1840-43 1845-48	*Hunt's Merchants' Magazine* (Vol. 18, p.518)

(continued)

TABLE A-10 (concluded)

Component Interpolated or Extrapolated		Interpolating and Extrapolating Series	Years[a]	Sources
		Bangor, Chicago Albany	1850	*Ibid.* (Vol. 40, p.229)
		Chicago, Albany	1851-53	*Ibid.*
		Chicago, Albany Baltimore, Florida	1855-56	*DeBow's Review* (Vol. 27, p. 105, Florida)
		Bangor, Chicago Albany, Baltimore, Florida	1857	
		Straight-line interpolation	1858	
5b.	Minor Group 13a	Crude iron consumed by the domestic iron industry	1840-43 1845-48 1850-53 1855-58	R. W. Fogel, *Railroads and American Economic Growth* (1964, p. 192)
5c.	Minor Group 14a	Raw wool prod. plus imports	See item 4a.(2)	
5d.	Minor Group 14b	Consumption of raw cotton by manufact.	See item 4a.(1)	
6.	Manufactured producer durables	Production of sewing machines *Minor Group 27:* Farm equipment *Minor Group 30:* Locomotives *Minor Group 31:* Ships & boats *Minor Group 32:* Business vehicles, horse-drawn, & railroad cars	1855-58	*DeBow's Review* (Vol.28, p.236)
6a.	Minor Group 27	Production of agricultural machinery	1850-53 1855-58	Unpublished series received from Paul David
6b.	Minor Group 30	Production of locomotives	1834-38	Unpublished series received from Albert Fishlow
		Railroad investment in equipment	1839-59	*Ibid.*
6c.	Minor Group 31	Prod. of ships less sales of ships to foreigners	1835,37 1838 1840-43 1845-48 1850-53 1855-58	Prod.: *Historical Statistics* (1949) Series K-120 Ship sales: North (*Trends in American Economy*, pp. 619-621)
6d.	Minor Group 32	Lumber trade (see item 5a)	See item 5a	

NOTES TO TABLE A-10

[a] These are years for which estimates were made by interpolation or extrapolation on the series listed. They are Census or fiscal years, identified by the first of the two calendar years over which the Census or fiscal year runs. Calendar year series were converted to approximations to Census year series by running two-year moving averages.

[b] See the discussion of the benchmark year estimates. Deflation was carried out by use of North's price indexes, shifted to the base 1860 without reweighting (North, *Economic Growth*, pp. 281-282). For imports, we used the index for manufactured foods. We deflated the two components of exports separately by the indexes for raw foods and processed foods.

[c] We assumed that about 15 per cent of wheat output and 80 per cent of corn output was used for seed and animal feed. See Gallman in *Trends in the American Economy*, p. 52. See *ibid.*, p. 50, for a discussion of the prices used.

[d] For an appraisal of the Patent Office estimates, see Robert E. Gallman, "A Note on the Patent Office Crop Estimates, 1841-1848," *Journal of Economic History*, June 1963, pp. 185-195.

[e] Pieced together from the following sources: David A. Wells, *The Year-Book of Agriculture*, Philadelphia, 1856, pp. 375, 377 (which includes estimates for the crop year 1855 from the *Cincinnati Price Current*, the *New York Herald*, and the *New York Times*, and estimates for crop years 1839-55 by Charles Cist); the *New York Times* for Sept. 22, 1855 (the original source of the Cist estimates, cited above, which also contains estimates for the crop year 1855 by the *Courier and Enquirer* and the *Economist*); *DeBow's Review*, Vol. 18, pp. 467, 471; Vol. 25, p. 575; *Hunt's Merchant's Magazine*, Vol. 41, p. 252; Vol. 43 (which contains estimates for the U.S. for each year of the decade, together with estimates for 1853-57 for Ohio and Indiana from Vol. 40, p. 762, the former evidently taken from the Annual Reports of the Auditor); *Annual Report of the Auditor of the State of Ohio*, 1860, p. 86 (data for Ohio for each of the years of the decade, collected by tax assessors); *Transactions of the California State Agricultural Society*, 1859, p. 325 (data for 1852, probably Census data, 1855-59). The estimates for 1855 vary widely, mainly because the estimators differ with respect to the output of Ohio (see Cist's article in the *Times*). Subsequently, the returns of the Auditor became available and, with these data in hand, it is simple enough to settle the issue. A second difference of some importance has to do with output in Indiana. The data later appearing in Hunt (see above) appear to be official, although we have been unable to locate the official source. If the official Ohio and Indiana data are substituted for the estimates, and if we assume that the Hunt U.S. total includes the official Ohio and Indiana data, the range of the estimates narrows to 140-160 million bushels. The Price Current estimates for the main producers suggest an even lower national total (perhaps about 130 million bushels), but there is no question that the Price Current estimates are low. For example, the Iowa Census of 1856 returns almost 5.5 million bushels for the crop year 1855, while the Price Current estimate is only 2.5 million. In addition, the Price Current gives Michigan only 6 million bushels, whereas the state produced a million more than this two years previously (Michigan Census of 1854). For the remaining years, the various sources are roughly consistent. Our estimates are as follows (in millions of bushels):

1850	107	1855	145
1851	118	1856	153
1852	122	1857	153
1853	135	1858	148

[f] Berry's series gives the number of hogs slaughtered. We assumed that live weight per hog was 200 pounds in "normal" years, 210 pounds, in years in which the hog crop was large, and 190 pounds, in years in which it was small. See Gallman in *Trends in the American Economy*, p. 49.

[g] The prices are those used in *ibid.*

[h] We assumed that live weight per animal slaughtered was 1000 pounds probably a little too high, but the rounding simplifies calculations and does not affect the interpolator adversely.

[i] We assumed a constant rate of change in the output of beef between between benchmark years, on the basis of postwar experience. As a test, we estimated beef production in 1870-79 by assuming a constant rate of increase between 1869 and 1880. The ratios of our estimates to actual production are:

NOTES TO TABLE A-10 (concluded)

1870	1.001	1875	1.041
1871	1.003	1876	1.052
1872	1.005	1877	1.054
1873	1.018	1878	1.033
1874	1.022	1879	1.017

That is, our maximum error was 5.4 per cent. (Actual production was taken from Frederick Strauss and Louis H. Bean, *Gross Farm Income and Indices of Farm Production and Prices in the United States, 1869-1937*, U.S. Department of Agriculture Technical Bulletin 703, Washington, 1940.) In the postwar years pork production fluctuated from year to year in the same direction as, but much less pronouncedly than, commercial production. The pattern can be approximated by a series composed of a component growing at a constant rate, weighted 2, and a component following the path of commercial production, weighted 1. We use such a series as our prewar interpolator and extrapolator.

^{j}See the discussion of the benchmark year estimates. We used North's export price index for manufactures and import price index for manufactures and semimanufactures to deflate (see note b).

^{k}Imports (amounting to between one-fifth and one-third of the total consumption of raw wool in American wool manufacturing) are from Chester W. Wright, *Wool Growing and the Tariff*, Cambridge, Mass., 1910, p. 340. Benchmark estimates of production were interpolated and extrapolated on the following series: imports of raw wool (1834, 1835, 1837, 1838), deflated estimates of output of the woolen industries (1840, 1841, from *Niles Register*, Vol. 66, p. 387), raw wool carried on New York and Pennsylvania canals (1842, 1843, 1845-48, from Wright, *Wool Growing*, p. 145), the number of sheep in Ohio (1850-53, 1855-58, from *Annual Report of the State Board of Agriculture*, 1865, p. 292), Illinois (1855-58, *Transactions of the Department of Agriculture, Illinois*, 1876), and California (1855-58, *Transactions of the State Agriculture Society*, 1859, p. 345).

attempted to dilute the effect of these oscillations by bringing the leather series into the interpolator.

2. While we have brought together a fairly large number of interpolating series (even more could have been used, but none could have borne a very heavy weight), in fact one or a few dominate the interpolating series for each major group: the flour series, perishables (less firewood and unmanufactured animal products); the cotton and woolen series, semidurables; the lumber series, consumer durables; the lumber and ship series, producer durables. This is especially true of the perishables and semidurables series, but is quite appropriate in these cases since flour and textile production do, indeed, dominate the final flows of these major groups. Many series carry some weight in the interpolation and extrapolation of the two durables major groups. However, important components of these groups—notably, industrial equipment—are inadequately represented. Additionally, the lumber series, a prominent member of the interpolating and extrapolating series, is probably not an especially good estimator, for a number of reasons.[100] However, one should bear in mind that the two durables groups are relatively much less important than the perishables and semidurables groups.

3. Estimates of flows of materials into production dominate the interpolators and extrapolators (e.g., wheat, corn, raw cotton and wool, lumber).

The above remarks will serve as general warnings. Lest they raise too many doubts, bear in mind that the interpolations and extrapolations generally carry over only four years, and frequently fewer years than this. The estimates produced are only used in decade averages. We use them to reduce our dependence on benchmark year estimates to establish prewar levels of performance. For this purpose they appear to be adequate.

VALUE OF IMPROVEMENTS TO FARM LAND, 1860 PRICES

Our construction estimates (except for railroad and canal construction) are limited to projects carried out with nonfarm materials. They do not cover farm building and the clearing, fencing, and draining of agricultural land carried out with farm materials. These agricultural construction activities were important. Martin Primack's work provides most of the material necessary to determine how important they were.[101] Our summary

---

[100] But it may not be a bad indicator of the domestic lumber trade. According to the Secretary of the Treasury (*Finance Report*, 1846) Maine, Maryland, and Florida accounted for about 76 per cent of the domestic trade in 1845.

[101] Martin Primack, "Farm Formed Capital in American Agriculture, 1850 to 1910," unpublished Ph.D. dissertation, University of North Carolina, 1962.

estimates appear in Table A-4. For convenience, we refer to them as estimates of the "value of improvements," but the reader should understand by that term the "value of improvements to farm land carried out with farm materials."

Primack's most important estimates are of the number of acres of land cleared and broken, distributed among the categories of virgin forest, abandoned forest, prairie, and plains; the number of rods of fence put in place, by type of fence (Virginia rail, post and rail, board, stone, hedge, straight-wire, barbed-wire, woven-wire); and unit (acre, rod) labor requirements for clearing and breaking the various types of land and constructing the various types of fence. The series cover the period 1850–1910. Labor requirements are given at decade intervals (1850, 1860, etc.). The estimates of land cleared and fence constructed are regional, decade (1850–59, 1860–69, etc.) estimates.

With Primack's data, one can compute total labor requirements of clearing and fencing for each decade. Labor was virtually the only input and, therefore, one can approximate the total cost of these improvements by valuing labor requirements. We were interested in constant price estimates. Therefore, we valued (regional) labor requirements by use of 1860 (regional) wage rates.[102] Additionally, the labor requirements we used were labor requirements *under 1860 techniques*.[103] That is, we estimated (for example) the cost of constructing the improvements of 1900–09 with 1860 techniques and wage rates. This is as close as we can presently come to the value, in 1860 prices, of improvements constructed during 1900–09. One should note that unit values of, for instance, clearing implicit in the estimates vary from decade to decade with changes in the distribution of cleared land among types of land (forest, prairie, etc.) and among regions, and only for these reasons.

We also derived the value of irrigation improvements to rice land and the value of construction carried out with farm materials on new farms. Neither of these series is quantitatively important. We assumed that irrigation cost $25 per acre, using 1860 techniques. Primack has data on the number of acres irrigated each decade. We assumed that frontier methods of farm building were used only in new farm creation in the southeast, south central, prairie, southwest, mountain, and Pacific regions. Primack has data on new farms created in each decade. We

---

[102] We produced two variants (see Table A-4). In Variant I, the wage rate includes the value of board; in Variant II, it excludes the value of board. The wage rates are from the agriculture volume of the *Census of 1860* (p. 512). We assumed that board was worth $.25 per day (see *Historical Statistics*, 1949, Series E-67 and E-68).

[103] Board and wire fence was constructed with manufactured materials and therefore did not figure in the estimates.

assumed that building costs (1860 prices) ran $100 per farm in all but the last three regions. In these regions we assumed costs of $150, $50, and $120, respectively. These judgments cannot be defended in detail here. They rest mainly on our interpretations of data in Primack's study.[104]

Primack's data do not extend back beyond 1850. In a previous study, we had estimated the number of acres cleared in the decade 1840–49, but we had no estimate of the regional distribution of this land nor of the magnitudes of the other components of improvements.[105] However, the principal factor influencing the average value of improvements per acre is the distribution of cleared acreage between virgin forest and all other land. According to Primack, 82 per cent of the land cleared in the decade 1850–59 was forested. Presumably a larger share was forested in 1840–49; we assumed 90 per cent, and estimated the value of improvements per acre cleared in 1840–49 by use of the following formula:

$$\frac{F^a \cdot Mf + N^a \cdot Mn}{F^b \cdot Mf + N^b \cdot Mn} \cdot \frac{V^b}{A^b},$$

where F^a is the share of forest land in cleared land for 1840–49; F^b is the same for 1850–59; N^a is the same for all other land in 1840–49; N^b is the same in 1850–59; Mf is man-days per acre to clear forested land, 1860 techniques; Mn is the same for all other land; V^b is the value of improvements made in 1850–59; and A^b is the number of acres cleared in 1850–59.

Since we wanted to run overlapping decade estimates, it was necessary to distribute the decade estimates among years.[106] The annual series used for this purpose were as follows: for 1840–50, federal land sales, two-year moving averages of calendar year data;[107] for 1850–60, federal land sales less graduation sales, two-year moving averages of two-year moving averages of fiscal year data;[108] for 1869–1909, homestead final entries, set back in time five years, two-year moving averages of fiscal year data.[109]

To get annual estimates for years before 1840, we extrapolated the estimates of the 1840's on two-year moving averages of calendar year

[104] Primack, "Farm Formed Capital," Chapters III, VI.
[105] See Gallman in *Trends in the American Economy*, p. 49.
[106] To get decade averages for 1844–53, 1874–83, etc.
[107] Benjamin H. Hibbard, *A History of Public Land Policies*, New York, 1939, pp. 103–106. Hibbard lists his data as fiscal year data, but apparently they really refer to the calendar year. See *Historical Statistics*, 1960, p. 233. The calendar year data were averaged to approximate Census year values.
[108] Hibbard, *Public Land Policies*. Graduation sales were associated with speculation. See Gallman in *Trends in the American Economy*, pp. 70–71. The averaging is intended to smooth out any remaining effects of speculation on the series.
[109] Hibbard, *Public Land Policies*, pp. 396–397. Presumably improvements were made when land was first entered, i.e., five years before final entry. The fiscal year data were averaged to approximate calendar year values.

federal land sales data.[110] Sales for the years 1835 and 1836 were heavily influenced by speculation. Therefore, we substituted interpolated estimates for the Land Office data for these years.

In a previous study, we had estimated the value of improvements in 1879 prices by multiplying the number of acres improved by the cost per acre of improvements in 1879.[111] Primack's work makes clear the fact that this procedure is defective. Nonforest land required much less labor to improve than did forest land. The composition of cleared land changed markedly over time. The cost of improvements per acre in 1860 prices (Variant I) declined from about $24 in the decade 1850–59 to $12 in 1870–79 and to roughly $9 in the full period 1880–1909. Consequently, the estimates we had previously put forward for the prewar period were too low, while the estimates for the postwar period, after the 1870's, were too high.

VALUE ADDED BY HOME MANUFACTURING

During the nineteenth century, the location of various kinds of manufacturing shifted from the home or the small establishment to the factory. One evidence of this process is the changing allocation of certain commodity flows between final and intermediate production. For example, over time, the ratio of flour that is final product to flour that is intermediate product declines. Flour flowing into the home is treated as final product, since the processing it receives in the home is not measured in the national product. Furthermore, flour flowing to small retail bakeries whose product is omitted from GNP, intentionally or otherwise, is also treated as final product, since the flow cannot be identified as intermediate (and should not be, as long as the product of retail bakeries is not counted in national product). A transfer of baking to factories would then reduce the share of flour going directly to final uses.

The transfer of activities from sectors whose product is not counted to sectors whose product is counted gives the GNP measure a rate of growth higher than that of the total product of the society, of course. It would be useful to know the extent to which the two rates of growth differ from each other. We have attempted to judge this by estimating the value added to so-called final product in benchmark years before 1899 by activities conducted in the home or in small establishments before 1899 but transferred to factories by that year. That is, we have not attempted to estimate all value added by home manufacturing in, e.g., 1879, but only the value added to that part of final product which would have flowed into factories had economic organization been the same in 1879 as in 1899.

[110] See footnote 107.
[111] Gallman in *Trends in the American Economy*, pp. 46–51.

The procedure is simple and can be explained best by an example—baking. We calculated the ratio of the value of flour flowing to factory baking to the value of the total flow of flour (final plus intermediate) in 1899. For each of the major benchmark years before 1899, we multiplied the value of the total flow of flour by this ratio and we subtracted the value of flour flowing into factory baking from the result. This gave us the value of final product that would have been intermediate product had the economy been organized along the lines of 1899. We then multiplied the result by the ratio of value added to value of flour consumed in factory baking. This gave us an estimate of value added by home and retail baking which was displaced by factory production by 1899.

The procedure obviously works best where production is largely confined to the processing of a single commodity and the flows of that commodity are readily traced. For example, we were able to make estimates relating to baking, slaughtering and packing, and clothing production (see Table A-5). But we were unable to do anything with certain activities which may have been important, at least relative to factory production of the same goods, but which involved operations on several commodities or on commodities whose flows were not easily traced. For example, we could not derive estimates of the production of tools, furniture, wagons, harness, etc., carried out in the home or by practitioners of hand trades which might not be adequately covered by the Censuses. However, our estimates do cover the activities which were of greatest quantitative importance.[112]

From 1839 through 1869, the Census returned value added by home manufacturing (perhaps value of output in 1839). The treatment of the item by contemporary observers (e.g., Seaman) suggests that the returns covered mainly textile production. However, virtually all raw wool and cotton available for processing was apparently used in factory production. No doubt flax and similar fibers were worked up in the home, but the magnitudes involved were probably smaller than the magnitudes implied by the returns of home manufacturing. It seems possible, then, that the Census covered home production of furniture, harness, etc., and perhaps even some components of the home manufacturing which we have measured

---

[112] Some components of "home production" are covered, of course, in series previously discussed. For example, farm construction on new farms carried out with farm materials is included in the "farm improvements" series. The gross new construction series is based, in part, on flows of construction materials and therefore includes gross new construction carried on outside the market, e.g., the construction of a shed by a home-owner or of a barn or fence by a farmer. The value of home production of intermediate goods (e.g., pot and pearl ashes, rags) is included in the value of final product (soap, paper), insofar as these intermediate goods were sold to commercial producers of final product.

through the commodity flow method. Nonetheless, we were anxious not to understate the significance of home manufacturing and, therefore, accepted the Census returns as accurate measures of value added by home textile production. We also included the value of textiles produced in the home in the input into home clothing manufacture. It appears likely, then, that if our estimates are in error they are too large, rather than too small. This is especially true in the case of the 1839 figure, since we have treated the Census return as though it referred to value added, whereas, as noted above, it may refer to value of output.

The estimates described above use commodity flow procedures (in the main) to measure value added by activities neglected by the GNP measure, manufacturing in the home and in small establishments. However, there remain other activities which are missed even by the commodity flow procedure. Among the most important is surely the raising of horses and mules for sale to commercial enterprises (producer durables) or to consumers (consumer durables) and the production of hay and other feeds for horses owned by consumers (consumer perishables).[113] However, the magnitudes missed are very small. For example, the value of all hay sold to nonagricultural sectors (i.e., including intermediate product) remained near, and generally below, 1 per cent of GNP during the full period.[114] According to Seaman, the total value of horses and mules sold to consumers and nonagricultural businesses plus the value of the increase of inventories of horses and mules amounted to about $14 million in 1839, or less than 1 per cent of GNP.[115]

COMMENT

Richard A. Easterlin, University of Pennsylvania

Robert Gallman's paper, like his previous one in Volume 24, provides a summary report on research in progress since 1953. The magnitude of his contribution, present and potential, is suggested in the brief but fascinating analysis contained in the first part, and the temptation to take up the issues touched on there is great. But my task, as I understand it, is to deal with the hard facts of life in the appendix. Since his work is unfinished, I trust I am spared the necessity of a definitive appraisal of the estimates (aside from the question of adequacy for the task), but the thought is a disturbing one, and in the end I shall come, if not to an appraisal of

---

[113] Kuznets, *Capital in the American Economy*, p. 516.
[114] Gallman in *Trends in the American Economy*, p. 46.
[115] Seaman, *Essay*, 2nd ed., p. 453.

Gallman's estimates, at least to the issue of the criteria appropriate for such an appraisal.

I should like first, however, to make a few comments specifically on the estimates, chiefly regarding possible further tests and areas of weakness. Regretfully, my remarks are confined almost wholly to the benchmark year estimates in current prices.

FLOW OF COMMODITIES

The Underlying Detail

In the estimates published in the present paper, the detail is confined to the major components of commodity flow. In itself, this is a major contribution, filling an important gap in our knowledge of midnineteenth century American development as well as suggesting important revisions of accepted views on the ensuing period up to World War I. However, I should like to point up also the immense potential value of the underlying industrial and commodity detail pieced together by Gallman in the construction of the estimates, but not published in the current paper. The nature of the detail is suggested by Table 1, which illustrates Gallman's estimating procedures. For the present purpose, the point to be noted is that Gallman worked with as many as 631 industries (1859) and forty classes of farm output in the process of establishing comparable classifications for each Census date in the period 1839–69 and developing distributions of output by type like those in the table. Indeed, the table understates the amount of underlying detail because, following Shaw, not six but around forty classes of commodity output were recognized. In addition, imports and exports as given in Treasury reports were allocated among these same commodity classes to obtain estimates of domestic consumption as well as production. One can only surmise the obstacles and discouragements surmounted by Gallman in this undertaking, but the time and effort consumed are self-evident and clearly have a bearing on the issue of appraisal.

Crude as these detailed estimates are—and they are significantly less reliable than the large aggregates since, as Gallman notes, errors due to misclassification often cancel out in the process of summation—they are of significant analytical interest. They give at least a rough idea of the degree to which different industries were sensitive to various classes of final demand. They provide also an important first step toward an interindustry flow table (which would further require distributing the estimated output of intermediate products according to industry of destination). Finally, these estimates, or more precisely, the detailed structural framework which they embody, provide an explicit link between recent work in

TABLE 1

VALUE OF PRODUCTS OF MANUFACTURING, BY TYPE
AND INDUSTRY GROUP, 1869
(million dollars)

Industry Group	Value of Products (1)	FINISHED GOODS				Intermediate Products (6)
		Consumer Goods			Construction Materials (5)	
		Perishables and Semidurables (2)	Durables (3)	Producer Durables (4)		
All manufacturing	3,794	1,446	245	296	320	1,486
Food and kindred products	692	606				85
Textiles and products	717	370	33	4		310
Iron and steel	600	7	26	147	115	305
Lumber and its manufactured products	419	5	68	12	108	226
Leather and its products	375	185	7	18		165
Paper and printing	158	28	8		2	120
Liquors and beverages	109	74				36
Chemicals and allied products	179	65		3		96
Stone, clay, and glass products	115	1	16		18	36
Metals other than iron and steel	123		38		60	69
Tobacco manufactures	72	72			15	0
Vehicles for land transportation	96		2	35	48	12
Miscellaneous	138	32	12	64	3	27

Source

Col. 1: Sum of cols. 2-6. Closely approximates group totals from Census shown by Shaw in *Commodity Output*, p. 200.
Cols. 2-5: *Ibid.*, pp. 108-135. Detailed entries under each major commodity group were classified by industry group according to the industry group number for each entry shown by Shaw.
Col. 6: *Ibid.*, pp. 138-151.

economic history emphasizing broad magnitudes of economic analysis and earlier research on individual industries and firms. Using this detail, the researcher at the microlevel can trace in rough fashion some of the connections between his special area of study and the economy as a whole. A significant feedback advantage for the income estimator himself will be noted subsequently.

Consistency Tests of Commodity Flow Data

An important part of Gallman's work involved testing whether estimates of farm output in physical terms derived from the agricultural Census were consistent after allowance for other uses with returns on materials consumed in the manufacturing Census. He reports that these tests were generally reassuring, but there was one big exception. In 1869, estimates of the uses of animal products developed by Shaw chiefly from the manufacturing Census fall far short of exhausting the farm output total. This is less true for Shaw's 1879 estimate and thereafter the check works out fairly well. Gallman infers that, with the adoption of the refrigerator car in the 1870's and 1880's, there was a major shift in slaughtering from the retail sector not covered in the industrial Censuses to large packing firms which were covered. This view seems plausible, but it would be strengthened if Gallman could show that the pre-Civil War Censuses exhibit the same inconsistency as those for 1869.

The significance of this testing procedure for assessing the reliability of the data in the industrial Censuses should be underscored. The midnineteenth century Censuses have been written off by some as virtually worthless. While inspection of such charges often shows them to be grounded on slim and (for the present case) irrelevant evidence, it is nonetheless reassuring to find that the returns from the nation's farmers check out fairly well with manufacturers' reports. Similar testing of mutual consistency might be attempted against data from the transport sector, especially for the earlier years when channels of commerce were fewer, but I am not sufficiently familiar with the source materials for transportation to evaluate this possibility.

In principle, outputs and inputs of different manufacturing industries can be checked against each other in the same way, but Gallman found this difficult with the data available. Hence his testing of and confidence in the estimates for the earlier stages of manufacturing is greater than for the later. However, even if such testing is not possible for individual industries and commodities in physical output terms, it may still be possible to develop a very rough test for entire industrial sectors in value terms. Such a procedure would add to Gallman's appraisal of the consistency of reports on raw material flows an impression of the consistency of sector reports on product values, including not only raw materials but also fabricated items.

Table 2 is an attempt to develop such a test for Shaw's commodity flow estimates, which were readily available in full detail. (And here I should like to second Gallman's appreciation of Shaw's lucid presentation.) In effect, lines 1–7a apply the commodity flow procedure to entire sectors to derive an estimate of the value of output of finished manufactures and manufactured construction materials. Thus, from the output of the extractive industries (line 1) are deducted the flow into inventory (line 2a—estimates were available only for livestock), the flow of nonmanufactured finished goods and construction materials to ultimate consumers and abroad (line 2b), and the flow of intermediate goods abroad (line 2ci). The residual provides an estimate of the flow of intermediate goods from the domestic extractive industries to the manufacturing sector (line 2cii). To this are added imports of intermediate goods for domestic manufacturing (line 3) to obtain the total input into manufacturing of intermediate goods (line 4). Adding to this the value added in domestic manufacturing (line 5) and deducting the flow of manufactured intermediate goods abroad (line 6), one obtains an estimate of the output by the manufacturing sector of finished goods and construction materials (line 7a). Shaw's estimates of the same, obtained from Census returns detailing manufacturing output by type (not used in the present calculation), are shown in absolute amount in line 7b and as a percentage of the present estimate in line 8.

There is some reason to expect Shaw's results to be lower since he tried to confine his estimate to factory production. However, a glance through the notes describing how the present estimate was obtained and some of its imperfections warns against expecting very close correspondence.[1]

[1] A partial check of the present procedure for one date is provided by the 1899 Census of Manufactures where separate returns on the raw materials and partly manufactured components of materials were obtained. See *Twelfth Census*, Vol. VII, Manufactures, Part I, p. cxxxvii. The estimate of $2,506 million obtained here (line 4) for the flow of raw materials to manufacturing is quite close to the direct report of manufacturers, $2,389 million.

It may be noted that Shaw used these 1899 Census figures to develop a check somewhat similar to the present one. See Shaw, *Commodity Output*, pp. 89–92. However, Shaw's check was only possible for the one date when the Census of Manufactures obtained direct reports on the value of raw materials consumed, while the present procedure develops the estimate of raw materials input from nonmanufacturing data and has the advantage not only that such data are to a substantial extent available throughout the entire period, but also of providing at the same time a rough test of the consistency of the Census returns for manufacturing with those for the extractive industries.

TABLE 2

ALTERNATIVE ESTIMATES OF OUTPUT OF FINISHED MANUFACTURES AND MANUFACTURED CONSTRUCTION MATERIALS AT PRODUCER PRICES AND COMPARISON WITH SHAW ESTIMATES, 1869-1919
(million dollars)

	1869	1879	1889	1899	1909	1919
1. Gross value added of extractive industries	2,417	2,538	2,923	3,743	6,973	18,159
2. Drafts on output of extractive industries						
a. Changes in livestock inventory	47	70	43	133	-106	-214
b. Flow of nonmanuf. finished goods and construct. mater. to ultimate consumers and export	736	756	1,017	1,247	2,263	5,213
c. Flow of intermediate goods						
i. To export	223	452	374	469	532	1,782
ii. To domestic manufacturing (line 1 minus sum of lines 2a, 2b, and 2ci)	1,411	1,260	1,489	1,894	4,284	11,378
3. Imports of intermediate goods for domestic manufact.	262	520	538	612	980	3,299
4. Total flow of intermediate goods to domestic manufact.	1,673	1,780	2,027	2,506	5,264	14,677
5. Gross value added of domestic manufacturing	1,631	1,962	3,727	5,044	8,160	23,842
6. Exports of intermediate goods by manufacturing	34	99	85	299	401	1,551
7. Output of finished manufactures and manufact. construction materials						
a. Lines 4 and 5 minus line 6	3,270	3,643	5,669	7,251	13,023	36,968
b. Shaw	2,308	3,079	4,807	6,373	11,330	34,067
8. Ratio of line 7b to line 7a (per cent)	71	85	85	88	87	92

NOTES TO TABLE 2

Line 1: Sum of estimates for agriculture, mining, and fisheries derived from sources below. Comparability with the scope of Shaw's final flow estimates would be improved by addition of an estimate for forestry.
Agriculture
1869-99: *Trends in the American Economy*, p. 47. Forest products (estimated at $100 million in 1889), improvements to land made by farm labor, and home manufactures were excluded in order to improve comparability with the scope of Shaw's final flow estimates.
1909,1919: Strauss and Bean, *Gross Farm Income*, p. 24. To improve comparability, total in source was reduced by arbitrary estimates for forest products of $150 million in 1909 and $300 million in 1919.
Mining
1869-99: *Trends in the American Economy*, p. 54. Comparability with Shaw's final flow figures would be improved if estimates were added for precious metals mining (omitted in source).
1909, 1919: *Historical Statistics*, 1960, Series M-4.
1869-1919: Figures derived as above were increased by value of production of natural mineral waters as given by Shaw, *Commodity Output*, pp. 247-248.
Fisheries
1869-1919: Shaw, *Commodity Output*, p. 252.
Line 2a, 1869-99: Strauss and Bean, *Gross Farm Income*, p. 23, by subtraction.
Line 2a, 1909, 1919: *Ibid.*, p. 24.
Line 2b: Sum of flows from agriculture (Shaw, *Commodity Output*, p. 247, cols. 1-8), mining (Shaw, *Commodity Output*, p. 262, col. 9; p. 247, col. 10; p. 264, cols. 3-4), and fisheries(*ibid.*, p. 252, col. 3).
Line 2ci: Exports of crude materials and crude foodstuffs (*Historical Statistics*, 1960, Series U-62 and U-63) for fiscal years 1870, 1880,1890, 1900, and 1909-10 average, and for calendar year 1919 were multiplied by a factor of 0.9 to place them on a valuation basis roughly comparable with Shaw (*Commodity Output*, p. 271). Shaw's estimates of exports of nonmanufactured food and construction materials (*ibid.*, pp. 30, 64) were then deducted from these to obtain exports of unfinished nonmanufactured goods.
Line 2cii: Line 1 minus lines 2a, 2b, and 2ci.
Line 3: Total imports (*Historical Statistics*, 1960 Series U-67) for same dates as crude exports in line 2ci minus Shaw's estimates of imports of finished commodities and construction materials (*Commodity Output*, pp. 62-65). Comparability with final flow figures would be improved if estimates were added for duties on these imports.
Line 4: Sum of lines 2cii and 3.
Line 5, 1869-99: *Trends in the American Economy*, p. 56.
Line 5, 1909, 1919: *Historical Statistics*, 1960, Series P-8.
Line 5: Exports of manufactured foodstuffs, semimanufactures, and finished manufactures (*Historical Statistics*, 1960, Series U-64, U-65, and U-66) for same dates as crude exports in line 2ci were multiplied by a factor of .85 to place them on a valuation basis roughly comparable with Shaw (*Commodity Output*, p. 271). Shaw's estimates of exports of finished manufactured commodities (*ibid.*, p. 62, less group 1b, pp. 30-31) and construction materials (*ibid.*, p. 63) were then deducted from these to obtain exports of unfinished manufactures.
Line 7a: Lines 4 and 5 minus line 6.
Line 7b: Shaw, *Commodity Output*, p. 152, sum of finished commodities and construction materials.

It is encouraging, therefore, to find that for 1879–1919 the two totals do not differ drastically, and that the trend in this period is generally consistent with expectations based on a shift of manufacturing production into the factory. Moreover, the 1869 figures, which Gallman on the basis of his detailed testing of materials flows concludes are deficient, are shown by this test also to be seriously in question.

The results seem encouraging enough to warrant fuller investigation of this testing procedure. Clearly, it would be interesting to replicate the test for all the Census year estimates presented here by Gallman, and particularly to see how his pre-Civil War and revised 1869 estimates fall in line. It would also be informative if, to the extent possible, such a comparison were presented using the returns from the industrial Censuses *before* adjustment for comparability over time. In this way, one might form some over-all impression of the degree of processing to which the original data were submitted. Finally, it should be noted in passing that the data in Table 2 are of analytical interest too, providing in a very summary fashion a notion of the changing interdependence between the extractive sector, manufacturing, and the rest of the world.

Table 3 is an attempt to develop a similar test for the Census year estimates in the pre-Civil War period, using only the figures published by Gallman here and in Volume 24. The procedure is much cruder, involving among other things, only an over-all adjustment for international trade flows and different levels of valuation for the direct estimate of manufactures of finished commodities and construction materials (line 5) and the implied estimate derived from the industrial data (line 4). There is a suggestion that the 1839 figures may be somewhat out of line, either on the industrial or final product side, but without more experience with this procedure and a longer series, it is difficult to draw any firm conclusion. The procedure is less precise than that employed in Table 2 and is noted here only to suggest a cruder but simpler alternative, and to provide a link with a similar comparison presented below in discussing the service estimates.

One final point should be noted regarding Gallman's revision of the 1869 figures for the flow of animal products into domestic consumption. While the testing procedure presented in Table 2 provides general support for Gallman, it does not of course identify the particular source or sources of inconsistency. In Volume 24, William N. Parker expresses the view that the estimates of hog slaughter by both Gallman and Towne-Rasmussen—which check with each other quite closely—may in fact be high, and Albert Fishlow, in his review of the volume, supports this position with the opinion that Gallman's method yields an estimate that constitutes in

TABLE 3

COMPARISON OF ESTIMATES OF VALUE ADDED BY INDUSTRIAL SECTOR
WITH ESTIMATES OF FINISHED COMMODITIES AND
CONSTRUCTION MATERIALS, 1839-59
(million current dollars, producer prices)

	1839	1849	1859
1. Gross value added	907	1,239	2,276
a. Agriculture	658	775	1,427
b. Mining	9	17	34
c. Manufacturing	240	447	815
2. Exports	112	135	316
3. Imports	98	174	354
4. Line 1 adjusted for trade flows (line 1 minus lines 2 and 3)	893	1,278	2,314
5. Finished consumer commodities, producer durables, and construct. materials (buyer prices)	1,115	1,759	3,141
6. Ratio of line 5 to line 4 (per cent)	125	138	136

Source

Line 1: Sum of lines 1a, 1b, and 1c.
Line 1a: *Trends in the American Economy*, p. 47. To improve comparability with the scope of estimates in line 5, improvements to land made by farm labor and home manufactures were eliminated from value added.
Line 1b: *Ibid.*, p. 54.
Line 1c: *Ibid.*, p. 56.
Line 2: *Historical Statistos*, 1960, Series U-61, figures for 1840, 1850, and 1860. Values as published were used.
Line 3: *Ibid.*, Series U-67 for same dates as Series U-61 in line 2. Values as published were used.
Line 4: Line 1 minus lines 2 and 3.
Line 5: Gallman's Table A-2, cols. 2-4, plus his Table A-3, col. 2, plus *Trends in the American Economy*, p. 63, Table A-10, lines 1 and 2.

effect an upper limit.[2] However, a comparison of meat consumption implied by Gallman's production estimate for 1849 with fragmentary direct information on consumption in 1848 seems to support the reasonableness of the level of the Gallman and Towne-Rasmussen estimates for that date.[3] But only about a dozen farmers' returns on consumption

[2] *Trends in the American Economy*, p. 284, editor's note. Albert Fishlow, "Trends in the American Economy in the Nineteenth Century," *Journal of Economic History*, March 1962, p. 78.

[3] See *Annual Report of the Commissioner of Patents for the Year 1848*, H. Exec. Doc. 59, 1849, pp. 660–663. Gallman's figures in pounds of live weight for hogs and cattle slaughtered were taken from his unpublished Ph.D. dissertation ("Value Added by Agriculture, Mining, and Manufacturing in the United States, 1840–1880," University of Pennsylvania, 1956, p. 346), converted to a per capita basis and then adjusted for the proportion that meat (and, in the case of hogs, lard) form of live weight. The estimates of these proportions, which Gallman kindly made available to me, were .625 for pork and .59 for beef. No attempt was made to adjust for international trade flows, which were relatively unimportant at the time.

provide the basis for this test and their comparability is uncertain; hence, not much confidence can be placed in it. The issue is a disturbing one since it is important quantitatively—Gallman's revision of Shaw's 1869 estimate on this score alone increases the value of finished consumer commodities by about one-fourth. It is to be hoped that further attention will be given to it by experts in the area.

Other Features

Brief mention should be made of additional aspects of the estimates on the commodity side, many of which draw considerably on other recent studies. In marking up output from producer to consumer prices in the period from 1869 on, Gallman utilizes margin data from Barger's recent work, not available when Kuznets first made his estimates. For the pre-1869 estimates, these margins are extrapolated and tested against related estimates by Seaman and Marburg for 1839. In the estimate of new construction, derived by marking up the input of construction materials, Gallman adopts a suggestion by Fishlow and makes separate estimates for (1) railroad plus canal building, using recent work by Fishlow, Ulmer, and Cranmer, and (2) all other construction. The materials-to-output ratio is significantly different for these two types of building and an important change is made in the 1869 estimate of new construction, a year in which the building "mix" involved a disproportionate representation of railroad building. Allowances for nonmarket activities typically excluded in conventional GNP estimates are made by using new estimates by Barger of firewood, and by Primack and Parker of land improvements by farmers. Also an ingenious estimate is developed by Gallman to test the significance of changes over the period in the relative importance of home and commercial manufacture. Figures for the annual net balance in the pre-Civil War period are obtained from North's work in Volume 24. Finally, in adjusting from current to constant dollars, Gallman uses new price indexes prepared especially for this purpose by Dorothy S. Brady plus some additional series developed by Hoover and Ulmer. I have the impression that the level of commodity detail at which the adjustment is carried out is significantly finer than that employed in the present official estimates.

FINAL CONSUMER SERVICES

As Gallman's own appraisal states, the service estimates, which account for somewhat more than one-fifth of GNP throughout the period, are the least reliable. Gallman accepts Kuznets' estimates for the period since 1869 and extrapolates them *in toto* to the pre-Civil War period, testing the result for 1839 against Seaman and also the movements during the period against Lebergott's labor force figures. He suggests that the 1839 estimate of final services may be somewhat high. This may be, since there is some evidence that the farm sector grew slower than the nonfarm in the 1840's, not at an equal pace as Gallman assumes in pushing the figures for the 1850's back to 1839. But quite aside from this specific issue, it is difficult to place much confidence in the service estimates in their current stage of development for *any* of the dates shown. The very rough and aggregative approach followed here by Gallman contrasts sharply with his detailed work on the commodity estimates. And while Kuznets' estimates clearly involved much labor and ingenuity,[4] certain aspects of them make one uneasy. For example, the final estimate for nonrent services exceeds that for rent, although thirty-seven out of the forty expenditure surveys for the pre-1914 period—the basic data used by Kuznets—show the difference running the other way and usually by a substantial amount.[5] This result is not impossible, since the surveys relate to urban low-income groups, but the key element in adjusting these to obtain an estimate for the total population is the expenditure differential by population class in 1935-36. This was the earliest date for which sufficient information was available; it is also a date when "other services" included to an important extent items hardly relevant to much of the nineteenth century, such as auto service and repair, movies, telephone, electricity, and other household utilities. While Kuznets' estimates may ultimately prove to be sufficiently reliable—a result which would surprise no one who has come to appreciate his uncanny feel for the use of data—still it does seem that more can be done to explore alternative approaches, several of which have been opened up by the development of new data unavailable to Kuznets at the time.

A few suggestions may be ventured about these alternatives. First, it would seem worthwhile to distinguish the individual service sectors rather than working with a broad aggregate. That the composition of services changed substantially during this period is suggested by Daniel Carson's labor force estimates, which show the share of domestic service in the service industries declining from 58 to 25 per cent between 1870 and 1930.[6] Second, fuller use should be made not only of income and expenditure data, which have been worked up in varying degree now for sectors such as government and education, but also of factor input data such as Carson's and, in the present volume, Lebergott's. Gallman's reluctance to use such data because it reduces the analytical value of the resulting estimates imposes stricter constraints on his methods than are employed in the current official estimates, where the procedures for the service

---

[4] See Kuznets, *National Product*, pp. 123–182.
[5] *Ibid.*, pp. 144, 132.
[6] *Studies in Income and Wealth*, *11*, New York, NBER, 1949, p. 47.

sectors often preclude significant productivity change. Finally, not only factor inputs but also material inputs into the service industries warrant investigation. Conceivably, it might be possible to develop a procedure parallel to the commodity flow technique in which the flow of material inputs into the service sector is adjusted in the same fashion as shown for

TABLE 4

COMPARISON OF ESTIMATES OF VALUE ADDED IN SERVICE INDUSTRIES WITH ESTIMATES OF FINAL CONSUMER SERVICES OTHER THAN RENT, ANNUAL AVERAGES FOR SPECIFIED PERIODS, 1869-1948

Period	Net Value Added in Service Industries[a] (billion dollars) (1)	Final Consumer Services (2)	Col. 1 as Percentage of National Income[a] (3)	Col. 2 as Percentage of Net National Product (4)	Ratio of Col.2 to Col. 1 or of Col. 4 to Col. 3 (per cent) (5)
			PANEL A[b]		
1. 1919-28	8.4	13.0	11.4	18.0	155
2. 1924-33	9.2	15.2	13.0	21.7	165
3. 1929-38	8.3	14.3	13.6	23.3	172
4. 1934-43			12.1	19.4	160
5. 1939-48			10.5	17.8	170
			PANEL B[c]		
6. 1879-89			13.6[d]	12.8	94
7. 1899-1908			9.6	14.8	154
8. 1919-28			9.4	17.5	186

Source

Cols. 1, 3, lines 1-3: Simon Kuznets, *National Income and Its Composition, 1919-1938*, New York, NBER, 1941, p. 163, cols. 8 and 11.
Cols. 2, 4, lines 1-3: Kuznets, *National Product*, p. 144, col. 2, and p. 119, col. 5.
Col. 3, lines 4-8: Kuznets in *Income and Wealth of the United States*, p. 89, col. 7.
Col. 4, lines 4-8: *Ibid.*, p. 168 (col. 5) times p. 155 (col. 5).

[a] For panel B, col. 1, aggregate payments instead of net value added; col. 3, aggregate payments instead of national income.
[b] Kuznets' estimates for all columns.
[c] Martin's estimates for col. 3; Kuznets' for col. 4.
[d] Average for 1879 and 1889 only.

the commodity sector in Table 1 to obtain the output of final consumer services. Table 4 is a very rough effort to explore this possibility, using a treatment paralleling, not Table 2 which I would have preferred, but Table 3. In panel A, column 1 presents Kuznets' estimates of net value added in the service industries, and column 2 presents his estimates of final consumer services other than rent for three dates. The ratio of the two is given in column 5 and is extended to two additional dates using more readily available data on shares of these two components in the income and product totals. Since, on the one hand, final consumer services includes the output of industries not covered in column 1 (such as public utilities), while, on the other, the industries in column 1 provide intermediate products to business as well as final services (e.g., legal services), one would not expect the levels of the two totals to correspond. But there is some basis for expecting correspondence in their movements, and, indeed, columns 3 and 4 of panel A show the percentage shares changing in rather similar fashion. However, this pattern does not appear in the figures for the earlier period in panel B, which compares Martin's industrial sector estimates with Kuznets' final product figures. Instead, the shares move in diverging fashion. The implication is that these two sets of figures may be seriously inconsistent in the earlier period. Faced with an immediate choice, one would certainly prefer Kuznets' estimates, which are the ones used by Gallman, to Martin's, but clearly the desirable course is a detailed re-examination of the estimates. The main point of the present illustration, however, is simply to suggest how it may be possible to test and perhaps develop estimates for the services following a two-pronged approach from both the industry and final product side.

GENERAL CONSIDERATIONS

In conclusion, I should like to offer three broader observations prompted by the general and understandable concern with the decline in basic data input relative to output as estimates such as these are extended back in time.

The first relates to the complementary nature of micro- and macroresearch in economic history. Mention has already been made of the manner in which the detailed structural framework underlying Gallman's estimates enables the microworker to relate his work to the economy as a whole. But the income estimator too may derive important gains from research at the microlevel. Obviously no estimator can have a comprehensive knowledge of an economy's operation, and as data become scarcer and sources more scattered, he must rely increasingly on the specialist and specialized studies for assistance. A number of examples of this in Gallman's work have already been given. I should like here merely to emphasize two obvious implications for income and wealth estimates. First, there is the need to present the estimates in full detail (I am not referring, of course, to summary reports of the type presented in this volume). The detail should be published (with appropriate warnings), even though crude and unreliable, because only in this way can the expert knowledge of the specialist be best brought to bear and improvements made. The other implication is that any historical income and wealth estimate is necessarily unfinished, for new knowledge will render obsolete various parts of the estimate. An established view that these estimates, though useful, are imperfect and mortal would perhaps reduce the dangers of misuse and casual criticism.

The second and related observation is the need for wider recognition by users of the interdependence of the construction of estimates, on the one hand, and their possible analytical uses, on the other. This point, made a decade ago in an excellent article by Stanley Lebergott,[7] applies with special force to historical estimates, where resort to analytical models to supply figures for components not covered by the basic sources is more widespread. Proper analytical use of such figures requires knowing the analytical models used in their derivation. Otherwise the analyst may find himself proclaiming as a finding a relationship built in by the estimator.

Finally, there is the fundamental question which has haunted these remarks from the start—how does one appraise the acceptability of estimates such as Gallman's and others attempted here and in Volume 24? Even for the current period, it is difficult to assess the reliability of estimates of economic magnitudes. But as one pushes such figures back into periods when basic data sources are increasingly deficient and the use of analytical models to plug gaps rises correspondingly, the danger grows that the outcome will be no more than a house of cards. It is this concern, of which the best estimators themselves are only too well aware, which is voiced in several of the discussions in Volume 24 and in Fishlow's thoughtful review,[8] and which Parker expresses pointedly in his introduction to Volume 24 when he notes that "at some point the game goes beyond the bounds of good scholarship."[9]

As one reflects on this question, the lack of widely accepted rules for appraising estimates of economic magnitudes becomes increasingly apparent. It is curious that, while in recent years highly sophisticated methods have been developed for the analysis of data, relatively little systematic thought has been given even to procedures for the construction of estimates, let alone their appraisal. Current quantitative work on long-term growth only throws into bolder relief the need for developing more formal procedures for judging, if not the precise reliability of the estimates, at least whether the bounds of good scholarship have been exceeded. Such a development would doubtless be welcomed by the serious estimators themselves, since it would make for quicker rejection of

---

[7] Stanley Lebergott, "Measurement for Economic Models," *Journal of the American Statistical Association*, June 1954, pp. 209–226.

[8] Fishlow in *Journal of Economic History*, March 1962.

[9] *Trends in the American Economy*, p. 9.

ad hoc pseudoestimates which on occasion clutter a field. (It would have the added advantage of protecting the estimators against the frequent tendency of reviewers, uncurbed by the need to observe an established set of rules, to seize on weak points, honestly exposed to full view by the estimator himself, and so magnify them as to descredit the entire work.) Indeed, it is the estimators who have wrestled most with this problem and to whom one can turn for some initial guidance.[10]

In this connection, it is relevant to note features of Gallman's work, even in its present unfinished state, which strengthen one's confidence in it. Attention has already been given to his use of the commodity flow technique to test the consistency of different data sources. There is also his effort to test the results against other estimates of the same or related magnitudes. An example of this is the comparison of the 1839 estimate for distribution with the figures developed by Seaman and Marburg. Other possibilities that come quickly to mind are comparisons of the construction figures with building permits series and of some of the commodity estimates with Frickey's series on manufacturing production. In addition, Gallman identifies key assumptions and data deficiencies, and tries to evaluate their quantitative significance—an example is the underestimate of imports. Finally, Gallman gives attention to possible omissions and conceptual variants, as in the case of home manufactures and land improvements. All of these practices have merit and are applicable in varying degrees to other types of historical estimate.

One should not leave the question of appraisal without noting that in part, of course, it has always involved a matter of personal quality—the amount of thought and effort that has gone into Gallman's estimates is well known to anyone who has followed his work in the last ten years. When two agricultural experts such as Towne and Rasmussen go over Gallman's figures for agricultural income and arrive at quite similar results, it is a real personal tribute and furthers acceptance of other aspects of his work as well.[11] But as estimates and estimators multiply (as I hope they will), objective bases of appraisal, such as the procedures mentioned above, must necessarily grow in importance, and more systematic standards must be developed.

[10] In the income and wealth field, see, for example, *National Income: 1954 Edition*, U.S. Department of Commerce, Washington, 1954, Part III, and the National Bureau studies by Kuznets, Shaw, Fabricant, Kendrick, and others on national income and productivity.

[11] See *Trends in the American Economy*, pp. 259–314, esp. pp. 279–280.

Price Deflators for Final Product Estimates

DOROTHY S. BRADY

UNIVERSITY OF PENNSYLVANIA

Nature of the Price Data

The price indexes presented in Tables 1a, 1b, 2a, and 2b, at the end of the text, were a contribution to Robert Gallman's estimates of the value of output in constant dollars. More than 200 classes of goods and services were employed in his basic compilations, and the price deflators were constructed for that level of disaggregation. The indexes shown in the tables are limited to classes for which the prices of a reasonably representative number of commodities could be compiled.

The publication of *Prices Current* as newspapers or in newspapers has given price historians a regular, continuing source of data for wholesale price histories since newspapers have generally been preserved in libraries. There is some evidence that printed price lists of consumer goods were common but, if so, they were treated like handbills and destroyed. Price lists of individual manufacturers which appeared in their catalogs may be found where trade catalogs have been collected for various historical purposes.

In the absence of a regular file of published price lists for consumer goods, information on prices can be compiled from advertisements in newspapers and magazines, but advertisements are in no way a source of continuous information nor do they cover all classes of goods and services. Secondary sources in the contemporary literature offer fairly extensive information on prices but, except for the instructional books on building homes, such sources are extremely irregular. Thus, in the main, information on the prices of consumer goods and producer durables has been sought in manuscript sources, mainly account books and records.

Prices from 116 account books and stores for the years 1752-1860 were collected by the Massachusetts Bureau of Statistics of Labor [A-2][1] and published in 1885. Prices from the account books of stores, including a

---

[1] Numbers in square brackets refer to the bibliography at the end of the paper.

disproportionate representation of company stores, were collected for a special report of the *Tenth Census* [A-5] for the period 1851–80. Similarly, food prices in the 1890's were obtained from store records consulted in 1902 and published by the Commissioner of Labor in 1903. There are some transcriptions from records for the years 1800–40 available at the University of Pennsylvania [F-4, 8], but apart from "Prices Paid by Vermont Farmers," there have been no recent compilations of retail prices from manuscript sources.

Data taken from a few accounts are naturally incomplete so that a large number of such sources must be used in order to construct a continuous price series. In addition, the records are likely to provide only a limited amount of information on the qualities of foods and other perishable goods. As Ethel D. Hoover's tables [A-1] demonstrate, foods could not be differentiated by quality in the source, the report by Weeks of the *Tenth Census* [A-5], except through the distinction of several cuts of meat, two types of flour, two grades of yellow sugar, and two brands of molasses. It was, apparently, only at the end of the nineteenth century that the records consulted provided some indications of the quality of such foods as butter, cheese, and cornmeal.

While the qualities associated with prices compiled from account books cannot be ascertained, the catalogs of manufacturers contain a wide range of information on quality. The matching of prices from the sources based on records for two dates to construct price relatives must be based on some assumption about their comparability. The matching of prices taken from catalogs depends on the scheme for averaging the prices from different manufacturers for the same year. Whatever the source, the price relatives over the span of years have to be chains in which new qualities are "linked in" and old ones are "linked out."

Prices of the particular commodities listed in Tables 1 and 2 and for classes shown in the outline at the end of this paper were assembled from the sundry sources not only for the years shown in the tables but also for various intermediate years. Quite generally, information for some year in the 1820's was needed to connect the series of price relatives from 1809 to 1834, and price data for 1865 were used to link the series between 1860 and 1869. A plan to accumulate a minimum of three sets of price relatives for each commodity and pair of years was carried out with a few exceptions.

Price Relatives from Price Distributions

The collection of prices from catalogs and advertisements was mainly limited to durable goods. Most of the series for 1809–59 were based primarily on prices in the Massachusetts report on wages and prices, 1752–1860 [A-2], the industrial censuses in Massachusetts [A-4], and transcriptions from manuscript sources in Pennsylvania [F-4, 8]. The series for 1869–99 were constructed from prices in the Weeks reports on wages and prices [A-5], the Aldrich report on retail prices [A-7], and the *Eighteenth Annual Report* of the Commissioner of Labor [A-8]. The industrial censuses in Massachusetts provide the quantity and value of output in each town for a large number of commodities. Frequency distributions of the unit values derived from these data were compared with the prices in the report on wages and prices which were recorded as "high, medium high, medium, medium low, low." All too frequently, that source gave only a high and low price so it was necessary to locate these apparent extremes in a complete distribution of prices. The distributions of unit values for shoes, skirts, straw hats, and bonnets were bimodal with modes that did not differ greatly from the high and low prices given by the report on wages and prices for the same year, 1854. This comparison suggested that the high and low prices found in account books represent two levels of quality.

The Pennsylvania records suggest why the real extremes of a distribution of prices might not appear in account books. Any one commodity was recorded only a few times a year in a given account. Hence many more account books than the compilers were able to consult would be required to extend the distributions of prices to their extremes.

The Weeks reports and the Massachusetts report on wages and prices both provide prices for the 1850's. A comparison of flour prices from the two sources supports the attribution of distinct qualities to the high and low prices reported. The low and high prices from the Massachusetts report corresponded fairly well with the prices for "extra-family" and "superfine" flour in three cities of Massachusetts from the Weeks report. This correspondence means that the differences from year to year in the seasons represented by the recorded prices did not completely obscure the price differences between qualities. The prices of the specifications given in the Weeks report are for particular towns and cities and, except in St. Louis and Pilot Knob, Missouri, the "superfine" was always associated with the higher price. On the basis of these comparisons of sources, price relatives were calculated for the high and low prices taken from the Massachusetts report on wages and prices and from the Pennsylvania transcription for the years before 1860.

The Massachusetts industrial censuses provided for a considerable expansion of the list of items offered by the other two sources. Where specifications seemed clear cut, as with maple sugar, the price relatives were based on simple averages. Where the price distributions were bimodal, the relatives were based on the two modes. This was the case for boats, scythes, spades, and hoes, but not for plows and axes. The price distributions for plows were peaked with around 40 per cent at a middle-price line, while those for axes were rectangular with a wide range. A comparison with advertised prices in the 1840's indicated that the greatest number of plows manufactured and sold before 1860 must have been fairly well-constructed one-horse wooden plows with iron points, so the modal price was used for the construction of the price relatives. Axes, though important, were not included in the list of price relatives principally because no way was found to break the wide range into separate distributions.

The unit values derived from the industrial censuses provide a useful insight into the enumeration process. For many articles such as hardware and tools the unit values were neat round numbers, transparently taken from some kind of price lists. Thus for plows, the unit values in 1854 were even dollars from $3 to $10 accounting for 5,572 plows and $6.50 for 894 plows. Since the averages were for towns, it appears that the same price was used for all the manufacturers of a given type within a town. That practice was probably universal, although it cannot be so easily detected in the unit values for broad classes such as all shoes, all boats, or all carriages. If physical output was as simply valued for the Census of Manufactures and other sources used for the estimates of output, the selection of one or two modal price lines for the deflators is entirely appropriate as long as there is evidence that these prices did not reflect substantial changes in quality over time.

The linking of the relatives can be explained best through an illustration. The price series for men's hats was based on the relatives of the high prices and the relatives of the low prices for the types of hats shown below for the different comparison dates:

1809–34	*1834–49*	*1849–60*
Beaver	Beaver	Wool
Youths' beaver	Wool	Straw
Straw	Straw	

The relative for 1809 was based on 1834; the relatives for 1834, 1836, 1839, and 1844 were based on 1849; and the relatives for 1854 and 1859 were based on 1860. The relative for 1809 on the 1860 base was thus determined by the chain from 1809 to 1834 to 1849 to 1860.

PRICE DEFLATORS FOR FINAL PRODUCT ESTIMATES 95

Relatives Based on Prices from Lists and Reports

The Weeks report on wages provides many series of prices for individual manufacturers of a wide range of commodities. When three or more manufacturers reported on prices for a specified commodity, such as the top buggy, the price relatives were based on the median price. When only one manufacturer reported prices—as for pianos and sewing machines—their reports were included with the price information compiled from advertisements, catalogs, and other sources. Advertisements proved most useful in the case of consumer durable goods—pianos, sewing machines, and stoves—all products for which the variation in prices could not be simply attributed to quality differences. The price relatives for these commodities were determined as the median of the price relatives calculated for the product of each manufacturer, and here again it was necessary to follow a chain of comparisons over time as the list of manufacturers was changed.

The library collections of catalogs used principally for prices of producer durable goods did not often include the catalogs of the same manufacturer for a number of years except for data late in the century. Hence it was necessary to average the prices for commodities defined as similar in terms of some technical characteristic. The measure of size commonly used was selected, for reasons given elsewhere [A-9], and the prices for a given size were averaged without regard to any other qualitative characteristic. Since the size range expanded over time, the price relatives based on these averages represented a different set of specifications for different spans of years. The chaining procedure was thus the same as for other commodities.

Derivation of the Indexes for Classes of Commodities

There were only five classes of commodities for which the prices collected were representative of subclasses—flours and meals, sugars and syrups, clocks, carriages and wagons, and reed organs. The price indexes for these classes were weighted averages of the price relatives determined for the subclasses, and the weights were changed to represent the historical trend in their relative importance. The weight of flours was increased from 20 per cent in 1809 to 80 per cent in 1899; the weight of sugars was increased over the same period from 30 to 90 per cent; the weight of brass shelf clocks was increased from 10 per cent in 1839 to 30 per cent in 1879; buggies, an innovation, were given a weight of 10 per cent in

1854 and their relative importance was increased to 50 per cent in 1879; reed organs, likewise an innovation, were given a weight of 5 per cent in 1854, which was raised to 20 per cent in 1879. The weights for the foods were based on the Pennsylvania records for 1809 and the cost-of-living surveys at the end of the century. The weights for the durable goods were based on estimates of the physical volume of output at one or two dates found in the commodity histories. Although the weights are very rough, better information on the output of the specific commodities would probably change their magnitude but not their direction.

The price indexes for the subclasses of these five classes and for all other classes were the medians of the price relatives when the class or subclass included three or more items, and the geometric mean when there were only two. The median was selected instead of the unweighted average not only because the products for which prices were located could not be considered equally important, but also because there were frequent gaps in the information that could not be filled by an imputation procedure. Wherever possible, a missing price for a given year, say, 1844, was estimated from the relation of its price to the price of a similar item in the year before or the year after. When there were several gaps over a span of years, the item was dropped instead of using two or more estimated prices. In the example of men's hats given above, the beaver (castor) hats were not included in the series for 1849–60 because the data were missing and not because the fine fur hat had disappeared.

Geographic Price Differences and Population Changes

The indexes for the years before 1860, based as they are mainly on data for Massachusetts and Pennsylvania, may not reflect national price changes if there were significant differences in prices and price changes among areas. Certainly the price trends in the new towns of the West declined over a long period relative to the price trends in the established settlements of the East. Comparisons of the price levels and price trends in the communities covered by the Weeks report suggest that differences were not so much a matter of state or region as of location in relation to transportation networks. The ranges of prices within states tended to be similar, and only on "imported" foods—flour, sugar, salt fish, coffee, and tea—did the absolute levels of prices vary systematically during the period between 1857 and 1880. Other food prices and textile prices varied as much among the communities in the central states as among those in the East.

The differences between the prices in Massachusetts and Pennsylvania

PRICE DEFLATORS FOR FINAL PRODUCT ESTIMATES 97

for the years before 1860 also suggest that transportation was the chief factor in community variation—apart, of course, from essential quality differentials that could always be present. The range in the price of flour, for example, was about $1 in both places, except at peaks when it reached or exceeded $2. The level in Pennsylvania was, however, consistently below the level in Massachusetts, by a factor varying between 12 and 20 per cent. Flour in Massachusetts came from the Carolinas, Maryland, and Pennsylvania and by the 1850's from New York.

Whatever the reasons for community differences, indexes constructed for deflating output estimates should reflect movements of the population as well as changing community differentials. It is possible that the failure to use changing population weights for communities of different types introduces larger biases than any other kind of procedural or measurement error. If so, the historical estimates are more seriously affected than contemporary series are, simply because the locality differences have been reduced over time. Thus, by the 1890's, there was no significant difference in the level of flour prices among the cities and towns of Massachusetts and Pennsylvania. The effect of reductions in the magnitude of community differences on the general level of prices is one of a number of studies that could be based on the excellent histories of wholesale prices.

The biases in the price indexes presented in this paper arise from the lack of information for localities outside Massachusetts and Pennsylvania and also from giving all observations equal weight, whatever their geographic distribution. In general practice, the bias, as in the current procedures for determining the gross national product in constant dollars, arises from the substitution of a Laspeyres for a Paasche index for the community or the class of commodities. If the three indexes are called P, L, and U, the biases are $(L - P)/P$ in the procedure used currently, and $(U - P)/P$ in the procedure used in this study, assuming that the medians are estimates of unweighted means. The size of the errors can be illustrated by a simple example based on the figures below.

	Regions of Settlement 1860		Other Regions	
	1860	1890	1860	1890
Population (millions)	8	23	25	40
Price of flour (cents)	9.0	2.2	2.5	2.3
Price of cornmeal (cents)	0.5	1.5	1.0	2.2

The social histories of new settlements suggest that the per capita consumption of cornmeal in 1860 was three times as great in the regions newly settled that year and the per capita consumption of flour no more than one-fourth that in the rest of the country. In 1890, the differences were small enough to be ignored for present purposes. With these assumptions, the biases in the Laspeyres and unweighted indexes compared with those in the Paasche are small, 2 and 1 per cent for cornmeal. The bias in the Laspeyres index for flour is very large, 100 per cent, but the unweighted index was in error by only 2 per cent. The use of the Laspeyres index as a deflator would result in a 50 per cent downward bias in the estimates of output in 1860 dollars.

This example may seem to exaggerate the problem since the results turn on the high price of flour attributed to regions newly settled in 1860 but it actually is an average for Salt Lake City, Denver, and Lawrence, Kansas. The long-period comparison is unrealistic, inasmuch as the Laspeyres indexes as compiled at the present time are chains with population and consumption weights changed about every ten years. Since the price of flour could and did change relatively as much as indicated in the example within a period of ten years, as communities were connected with transportation systems, the illustration serves to underscore the need to explore this problem for historical studies in particular. The variation in changes in population, in consumption, and in prices among localities must, nevertheless, be recognized as much for current estimates as for historical studies. While the population distributions are not changing as dramatically as in the nineteenth century, consumption levels, which are a function of income, have changed enough to reinforce the effect of population changes on the differences between the price indexes.

The unweighted indexes fared better than the Laspeyres indexes because of the assumptions made about the relative consumption in the two sets of regions. There is no doubt that the per capita consumption of goods produced locally was higher in small and isolated communities, while the per capita consumption of imported commodities was lower than elsewhere. The price information for different localities could be used to estimate the relation between price and consumption between communities if the aggregate physical output, as well as the aggregate value of output, is known. By distributing the population among the communities for which the price information is available, the parameters of a linear demand function between communities could be ascertained. Such studies for various types of specific commodities might lead to a systematic procedure for estimating the bias in price indexes that do not reflect the changes in the population distribution among communities.

Chain Indexes, Quality Changes, and New Goods

New qualities of goods and new types of goods must be linked into a chain of comparisons, regardless of the formula for the index. In the

PRICE DEFLATORS FOR FINAL PRODUCT ESTIMATES

Laspeyres index and others of that type in general use, new qualities are linked in as substitutions at any date, but new types of goods are introduced only when the indexes are revised. The chaining is at the level of the specific commodity and, as a result, the indexes are averages of chains. In the present study, the indexes for classes of goods can be described as chains of average prices or price relatives. The procedures used avoid a very real possibility of bias through the correlation in the price movements of the new qualities over a given period with the price movements of the "old" commodities over the preceding period of comparison. When technical progress was affecting the prices of all the products within a class in the same way, these correlations were certainly positive and, perhaps, of significant magnitude.

Avoiding the effect of the correlations removes only part of the biases arising from the introduction of the prices of new qualities of goods into the index calculations. The timing of the introduction of new qualities determines the behavior of the indexes, and the accumulation of price data for many more years than selected for this study would, without doubt, lead to different patterns of substitution and hence to different estimates for the price indexes. The nature of the problem can be described by the example of an index based on two qualities which are changed over time in the way shown below.

Quality	1834	1836	1849	1854	1860
1	P_{41}	P_{31}	P_{21}		
2		P_{32}	P_{22}	P_{12}	
3			P_{23}	P_{13}	P_{03}
4				P_{14}	P_{04}

The price index for 1834 on the 1860 base is determined by the chain

$$\frac{P_{41}}{P_{31}} \frac{Av(P_{31}, P_{32})}{Av(P_{21}, P_{22})} \cdot \frac{Av(P_{22}, P_{23})}{Av(P_{12}, P_{13})} \cdot \frac{Av(P_{13}, P_{14})}{Av(P_{03}, P_{04})}.$$

Rearranging the chain reveals the assumption made about the quality difference between the 1834 quality and the 1860 qualities. The second, third, and fourth links in the chain,

$$\frac{P_{41}}{Av(P_{03}, P_{04})} \frac{Av(P_{31}, P_{32})}{P_{31}} \frac{Av(P_{22}, P_{23})}{Av(P_{21}, P_{22})} \frac{Av(P_{13}, P_{14})}{Av(P_{12}, P_{13})},$$

are comparisons of the prices of different qualities at the same dates. The price indexes thus reflect an accumulation of such estimates of quality differences. Examination of many price schedules indicates that the price for different qualities is by no means simply proportionate to the physical measure of the quality. Thus, the accumulated estimates of the quality differences are very sensitive to the selection of qualities for the price index. Since each link compares two parts of a price-quality schedule, the results of the selection will vary according to the nature of the schedule.

The schedules changed significantly over time, and fixing the links at different dates would yield quite different estimates of the quality component of the price index calculation. The whole problem needs thorough investigation, and the abundance of price information of the kind used in this study for the early part of this century suggests that methodological studies with data for 1900–20 could contribute to the improvement both of methods used with historical data and of the current practices in the determination of index numbers.

New products and old products do not present the same kind of problem for the determination of a set of price deflators for the construction of a general price index. The output estimates include the new products as they appear and leave out the old products when they disappear. The price deflators can be constructed accordingly with some assumption about nonexistent base-period prices. It is hard to devise a procedure for estimating a hypothetical price, and thus the attribution of the price, at the date of introduction or disappearance, to the base date seems to be the only operationally feasible procedure. Such a conclusion simply invites an examination of the problem that will prove the contrary.

Sources and Notes

The censuses, surveys, and compilations listed in the bibliography were a main source of information on prices, particularly for perishable and semidurable goods. Between the Weeks report on wages and prices, which covered prices of a wide variety of durable goods, and the industrial censuses in Massachusetts, it was possible to construct series for a number of durable goods after identifying the price lines with specifications through the descriptive literature. Thus, without much additional information, it would have been possible to construct price indexes for such products as agricultural machinery, furniture, pianos, locomotives, and carriages from the data in these publications. They do not, however, provide the necessary price information on heating equipment and industrial machinery, clocks, watches, optical goods, and the like. Heating equipment was reported mainly as hollow ware, per ton, machinery as castings, per pound, while the smaller articles were reported in mixtures such as "all philosophical apparatus."

The histories, the books of instruction, advertisements, and trade catalogs were used to fill in the series of prices and extend them to 1809, 1889, and 1899. Among the histories, Bishop [B-3], Bolles [B-4], and Clark [B-6] quote many prices for early years in the century generally without citing the original source. It was possible to trace some of the quotations to Fearon [B-14] and Cobbett [B-7], but the detail in some cases suggests that the authors had access to price lists or tabulated cost estimates that were widely used. With few exceptions, the histories of particular products are time-consuming references for the collection of price statistics.

The instructional literature is much more rewarding. In such books as Sloan [C-17], the cost estimates for the construction of houses and churches are itemized in detail and thus provide prices for equipment such as stoves, furnaces, and pumps. These books evidently developed, to begin with, as the result of popular articles in such magazines as the *American Farmer*, the *Cultivator*, and *Godey's* [D-2, 8, and 10].

Advertisements, particularly in the directories, registers, and illustrated magazines like Frank Leslie's [D-9], are the best source of information on such things as sewing machines, refrigerators, patent medicines, cosmetics, watches, paper products, magazines, books, and surgical appliances. There are other articles widely advertised, like playing cards, sheet music, and daguerreotypes, for which price indexes could not be calculated for want of specifications of size or other aspects of quality.

The material from various account books supplemented the compilations for all types of goods except clothing, furniture, housefurnishings, and toys. The Irving and Leiper accounts [F-2] provide information not found elsewhere on installation costs for various types of machinery, and construction costs for such structures as stables and bathhouses. Account books could be particularly useful in assigning weights to different qualities of a product and to various products within groups. They could also eventually lead to an expansion of the coverage of commodities classified as consumer goods for the years before the Civil War and even for later years. Households bought sand for floors, earth for the earth closets, and crushed stone for walks, until rugs and carpets, water and sewage systems, and brick sidewalks supplied such needs. Housewives used saltpeter and potash for preserving meats, and beeswax and tallow for making candles, as late as the 1890's. Such products might represent an aggregate at least comparable with the textiles that represent the volume of clothing, mattresses, and pillows manufactured in households throughout most of the century.

Price statisticians have to deal with a level of disaggregation which is discomfiting to those engaged in the study of large aggregates. As long as our concept of price change relies on the identification of the same product in different places or at different dates, the collections of price data will necessarily be detailed and voluminous and not susceptible of the kind of annotation now customary in estimating various aggregates.

Price indexes have to be some kind of a sample of items within classes and of qualities within items. For historical studies, the sample of items is determined by the information found in records, and the recorded prices, by and large, represent the kind of products that could be transported some distance and sold in a market, and do not necessarily represent the range of products within a class in any meaningful way. Peaches, melons, strawberries, raspberries, cherries, and grapes were all important in the localities very near where they were produced, but no one ever kept records in farmers' markets. Thus, a price index for fruits has to be based on the hard-skinned fruits that could be transported to and stored in the general store for a reasonable period of time. In some classes of products, such as furniture, it is quite possible that the output estimates were similarly limited but, in general, the items included in the price indexes must be a poor representation of the complete range. It would take many more items than are shown in the outline below to assure that the sample even approached a cross section of all the diversity in output.

A glance at Tables 1 and 2 should lead the reader to ask why there is so much methodological discussion here of series characterized mainly by blanks. This question has concerned the author because there seems to be no good reason for presenting these poor estimates of price developments without a simultaneous publication of the output estimates in equivalent detail.

Outline of Items Included in Price Indexes

Class of Product	Items in Price Index
Foods, manufactured	
Bakery products	Bread, biscuits, crackers
Canned, smoked fish	Cod, mackerel
Canned, dried fruits, vegetables	Dried peas, beans, canned corn, tomatoes
Cheese	Cheese
Chocolate, cocoa	Chocolate
Coffee and spices	Coffee, pepper
Hominy, flour, meals	Wheat flour, corn and rye meal
Other cereal products	Macaroni

PRICE DEFLATORS FOR FINAL PRODUCT ESTIMATES   103

Class of Product	Items in Price Index
Distilled spirits	Rum, whisky
Malt liquors	Beer
Vinous liquors	Grape wine
Rice	Rice
Meat products	Salt pork, beef, ham, sausage
Syrups, sugars	Molasses, yellow, brown, loaf sugar
Vinegar, cider	Vinegar
Baking powders, yeast	Cream of tartar, saleratus, soda
Butter	Butter
Canned milk	Condensed milk
Lard	Lard
Salt	Salt
Foods, nonmanufactured	
Orchard fruits	Apples, quinces
Citrus fruits	Lemons
Small fruits	Cranberries
Potatoes	Potatoes
Sweet potatoes	Sweet potatoes
Peas and beans	Peas, beans
Vegetables	Cabbages, onions, turnips
Cereals	Corn, rye meal
Fish, fresh	Cod, halibut, haddock, mackerel
Poultry, eggs	Chickens, eggs
Meat, fresh	Beef, pork, mutton
Tea	Suchong
Tobacco products	
Tobacco	Tobacco, snuff
Drugs, household preparations	
Bluing	Bluing
Patent medicines	Herb remedy
Perfumes	Rosewater, oil of bergamot
Soap	Soap
Blacking, stains	Blacking
Cleaning preparations	Sapolio
Castor oil	Castor oil
Mazagines, paper products	
Envelopes	Envelopes
Writing paper	Writing paper
Newspapers, magazines	Newspapers, magazines
Ink	Ink
Mucilage, paste	Mucilage

Class of Product	Items in Price Index
Fuel, lighting products	
Candles	Tallow candles
Matches	Matches
Oil, sperm whale	Whale oil
Oil, coal	Kerosene
Firewood	Firewood
Coal	Coal
Dry goods and notions	
Needles, pins, etc.	Needles, pins, hooks, and eyes
Pocketbooks, purses	Pocketbooks
Buttons	Metal buttons
Cotton thread	Cotton thread
Cotton woven goods	Calico, muslin sheeting, shirting, gingham, cambric
Silk ribbons, cloth	Ribbons, handkerchiefs
Woolen worsted goods	Broadcloth, cashmere, flannel
Mixed textiles	Plaid linsey
Clothing and personal furnishings	
Clothing, men's, factory	Pantaloons, overalls, vests, coats
Clothing, women's, factory	Balmoral skirts, shawls, hoopskirts
Gloves, mittens	Gloves
Hats, men's	Beaver, wool hats
Bonnets, trimmed	Straw bonnets, trimmed
Underwear	Hose, stockings
Umbrellas	Umbrellas
Shoes, etc.	Boots, shoes, slippers
Rubber footwear	Rubbers
Housefurnishings	
Brooms	Brooms
Bedding	Bedspreads, sheets
Linen woven goods	"Linen"
Towels	"Linen"
Furniture	Tables, chairs, bedsteads
Refrigerators	Refrigerators
Sewing machines	Sewing machines
Stoves	Stoves
Washing machines	Washing machines
Floor coverings	Rugs
Feather beds, pillows	Feather pillows
Mattresses, springs	Mattresses
Blankets	Blankets

PRICE DEFLATORS FOR FINAL PRODUCT ESTIMATES 105

Class of Product	Items in Price Index
Cutlery	Forks, spoons
China, earthenware	Cups, teapots
Woodenware	Bowls
Glassware	Drinking glasses
Lamps	Globe, kerosene lamps
Other consumer goods	
Musical instruments	Pianos, reed organs
Clocks	Grandfather, brass clocks
Carriages	Gigs, carriages, buggies
Artificial limbs	Wood, reed, metal
Eyeglasses	Spectacles
Books	Bible, spelling book
Billiard tables	Billiard tables
Pocketknives	Pocketknives
Firearms	Muskets
Cameras	Magic lanterns, cameras
Children's wagons, sleighs	Children's wagons
Toys	Cast-iron wheel toys
Producer durable goods	
Machine-shop products	Steam engines, waterwheels, lathes, looms
Electrical apparatus	Motors, arc-lighting equipment
Agricultural machines	Corn shellers, threshing machines, mowers
Agricultural implements	Plows, scythes, spades, hoes
Pumps	Pumps
Barbed wire	Barbed wire
Woven wire fence	Wire fence
Cash registers	Cash registers
Scales	Scales
Typewriters	Typewriters
Office furniture	Physicians' chairs
Railroad equipment	Locomotives, cars
Ships and boats	Sailing vessels, river steamers
Farm, freight wagons	Freight wagons
Optical goods, etc.	Chronometers, barometers, microscopes, compasses
Edge tools	Augers, planes
Other tools	Saws, files
Buildings	
Housing, community services	Houses, churches, schools
Nonfarm business	Factories, office buildings

TABLE 1a

PRICE INDEXES FOR CONSUMER GOODS, SELECTED YEARS, 1809-49
(1860=100)

Product	1809	1834	1836	1839	1844	1849
Perishables						
Manufactured foods, etc.						
Bakery products	131	113	127	102	85	84
Canned, smoked fish, etc.	120	90	91	96	70	83
Canned, dried fruits, vegetables	93	54	67	62	57	66
Cheese	93	89	120	124	65	85
Chocolate, cocoa	139	80	87	92	92	90
Coffee, spices	158	90	74	81	67	98
Hominy, flour, meals	136	103	123	120	80	103
Other cereal products	117	103	135	99	87	93
Rum, whisky	154	86	112	149	94	95
Malt liquors	130	104	100	100	73	87
Vinous liquors	--	60	67	113	63	73
Rice	83	75	85	98	68	82
Meat products	130	86	120	93	64	79
Syrups, sugars	161	100	120	102	97	103
Vinegar, cider	137	102	115	102	84	84
Baking powders, yeast, etc.	125	126	116	114	113	84
Butter	117	87	114	116	75	91
Condensed milk	--	--	--	--	--	--
Lard	107	71	104	92	53	61
Salt	202	107	118	113	100	99
Nonmanufactured foods, etc.						
Orchard fruits	90	120	172	138	63	93
Citrus fruits	--	--	--	--	--	--
Small fruits	51	62	70	60	51	110
Potatoes	71	52	89	74	119	130
Sweet potatoes	--	--	--	77	83	98
Peas and beans	97	85	98	109	74	76
Vegetables	122	87	118	98	61	75
Wheat	143	103	114	110	71	96
Fish, fresh	57	50	63	70	60	72
Poultry, eggs	83	70	--	88	81	86
Dairy products	98	86	95	100	72	90
Maple sugar, honey	89	60	84	76	63	63
Meat, fresh	78	63	75	88	64	76
Tea	136	72	58	67	68	61
Tobacco	90	70	82	88	80	100
Drugs, household preparations						
Bluing	--	--	--	--	--	--
Patent medicines	200	--	--	146	100	76
Perfumes	100	--	--	100	90	70
Soap	172	125	115	113	100	86
Blacking, stains	259	174	196	202	150	120
Cleaning preparations	--	--	--	--	--	--
Castor oil	--	--	--	--	59	84
Magazines, paper products, etc.						
Envelopes	--	--	--	--	--	110
Writing paper	449	234	225	216	178	125
Newspapers, magazines	245	146	--	135	122	--
Ink	--	168	164	131	121	103
Mucilage, paste	154	128	112	109	97	120
Fuel, lighting products						
Candles	113	91	94	106	95	81
Matches	--	--	--	307	267	106
Oil, sperm whale	104	100	110	121	110	90
Oil, coal	--	--	--	--	--	--
Firewood	85	67	77	88	92	100
Coal	--	121	158	141	84	100

(continued)

PRICE DEFLATORS FOR FINAL PRODUCT ESTIMATES 107

TABLE 1a (concluded)

Product	1809	1834	1836	1839	1844	1849
Semidurable goods						
Dry goods and notions						
Needles, pins	800	262	--	163	147	--
Pocketbooks	--	--	--	--	150	--
Buttons	300	240	161	159	134	--
Cotton thread	290	157	136	123	101	85
Cotton woven goods	376	155	150	141	105	106
Silk ribbons and cloth	193	--	140	--	155	133
Woolen worsted goods	202	135	146	167	159	133
Mixed textiles	284	253	--	168	--	--
Clothing and personal furnishings						
Clothing, men's, factory	183	--	--	117	--	--
Clothing, women's, factory	185	90	111	150	94	88
Gloves, mittens	83	93	104	83	95	80
Hats, men's	185	153	132	140	126	97
Bonnets, trimmed	243	--	140	--	133	--
Underwear	487	359	195	151	115	95
Umbrellas	300	--	141	--	--	--
Shoes	185	153	153	163	114	111
Rubber footwear	--	180	--	143	107	--
House furnishings						
Brooms	140	108	--	102	120	104
Bed spreads, sheets	240	161	133	149	101	95
Linen woven goods	201	150	137	137	120	123
Towels	182	167	138	141	114	118
Toys, games, sporting goods						
Billiard tables	--	--	--	--	--	--
Pocketknives	--	--	--	--	--	152
Firearms	192	--	152	--	146	--
Cameras	--	--	--	--	--	145
Children's wagons, sleds	--	--	121	--	--	--
Toys	--	--	--	--	--	--
Durable goods						
Furniture						
Tables, chairs, bedsteads	289	--	--	--	--	111
Heating, cooking, appliances						
Refrigerators	265	--	--	137	--	--
Sewing machines	--	--	--	--	143	132
Stoves	154	115	119	117	113	116
Washing machines	120	--	·120	--	--	--
Floor coverings						
Rugs	301	--	--	241	151	139
Miscellaneous						
Feather beds, pillows	126	73	102	95	65	75
Mattresses, springs	302	--	--	--	--	--
Blankets	197	--	156	145	113	111
Looking glasses	300	135	138	--	150	117
China, utensils						
Cutlery	250	--	156	--	142	--
China, earthenware	172	--	--	140	--	--
Woodenware	194	--	--	--	129	132
Glassware	256	--	--	147	149	--
Lamps	180	--	--	120	120	154
Musical instruments, books						
Pianos, reed organs	143	71	108	--	93	96
Clocks	517	270	--	157	102	--
Carriages, buggies, wagons	234	--	193	--	137	91
Artificial limbs	156	--	140	--	120	--
Eyeglasses	231	117	125	--	117	118
Books	267	181	--	--	167	--

TABLE 1b

PRICE INDEXES FOR CONSUMER GOODS, SELECTED YEARS, 1854-99
(1860=100)

Product	1854	1859	1869	1879	1889	1899
Perishables						
Manufactured foods, etc.						
Bakery products	128	114	156	93	111	110
Canned, smoked fish etc.	106	100	151	122	124	112
Canned, dried fruits, vegetables	84	80	184	117	116	91
Cheese	109	97	143	98	167	181
Chocolate, cocoa	92	112	154	164	145	92
Coffee, spices	97	98	186	123	126	105
Hominy, flour, meals	125	105	151	111	105	83
Other cereal products	103	100	154	112	105	92
Rum, whisky	98	114	326	295	285	314
Malt liquors	100	102	143	114	106	93
Vinous liquors	79	119	210	138	137	136
Rice	100	95	225	167	158	158
Meat products	134	100	158	98	98	104
Syrups, sugars	92	99	159	97	68	65
Vinegar, cider	100	98	250	182	112	108
Baking powder, yeast, etc.	127	99	140	93	80	65
Butter	104	104	150	96	114	125
Condensed milk	--	--	144	62	55	43
Lard	90	102	164	92	77	83
Salt	137	100	176	61	56	45
Nonmanufactured foods, etc.						
Orchard fruits	170	111	233	208	117	136
Citrus fruits	--	100	--	83	73	70
Small fruits	157	80	190	90	130	--
Potatoes	133	107	137	140	111	125
Sweet potatoes	--	100	258	117	141	--
Peas and beans	98	119	187	100	137	--
Vegetables	113	108	118	72	90	--
Wheat	108	104	155	109	107	103
Fish, fresh	104	127	220	119	120	121
Poultry, eggs	101	96	143	102	--	--
Dairy products	104	102	170	107	118	116
Maple sugar, honey	88	107	124	119	130	110
Meat, fresh	125	98	162	114	122	127
Tea	93	101	179	101	104	109
Tobacco	100	104	267	124	92	75
Drugs, household preparations						
Bluing	--	--	223	84	98	100
Patent medicines	90	100	117	72	83	90
Perfumes	77	104	129	111	67	70
Soap	89	101	128	88	72	91
Blacking, stains	128	104	--	--	--	--
Cleaning preparations	--	105	212	53	52	48
Castor oil	59	91	100	45	52	51
Magazines, paper products, etc.						
Envelopes	120	90	--	85	--	--
Writing paper	--	104	121	80	--	--
Newspapers, magazines	--	--	--	107	--	95
Ink	--	100	--	91	--	--
Mucilage, paste	112	101	--	--	--	--
Fuel, lighting products						
Candles	101	106	134	65	51	44
Matches	110	100	--	--	--	--
Oil, sperm whale	86	97	194	120	83	74
Oil, coal	120	100	67	29	16	15
Firewood	112	102	176	147	108	--
Coal	109	93	143	102	88	90

(continued)

PRICE DEFLATORS FOR FINAL PRODUCT ESTIMATES

TABLE 1b (concluded)

Product	1854	1859	1869	1879	1889	1899
Semidurable goods						
Dry goods and notions						
Needles, pins	122	103	124	95	88	75
Pocketbooks	101	100	117	--	--	--
Buttons	120	--	145	--	--	--
Cotton thread	100	102	181	91	80	76
Cotton woven goods	100	101	173	90	83	67
Silk ribbons and cloth	110	113	225	171	151	--
Woolen worsted goods	104	102	119	87	63	--
Mixed textiles	125	100	80	55	50	--
Clothing and personal furnishings						
Clothing, men's, factory	106	97	153	130	--	--
Clothing, women's, factory	106	104	107	104	98	--
Gloves, mittens	70	90	--	126	--	--
Hats, men's	112	105	122	131	127	--
Bonnets, trimmed	79	--	174	181	--	--
Underwear	100	108	100	57	45	--
Umbrellas	--	--	138	--	--	--
Shoes	107	100	148	108	103	95
Rubber footwear	107	100	--	80	--	--
House furnishings						
Brooms	108	106	--	--	--	--
Bed spreads, sheets	100	98	155	85	100	75
Linen woven goods	100	102	124	113	110	95
Towels	103	100	124	92	82	63
Toys, games, sporting goods						
Billiard tables	--	100	90	50	57	--
Pocketknives	121	100	117	80	66	--
Firearms	117	--	83	--	--	--
Cameras	122	--	104	83	76	54
Children's wagons, sleds	109	100	102	89	73	--
Toys	--	100	83	60	60	46
Durable goods						
Furniture						
Tables, chairs, bedsteads	109	100	108	79	70	--
Heating, cooking, appliances						
Refrigerators	--	100	105	73	65	--
Sewing machines	125	100	83	66	62	53
Stoves	117	100	146	114	71	68
Washing machines	112	100	92	83	73	--
Floor coverings						
Rugs	123	100	189	91	78	71
Miscellaneous						
Feather beds, pillows	96	99	--	--	--	--
Mattresses, springs	--	100	--	65	43	53
Blankets	109	106	95	71	59	62
Looking glasses	106	100	120	133	100	--
China, utensils						
Cutlery	120	100	115	116	71	66
China, earthenware	123	--	112	89	70	--
Woodenware	130	108	129	86	76	72
Glassware	160	--	84	47	37	--
Lamps	105	96	88	50	65	50
Musical instruments, books						
Pianos, reed organs	91	99	112	95	92	82
Clocks	100	--	105	93	--	75
Carriages, buggies, wagons	95	--	148	89	76	77
Artificial limbs	--	--	75	--	50	--
Eyeglasses	87	--	117	96	--	--
Books	115	100	--	86	--	92

TABLE 2a
PRICE INDEXES FOR PRODUCER DURABLE GOODS, SELECTED YEARS, 1834-49
(1860 = 100)

Product	1834	1836	1839	1844	1849
Industrial machinery, equipment					
Machine-shop products	156	162	149	152	138
Sewing machines	--	--	--	--	--
Electrical apparatus	--	--	--	--	--
Farm equipment					
Agricultural machines	157	150	130	118	127
Agricultural implements	--	240	--	142	--
Pumps	--	200	--	110	--
Windmills	--	--	--	--	137
Barbed wire	--	--	--	--	--
Woven wire fence	--	--	--	--	--
Office, store equipment					
Cash registers	--	--	--	--	--
Scales	--	242	--	190	--
Typewriters	--	--	--	--	--
Furniture	--	--	--	137	137
Railroad equipment					
Locomotives, cars	--	163	--	142	--
Ships and boats					
Sailing vessels, river steamers	189	--	--	150	--
Conveyances					
Farm, freight wagons	--	170	--	83	--
Professional, scientific equipment					
Optical goods, etc.	--	337	--	--	244
Carpenters, mechanics tools					
Edge tools	--	230	--	185	--
Other tools	--	270	--	242	--
Construction costs					
Houses, churches, schools	--	--	--	109	95
Factories, office buildings	--	147	--	76	107

PRICE DEFLATORS FOR FINAL PRODUCT ESTIMATES

TABLE 2b

PRICE INDEXES FOR PRODUCER DURABLE GOODS, SELECTED YEARS 1854-99
(1860=100)

Product	1854	1859	1869	1879	1889	1899
Industrial machinery, equipment						
Machine-shop products	115	107	113	71	32	28
Sewing machines	120	120	135	72	64	50
Electrical apparatus	--	--	100	88	78	69
Farm equipment						
Agricultural machines	104	105	116	77	65	62
Agricultural implements	100	--	143	75	63	47
Pumps	115	--	120	91	65	60
Windmills	--	--	86	49	48	45
Barbed wire	--	--	--	100	80	65
Woven wire fence	--	--	79	50	34	27
Office and store equipment						
Cash registers	--	--	--	75	70	70
Scales	175	--	--	83	67	65
Typewriters	--	--	--	100	95	79
Furniture	--	--	--	92	86	--
Railroad equipment						
Locomotives, cars	123	--	67	57	49	43
Ships and boats						
Sailing vessels, river steamers	118	--	84	72	57	51
Conveyances						
Farm, freight wagons	100	--	138	113	86	72
Professional, scientific equipment						
Optical goods, etc.	--	--	132	87	85	77
Carpenters, mechanics tools						
Edge tools	124	--	128	94	83	64
Other tools	118	--	142	85	85	77
Construction costs						
Houses, churches, schools	87	98	134	122	132	--
Factories, office buildings	107	--	94	107	89	--

BIBLIOGRAPHY

A. CENSUSES, SURVEYS, AND COMPILATIONS

1. Ethel D. Hoover, "Retail Prices after 1850," in *Trends in the American Economy in the Nineteenth Century*, Studies in Income and Wealth 24, Princeton for NBER, 1960, pp. 141–190.
2. *Sixteenth Annual Report*, Massachusetts Bureau of Statistics of Labor, Boston, 1885.
3. *Thirty-first Annual Report*, Massachusetts Bureau of Statistics of Labor, Boston, 1901.
4. Massachusetts Secretary of State, *Statistical Tables Exhibiting the Condition and Products of Certain Branches of Industry in Massachusetts for the Year Ending April 1, 1837*, Boston, 1838. The same for years ending April 1, 1845, April 1, 1855, and April 1, 1865.
5. *Tenth Census of the United States, 1880*, Special Reports, *Statistics of Wages in Manufacturing Industries with Supplementary Reports on the Average Retail Prices of the Necessaries of Life and on Trades Societies and Strikes and Lockouts*, by Joseph D. Weeks, Washington, 1886.
6. Senate Committee on Finances, Rept. 1394, *Wholesale Prices, Wages and Transportation*, by Mr. Aldrich, Washington, 1893.
7. Senate Committee on Finances, Rept. 986, *Retail Prices and Wages*, by Mr. Aldrich, Washington, 1892.
8. *Eighteenth Annual Report of the Commissioner of Labor*, Department of Commerce and Labor, Washington, 1903.
9. Dorothy S. Brady, "Relative Prices in the Nineteenth Century," *Journal of Economic History*, June 1964.

B. HISTORIES

1. *History of the Baldwin Locomotive*, Baldwin Locomotive Works, Philadelphia, 1913.
2. William Banning and Hugh George, *Six Horses*, New York, 1930.
3. J. Leander Bishop, *A History of American Manufactures from 1806 to 1860*, Vols. II and III, Philadelphia, 3rd ed., 1868.
4. Albert S. Bolles, *Industrial History of the United States*, Norwich, Conn., 1881.
5. Howard I. Chapelle, *History of American Sailing Ships*, New York, 1935.
6. Victor S. Clark, *History of Manufactures in the United States*, New York, 1929.
7. William Cobbett, *A Year's Residence in the United States of America*, London, 1828.
8. Carl C. Cutler, *Greyhounds of the Sea, The Story of the American Clipper Ships*, New York, 1930.

9. *One Hundred Years of American Commerce*, Chauncey M. Depew, ed., New York, 1895.
10. Everett Dick, *The Sod House Frontier, 1854–1890*, New York, 1937.
11. Henry Disston and Sons, *The Saw in History*, Philadelphia, 1916.
12. James Dredge, *The Pennsylvania Railroad, Its Origin, Construction and Maintenance*, New York, 1879.
13. Seymour Dunbar, *History of Travel in America*, Indianapolis, 1915.
14. Henry Bradshaw Fearon, *Sketches of America*, London, 1818.
15. Ruth E. Finley, *The Lady of Godey's*, Philadelphia, 1931.
16. Charles Flint et al., *One Hundred Years Progress of the United States*, Hartford, 1872.
17. Edwin T. Freedley, *Philadelphia and Its Manufactures*, Philadelphia, 1858.
18. George S. Gibb, *The Saco-Lowell Shops, Textile Machinery Building in New England, 1813–1849*, Harvard Studies in Business History 16, Cambridge, Mass., 1950.
19. *The Great Industries of the U.S.*, Horace Greeley, ed., Hartford, 1872.
20. Harper and Burtas, *The First Century of the Republic*, New York, 1871.
21. Arthur E. James, *Chester County Clocks and Their Makers*, West Chester, Pa., 1942.
22. Arthur E. James, *The Potters and Potteries of Chester County*, West Chester, Pa., 1945.
23. Chauncey Jerome, *History of the American Clock Business for the Past Sixty Years*, New York, 1860.
24. Helen LaGrange, *Clipper Ships of America and Great Britain*, New York, 1936.
25. James Mease, *Picture of Philadelphia*, Philadelphia, 1831.
26. *History of Transportation in the United States before 1860*, B. H. Meyer, ed., Washington, 1917.
27. Harold C. Passer, *The Electrical Manufacturers, 1875–1890*, Cambridge, Mass., 1953.
28. J. L. Ringwalt, *Development of the Transportation of the United States*, Philadelphia, 1888.
29. Jack D. Rittenhouse, *American Horse Drawn Vehicles*, Los Angeles, 1953.
30. Robert H. Thurston, *History of the Growth of the Steam Engine*, New York, 1886.
31. Lyman Horace Weeks, *History of Paper Manufacturing in the U.S., 1690–1916*, New York, 1916.
32. George S. White, *Memoir of Samuel Slater*, 2nd ed., Philadelphia, 1836.

C. INSTRUCTIONAL LITERATURE

1. Daniel T. Atwood, *Atwood's Country and Suburban Houses*, New York, 1882.
2. Catherine E. Beecher and Harriet Beecher Stowe, *The American Woman's Home*, New York, 1870.

3. John Bullock, *The American Cottage Builder, A Series of Designs from $200 to $20,000*, New York, 1854.
4. A. W. Chase, *Practical Recipes*, Ann Arbor, Mich., 1867.
5. C. T. Chase, *A Manual on School Houses and Cottages for the People of the South*, New York, 1863.
6. Andrew Jackson Downing, *Designs for Cottage Residences*, 2nd ed., New York, 1846.
7. *Ibid.*, New York, 1857.
8. *Idem, Rural Essays*, New York, 1853.
9. *Idem, The Architecture of Country Houses*, New York, 1851.
10. C. P. Dwyer, *Economic Cottage Builder*, New York, 1856.
11. M. Field, *City Architecture; on Designs for Houses, Stores, Hotels*, New York, 1853.
12. William Queron Force, *The Builder's Guide*, Washington, 1851.
13. Henry Hudson Holly, *Holly's Country Seats*, New York, 1863.
14. F. C. Hussey, *Home Building, New York to San Francisco*, New York, 1875.
15. Palliser, Palliser & Co., Architects, *Palliser's Model Homes*, Bridgeport, Conn., 1878.
16. John Riddell, *Architectural Designs for Model Country Residences*, Philadelphia, 1864.
17. Samuel Sloan, *The Model Architect*, Philadelphia, 1854.
18. Alfred Spitzli, *A Manual for Managers, Designers, Weavers*, West Troy, N.Y., 1881.
19. Gervaise Wheeler, *Homes for the People*, New York, 1867.
20. *Idem, Rural Homes*, New Orleans, 1854.
21. George E. Woodward and F. W. Wheeler, *Woodward's Country Houses*, New York, *The Horticulturist*, 1865.

D. PERIODICALS AND DIRECTORIES

1. *American Agriculturist*, New York, 1857.
2. *American Farmer*, Baltimore, 1824.
3. W. W. Atwater, *The Vermont Directory and Commercial Almanac for 1860*, Rutland, Vt., 1860.
4. *Americana Review, American Advertising*, 1800–1900, Scotia, N.Y. (no date).
5. *Americana Review, Locomotive Advertising*, Scotia, N.Y. (no date).
6. *Antiques Magazine*, New York, 1946–50.
7. *Compiler's List, A Visiting and Shopping Directory*, New York, 1880.
8. *Cultivator*, later *The Country Gentleman*, Albany, N.Y., State Agricultural Soc., 1834–36, 1838–41.
9. *Frank Leslie's Illustrated Weekly*, New York, 1857–66.
10. *Godey's Magazine*, New York, 1830, 1856–59.
11. *Harper's Illustrated Weekly*, Boston, 1857–90.
12. O. L. Holley, *New York State Register for 1843*, Albany, 1843.
13. *Norton's Literary Register*, New York, 1852–53.

14. *Paxton's Philadelphia Directory and Register*, Philadelphia, 1832.
15. *Philadelphia Almanac and General Business Directory for the Year 1848*, Philadelphia, Downes, 1848.
16. *Scientific American*, New York, 1857–59.
17. Edwin Williams, *New York Annual Register for the Year of Our Lord 1830*, New York, 1831; the same for 1834, 1836-37.

E. TRADE CATALOGS

1. *Circular and Price List*, Cambridge Scientific Instrument Co., Cambridge, Mass., 1882.
2. *Illustrated Catalog of Surgical and Dental Instruments*, Codman and Shurleff, Boston, 1859.
3. Morris Fund Collection, Library of the Franklin Institute, Philadelphia, (hardware, locomotives, machinery mills, pumps, steam engines, water wheels, windmills).
4. Benjamin Pike, Jr., *Pike's Illustrated Catalog of Optical, Mathematical and Philosophical Instruments*, New York, 1848, 1856.
5. *Catalog of Mathematical, Optical, Philosophical and School Apparatus and Furnishings*, Philadelphia, 1856, 1868, 1871, 1878, and 1882.
6. *Illustrated Catalog of Hardware*, Sargent & Company, New York & New Haven, 1880.
7. *The Averill Paint Catalog*, Seeley Brothers, New York, Boston, Chicago, 1882.
8. *An Illustrated Catalog of Instruments*, A. and A. F. Spitzli, Troy, N.Y., 1881.

F. MANUSCRIPTS

Transcriptions from original records in the Industrial Research Unit, University of Pennsylvania. The records are in the possession of the Historical Society of Pennsylvania unless otherwise noted.

1. Brown and Sharpe Co., sales records, Purdue University, 1869–99.
2. J. Washington Irving and B. M. Leiper, account books, 1863–99.
3. Lippincott and Company, account books, 1809–19, 1880–85.
4. J. Krauss, farm account books, 1810–15.
5. Pennsylvania General Hospital, day books, 1809–40.
6. Trotter, Nathan, and Company, account books, Baker Library, Harvard University, 1821–60.
7. Westtown School, account books, 1833.
8. Wetherill and Brother, account books, 1809, 1830–60.

Labor Force and Employment, 1800–1960

STANLEY LEBERGOTT

WESLEYAN UNIVERSITY

I

Historical Comparison of U.S. and U.K. Employment

The full meaning of the employment trends shown in Tables 1 and 2 for this lengthy period can be understood best by reviewing the entire span of American history. So laudable an enterprise must be left to others. Here we seek only to consider a few obvious implications. In this section we make some contrasts with the concurrent employment changes in the United Kingdom—that colonial power once dominating this country, our competitor in third country markets, and perhaps our closest ally (Table 3). To do so we telescope our history into five periods.

1840–60

From the late 1830's, with Jackson's frigid treatment of joyous entrepreneurial expectations in banking, down to the eve of the Civil War, the United States decisively expanded its home market, while the United Kingdom extended its outward markets even more than those at home. The 60 per cent rise in U.S. farm employment was twice the rate of gain for the U.K. But exports were not the key. U.S. grain exports constituted an undistinguished footnote to the rise: wheat exports rose from $2 million to a mere $4 million; and while cotton exports gained from 744,000 to 1,768,000 pounds, tripling in value, neither category accounted for the bulk of the rise in farm employment. Even were we to attribute all the rise in farm slave employment to export sales—and a large segment was surely attributable merely to maintenance and expansion of the slave capital stock—the rise of over 50 per cent in the free farm labor force was another matter. That gain derived primarily from the support of a massive population increase—in city slums, in the open country, on frontier farms.

TABLE 1

THE LABOR FORCE, BY INDUSTRY AND STATUS 1800-1960
(thousands)

	Total	Free	Slave	Agriculture	Fishing	Mining	Construction	Manufacturing Total Persons Engaged	Cotton Textile Wage Earners	Primary Iron and Steel Wage Earners	Trade	Transport Ocean Vessels	Transport Railway	Teachers	Domestic Service
1800	1,900	1,370	530	1,400	5	10			1	1		40		5	40
1810	2,330	1,590	740	1,950	6	11			10	5		60		12	70
1820	3,135	2,185	950	2,470	14	13		75	12	5		50		20	110
1830	4,200	3,020	1,180	2,965	15	22			55	20		70		30	160
1840	5,660	4,180	1,480	3,570	24	32	290	500	72	24	350	95	7	45	240
1850	8,250	6,280	1,970	4,520	30	102	410	1,200	92	35	530	135	20	80	350
1860	11,110	8,770	2,340	5,880	31	176	520	1,530	122	43	890	145	80	115	600
1870	12,930			6,790	28	180	780	2,470	135	78	1,310	135	160	170	1,000
1880	17,390			8,920	41	280	900	3,290	175	130	1,930	125	416	230	1,130
1890	23,320			9,960	60	440	1,510	4,390	222	149	2,960	120	750	350	1,580
1900	29,070			11,680	69	637	1,665	5,895	303	222	3,970	105	1,040	436	1,800
1910	37,480			11,770	68	1,068	1,949	8,332	370	306	5,320	150	1,855	595	2,090
1920	41,610			10,790	53	1,180	1,233	11,190	450	460	5,845	205	2,236	752	1,660
1930	48,830			10,560	73	1,009	1,988	9,884	372	375	8,122	160	1,659	1,044	2,270
1940	56,290			9,575	60	925	1,876	11,309	400	485	9,328	150	1,160	1,086	2,300
1950	65,470			7,870	77	901	3,029	15,648	350	550	12,152	130	1,373	1,270	1,995
1960	74,060			5,970	45	709	3,640	17,145	300	530	14,051	135	883	1,850	2,489

[a] Persons engaged (employees, wage earners, salaried, self-employed, and unpaid family workers), unless otherwise specified. Aged ten and over.

TABLE 2

PERCENTAGE DISTRIBUTION OF THE LABOR FORCE, BY INDUSTRY AND STATUS

	Total Labor Force	Free	Slave	Farm	Nonfarm	Primary (Farm, Fishing, Mining)	Construction	Manufacturing	Trade	Ocean and Rail Transport	Domestics
1800	100.0	72.1	27.9	73.7	26.3	74.5				2.1	2.1
1810	100.0	68.2	31.8	80.9	19.1	81.6				2.6	3.0
1820	100.0	69.7	30.3	78.8	21.2	79.6		2.8		1.6	3.5
1830	100.0	71.9	28.1	68.8	31.2	69.7				1.7	3.8
1840	100.0	73.8	26.2	63.1	36.9	64.1	5.1	8.8	6.2	1.8	4.2
1850	100.0	76.1	23.9	54.8	45.2	56.4	5.0	14.5	6.4	1.9	4.2
1860	100.0	78.9	21.1	52.9	47.1	54.9	4.7	13.8	8.0	2.0	5.4
1870	100.0			52.5	47.5	54.1	6.0	19.1	10.1	2.3	7.7
1880	100.0			51.3	48.7	53.1	5.2	18.9	11.1	3.1	6.5
1890	100.0			42.7	57.3	44.8	6.5	18.8	12.7	3.7	6.8
1900	100.0			40.2	59.8	42.6	5.7	20.3	13.7	3.9	6.2
1910	100.0			31.4	68.6	34.4	5.2	22.2	14.2	5.4	5.6
1920	100.0			25.9	74.1	28.9	3.0	26.9	14.0	5.9	4.0
1930	100.0			21.6	78.4	23.8	4.1	20.2	16.6	3.7	4.6
1940	100.0			17.0	83.0	18.8	3.3	20.1	16.6	2.3	4.1
1950	100.0			12.0	88.0	13.5	4.6	23.9	18.6	2.3	3.0
1960	100.0			8.1	91.9	9.1	4.9	23.2	19.0	1.4	3.4

TABLE 3

U.S. AND U.K. EMPLOYMENT BY INDUSTRY: 1840-1960[a]

	Agriculture		Fishing		Mining		Construction			Cotton Textiles			Ocean Transport		Railway		Trade U.K. Commercial Occupations[c]	
						U.K. (Coal Only)			U.K.[b]			U.K.[b]						
	U.S.	U.K.	U.S.	U.K.	U.S.		U.S.	A	B	U.S.	A	B	U.S.	U.K.	U.S.	U.K.	U.S.	A	B
1840-41	3,570	1,515	24	24	32	n.a.	290	377		72	260		95	76	7	2	350	95	
1850-51	4,520	2,017	30	37	102	n.a.	410	497		92	331		135	156	20	29	530	91	
1860-61	5,880	1,942	31	40	176	n.a.	520	594		122	450		145	203	80	60	890	132	
1870-71	6,790	1,769	28	48	180	351	780	716		135	450		135	192	160	96	1,310	217	
1880-81	8,920	1,633	41	61	280	485	900	877		175	485		125	206	420	158	1,930	363	
1890-91	9,960	1,502	60	54	440	632	1,510	902		222	529		120	236	750	213	2,960	475	
1900-01	11,680	1,425	69	51	637	780	1,665	1,219		303	524		105	264	1,040	320	3,973	673	
1910-11	11,770	1,553	68	53	1,068	1,049	1,949	1,145		370	580		150	293	1,855	373	5,320	896	
1920-21	10,790	1,449	53	51	1,180	1,248	1,233	899	739	450		560	205	314	2,236	357	5,845	1,491	1,759
1930-31	10,560	1,353	73	40	1,009	931	1,988		987	372		564	160	305	1,659	305	8,122		2,323
1940-41	9,575	n.a.	60	n.a.	925	n.a.	1,876		n.a.	400		n.a.	150	n.a.	1,160	n.a.	9,328		n.a.
1950-51	7,870	1,219	77	26	901	791	3,029	1,282		350	322		130	218	1,373	318	12,152		2,213
1960-61	5,970	n.a.	45	n.a.	709	n.a.	3,640		n.a.	300		n.a.	135	n.a.	883	n.a.	14,051		n.a.

Source: For the U.S., present estimates. For the U.K., B. R. Mitchell and Phyllis Deane, *Abstract of British Historical Statistics*, 1962, pp. 60-61, 118, 188; and United Kingdom, *Annual Abstract of Statistics*, 1961, p. 106. For U.K. vessel transport we use "sea, canals, and docks"; for agriculture, we use agriculture, horticulture, and forestry; for coal, 1940 figure is that for 1938.

[a] Data in column A based on Factory Inspectors returns; in B, for insured employees.

[b] Data in column A are for "commercial occupations"; in B, for "commercial finance and insurance occupations (excluding clerical staff)."

Intimately linked to the advance was the concurrent rise in railroad employment: 300 per cent for the United States, compared with 100 per cent for the United Kingdom.[1] For the United Kingdom, railways offered only a superior means of transport, competitive with existing roads and canals; for the United States, they constituted the very conditions for opening new territory, breaking into areas that had virtually no transport worthy of the name.[2]

Linked to the population advance was the 150 per cent rise in U.S. trade employment, compared with a mere 30 per cent for the U.K. London, Glasgow, Bath, and Barset had long since acquired their complement of drapers, greengrocers, and apothecaries. New London, Chicago, and Etruria had still to develop such a network of shops. Why, one may ask, if extensive development were so characteristic of the U.S., did construction employment in the U.S. gain 80 per cent—not much more than the U.K. 60 per cent? It is likely that the answer lies in the nature of our measures. A substantial amount of construction for the new U.S. population was of the crudest sort, done by farmers themselves with the help of their laborers or slaves. Performed in this way, it created fewer opportunities for full time construction employees than the mere volume of construction would suggest.

Finally, for both fishing and vessel employment, the rate of U.S. rise (50 per cent) was below the U.K. (70 per cent). For both industries, 1860 was a U.S. peak, the war then breaking permanently the U.S. rate of advance in these industries.

1860–80

The most decisive contrast for these decades is in agriculture, where U.S. employment increased 100 per cent, while in the U.K. it decreased 15 per cent. The forceful U.S. advance in agriculture did far more than surpass the 1840–60 rate; it was of a different character. American wheat had begun flooding into markets from Wales to Sicily, successfully competing with exports from Devon, Cawnpore, and the Ukraine. The greater 1840–60 rise in U.S. than U.K. farm employment had reflected the extensive development of the U.S. and its home market. The 1860–80 rise now reported the swelling U.S. competitive advantage in world export markets. Concurrent export strength in mining (a 60 per cent employment

---

[1] For railroads we compute an 1850–60 change as being a more helpful basis for contrasting the two nations than the astronomical 1840–60 change.

[2] We are not designating the railways as a sine qua non in development, but simply noting that the first transport network, whether road, rail or canal, had a role in cutting the cost of importing population, as well as of exporting goods, that was so significant as to be different in kind from a merely cheaper means of transport.

gain compared with the U.K. 40 per cent) and cotton textiles (40 per cent compared with the U.K. 10 per cent) was apparent.[3] In the less export-oriented activities for which we show data, the U.K. rise was either greater (fishing: U.S.—30 per cent, U.K.—50 per cent; vessels: U.S.—minus 15 per cent, U.K.—no change), lesser (construction: 80 per cent compared with 60 per cent), or the same (trade: 200 per cent). For railways alone there was substantial growth for each, but the 400 per cent rise for the U.S. was much more dramatic than the 200 per cent for the U.K. And here, of course, it was the interaction between government subsidy, export market possibility, and the attractive powers of mineral wealth and the soil that conjointly brought the growth of agricultural exports and railroads. In Bernard's apt phrase, every mile of railroad in the new nation was "a kind of centrifugal pump furnishing for exportation hundreds of tons of the products of such country."[4]

1880–1910

The third of a century from James Garfield to William Howard Taft undoubtedly lacked some of the more florid and grandiose excursions in political life that characterize earlier decades. But for these decades a common character of significant aspect marks the employment changes (shown in Table 3), and presumably the underlying output changes as well. Substantially greater gains by the United States than by the United Kingdom appear in every major group shown, and indeed in every category shown, except vessels.

	U.S.	U.K.
	(per cent)	
Agriculture	30	−5
Fishing	60	−10
Mining	280	120
Construction	115	31
Textiles	100	20
Vessels	20	40
Railway	230	133
Trade	180	144

Let us particularly note the construction rise, nearly four times as great for the United States as for the United Kingdom. This differential reflects the differential stimuli to population growth apparent in each. From 1880 to 1890 alone, over 6 million immigrants entered the United States (on a 50-million population base). Concurrently, the United Kingdom lost 2 million emigrants from a population half our size. Between

---

[3] For mining we use an 1870 base because of the absence of a U.K. figure for 1860.
[4] Quoted in David A. Wells, *Recent Economic Changes*, New York, 1890, p. 176.

1900 and 1910 the United States gained 8 million immigrants, while the United Kingdom lost 1.5 million emigrants.[5] These contrasting migration flows plus variations in the rate of natural increase generated differing manpower requirements in residential construction. The induced effects on highway and public building construction, on plant for making steel, brick, and lumber can be surmised, although not measured at present.

1910–30

We note three generalizations about the employment changes shown for these two decades. (1) Both nations had reached a peak of agricultural employment in 1910—the United States clearly, the United Kingdom somewhat less clearly—and both then began an uninterrupted descent from that peak by a 10 per cent decline. (2) For most other categories shown no significant employment change occurred for the United States, whereas the United Kingdom showed declines for nearly all. The long weakness of the United Kingdom after the effort of World War I is particularly apparent in the declines for cotton textiles, mining, fishing, and railway employment. (3) The one marked increase in labor requirements (Table 3) was for trade, with a 60 per cent gain for the United States and a more than 100 per cent rise for the United Kingdom. (Data for service and government in the United States, and presumably the United Kingdom, would show marked gains.)

1930–50

In the two decades from the beginning of the Great Depression to the more durable cold war, declines took place in virtually all industries except those linked to the lively postwar population increase. Marked declines in agriculture for both nations reflect a cut in disguised unemployment, a rise in alternative opportunities. A 20 per cent decline in cotton textiles for the United States and a 30 per cent decline for the United Kingdom indirectly reports the competition of new nations and new fibers. The 10 per cent further declines for mining likewise reflect the fresh availability of alternative jobs, competitive fuels from abroad, productivity advance. International competition also helps explain the decline of vessel employment in both countries (20 per cent and 30 per cent respectively) despite active U.S. subsidy programs, while the decline for U.S. railroad employment (contrasting with stability for the United

---

[5] Data for the United States from Bureau of the Census, *Historical Statistics of the United States, Colonial Times to 1957*, 1960, pp. 8, 56; and for the United Kingdom, from B. R. Mitchell and P. Deane, *Abstract of British Historical Statistics*, 1962, pp. 9, 50.

Kingdom) reflects in no mean measure a livelier U.S. subsidy program for nonrail transport than that in which the United Kingdom indulged.

The intense rise in construction employment (50 per cent for the United States and 30 per cent for the United Kingdom) presumably reflected needs of a growing population, as did the U.S. trade growth of 50 per cent. The absence of any U.K. gain in trade growth contradicts this inference only in part: the remarkable U.K. gain for the trade group in the 1951–61 decade suggests that it was the varied U.K. manpower and investment controls after World War II that limited such expansion as the free market would have generated.[6] The advance of U.S. trade employment, despite such productivity coadjutators as supermarkets and vending machines, suggests that the redistribution of the population (to suburban areas), as well as lack of investment constraints as in the United Kingdom, led to a proliferation of shops, stores and distributive convenience in general.

The Role of Education in U.S. Economic Growth

Publications on the economic effects of education have proliferated in the past decade. Disagreeing on many other issues, their authors all seem in cheerful concert on one point: formal education has made massive contributions to our economic growth. Now education per se, communication, or learning in general are not at issue. Such activities may encompass much that is labeled investment and more that is definable as consumption.[7] But to treat education in so broad a sense is to fashion a tool without a cutting edge. We are then involved with an amorphous totality, encompassing both our cultural values and our economic way of life.

Most of the discussions focus on formal education, but estimates for the count of teachers reach out to include not merely Millard Fillmore and Alfred North Whitehead, but John Sloan, Isadora Duncan, and James Smith IV—plus every errant instructor in art, eurhythmics, or tatting. But if we add them all together, higgledly piggledy, they account for no more than 2 per cent of the labor force during the first century and more of our national existence.

---

[6] Although no comparable Census data are available, we judge from the annual series for distribution trades, insurance, banking, and finance, as presented in the United Kingdom's *Annual Abstract of Statistics*, 1961, p. 108 and 1962, p. 109—which report a rise of 30 per cent for distributive trades in this decade and 21 per cent for finance, insurance, etc.

[7] Cf. Fritz Machlup's stimulating *The Production and Distribution of Knowledge in the United States*, Princeton, 1962.

Can so few candles have cast their beams so far? True, any handful of great souls could have offered Matthew Arnold's "unum porro necessarium." And this noble band might have provided vigor for a swiftly developing economy. But is there a stronger basis than some wishful guild-thinking to suggest this?

What of the quality of the education offered?

> a spirit yearning in desire
> To follow knowledge like a sinking star
> Beyond the utmost bound of human thought.

Was this the mighty spirit at work? Contemporary reports do not discover any inordinately high quality pervading the instruction. President Duer, on the New York state school system in 1837, found the "teachers inexperienced and transitory, snatched up for the occasion ... paid by salaries which hardly exceed the wages of a menial servant; and as a necessary consequence, ignorant and disqualified ...."[8] James Carter in 1826 reported that "The country schools are everywhere degraded .... It is thought a mean thing for a man of competent estate, or for any but the mechanic, the artisan or the laborer, to send their children to them for their education .... The teachers of the primary schools have rarely had any education beyond what they have acquired in the very schools where they begin to teach."[9]

It is hardly as though the quantity of education made up for its lack of quality. The South Carolina Legislature appropriated $40,000 for 840 schools in 1828.[10] Given the usual contemporary ratio of one teacher per school this would mean about $45 a year. Since teachers would have received somewhat more than the $7 a month paid for hired slaves, the school year must have averaged less than five months.[11] In 1838 the Superintendent of Schools for Ohio computed that New York funds will "pay for teaching the whole [of the enrolled student body] less than 3 months in the year."[12] His own report leads to an Ohio figure only slightly higher, at under four months.[13] In 1868 it was asserted with some

---

[8] Quoted in Samuel Lewis, *First Annual Report of the Superintendent of Common Schools of the State of Ohio*, Columbus, 1838, p. 13.

[9] James Carter, *Essays on Popular Education*, 1826, quoted in Newton Edwards and H. G. Richey, *The School in the American Social Order*, Boston, 1947, p. 269.

[10] *An Accompaniment to Mitchell's Reference and Distance Map of the United States*, Philadelphia, 1835, p. 221.

[11] The $7 figure is from Table A-23 in the writer's *Manpower in Economic Growth*, New York, 1964.

[12] *First Annual Report, Superintendent of Schools, Ohio*.

[13] *Ibid.*, p. 46, comparing his figures for children attending school with "the number of months scholars have been taught" and the "number in usual attendance."

pride to would-be emigrants that "Alabama has made ample provision for the education of her children, the poll-tax of $1.50 being set aside for this object exclusively," yet this appeal anticipated that the migrant would bring even superior skills: "under your superior knowledge of cultivating the soil, the fields that have been scratched by a lazy nigger and a poor mule will yield abundantly. Go to Alabama."[14] And, in fact, it appears that the Alabama school year for the children enrolled averaged four months, compared to only three in North Carolina.[15]

As late as 1870, when we have our first comprehensive figures on attendance, the average child enrolled in school attended less than four months out of the year, while a reckoning which included the numbers not enrolled would pull the average down close to three months.[16] They are well below figures from the earlier Censuses, which appear to refer to enrollment rather than attendance. But even the three-month average applies only to native students in the northern and midwestern states. Most southern states had only recently instituted public schools. And the bulk of the foreign born and Negro labor force had had no formal schooling whatever. It is a fair inference, therefore, that from 1800 to 1870 formal education per person in the U.S. labor force came to less than two months a year during their years of schooling. Nor could it have risen much above three months prior to 1900. It is scarcely likely, therefore, that the quantity of schooling compensated for the limitations on its quality. Taken together they do not suggest that formal education was anything like a significant factor in raising the quality of the American labor force, or in stimulating economic growth.[17]

The Relative Contribution of Agriculture to U.S. Economic Growth

Obtruding through this motley array of statistics is a single, overwhelming fact about American economic growth. Brilliantly obvious though it may

[14] *To the Emigrant. The Descriptions of the Lands on the South and North Alabama Railroad Are Not Overdrawn*, Louisville, 1878.

[15] North Carolina data on duration and teacher pay from *Report of the Commissioner of Education . . . for the Year 1870*, 1870. Assuming the same rate of pay, the teacher-school income ratios (Ninth Census, Vol. I, p. 452) suggest a third longer duration than for North Carolina.

[16] *Report of the [U.S.] Commissioner of Education*, p. 504, gives data by state on school population, enrollment, attendance, and average duration of school year from which we compute these averages.

[17] By definition the extent of education that a labor force member received at his mother's knee, by consulting his soul while fishing at the brook, or on the job, is not at issue. And the presumably greater contributions for such specialists as doctors or lawyers seem some distance away from basic factors in economic advance.

be, it is nonetheless usually lost sight of in the pursuit of theorems on fascinating, but yet undiscovered, lesser points. That fact is simply the overwhelming importance of agriculture during the many decades when the structure of our present economy and social order was being shaped. Occupying nearly 75 per cent of our labor force in 1800, farming occupied over half the labor force until some time between 1880 and 1890. (It is, of course, no coincidence that it was at the end of the same decade that the Census superintendent found that continuity in the line of settlement could no longer be observed—an event defined more pungently by Turner as "the end of the frontier.") The Kuznets and Gallman estimates of national income are consistent with this conclusion.

But if this is so we may draw one rather significant inference with respect to the factors that dominated our growth. Recent discussions of backward and forward linkages in economic development have tended to emphasize the contributions made by nonfarm sectors to economic advance. Yet in our own history the gross dominance of agriculture suggests, by mere probability, that it accounted for a greater portion of these effects than all other industries put together. An industry that used from 50 to 90 per cent of the nation's labor input over the first century of our national existence was surely more likely to make greater demands on most supplying industries than one that merely accounted for 5 or 10 per cent.[18]

Yet this proposition is surely not true with respect to every particular supplying industry. Moreover, the difference between input coefficients can be such that the marginal effects of an advance in nonfarm industries could be much more potent than our argument suggests, resting as it does on an assumption of equal average effects.

Can we throw any light on the net results by reference to empirical results? Some interesting data are available for the textile industry, characteristic leader in the industrialization of so many nations. As of 1831, an estimated 3,200 men and 11,000 tons of iron and steel were required to produce machinery at the rate the textile industry was then installing it, according to the reports of the New York Convention.[19] Since the Convention was busily emphasizing the importance of manufacturing industries as markets for native industries, the figure is not

[18] The labor force totals represent a minimum statement of input to agriculture: (1) hours were longer in farming than the average for nonfarm industries after 1830, (2) work by female family workers is incompletely recorded. Offsetting, in part, is the possibly higher quality of some nonfarm labor input. But this fact would be largely irrelevant to the point being made here.

[19] Quoted in *Niles' Weekly Register*, 1832, Vol. 42, Addendum, p. 8. The ratio is consistent with, and probably rests on, data for the Lowell, Concord, and Merrimac River Company.

likely to be an underestimate. A decade later, the Locks and Canal Machine Shop, working for all the Lowell mills, employed between 1,000 and 1,200 hands "directly and indirectly" when actually building mills.[20] Given the ratio of Lowell textile output to that of the United States, it is unlikely that the indirect manpower requirements for the nation as a whole would mount as high as 10,000.

For the woolen industry the figures are so small as to be trivial. In January 1827, Representative Davis, trumpeting the substantial role of the industry, gave figures on its total input of wool, while in July the pro-tariff General Convention estimated iron and steel requirements per 100,000 pounds of wool consumed.[21] Combining these figures leads to a requirement by the entire industry of 160 tons of iron and steel annually —or less than a tenth of 1 per cent of total input.[22]

But if only 10 per cent of iron and steel output went to the major textile industries, where did the bulk of U.S. iron and steel production at this period go? A detailed breakdown for the output of Litchfield, Connecticut, in 1831 may be suggestive:[23]

Total		$178,000
Scythes	56,000	
Pitchforks	20,000	
Ploughs	3,800	
Hoes	7,150	
Shovels	6,500	
Subtotal		93,450
Axes	26,500	
Rat traps	9,500	
Shoetacks and sparables	40,000	

The share of agricultural items in the total output is clearly overwhelming even if one excludes axes, the bulk of which must have been used in clearing land for farming.

Of the total investment requirement for agriculture, of course, much went into land clearing, much to current agricultural production. The

---

[20] *Hunt's Merchants Magazine*, Vol. 9, November 1843, p. 426.

[21] Gales and Seaton, *Register of Congressional Debates*, January 31, 1827. *Proceedings* of the General Convention of Agriculture and Manufactures, July, 30 1827, p. 42.

[22] An output of 112,866 tons of bar iron equivalent was reported by the *General Convention of the Friends of Domestic Industry; Report of the Committee on Iron*, New York, 1831, pp. 19, 28.

[23] *Ibid.*, p. 28.

Parker-Primack estimates for the 1850's report 450,000 equivalent persons in land clearing.[24] This "industry" accounts for less than 10 per cent of our total farm labor force average for the decade but undoubtedly required well above that proportion of supporting investment per farm worker. (If one adds in roads, canals and other social overhead capital—more accurately social underfoot capital—the proportion will zoom. However, the growth of such capital reflects broader forces than mere market forces originating in agriculture.)

For farm operations we may at least refer to very recent figures which Leontief has given.[25] These indicate that the direct capital coefficients per million dollars of output ran as follows in recent years:

Agriculture	1.61
Iron ore mining	1.48
Spinning, weaving, dyeing	0.31
Sawmills, planing and veneer mills	0.53
Blast furnaces	0.96
Footwear	0.10

(These coefficients measure not stock of capital but flow of current services.) They suggest that an equivalent dollar volume of farm output requires far more iron and steel output, machinery production, than does an equivalent dollar volume of output generated by (a) the textile and footwear industries that bulked large in our early growth, or (b) by the omnipresent mills that employed the largest single group of manufacturing employees before 1860, or even (c) iron and steel production per se. Technological forces joined with institutional ones to generate significant agricultural demands for output by the metals industries. Even the rudest of techniques required axes to clear the land and breaking ploughs to open the plains, scythes to cut wheat, and hoes to chop cotton. Such demands burgeoned when wage-price trends made advantageous a shift to the more capital-intensive techniques of drill and harvester. A complementary stimulus to nineteenth century growth was the influence of widespread land ownership—widespread, say, as compared with Ireland or Italy at the same time. Such ownership made it possible for cultivators to reap the financial benefits of investment, inducing more farming and educing more farm investment than the volume in which a rational crew of monopolists, hiring employees, might have indulged themselves.

[24] William Parker and Martin Primack, "Land Clearing Under Nineteenth Century Techniques" (unpublished paper), p. 12.
[25] Wassily Leontief, "Factor Proportions and the Structure of American Trade," *Review of Economics and Statistics*, November 1956, Appendix III.

In a world of intensified nationalism, with every new nation seeking immediate advance, it is not surprising that extended attention has been given to a generalization stated by Colin Clark, but described by him as stemming from Petty.[26] That generalization refers to a tendency for the proportion of the labor force in agriculture to decline when an economy develops. Any predictive law is, of course, of interest. One that predicts a more or less irresistible trend of such arrant significance is even more so.

What is of greater interest, however, is the question of what forces work to produce such a trend. To the extent that developing nations refer to our experience as precedent they may, however, find no impelling requirement to shift out of agriculture (or mining) into manufacturing and tertiary industries. Table 2 indicates fairly flatly that the proportion of the labor force in agriculture in this country hardly shifted at all from 1850 until some time after 1880. Growth during this period—and growth there was by a dozen criteria—came from no disembodied force for industrialization.

The engine of advance during these decades was comparative advantage on a global scale; the midwestern states became producers of wheat far superior to Italy, France, and England, while the southern states continued to be the still dominant world suppliers of cotton and tobacco. The central states readily secured labor by immigration. Of her central labor force group (males, age 25–44) in 1880, for example, Kansas had received just under two-thirds by immigration, immigrants flowing from other states and countries as distant as Russia.[27] And with prospects so bright that interest rates of 17 per cent were being paid (in Kansas in the 1870's) capital, too, flowed in from eastern and foreign sources.[28] Indeed, Easterlin estimates that approximately 18 per cent of the total wealth of the six rapidly growing central states in 1880 was owned outside the state.[29]

Another way of looking at the long-term trend is to separate the figures for the North and West from those for the South, treating them as separate nations. Major labor force shifts occurred within each. Those for the North and West were well publicized. But those within the South were just as significant. Of those persons numbered in the 1850 Census,

---

[26] *Conditions of Economic Progress*, 1940, pp. 6–7. A more recent precursor was Allen Fisher in the *International Labor Review*, January 1920.

[27] Everett Lee, Ann Miller, Carol Brainerd, and Richard Easterlin, *Population Redistribution and Economic Growth, United States, 1870-1950*, 1957, pp. 142–144.

[28] Kansas rates in Allan Bogue, *Money at Interest*, 1955, pp. 116–117, 272.

[29] Lee *et al.*, *Population Redistribution*, p. 729. We combine his data for Minnesota, Iowa, Missouri, Nebraska, Kansas, and the Dakota Territory.

for example, the percentages then living outside the state of their birth were:[30]

South Carolina	42
North Carolina	34
Virginia	31
Alabama	26
Georgia	24
Mississippi	18
Texas	5

We take the 5 per cent for Texas, with its heavy immigration, as a practical minimum. The proportions for the old south ran six to eight times as great as that minimum. For the Gulf states (whose out-migrations in volume began perhaps a decade later) the rates were three to five times as large. We know, too, that between 1800 and 1860 indigo disappeared from the list of the major crops, tobacco had spread its domains, while cotton and cane had risen from almost nothing to a dizzying eminence. But so far as the trend of the farm proportion in the labor force was concerned, none of this is in evidence. The labor force proportions changed as follows:[31]

	Percentage of Labor Force in Farming	
	1800	1860
South	82	84
North and West	68	40

Had the South from 1800 to 1860 been the separate nation it sought to become in 1860 the labor force figures would indeed have reported massive growth during sixty years—but with no taint of decline in the farming share of the labor force.

We take the above data to indicate only one thing: U.S. experience reveals no higher law at work forcing a decline of the share of the labor force in agriculture during economic growth. Our experience suggests, instead, that the optimal alternatives change from time to time, depending on the marginal efficiency of capital among regions, products, and activities, and—just to complicate matters—depending on the incentives and limitations that the social order laid upon one or another set of alternatives. One inference for those of today's underdeveloped nations whose economies are primarily based on one export product is that U.S. experience does not prove that the proliferation of manufacturing employment is prerequisite to growth.

[30] James De Bow, *Compendium*, pp. 116–117.
[31] Estimated in the writer's *Manpower and Economic Growth*.

Derivation and Explanation of Estimates

The present estimates were derived to permit analysis of long-run changes in the structure of the American economy, and to assist in projections of employment and GNP in the years ahead.[32] To facilitate such work it is most desirable that the series developed be made comparable with the major series on the labor force and its distribution currently available. Hence the totals, with exceptions noted below, are generally comparable with the explicit or implicit series compiled by the Bureau of the Census and published by the Bureau of Labor Statistics as part of the *Current Population Survey, Monthly Report on the Labor Force*. Since the CPS figures are not comparable with the Decennial Census results, our series will automatically also differ from the Census results for 1940–60, when both sets of figures are available. Hence, too, they will differ from the figures of Clarence Long, Daniel Carson, Solomon Fabricant, and John Durand—all of whom adopt the Population Census levels for the 1940 labor force, and (in some instances) its components.[33] Since the CPS figures are not comparable with the results of the Census of Agriculture, our figures for agricultural employment will necessarily differ also from those of the U.S. Department of Agriculture, and from series such as those of Kendrick, Knowles, Barger, and Landsberg and others who link to the USDA series.[34]

A basic difference between CPS and USDA series (as between various series for construction, teachers, fishermen) and our own is that we seek to include each person only once, in the industry to which he is primarily attached, rather than to develop industry series that include everyone who may devote some time to a particular industry.

---

[32] Differences between the present series and the major alternatives for the years 1930 ff. are discussed in some detail in *ibid.*, Part III.

[33] Clarence D. Long, *The Labor Force under Changing Income and Employment*, Princeton University Press for NBER, 1958; Solomon Fabricant, "The Changing Industrial Distribution of Gainful Workers" and Daniel Carson, "Changes in the Industrial Composition of Manpower since the Civil War" in *Studies in Income and Wealth*, 11, New York, NBER, 1949; John Durand, *The Labor Force in the United States, 1890–1960*, 1948, Table A-6. Durand (p. 207), however, gives ratios for converting to CPS levels.

[34] John Kendrick, *Productivity Trends in the United States*, Princeton for NBER, 1961, Table A-VI. Kendrick adjusts his series down to a full-time level by 1940–55 ratios. James Knowles, *The Potential Economic Growth in the United States*, Joint Economic Committee, Washington, 1960, uses the Kendrick estimates. Harold Barger and Hans Landsberg, *American Agriculture, 1899–1939: A Study of Output, Employment and Productivity*, New York, NBER, 1942.

Differences for the years prior to 1940 between our labor force totals and the gainful workers estimates of Kaplan and Casey, Edwards, Whelpton, Carson, and Miller arise for a number of reasons. Most generally we differ for two reasons. (1) We accept the original Census figures for 1900, 1920, and 1930 without adjustments proposed by various writers, and the 1870 and 1890 Censuses only as adjusted in official reports of the Census Bureau. Hence we will differ from Carson, Miller, Long, and Kendrick. (2) We adjust the 1910 Census by procedures differing from those pioneered by Edwards, and hence will differ from Kaplan and Casey and others.

For the major categories of construction, trade, and domestic service, the CPS totals differ from those of the Population Census, and the National Income Division, and our figures will differ correspondingly. These differences for 1930–60 are necessarily associated with differences in level for prior years in order to prevent the series from taking an abrupt and unreasonable path from 1920 to 1930. Hence, we differ from Carson, Kuznets, and others.

For total manufacturing (and railway and ocean vessel employment), we link to the BLS series for employees in nonagricultural establishments, since this is the series used for current analysis of trends in manufacturing, etc. Because that series is adjusted to social security benchmarks, its level will differ from that of the Census of Manufactures and the infinite number of series that depend on the Census of Manufactures for current and/or earlier years—including the well-known series of Kuznets, Fabricant, Easterlin, and others.

For cotton textiles and iron and steel we similarly link to BLS totals except that we have had to make new estimates for cotton textiles in recent years, no official series being available.

For railways also we link to BLS, and hence differ from ICC totals for 1890 and after (which omit switching and terminal companies in some years) and from NID totals for 1929 and after (which differ slightly from BLS). For earlier years we rely on direct ratios of employment per track miles and estimates of trends in such ratios over time, whereas the forthcoming important study by Fishlow applies cross-section relationships between employment and activity measures from one region to others.

For fishing and teachers we accept the Population Census as having the most comprehensive coverage, measuring the numbers of persons who consider each as their primary activity. Hence, we arrive at different results from the estimates, respectively, of the Bureau of Wildlife and Fisheries and the Office of Education, not to mention the many series which use these estimates as benchmarks.

II

Methods of Estimate

GAINFUL WORKERS: 1800

The size and distribution of the 1800 labor force was estimated in a number of steps. The key steps and assumptions are outlined below.

1. Estimates were made of the total number of free males aged ten and over.[35] It was assumed that the proportion of these males gainfully occupied was 87.2 per cent—the same rate as given by the 1850 Census, the first providing such data, and much the same as that for all subsequent Censuses of the gainfully occupied.[36]

2. The number occupied in navigation and fishing was estimated by procedures described in the section on navigation employment.

3. The number of farmers was estimated primarily on the basis of the number of heads of rural white families:

a. Estimates of the number of white families had been previously made by utilizing Census counts of white families in 1790 and 1850 ff. in a regression relationship against the white population divided by a series for estimated average size of family.[37]

b. The proportion of white families in 1800 that was rural is estimated at 93.8 per cent—the ratio implicit in the U.S. white population data. The number of urban slaves was then deducted from the total slave population—thus, by further subtraction from the total Census figure for rural population—giving the required figure.

Urban slaves were estimated as the sum of slaves living in the major slaveholding cities. The ratio of this group to total slaves can be computed for 1790 at 5 per cent and was assumed the same for 1800. For 1790, the slave population for Henrico County, Virginia, Baltimore County, Maryland, New Hanover County, North Carolina, Jefferson County, Kentucky, and Chatham County, Georgia, was assumed to equal the total urban slave population for those states. For South Carolina, 22

---

[35] The white population aged ten and over is reported in *Historical Statistics of the United States, 1789-1945*, 1949, p. 28. The ratio of males to total white population aged sixteen and over in 1790 was applied to this figure. Census, *A Century of Population Growth*, 1909, p. 208.

[36] *The Seventh Census of the United States: 1850*, 1853, pp. xlii, lxxx. For the stability over the 1870-1930 period of the rates for all males aged sixteen and over, compare Alba Edwards, *Comparative Occupation Statistics*, Washington, 1943, p. 92.

[37] See the author's "Population Change and the Supply of Labor," in *Demographic and Economic Change in Developed Countries*, Princeton University Press for the National Bureau of Economic Research, 1960, p. 414.

per cent of the Charleston County slave total in that year was assumed to be urban—that being the ratio of city to county slave population in 1840.[38]

c. From this 708,000 total for white rural heads we deduct 70,000 for those engaged in handicrafts and trades in rural areas, leaving 638,000 as the estimate of white heads of families who were farmers and farm laborers.[39]

Data from the 1790 Census on the distribution of the free colored population gives us the basis for estimating 5,000 gainfully occupied in urban areas, and 62,000 in rural—the latter then arbitrarily split into 50,000 farm labor and 12,000 farmers.[40]

4. The number occupied in nonfarm occupations other than navigation and fishing was estimated as the sum of those in urban areas and those outside urban areas.

a. Of all whites aged ten and over in 1790, 50.2 per cent were males according to the 1790 Census.[41] The same ratio was applied for 1800. Of the male group thus computed, it was estimated that 9.1 per cent (or 129,000) were in urban areas—87.2 per cent of whom (using the same gainfully occupied rate as above for the total males) were gainfully occupied, or 112,000.

The proportion of males in urban areas was based in turn on 1790 Census figures for cities that accounted for nearly half the urban population in that year—Baltimore, Boston, Charleston, New York, and Philadelphia. For these cities taken together, the ratio of white males aged sixteen and over to total free population was computed, and applied to the total urban population. Thus we derive a figure for free white males sixteen and over in urban areas. The ratio of this figure to the total U.S. population for this group is the 9.1 per cent noted above. This leaves only a single gap—namely, the total free urban population in 1790. This was estimated by deducting from the reported totals the number of urban slaves, as estimated above in step 3. Reference to the Charleston,

---

[38] The various county figures for 1790 are from Census, *A Century of Population Growth*, Table 104, while totals for free population and urban population in 1800 are derived from data in *Historical Statistics*, 1949, pp. 25, 29.

[39] In section 4b we estimate 227,000 whites gainfully occupied in rural areas but not in farming. Now, over-all, we have 1,240,000 gainfully occupied whites and 755,000 white families, giving a ratio of 61 per cent. This ratio was cut down to 30 per cent to reflect the fact that among the 227,000 (who were primarily in handicrafts and trades) a larger proportion of secondary workers would appear, the multiplication then giving 70,000.

[40] Census, *A Century of Population Growth*, Table 104, gives city data.

[41] The basic source materials used in the following estimate appear in Census, *A Century of Population Growth*, Table 104, and in *Historical Statistics*, 1949, pp. 25-29. In addition to free males, however, others were employed in urban areas.

Philadelphia, and New York City directories for 1790 and 1800 indicates many listings of female tavern keepers, in addition to which a significant number of domestic servants were employed in urban areas. To the 116,000 white males aged ten and over, therefore, 40,000 was added for female domestic service (estimated below) plus an arbitrary 10,000 addition for other females. This brought the total up to 166,000.[42]

b. For the group outside urban areas but in nonfarm pursuits, we begin from the certainty that the extent of local handicrafts (blacksmiths, saddlemakers, etc.) and traders and learned professions changed at a different rate in different parts of the country from 1800 to 1840—the latter year providing our first occupational data for these groups. For the big cities, however, as for New England with its rising factory system, the 1840 data are certainly irrelevant. On the other hand, for the most populated southern states in 1800 (Virginia, the Carolinas, Georgia, Tennessee, and Kentucky) plus Alabama and Mississippi, this is less true. Exclusive of the towns in these states, it was assumed that the ratio in 1840 of (a) persons in such occupations to (b) total white male rural population aged ten and over applied equally well in 1800. (This amounts to assuming that the supporting service personnel required per 1,000 population in 1840 for such areas was the same as required in 1800. Since we have excluded all towns, and the northern areas of nascent industrialization, the resultant ratio estimate of saddlers, shopkeepers, school teachers should apply reasonably well.[43]) In step 4a the number of free urban males (aged ten and over) was estimated at 129,000, which, by deduction from the total, leaves 1,290,000 free male population in rural areas. Taking 1.41 per cent of these as gainfully occupied in learned professions, 2.76 per cent in commerce, and 13.4 per cent in hand trades and trade gives a total in nonfarm pursuits, but outside urban areas, of 227,000. Allowing for 13,000 free domestic servants of the 694,000 rural families (the South, of course, must be largely excluded from this estimate) gives 240,000.

Combining the 130,000 gainfully occupied in urban areas with the 240,000 in nonfarm pursuits outside urban areas gives a total of 370,000.

5. White farm laborers were estimated as (a) the 210,000 difference

---

[42] *The New York Directory for 1799;* Jacob Milligin, *The Charleston Directory and Revenue System*, 1790; and *The Philadelphia Directory for 1800* report tavern keepers. For the domestic service estimate, see section on domestic service.

[43] From the 1840 Census *Compendium* (Allen edition) the ratio of persons in the learned professions, in manufacturing and trade, and in commerce to the white male population aged fifteen to seventy was computed for the above states exclusive of the towns reported for each state. Data for the Middlewestern states were not used on the assumption that the density and pattern of settlement was so much different from that of the coastal states in 1800 as to be an irrelevant guide. The 1800 urban data were estimated above and the rural figure is simply the difference from the total.

between all white males gainfully occupied and those in the specified occupations noted above plus (b) an arbitrary 50,000 of the 638,000 farmers and farm laborer family heads estimated above. Since this important group was estimated as a residual it is particularly necessary to assess its reasonableness. Our best test for this purpose is the ratio of all free laborers (260,000 whites plus 50,000 nonwhites, estimated below) to the number of farmers. For 1800 it proves to be 52 per cent of that number, while for 1860 the Population Census data (also for the free population) give a 41 per cent ratio. The decline in the ratio is consistent with the opening of the midwest and the greater consequent rise in self-employed farmers than farm laborers.

6. Added to the above estimates for the white population were those made for slaves and free colored. For both groups the numbers aged ten and over were estimated, and 87.2 per cent were estimated as gainfully occupied.[44] The basis for the latter ratio has been discussed elsewhere, and rests basically on the fact that an examination of 1820 and 1840 Census unpublished schedules for various southern counties reveals that planters commonly reported all their slaves aged ten and over (both men and women) as gainfully occupied.[45]

The allocation of the slaves between those in rural and urban areas has been described above, and that between farm and nonfarm occupations is assumed as identical. The many individual examples of slave blacksmiths, turpentine tappers, carpenters, etc., living in rural areas (and used on the plantation or hired out) suggest some distortion here. But undoubtedly it is a trivial one. The allocation of the free colored population is more in doubt, but fortunately few are involved.[46]

7. For 1805, Samuel Blodget estimated the total active population at 1,866,000, compared with the present estimate of 1,900,000 for 1800.[47] This similarity conceals differences in the components: his figure for slaves is 400,000 compared with the present 530,000. For seamen and fishermen, his 116,000 compares with the present 45,000. (He apparently counts both entrances and clearances of seamen.) The major difference, however, appears in his estimate of artisans (100,000) and professionals

[44] The total population for the groups is from *Historical Statistics*, 1949, p. 27. The ratio aged ten and over for slaves was 65 per cent in 1830, 66 per cent in 1840, and was taken as 65 per cent in 1800. For free colored the ratio was 70 and 72 per cent in 1830 and 1840, respectively, and was assumed at 70 per cent for 1800.

[45] Compare the discussion below in connection with the estimates for 1820 and 1840, and the historical materials in my *Manpower in Economic Growth*.

[46] The 68,000 involved were allocated as follows: 5,000 to urban areas based on the numbers shown for the larger cities; 53,000 of the remainder to farm laborers and 10,000 to farmers.

[47] Samuel Blodget, *Economica, A Statistical Manual for the United States of America*, Washington, 1806, p. 89.

and traders (50,000), compared with the present 300,000 for both groups. The present figure of 137,000 males in urban areas is not far from his 150,000 total. It is likely, therefore, that he made a much smaller allowance for those in the hand trades and in general stores throughout the nation that were *not* in the big cities. Because of the subtraction procedure used, this in turn is a major factor in producing the difference between the present figure of 910,000 free persons in agriculture and Blodget's 1,200,000 for "free planters and agriculturists." The major difficulty with Blodget's estimate lies in precisely this figure. For if one takes the present estimate of 600,000 free farmers, his figure implies some 585,000 free farm laborers —giving a ratio of 98 per cent, which is unreasonably high. (His total in agriculture is, of course, much closer to our 1,400,000.)

One may also mention an *ad hoc* assertion of William Duane to the effect that 17/20 of the free occupied population were "farmers and those who acquire support from labor"—a figure to be compared with the 75 per cent implicit in the present figures.[48] Duane, as a brilliant and bitter journalist and politician of the time, provides a figure that presumably lacked pedestrian accuracy, but does distinguish the orders of magnitude involved.

Since the composition of the labor force did not change greatly over the years, it may be relevant to cite two estimates for slightly later dates. Representative Pearce of Rhode Island stated in January 1827 that "83 in every 100 are engaged in agriculture"—or much the same as Duane's statement.[49]

In 1820, Matthew Carey estimated that there were 5 million agriculturists, 1.5 million artists, mechanics, manufacturers, etc., plus 1.5 million professors of law and physic, gentlemen who live on their income, merchants, tradesmen, seamen, etc.[50]

GAINFUL WORKERS: 1810

The 1810 gainful worker total was estimated by procedures similar to those used for 1820–40 estimates. For free males and slaves we apply the same worker rates as in 1820 to the appropriate 1810 population figures.[51]

For slaves, we estimate those ten and over at 31 per cent of the total

---

[48] William Duane, *Politics for the American Farmer*, Washington, 1807, p. 3. Tench Coxe in 1787 had guessed that nine out of ten persons were engaged in agriculture. Harold Hutcheson, *Tench Coxe*, 1938, p. 79. For a similar assertion (by Franklin)— that "calculations carefully made do not raise the portion of property or the number of men employed in manufactures, fisheries, navigation and trade to one-eighth" of that in agriculture, for New England, see H. C. Adams in *John Hopkins Studies in Historical and Political Science*, 1884, p. 10.

[49] *Annals of Congress*, January 30, 1827, col. 863.

[50] Matthew Carey, *The New Olive Branch*, Philadelphia, 1820, pp. 151–152.

[51] Population data from *Historical Statistics*, 1949, p. 28, for white males ten to fifteen, and sixteen and over.

(as in 1820), that rate in turn being based on the 31.8 per cent for 1850 and the 31.1 per cent for 1860. The 87.2 per cent worker rate for adult males, the 25 per cent for boys, and the 90 per cent for slaves are discussed in connection with the 1840 estimate.

For free females we estimate three components. Domestic servants were estimated at 70,000 on the basis of a regression against the number of white families (as discussed in the section on employment in domestic service). Employment in textiles, including wool, was estimated at 10,000 as outlined in the section on cotton textile employment.

For employment in the clothing trades and all other industries we add an arbitrary sum equal to domestic service. The first year in which we can estimate with some likelihood of reason, 1860, shows this group to be under 300,000, or half the estimate for domestics in that year. In the infancy of the factory system, before "boughten clothing" was at all common, we can hardly estimate the count for this group at greater than the 70,000 in domestic service. One might reasonably dispense with an estimate altogether and get about the same totals, but we follow tradition to show that a moderately rational estimate has been made for this category. We make no allowance for females in agriculture, since examination of the unpublished Census schedules for 1820 and 1840 indicates they were not included in those years, while the county data for 1860, 1870, 1880 show that only nonwhite women in the South were included in the agriculture counts. The latter category, for 1810, is comprehended in our estimate of the slaves gainfully occupied.

GAINFUL WORKERS: 1820

The 1820 Census secured data on the occupation of those gainfully employed in agriculture, commerce, and manufactures.[52] Long has pointed out the unreasonably low proportions gainfully occupied in fourteen cities, and Whelpton has adjusted for the inclusion of professional service and other urban occupations that were not in principle included, as well as for the inclusion of navigation, lumbering, etc.[53]

In the present estimates we begin from two premises. The first is that the coverage in the important rural areas of the nation was irregular, and

---

[52] These data are conveniently summarized, with minor corrections from the original, in 1900 Census, *Occupations*, p. xxx.

[53] Clarence D. Long, *Labor Force*, p. 407. P. K. Whelpton's pioneering and still standard study is "Occupational Groups in the United States, 1820-1920," *Journal of the American Statistical Association*, September 1926. Fabricant has shrewdly noted the relationship between a possible 1820 understatement and the schedule sequence in which occupation questions were asked before the slaves and free colored were enumerated. See Solomon Fabricant in *Studies in Income and Wealth*, *11*, p. 31. However, as noted below, examination of the schedules suggests that slaves were rarely omitted en bloc for any family report, though a few such examples do turn up.

in some urban areas it was uncertain. Second, that to adjust for omitted industries directly is the least satisfactory way of arriving at an adjusted total, though eminently desirable for industry information. (That rural census canvassing was irregular can be seen from an examination of a sample of the original 1820 county returns.)[54]

The premise that adjustment for omitted industries should not be made directly if we seek a reliable total simply turns on the difficulty of making such estimates in this volatile growth period. At what rate did the ratio of navigation employment to population, of learned professions to population, etc., actually change? Slight errors in estimating the four or five omitted industry groups can cumulate, in estimating the total.

Since we know that the proportion of adult males working changes less from decade to decade than does the proportion engaged in navigation, lumbering, etc., we prefer to work from the former ratio. Doing so, of course, likewise helps compensate for the irregularity in Census coverage of the industries that it purports to cover. We therefore compute separate estimates of the gainfully occupied among free males sixteen and over, free males ten to fifteen, slaves ten and over, and free females ten and over. We apply rates for other years to the population data for each group. In some instances minor adjustment in the Census totals is needed. For free males sixteen and over—the vast bulk of the labor force in those years—we adopt the 87.2 per cent participation rate derived from 1850 data and the same rate for slaves ten and over. (The applicability of these rates for earlier years is discussed in connection with the section on slaves in agriculture, 1800–60, below.)[55]

[54] The 1820 returns are now in the U.S. National Archives. Some examples follow. Granville County, North Carolina: Chisley Davis family, no occupational entries; Howel Frazier, six in agriculture in family with two white males aged ten and over, and four slaves aged fourteen and over; Halifax, North Carolina: Gideon Allston, thirty in agriculture and thirty slaves aged fourteen and over; Williamsburg, South Carolina: Francis Cordes, eighty in agriculture; Ashe, North Carolina: David Edwards, three in agriculture in a family with two white males, aged sixteen and twenty-six, one aged 45, but eight male slaves over 14; Salisbury, Massachusetts: occupational entries for males but not females; Indiana and Pennsylvania: many occupational entries of one, though females were present. These and other examples show instances where slaves and white adult males in rural areas were omitted, though not commonly. What a sizable random sample would show is another matter.

[55] Population data from *Historical Statistics*, 1949, p. 28. The population of slaves under ten was estimated at 70 per cent of those under fourteen—the ratio prevailing in both 1850 and 1860. For free males, aged ten to fifteen, we assume a 25 per cent rate (see section on 1850 gainfully occupied). For free females, aged ten and over, we estimate 110,000 domestics (see section on domestic service) and an arbitrary 50,000 in other occupations. (A total of 12,000 persons in cotton textiles in 1820, and the 1818–20 decline of all textile hats and clothing industries suggests this as a reasonably generous figure.) Our estimate of 3,135,000 is within 10 per cent of Whelpton's, and those who essentially adopt his figures (i.e., Edwards, Fabricant, and Long).

LABOR FORCE AND EMPLOYMENT, 1800–1960   141

To allocate omissions between farm and nonfarm pursuits we apply the proportion of farm to total implicit in the reported Census figures.[56]

GAINFUL WORKERS: 1830

The procedures used for 1830 were much the same as for 1820. For free males aged fifteen and over the proportion gainfully occupied as reported in the 1850 Census was used, while for slaves aged ten and over a 90 per cent rate was used, both ratios having been discussed above in connection with the 1820 estimates.[57] For free males aged ten to fifteen a 25 per cent rate was used (as in 1840), the rate for native whites ten to fifteen in 1900 being 22.6 per cent. The total for females was estimated as the sum of those engaged in individual industries and occupations. Female domestic service employees were estimated by procedures outlined in the section for employment in that group. An aggregation of establishment reports in a large-scale survey by the U.S. Department of State in 1832 leads to a total of just under 50,000 female factory workers in cotton, wool, shoe, and palm-leaf hat manufacturing, to which we add 25,000 for mantua makers, etc.[58]

GAINFUL WORKERS: 1840

The total was estimated as the sum of four major categories.

1. Free males sixteen and over: 4,075,000. It was assumed that 87.2 per cent of these were gainfully occupied, the same rate as shown by the Census data for the free males sixteen and over in 1850.[59]

2. Free males ten to fifteen: 235,000. An arbitrary 25 per cent of this group was taken as gainfully employed, the rate for native whites in 1900 being 22.6 per cent.

3. Slaves: 1,430,000. Following the procedure used for 1850 and 1860 estimates, the proportion of all slaves aged ten and over gainfully occupied was assumed the same as the 87.2 per cent for white males aged sixteen and over.

[56] Any passionate arithmetician will compute a different figure than the one we derive here because we assume that the entire female labor force, as we have estimated it, was omitted, and was, in addition, in domestic service or other nonfarm occupations. Hence the 83 per cent farm ratio applies only to the estimate for males omitted.

[57] Population data from *Historical Statistics*, 1960, pp. 10–11.

[58] Reports presented in *Documents Relative to the Manufactures in the United States*, 22d Congress, 1st Session, House Doc. 308, 1833. For cotton textile manufactures in some states the data used were instead from the New York Convention survey, reprinted in *Niles' Register*, Vol. 41, Addendum, p. 7. See also the writer's "Population Change and the Supply of Labor," *Demographic and Economic Change in Developed Countries*, Special Conference 11, Princeton for NBER, 1960.

[59] De Bow, *Compendium*, pp. 55, 69, 128. The same source was used for other population data below.

4. Free females sixteen and over. The number in industrial pursuits is estimated for 1830 at 75,000 (see note 58), and for 1850 (below) at 220,000. The 1840 figure was estimated by proportionate interpolation, using cotton textile employment 1830–40 and 1830–50.[60] The number in domestic service—240,000—is estimated by procedures described in the section on domestic service.

GAINFUL WORKERS: 1850

The total was estimated as the sum of major components outlined below.

1. Free males sixteen and over: 5,330,000. The reported Census total, minus the number of students included therein, is used.[61]

2. Free males aged ten to fifteen: 280,000. The proportion of white males in this age group that were gainfully occupied in the first Census providing such information—that of 1900—was used.[62] The trend for the combined white-nonwhite group from 1870 to 1900 is not such as to suggest that the passage of these decades changed the rate for this group.

3. Free females: 675,000. This group is estimated as the sum of women in a number of specified occupations. This procedure was tested for 1860, and gave results that were close to the sum estimated by the procedure actually used. We therefore have a basis for using it for 1850 beyond its general reasonableness.

a. Domestic servants were estimated at 350,000, by using the relationship to white families, as described in the section on domestic servants.

b. Population Census data for dressmakers, milliners, and tailors are available for 1860 and subsequent decades, and for males (only) in 1850. Taking the total for males in 1850 (52,000) and for 1870 (67,000) we estimate males in 1860 by the trend for males employed in clothing industries as reported by the Censuses of Manufactures.[63] (Virtually all males in these occupations were employed in factories.) Deducting from the reported 1860 Population Census total for both sexes gives us 190,000 for females.

We now compute the ratio of males to total gainfully occupied in these occupations, which proves to be identical in 1860 and 1870. It is therefore assumed the same in 1850, applied to the Census total for males to give 165,000 for females.

---

[60] See section on cotton textile factory employment.
[61] *The Seventh Census of the United States: 1850*, pp. lxxix, lxxvii.
[62] 1900 Census, *Occupations*, p. cxviii.
[63] Occupation Census data from Edwards, *Comparative Occupation Statistics*, Tables 8–10, and 1890 Census, *Population*, Part II, p. cviii. Manufactures Census data from 1870 Census, *Industry and Wealth*, pp. 394, 400, 406.

c. Mill and factory operatives, not specified. The bulk of female factory operatives, other than those included in the dressmaker, etc. category were returned under this heading. We therefore take the Census of Manufactures trend for the employment of males in factories, deducting clothing factories, to extrapolate the Population Census 1860 total for this occupation to an 1850 level.[64] The resultant figure, minus the males reported in this occupation in 1850, gives a 50,000 estimate for females. Alternatively, we assume that the Population Census enumerators enumerated the same proportion of all male factory operatives in 1850 as they did in 1860. By then taking the Census of Manufactures as a control, we assume that the proportion of male to total employment indicated in the Census of Manufactures data (for all industries but clothing) can apply to the Population Census figure for males, giving a 60,000 figure for females. We arbitrarily average the two figures since no basis exists for preferring one to the other.

d. Teachers. A comprehensive Census enumeration of educational institutions in 1850 gives us the total for that year, the female proportion being taken as the same as that for 1870 (as the ratio changed little in subsequent years).[65] The resultant figure is 80,000.

e. Nurses, boardinghouse keepers. For these small groups, whose numbers changed little over the 1860–80 period, the difference between the 1850 male total and the 1860 male plus female total was assumed to give the correct 1850 figure—of 18,000.

Adding the number in these occupations gives a figure of 662,000. This was rounded to 675,000 to allow for minor omitted occupations.[66] No estimate is made for females in agriculture since the Censuses of 1860, 1870, and 1880 did not include females, except nonwhites in the South—a group which was included under the slave total in 1850.

4. Slaves gainfully occupied: 1,890,000. The number of slaves aged ten and over was derived from Census data for each state in 1850.[67] It was assumed that the worker rates for slaves in 1850 averaged 90 per cent, or above the relatively low rates (of about 80 per cent) for planters and white persons. We rely on Jefferson's statement that numbers of planters

[64] Manufactures and Population Census sources as for dressmakers.

[65] 1850 Census, *Compendium*, pp. 141–143. As can be seen from the similar 1860 enumeration the figures are well above the Population Census counts. They are preferred as being clearly more comprehensive and reliable. The 1870 percentage is based on data from 1870 Census, *Population and Social Statistics*, pp. 676, 638.

[66] One occupation of consequence in later years is laundresses. These are included in the estimate of servants. Another group—clerks, bookkeepers, and saleswomen—included very few women in 1860 and must have had well under 10,000 in 1850.

[67] *The Seventh Census of the United States: 1850*, p. xliv; De Bow, *Compendium*, p. 89.

were not in the labor force, assuming that the same causal factors did not minimize the slave labor force. (Using the adult white male rates by state, however, would only reduce the total by 70,000.)

GAINFUL WORKERS: 1860

The 1860 Census secured data on the number of gainfully occupied persons aged sixteen and over. It did not cover slaves, nor did it distinguish between males and females. To expand the Census figures to total, and to provide a basis for linking with the 1850 Census, it was necessary to estimate for 1860 the number of women who were included by the Census as gainful workers, to estimate the number of slaves, and to estimate the number of children ten to fifteen, who were gainful workers but were not included by the Census.

The number of women included in the occupation totals for 1860 was estimated as 895,000 (a check made by a completely different procedure led to an estimate of 950,000). The basic estimating procedure was the following.

1. The ratio of males aged fifteen and over who were gainfully occupied in 1850, 1880, and 1890 was computed for each state.[68]

For 1850 these ratios relate to the free population (which was almost all white). For 1880 and 1890 they relate to the free population, which then included a large number of ex-slaves. It will be seen that the ratios changed little from Census to Census despite this major social change, and appear to have varied as a result of differences in enumerator efficiency rather than because of any discernible trends in labor force rates. The 1850 ratios for free males were therefore assumed to apply equally well to the number of free males in 1860, except for Connecticut and Missouri. For Connecticut the 1880 rate was used instead—the 1850 rate being unreasonably low in the light of rates for other New England states in 1850 and every New England state in 1880 and 1890. The 1850 Louisiana rate was used to raise the Missouri rate so that it was not below that for any other slaveholding state in 1850. While it is possible that even this is not enough of an adjustment, the Louisiana rate was taken as a working indication of the minimum rate under 1850 slaveholding conditions.

If we apply these 1860 rates to the 1860 white male population aged fifteen and over, we get an estimated 7,075,345 white males in this group that were gainfully occupied. Increasing this total by 210,935 for male free colored and free Indians, and deducting from the reported occupations

---

[68] *The Seventh Census of the United States: 1850*, p. xlii; 1890 Census, *Population*, Part II, p. lxxxvii.

count, gives an estimate of 894,697 for gainfully occupied females in 1860, as implied in the Census count for that year.[69]

2. A check estimate of 950,000 was arrived at by an entirely different procedure. The female gainfully occupied total was estimated directly from data on occupations with substantial numbers of females in them in later years. Thus, four occupation groupings accounted for 80 per cent of all female nonfarm employment in 1880. The proportion of females in these occupations in 1880 and 1890 are shown below, together with the total number in 1860 and the estimated proportion of females in the same year.[70]

	Number in Occupation in 1860	Percentage Female			
		1890	1880	1870	Est. 1860
Teachers	112,969	72	68	66	65
Servants and laundresses	627,068	84	81	88	85
Clerks, bookkeepers, and salespersons	185,549	15	7	3	3
Dressmakers, milliners, and seamstresses	151,955	99	97	99	100
Tailors	101,868	34	39	—	36

Estimates were made for these occupations and, in addition, for housekeepers, boot and shoemakers, laborers, mantua makers, nurses and weavers, aggregating a total of 869,000. Total nonfarm females aged fifteen and over and occupied in 1880 were 118 per cent of the aggregate for these or comparable occupations in that year. To allow for the more limited entrance of women into the various occupations in 1860, this percentage was reduced to 110 per cent, and an estimate of 950,000 arrived at.

It was assumed that no women were reported employed in agriculture by the 1860 Census. This assumption was dictated by an examination of the individual state figures for 1880. Only a trivial number of females in agriculture were reported for that year in the great central and northern states—implying that the numbers reported in the southern states reflected primarily Negro women in that year. A fortiori, in 1860, when

[69] Given the small number of free colored and Indians, separate state estimates were not made. Instead, the U.S. number of free colored, aged fifteen and over (1860 Census, p. 595), was assumed to have an 80 per cent worker rate and the free Indians (1860 Census, pp. 596–597) a rate of 60 per cent—a lower rate to reflect Indian family relationships indicated in travelers' tales. But it clearly makes very little difference whether one increases the white male total by 2, 3, or 4 per cent.

[70] Census data from 1890 Census, *Population*, Part II, pp. civ ff.

slaves were not enumerated, the number of white or free colored females reported in agriculture would have been trivial at best.

The number of gainfully occupied slaves in 1860 was estimated at 2,340,000. This estimate was arrived at by applying to the slave population of each state, aged ten and over, the proportion of the free male population (white and colored), aged fifteen and over, that was reported as gainfully occupied in 1850. Several assumptions lying behind this procedure should be noted. First, it is assumed that the proportion of female slaves gainfully occupied was the same as that of males. Examination of a sample of individual Census schedules for 1820 and 1840, the only Censuses when slaveholders did not report on the proportion of their slaves gainfully occupied, showed no distinction of rates resultant from the sex composition of their holdings. Plantation accounts and literature of the time confirm the buying and selling of females and their use for activities productive of market values—the common sufficient indication of gainful occupation.[71] Secondly, the 1860 data could not be used as satisfactorily as the 1850 data for establishing worker rates for this particular group. The 1860 data include not merely men but women gainfully occupied; and while the slave holding states generally had small numbers in this group, some, such as Virginia, Missouri, Maryland, and Delaware, did have a significant number.[72] On the other hand, the 1850 data relate only to males, and we can compute in each state the ratio of the number gainfully occupied to the population base. These proportions generally ran from 84 to 88 per cent in most southern states. Except for the adjustment of the Missouri rate noted above these rates were therefore applied to the slave population aged ten and over in 1860 to estimate gainfully occupied slaves in that year.

Free children aged ten to fifteen were estimated at 535,000. Since virtually all of this group was composed of white children, we take the proportion of white children ten to fifteen gainfully occupied in 1900 as our guide—earlier Censuses being unsatisfactory on various grounds.[73]

---

[71] Different treatment is not warranted because periods of absence from regular field or house duties for parturition occurred, given the monetary values set on the results of such activity.

[72] We deduce the number of women by applying to the 1860 free male population the worker rate for the same group in 1850, deducting from the reported 1860 total. The result in most slaveholding states is under 10 per cent of the 1860 total in most southern states.

[73] The 1870 Census underenumerated gainfully occupied children; the 1880 Census does not give us the distinction by color; the 1890 Census does not give us that by age. The 1900 Census, *Occupations*, p. cxviii, gives data for deriving rates of 22.1 and 6.4 for boys and girls with native white parents. Those with foreign parents were not included in the weighting, first because the foreign group was much smaller in 1860, and second, as a means of offsetting any slight decline in rates from 1860 to 1900 that may have flowed from rising incomes or the spread of education.

Adding the minor adjustment for the fifteen year olds, we arrive at an estimate for the ten to fifteen group of 535,000.[74]

Adding together the number of free persons sixteen and over reported by the Census as gainfully occupied, the slaves and free children estimated as belonging to this group, and deducting students included in the Census total[75] gives us 11,110,000.

GAINFUL WORKERS: 1870–1900

For these years we adopt the figures from the recent Census report by Kaplan and Casey.[76] These figures are essentially the same as the ones given by Edwards who adopted the original 1880 and 1900 Census results which were the official revisions for 1870 and 1890. The only alternative for 1870 is an estimate[77] which adjusts the Edwards figure by one-half of 1 per cent. We prefer to adhere to the official Census results instead of following this relatively minor revision. For 1890 Clarence Long has adjusted the Census on the assumption that school attendance rates in later years applied in 1890, and that labor force rates in 1890 should equal the average of 1900 and 1920 for certain groups.[78] His net adjustment is about 1 per cent of the Census gainful worker total. We prefer to accept the official Census results rather than adopt this speculative and minor adjustment.

GAINFUL WORKERS: 1900–60

Estimates for the ten and over group were derived as initial steps in estimating the annual labor force figures 1900–30, and their derivation is outlined in connection with the latter series.

For 1940–1960 we increase our estimates for the labor force as currently defined (including those fourteen and older) by making an allowance for those aged ten to thirteen.

For 1950 we estimate the proportion ten to thirteen in the labor force on the basis of two special Census enumerations of those employed in that

---

[74] 1860 Census, *Population*, pp. 592–594, gives data for the population aged ten to fifteen. The 1880 Census, *Population*, p. 548, indicates that the ten to fifteen male group is 116 per cent, and the female, 116.5 per cent of the ten to fourteen group.

[75] The Census included precisely 49,993 students, no more, no less (*ibid.*, pp. 676–677).

[76] David L. Kaplan and M. Claire Casey, *Occupational Trends in the United States, 1900 to 1950*, 1958, p. 9.

[77] Simon Kuznets, *Capital in the American Economy: Its Formation and Financing*, Princeton for NBER, 1961.

[78] Long, *Labor Force*, Appendix G.

year.[79] Earlier percentages in the labor force were as follows.[80]

	1920	1930	1940 (March)	1950
Age 10–13	4.4	2.4	n.a.	7.9
Age 14–15	17.5	9.2	5.2	20.8
Percentage ratio	25	26		38

We assume that the relative rise in the rate for the youngest age group from 1930 to 1950 reflects the greater availability of part-time work during the war and postwar years, and hence take a 25 per cent ratio for 1940. Applying that ratio to the 5.2 rate for fourteen to fifteen, and rounding to an annual average rate for the ten to thirteen group, gives 105,000 for 1940, with 35,000 in agriculture.[81] For 1950 we take the data for the fourteen to fifteen year olds as a guide to estimate that half the annual average labor force ten to thirteen was in farming.[82] For 1960 we adopt the same 38 per cent ratio as given by the 1950 data, and this, applied to the 17.5 per cent reported rate for the fourteen to fifteen group, gives us 6.7 per cent for 1960.[83]

PREVIOUS GAINFUL WORKER ESTIMATES: 1820–60

Prior gainful worker estimates for these decades are essentially those of P. K. Whelton. These constitute the primary basis for the estimates made for the Census Bureau by Alba Edwards, as well as the combination

---

[79] Estimates of 1,095 for August and 719 for October are available. The average of these two was adjusted to an annual average of 718 by the parallel ratio for the fourteen to fifteen year olds. (August data from *Current Population Reports*, Series P-50, No. 83. October data kindly provided by Miss Gertrude Bancroft of the Bureau of Labor Statistics. Data for fourteen to fifteen year olds from Series P-50, No. 31, Tables 3, 8.) Estimates of 132,800 and 201,300 for 1940 and 1950 have been made by Ann Miller (in Simon Kuznets, *Capital in the American Economy*, Table 38). The striking difference in 1950 reflects her use of 1930 ratios for 1950 whereas we rely on the direct special Census enumerations.

[80] 1920 and 1930: 1930 Census, *Occupations*, p. 347. 1940: 1940 Census, *The Labor Force*, Vol. III, Part 1, p. 19. 1950 data: see above.

[81] The resultant rate of 1.3 per cent was rounded to 1.5 to give an annual average rate, and this, applied to the population count (1940 Census, Vol. 4, Part 1, p. 8) gave 104,000, or 105,000 allowing for population growth. The 1940 Census, *Industrial Characteristics*, Table 3, showed that 32 per cent of the fourteen to fifteen year olds were in agriculture and we assume that one-third of the ten to thirteen year olds were.

[82] Some 46 per cent of the fourteen to fifteen year olds employed in October were in farming, compared to 38 per cent for the year. We therefore reduce the October 62 per cent rate for ten to thirteen year olds to 50 per cent for the year, giving 360,000 in farming.

[83] Bureau of Labor Statistics, Special Labor Force Reports, *Monthly Labor Review*, March 1961, P. A-13.

of both series by Solomon Fabricant in a comprehensive and characteristically lucid review of these data.[84]

The gainful worker totals for 1850 and 1860, in either set of estimates, can be viewed simply as the sum of component estimates—white male population times the proportion of that group in the labor force, white female population times the proportion of that group in the labor force, and so on. The present estimates use the actual Population Census counts in each of the four major sex-color groups, and stipulate the proportions of each in the labor force. The alternative estimates stipulate that the proportion of the total population in the labor force in 1850 and 1860 was given by interpolation between 1840 and 1880 rates. This process fails to make use of the actual population counts of the changing numbers of each of these sex-color groups, and fails to use the Population Census records for the proportion of white males gainfully occupied—the largest single group in the labor force. Under the circumstances, interpolation between over-all rates based on the inadequate 1840 Census (when slavery prevailed and manufacturing industries had just begun employing females in significant numbers) and the 1880 Census (when slavery had ceased, a great number of immigrants had entered, and manufacturing was a significant employer of female workers) is hardly a preferred procedure.

A different consideration appears in the 1840 estimates. The procedures used for the present 1840 estimates are similar to those for 1850–60. Whelpton and Edwards, rejecting the improved Censuses of 1850 and 1860, take as the rock on which to rest their estimates that for 1840. Aside from Congressional investigations (exculpations by Webster and Calhoun joining in support of the most incompetent Census superintendent ever in office), we have the internal evidence of the 1840 Census testifying to its limitations. These are discussed in connection with the estimates for the agricultural labor force. These suggest both under- and overenumeration in different portions of the Census. It is impossible from the Census materials themselves to decide just where the net adjustment falls. We therefore reject that Census as a measure of absolute level of the gainfully occupied in the industries it purports to cover, and do not follow Whelpton's procedure of adding about 15 per cent for industries it did not purport to cover. Instead, we return to direct estimating of worker rates

---

[84] Whelpton, "Occupational Groups"; Edwards, *Comparative Occupation Statistics*, p. 142; Fabricant, "Changing Industrial Distribution." Fabricant adopts Edwards' minor revisions of Whelpton's total gainfully occupied (*ibid.*, p. 43). The Edwards estimates are reproduced, with minor changes, in the important Census monograph by David L. Kaplan and M. Claire Casey (*Occupational Trends*, p. 6). They also appear in Long, *Labor Force*, Appendix A.

based on the somewhat more satisfactory population counts. These, summed as above, suggest that the true total would be about 10 per cent above that estimated by Whelpton. Our difference is largely a result of our adjustment for Census undercoverage in agriculture, Whelpton accepting the Census totals for this group.[85]

GAINFUL WORKERS IN AGRICULTURE: 1800–60

Varieties of Census data are available for 1820, 1840, and 1860. The basic procedure adopted was to build upon these data, using relationships to other comprehensive figures to compute the major components of the gainful worker total. The various incomparabilities in coverage and definition make the estimates a laborious matter, and in some instances, quite chancy. But two elements tend to make the 1800–60 figures solider than might at first appear. One is the pattern of stable relationships for the key components of the labor force, discussed below. The second is the fact that in this period the share of the slave population in the agricultural total, and of the farm population in the U.S. total, is so great that the possibilities of differing estimates are much more limited than would be the case if we had to estimate, in the same fashion, farm employment today. The specific procedures used are outlined below.

Slaves

Reported Census totals for the slave population in each decade constitute our starting point. For 1830–60 the data also give us the population aged ten and over, while for 1820 we have figures on those aged fourteen and over.[86] The ratio of the under ten group to the under fourteen group was 70 per cent in 1850, and 69.5 per cent in 1860, and was assumed to be 70 per cent in 1820. The 1820 figure for the under ten group as thus derived proves to be 30.5 per cent, compared with 31.8 per cent for the comparable group in 1850 and 31.1 per cent in 1860. We therefore assume 31 per cent for 1800 and 1810. The reported Census figures for 1830–60, and those thus estimated for 1800–20, were then distributed between rural and urban residents. For 1790 we took the slave population of counties which included the five major cities as equivalent to the urban slave population at that early date (even in later decades these counties accounted

---

[85] The limitations of the 1840 Census figures are discussed by Richard A. Easterlin in "Interregional Differences in per Capita Income, Population, and Total Income, 1840-1950," in *Trends in the American Economy in the Nineteenth Century*, Studies in Income and Wealth 24, pp. 126 ff. He allocates these, as we do, to agriculture.

[86] *Historical Statistics*, 1960, pp. 9, 11.

for an overwhelming proportion of total urban slaves).[87] For 1850[88] and for 1860[89] we added up the slave population of all cities and towns reported in the Censuses of those years. The result of these calculations is that approximately 5 per cent of the slaves were in the urban population in 1790, in 1850, and again in 1860, from which we assume 5 per cent in the intervening years.

For 1850 we do have an estimate by De Bow to the effect that 400,000 slaves were urban—or about 13 per cent.[90] De Bow's figure comprehends "the slaves who are known to be residents of towns, and approximating for those towns that are unknown." However, as noted above, even if one adds up all reported towns in the 1850 Census, exclusive of the unreasonable entries for Missouri, one arrives at perhaps half that number, suggesting that De Bow simply doubled his figure to allow for "the unknown towns." Moreover, one might note a contrary bias—namely, the practice of including with the town population slaves in rural areas owned by those residing in towns. We therefore reject his figure as being less reliable than a direct summation of Census reports.

The proportion of both rural and urban slaves aged ten and over that was gainfully occupied was taken as 87 per cent. The basis for this ratio was discussed at length earlier; essentially it rests on a review of the original 1820 and 1840 Census schedules in which the slave-owners themselves reported on the number of their slaves that were gainfully occupied. Comparison with the numbers of persons (by age) in their families indicated that the slave-owners fairly consistently counted at least those ten and over, and frequently some at younger ages. However, a 100 per cent ratio was not used; we attempted to make a conservative estimate, allowing for sickness and disability. As a guide, the 87 per cent ratio derived for white males fifteen and over in the slave states in 1850 was employed. One may assume that the proportion for slaves ten and over

[87] The counties are those including Richmond (Henrico), Baltimore (Baltimore), Savannah (Chatham), and Louisville (Jefferson). For Charleston County, 22 per cent of the county total was used, that being the city-county ratio for Charleston slaves in 1850. The 1790 data is from Census, *A Century of Population Growth*, Table 104.

[88] *The Seventh Census of the United States, 1850*. Individual city and town totals appear in Table 111 for each state.

[89] 1860 Census, *Census of the Population*, Table 3, gives individual city totals. Since the city total for Augusta was lacking, that for Richmond County was used. The reports for Missouri apparently considered every inhabited area a town in 1850. The 1860 figure of 3,000 for St. Louis was therefore used instead of the summarized figure of 45,000. The population of the twenty-eight major cities plus Richmond County was computed as a check, since it is less subject to the whims of changing definition. This confirmed the essential identity of 1850 and 1860 totals.

[90] De Bow, *Compendium*, p. 94.

would be at least as great.[91] The use of the 87 per cent figure produces a mild underestimate of those employed in agriculture, thus partially compensating for the inclusion of domestic servants, carpenters, etc., employed on the plantations and small slaveholdings.[92]

For 1840 it appears that the Census enumeration must have counted all slaves aged ten and over in rural areas as engaged in agriculture, in contrast to the present estimates that make a 13 per cent reduction for those disabled, ill, or in purely household tasks. This inference rests upon the fact that if the white labor force in agriculture, estimated below, is deducted from the Census count for all in agriculture, the result is 1,550,000—in contrast to the present estimate of 1,563,000 slaves in rural areas (which constitutes the basis for a 1,410,000 estimate of those actually employed in agriculture). While the white labor force may have been estimated incorrectly, the various relationships of that group to the rural white families, of farmers to farm laborers, make it unlikely that a significant error was made. Moreover, the undercoverage of that Census in both northern and southern counties makes it likely that the white estimate, as does the total, errs by being too small rather than too large. On the whole, it seems more likely that the entire slave population in rural areas was classified as engaged in agriculture than that the free farm labor force was underestimated by about 8 per cent.

Farmers and Free Farm Laborers

For 1850 and 1860 we have fairly inclusive data; for 1820 and 1840, somewhat less so. These data were adjusted to a comprehensive coverage, then related to the number of rural white families as the basis for extrapolating to other Census years. Because of this sequence of building from the solider materials of the later Censuses, the estimating procedure will be described in the order actually followed.

The 1860 Census provides a count of farmers and free laborers in agriculture, aged sixteen years and older.[93] The population count for free males aged ten to fourteen was adjusted to include those aged fifteen, and 17 per cent of this ten to fifteen group was taken as the number in

---

[91] The practice of manumission removed from the slave population some of the oldest slaves and those least likely to be gainfully occupied, again tending to make this a conservative estimate.

[92] For 1860 a test was made of the effect of not using an over-all 87 per cent figure, but ratios were computed for each state from the 1850 data. The result, 2,339,685, was actually used instead of the 2,450,000 to be derived from the 87 per cent figure, but is not significantly different.

[93] 1860 Census, *Population*, p. 592.

agriculture.⁹⁴ (No allowance was made for females as is explained below.)

The 1850 Census provides a count of males, sixteen and over, gainfully occupied in agriculture.⁹⁵ Males in agriculture, ten to fifteen, were estimated from the reported ten to fourteen population totals, as was done for 1860. No allowance was made for free females engaged in agriculture for 1850, nor for 1800, 1810, or 1830. It is perfectly clear that many farm women did help with the crops, certainly with garden chores, taking care of chickens, etc. The reason why no adjustment was made, when many of lesser consequence were attempted, is simply that there is no evidence that this group was counted in the Censuses of 1820, 1840, or 1860–80 when the Census purported to cover all women, aged ten and over, gainfully occupied. We may make our own judgments as to the meaning of this exclusion. But if we were to include farm women in the agricultural count, it would be necessary to adjust every one of our nineteenth century Censuses.

If one examines the individual state totals for 1870–90, when the published data give a breakdown by sex, and if one examines the unpublished schedules for a random selection of counties in 1820 and 1840, it becomes clear that virtually no women were counted as employed in the midwestern states. The few in the border states of the Middle Atlantic and in New England are readily accounted for as employed in manufacturing, etc. The substantial number of women counted in those Censuses proves to be a phenomenon restricted to the southern states and clearly is a count of Negro females in agriculture—slaves prior to 1860 and free workers afterwards.

As late as 1880, for example (when we do not have the same problem of undercoverage as in 1870, but do purport to include females in agriculture), a grand total of less than 3,000 are reported for Illinois (as compared with 434,000 males). Less than 2,000 each are reported in Massachusetts, Michigan, Minnesota, Ohio, Pennsylvania—as compared with from 100,000 to 400,000 males in farming in each of those states.⁹⁶

⁹⁴ From the 1880 Census, *Population*, p. 548, we compute the ratio of white males aged ten to fifteen to those ten to fourteen, and apply it to the 1860 white plus free colored aged ten to fourteen. Edwards (*Comparative Occupation Statistics*, p. 97) summarizes Census data showing about 16 per cent for 1870 and 1910, and about 18 per cent for intervening years. An arbitrary 17 per cent was used, considering all of these as samples of a fairly constant average.
⁹⁵ *The Seventh Census of the United States: 1850*, p. lxx.
⁹⁶ 1880 Census, *Population*, p. 716. In the 1870 Census we find only 244 females in farm labor compared with 134,000 males (1870 Census, *Population*, Table XXVII); and in 1900, 835 to 79,000 in family farm labor, and 764 to 102,000 for hired (1900 Census, *Occupations*, Table 33). On the other hand, 64,000, 78,000 and 30,000 are reported for Mississippi.

In the 1840 Census, triumphant incompetence succeeded in simultaneously producing an undercoverage as well as an overcoverage of those in agriculture. Our problem is to decide which of these biases was greater, and by how much.

The undercoverage has been noted by careful students, who point to unreasonably low participation rates.[97] Reference to the unpublished schedules now in the National Archives confirms this fact, and suggests that the failure began with the actual enumeration. By making estimates for every county with no, or trivial, agricultural entries, we would increase the free labor total by only 30,000, however.[98] Most of the omissions were made in the slave states, and were therefore assumed not a problem, since the slaves in agriculture were here estimated independently.

But along with undercoverage in some counties was still greater overcoverage in others. Even the proudest Michigander could hardly believe that men, women, and babes in arms in Hillsdale and Livingstone counties were all occupied in farming.[99] And, more substantially, it is impossible to believe that corn production in the midwestern states rose as enormously as it did from 1840 to 1850 while farm employment rose only trivially—as the Census implies.[100] Since the 1850 Census was certainly better than the 1840, this computation suggests that on balance the 1840 Census overenumerated. Such assimilation of entries to the dominant classification group is characteristic of bad enumerative practice.

These limited tests suggest that overenumeration was probably greater than underenumeration. But do we have a more general control? One guide could be the change in the rural population. But unfortunately we have no assurance that the population enumeration of urban areas in 1840 was any better than the occupational check. Hence the estimate of rural population (made by subtraction) would be equally suspect. We may start, however, with certain outside limitations. (a) The ratio of farm laborers to farmers in 1840 cannot have differed greatly from the 1850 ratio. (b) The rate of change in free farm labor, and in free farmers, over

[97] Cf. the analysis of Richard Easterlin, "Interregional Differences."
[98] As a method of estimation, the ratio of agricultural gainful workers to total population in the other counties in the same state was used. (Counties with major cities were, of course, excluded.) The Allen edition was used, and the ratio for most states proved to be near one-third.
[99] 1840 Census, *Compendium*, pp. 94–95.
[100] The Illinois figures show a reasonable proportion of change. Indiana, Iowa, Michigan, and Ohio do not. If one assumes for each of these states that the 1850 ratio of horses and mules to those gainfully occupied in agriculture should have applied in 1840, an adjustment of 70,000 might be made for the four states. If we contend that mechanization would have changed the ratio the adjustment would be still greater.

this decade must not be markedly out of line with the change in prior and subsequent decades. (c) Examination of unpublished schedules shows that all slaves aged ten and over were included as gainfully occupied in many counties. We assume that in rural areas they were all classified in agriculture.

If we assume that the 1,563,000 rural slaves aged ten and over were all allocated to agriculture (instead of merely the 1,410,000 estimated as gainfully occupied), and deduct from the Census total for agriculture, we have a residual of 2,160,000 free persons in farming. The ratio of free persons in farming to farmers was 148 per cent in 1850, 131 per cent in 1860 (and 152 per cent in 1800). If we assume it at 150 per cent for 1840, the number of farmers can then be estimated at 1,440,000. Neither figure is grossly out of line. Any great amount of manipulation is impossible because it distorts one or more relationships, and it is therefore assumed that these are tolerable approximations—including the supposition of a 150,000 overestimate (net) by the Census in its count of slaves in agriculture.[101]

On the other hand, the 1840 data were not adjusted because of the trivial rise in agricultural employment shown, from 1840 to 1850, for Indiana, Iowa, Michigan, and Ohio. It is impossible to believe (a) that enormous gains of corn production occurred in these states, (b) that simultaneously, marked employment gains occurred in Illinois; but (c) that only trivial gains took place for these four states. An adjustment of about 70,000 might be made—but being so dubious, is omitted, though the reader is warned as to this probable bias.[102]

For 1820, the total for those gainfully occupied in farming, as derived in the section on gainful workers in 1820, was used; the estimated slave count was then deducted to obtain free persons in agriculture.[103]

The total of those gainfully occupied in agriculture in 1810 and 1830 was estimated as 150 per cent of farmers in 1810 and in 1830, the ratio to

---

[101] If the ratio is changed much, the time series for farmers shows an unreasonable hump (or dip) in 1840, particularly in relation to the long-run rural population trends. If one assumes a correct enumeration of the slave population and an error only in the free population—which is unreasonable *per se*—the ratio proves to be 157 per cent.

[102] Production and employment data from the 1840 Census. The adjustment suggested used the ratio of horses and mules to gainfully occupied in agriculture, by state, in 1850 and in 1840; the ratios for most midwestern and northern states changed little, but these changed markedly, and improbably. Assimilation of entries to the dominant classification group being characteristic of bad enumerative practice, it was assumed that the error lay in the well-known inadequacies of the 1840 Census, rather than undercoverage in 1850.

[103] The 1900 Census, *Occupations*, p. xxx, makes minor corrections in the printed 1820 Census figures. We adjust the total there shown for omissions.

be computed from our 1800 estimates being 152 per cent, that for 1850, 148 per cent.

If we look to the rate of admission of immigrants who called themselves farmers or farm laborers, and consider them as the major volatile source of farm labor, the trend is consistent with that of these ratios.[104] Farm labor was then estimated by deducting farmers from the total.

The derivation of the 1800 data is discussed in the section on 1800 labor force.

How do the present estimates compare with the Whelpton-Edwards figures for the same period?[105] For 1820 both are tied to reported Census totals. For 1840 Edwards adopts the Census figure while the present estimates assume that all slaves aged ten and over in rural areas were allocated to agriculture. An adjustment for this (despite a mild compensatory adjustment for counties completely lacking in agricultural entries) produces a total below Census, and therefore below Edwards.

Significant differences arise for 1850 and 1860 largely because Edwards ignored the Census reports for those years. Although this is understandable because these Censuses were limited to free persons (and that of 1850 to free males sixteen and older), we are not well advised to ignore an adequate Census of persons who constitute some 60 per cent of the agriculture total. This is particularly so if our alternative is the Edwards one of interpolating at one fell swoop between a ratio derived from the inadequate 1840 Census and one based on the 1910 Census—after the latter figure had been adjusted by Edwards himself. Edwards implicitly assumes that the trend in the proportion of slaves in rural areas declined as did the U.S. totals, whereas the present estimates, based on totals from reported Census data for cities in 1790, 1850, and 1860, indicate that the proportion of slaves in towns did not rise significantly over the decades.

AGRICULTURE: 1870–1900

The bourne from which no traveler has ever returned unscathed is the region where lie the Censuses of 1870–1900, with their indefinite estimation of "laborers." For by a "house that Jack built" process, the inability, and/or unwillingness of respondents, enumerators, and coders to classify laborers with adequate precision left a large group of "laborers, not specified." An unknown portion of these belong in agriculture. The

---

[104] Immigration data from 1860 Census, *Preliminary Report* of the Census, pp. 12, 17, 18.

[105] Edwards, *Comparative Occupation Statistics*, p. 142, based in large measure on procedures developed by P. K. Whelpton. These figures form the essential basis for those of Solomon Fabricant that appear in *Historical Statistics*, 1960, p. 57.

group is so substantial that we can have no reliable figure for agriculture without estimating them. The proper method of allocating them, however, is difficult to discover. By *force majeure* the laboriously detailed estimates of Alba Edwards have been widely used.[106] But other attempts have not been wanting. We therefore consider the major serious estimates in the field, then describe the present approach.

Previous estimates

Edwards estimates agricultural employment from the trend in the rural population.[107] By heroic interpolation, he estimates agricultural employment for 1850–1900, completely ignoring the Census information for those years. Specifically, he interpolates between the Harrison (1840) and Taft (1910) ratio of gainful workers in agriculture to the rural population, then applies his sixty years of ratios to the rural population totals for these decades to estimate gainful workers in agriculture. To assume a straight-line progression in a set of ratios, moving with relentless precision through years of war, cyclical change, and the spreading factory system, is difficult enough. To use this steady line for interpolating between one ratio based on an unsatisfactory Census (1840) and another based on a Census with an acknowledged substantial overcount of farm workers is an even more troublesome procedure. Edwards was clearly not happy with the procedure but "after much experimentation" adopted it failing any better alternative.

Daniel Carson pointed out, quite properly, the disturbance in the manpower-population ratio that results from wars, from the shifting of worker to nonfarm industries during the depression of the 1870's, and so on.[108] Having pointed to this irregularity, Carson decided to treat agricultural manpower as a function not of rural population, but of improved farm acreage. This procedure displaces the difficulty by an infinitely small amount. Adopting the Census of 1860 count of farm workers, and adjusting the 1920 Census count of farm workers, he interpolates between the two by applying to the improved farm-acreage figures interpolated ratios of workers to acreage.

What does Carson's procedure amount to? The slope of relationship is determined at one end by the extent of his debatable adjustment in the 1920 Census—and at the other, by the level of the 1860 figure, which is adopted from Edwards who adapted it from Whelpton. Whelpton in turn

---

[106] Edwards, *Comparative Occupation Statistics*, p. 144.
[107] *Ibid.*, p. 142.
[108] Daniel Carson, "Changes in the Industrial Composition of Manpower Since the Civil War," *Studies in Income and Wealth, 11*, New York, NBER, 1949, pp. 128 ff.

created an 1860 figure by interpolating between the 1840 and the 1910 ratio of farm workers to rural population. What matters in all this is the rate of change between 1860 and 1920—*the* consideration making Carson reject Edwards' detailed study. Having stigmatized Edwards' procedure because it produced unreasonable changes from 1860 to 1870 and 1880, because it disregarded the meaning of the changes in its ratios, Carson goes on to use a ratio that moves with the steady undeviating course of a sleep-walker—being either 3.2 or 3.1 acres per farm worker in any decade, depending on how the rounding falls. While some technological coefficient does relate acreage to farm employment, it is hardly this invariant. Since a large share of farm income derives from livestock (variably related to acreage), and since the men-machinery coefficients also changed over the period (as price-cost relationships changed), we cannot hope for too much from such ratios unless they are derived from some empirical sample or Census data. To derive them merely by assuming undeviating growth over this sixty-year period is unsatisfactory.

A third substantial attempt to estimate agricultural employment was recently made by Ann Miller and Carol Brainerd.[109] These authors, concerned with deriving state estimates, found previous procedures unacceptable, and assumed that "laborers not specified" could be split between farm and nonfarm occupations "in the same proportions as the two industries constituted of the total labor force excluding 'laborers not specified' for each sex."[110] In a helpful analysis they point out that this gives about a fifty-fifty split in 1870 and 1880, and a sixty to forty split in the next two decades. This differs from previous estimates, particularly in allocating a much higher proportion to agriculture in 1900. Their procedure has the considerable merit of simplicity. The major limitation of their treatment is not that the results come from a quite arbitrary assumption, but rather that the reasonableness of their resultant estimates is not tested, checked, or controlled—either by Edwards' method of control (to rural population trends), by Carson's use of farm acreage data, or some *tertium quid*. However, one may note that their figures are not notably distant from those of Edwards' for 1870–90. For 1900, when they are substantially higher than Whelpton's, Edwards', and Carson's, there is some warrant for the point they shrewdly quote Edwards himself as making, that a large rise from 1890 to 1900 appears to have taken place in the southern farm areas—a rise apparent in their estimates but not in Edwards'. We note below further reason for the assumption of a substantial 1890–1900 rise.

[109] In Everett Lee *et al.*, *Population Redistribution and Economic Growth, 1870-1950*, 1957, Part I, pp. 383 ff.
[110] *Ibid.*, p. 384.

Present Estimate

Agricultural employment in 1870–1900 is estimated by making fuller use of the direct Census materials for these years than previous estimators have done. We start from the actual reported Census figures for the gainfully occupied in agriculture,[111] instead of by-passing them as do Whelpton, Edwards, Carson, and Miller and Brainerd.

To these figures we add the number of farm laborers who were not tabulated as such but included in the general rubric "laborers not specified." We do not estimate their numbers by first estimating total agricultural gainfully occupied and then subtracting, as do Edwards, Carson, and Whelpton. Instead we tabulate for the cities for which data are available at each Census the ratio of "laborers not specified" to total population in those cities.[112] These cities had roughly half of all the U.S. urban population at each Census. We may therefore consider the ratios of laborers to population that we thus derive to be based on an enormous sample. But while the sample may be great enough to make sampling error no consideration, the possibility of bias does exist. To minimize that bias, the ratios in the different size groups were analyzed, and the three cities of New York, Chicago, and Philadelphia were excluded from the ratio computations. The ratios for 1870–1900 were then 5.13, 4.44, 4.60, and 4.05. Applying these ratios to the total urban population gives us our estimate of urban laborers not specified.[113]

Because we have an estimate of laborers in towns down to 2,500 population, we may have some confidence that we include all laborers, with trivial error, except forestry, mining, and agriculture. Reference to Edwards' analysis indicates that even he allocates a trivial number to forestry, which leaves mining and agriculture. Study of earlier Censuses indicates, as one would expect, that the misallocations in mining were not likely to be between mining and "labor not specified," but rather between mining labor and miners, the relationship to a mine being important and

---

[111] 1900 Census, *Occupations*, p. lxxxviii. These figures include the official Census revision of the 1890 count to allow for the omission of 582,522 children, aged ten to fifteen, from the total gainful worker count—an adjustment made in the 1900 Census for reasons outlined with considerable precision and persuasiveness (pp. lxvi ff.).

[112] 1900 Census, *Population*, Part I, pp. 430 ff. Laborers, not specified: 1880 Census, *Population*, Table XXXVI. 1890 Census, *Population*, Part II, Table 18. 1900 Census, *Occupations*, Table 42. For 1870 the 30 largest cities are shown; for 1880 and 1890, 58 cities with more than 25,000 population; and for 1900, the 161 cities with 25,000 or more. The estimating procedure computed ratios of laborers not specified to total population in these cities; the count for New York, Chicago, and Philadelphia was added at the end.

[113] The ratios, of course, were applied to the total urban population exclusive of the three cities, with laborers in the latter then added. Urban population from *Historical Statistics*, 1960, p. 14.

apparent to the respondent and enumerator. A fortiori, if mining laborers were thrown into this category on any substantial basis we should expect the ratios to population for Wilkes Barre and Scranton to be unusually high—which is not the case.[114] We conclude, therefore, that within the margin of significant adjustment we may take urban "laborers not specified" as including all those belonging to nonfarm industries, while the balance of "laborers not specified" belong in agriculture.[115]

We now compare the number of laborers not specified which was allocated to agriculture by Edwards, and Miller and Brainerd.[116] The most useful comparison is not one of mere differences but in relationship to a base. The entire laborious discussion of laborers not specified really turns on how many gainfully occupied persons listed by Census enumerators were not classified in agriculture but tossed into the laborers category. For this purpose we take as the base for each estimate the total reported originally by the Census, and add to it the number estimated by each investigator from the labor n.o.s. group.[117]

	1870	1880	1890	1900
		(per cent)		
Present estimates	16	28	24	25
Miller-Brainerd	15	22	21	19
Edwards	18	22	23	13
Carson	6	22	13	9

Aside from the extremely low proportion implicit in Carson's figures for 1870, what is most striking in the table is the signal 1890–1900 decline in the Edwards and Carson ratios. Such a decline is not only unwarranted in the abstract, but in the light of the specific statements made in the 1900 Census by the distinguished scholars, Joseph Hill and W. R. Rossiter. While every attempt had been made, they said, to classify laborers in agricultural districts to farm labor, and others to manufacturing and

[114] Edwards, in fact, does not attempt a direct estimate for laborers in mining. He simply divides a group of laborers that he has been unable to allocate to other industries, as between forestry, mining, manufacturing, and trade, in proportion to employees directly allocated to those industries.

[115] We assume that the small number of farm workers resident in towns that are allocated to nonfarm jobs by this procedure offsets the number of nonfarm workers who, being resident in open country, were assimilated to the prevailing classification and labeled as farm labor by the enumerators.

[116] Edwards, *Comparative Occupation Statistics*, p. 144; Carson, in *Studies in Income and Wealth*, 11, p. 126; Lee et al., *Population Redistribution*, p. 384.

[117] We intentionally exclude the subsequent 1870 and 1890 Census adjustments: our problem here is not that of a correct total but misclassification of persons recognized as gainfully occupied. The gainful worker data are from 1900 Census, *Occupations*, p. 1.

mechanical pursuits wherever the evidence on the schedule warranted it, "this effort did not prevent a very large increase in 1900 in this class of workers over the number reported in 1890."[118] We therefore infer that a reasonable estimate must assume that 1900 enumerators did not signally improve over the performance of their 1890 peers in the precision with which they classified farm laborers between "farm labor" and "labor not specified." We therefore conclude that the Miller-Brainerd and the present estimates have greater consistency with the known characteristics of enumeration. In addition, the present procedure, unlike previous ones, utilizes the information actually available from the Censuses on the geographic location of laborers not specified, to estimate by a systematic method the number in farm and in nonfarm pursuits.

In the most recent comprehensive Census publication on historical occupational data, Kaplan and Casey have made various revisions in Edwards' figures to provide a more precise classification.[119] While the differences from Edwards are small, they are improvements and we adopt these 1870–1900 figures as a starting point. We then adjust them by the difference between the Edwards count of "laborers not specified" in agriculture (which they implicitly adopt) and our present estimates. The results (in thousands) are the following:

	Kaplan and Casey	Present Estimates	Difference
1870	6,849	6,790	−59
1880	8,584	8,920	+336
1890	9,938	9,960	+22
1900	10,888	11,680	+792

FARMERS

For 1940–60, the Current Population Survey totals (mildly different from the Decennial Census figures) are accepted as the most carefully reported and classified figures.[120]

For 1900–30, we adopt the reported Census of Population counts for these years, as adjusted by Kaplan and Casey.[121]

The reported figure for 1890 overallocates Negro laborers and other

---

[118] 1900 Census, *Occupations*, p. xxvi.
[119] Kaplan and Casey, *Occupational Trends*. Edward Atkinson surmised in 1901 that "nearly 50%" of the 17.4 million gainful workers in 1880 were really in agriculture. Cf. 57th Census, 1st Sess. House Doc. 182, *Report of the Industrial Commission*, p. 522.
[120] Bureau of the Census, *Current Population Reports*, Series P-50, No. 2, p. 19, and No. 31, p. 23. Bureau of Labor Statistics, *Special Labor Force Report*, No. 14, p. A-21.
[121] Kaplan and Casey, *Occupational Trends*.

laborers, primarily in the southern states, to this occupation. While the 1900 Census corrected the 1890 figure for laborers, the corresponding correction that must flow for the count of farmers was not made. It is, however, made here.[122]

For 1860–80, Census occupational entries now grouped under the heading "farmers and farm operators" were combined to give total farmers and farm managers.[123]

For 1820–40, farmers were estimated by applying ratios to the total gainful workers in agriculture. The ratio of total to farmers was as follows: 1800, 152 per cent; 1850, 147 per cent; and 1860, 131 per cent. Interpolated values of 150 per cent were used for 1820 and 1840.

A regression of the number of farmers in 1800, 1820, 1840, 1850, and 1860 against the number of rural white families shows a very close relationship. The slope was used for estimating 1810 and 1830 values.

The derivation of the 1800 total for farmers is discussed in the section on 1800 labor force.

EMPLOYMENT IN NAVIGATION: 1800–1960

For estimating the trend of employment in navigation back to 1800 we have three bodies of information. Each has its own limitations. Taken as a whole, however, they help indicate the likely trend. The first set of data derives from the reports of the Treasury Department in entrances and clearances of American vessels. A second is the information from the Population Census of 1840, and subsequent years, on employment in certain navigation occupations. A third set of data, easily the most confusing, encompasses a variety of contemporary estimates. Let us review each in turn.

The port entrance and clearance data appear in the annual reports of the Secretary of the Treasury on Commerce and Navigation. We shall attempt to establish the level of employment, at decennial dates 1800–70, for each major type—fishing, whaling, coastal trade, foreign trade.

---

[122] 1900 Census, *Occupations*, p. lxxiii, discussed the overcount of farmers and, on page lxxi, an adjustment of 491,000 in the count of male laborers is made. If one interpolates between the ratio of farms to farmers in 1880 (107.5 per cent) and 1900 (101 per cent), and uses a ratio of, say, 105 per cent, an adjustment of 523,000 in farmers is indicated. To minimize differences from the Census, however, the 491,000 adjustment is used as the measure both of the required change in laborers and in farmers.

[123] 1900 Census, *Occupations*, pp. xxxiii, liii. These include farmers and planters, overseers, gardeners and florists, stock raisers, apiarists. For 1870 the Census total was increased by 107,000, which is Edwards' estimate of the number of farmers included in the undercount of that year. See Edwards, *Comparative Occupation Statistics*, pp. 104, 141.

Fishing Employment: 1870–1960 (decennial)

For 1870–1960 we adopt the Population Census totals for fishermen and oystermen.[124] Two other sources of data were rejected as indications of trend for the group. One is the annual report on documented tonnage in cod, mackerel and whale fisheries. This tonnage total was cut almost in half from 1860 to 1870. Hence, had our procedure for 1800–60 been used for 1870 ff., a substantial decline in employment would have been indicated. The Population Census does show an 1860–70 decline for the northern states, where much of the employment associated with documented tonnage appeared. However, a substantial 1860–70 rise for Virginia, Maryland, and other states with ex-slaves suggests a contrary factor not encompassed in the documented tonnage-employment relation. We, therefore, adopt the occupation Census reports beginning with 1870. (The result is to indicate a slight 1860–70 decline.)

A second possible source is the fishing employment total reported in the Censuses of Transportation beginning with 1906.[125] Their totals, however, include only wage earners working for firms with ships above a minimum tonnage figure—thus, omitting other wage earners as well as the far more substantial group of self-employed. It is, therefore, reasonable that they should be substantially below the more comprehensive total for the entire labor force associated with fishing.

Fishing

Benchmark estimates of employment were made for 1800 and 1860, with interpolation by tonnage engaged in the cod and mackerel fishery. For 1800 we have estimates by the House of Representatives, based on a Treasury report, that there were 3,841 men and 25,787 tons of shipping engaged in the fisheries.[126] Seybert's data on the enrolled and licensed tonnage shows that Massachusetts tonnage came to 94 per cent of the U.S. total, from which one can inflate the Treasury employment total to 4,100 (or 4,000 rounded). For 1860, the Census lists 15,579 employed in cod and related fishing as of June 1.[127]

---

[124] Kaplan and Casey, *Occupational Trends*, p. 14. 1940 Census, *The Labor Force*, Vol. III, p. 180. 1950 Census, Vol. II, Part 1, *Characteristics of the Population*, pp. 1–283. 1960 Census, *Occupations by Industry*, p. 7. In each of these last three Censuses the industry total, used here, is about 7,000 above the occupation total for fishermen. The latter is most directly comparable with the Census data for earlier years. No adjustment is made—*de minimis*.

[125] 1926 Census, *Water Transport*, p. 8.

[126] *American State Papers, Commerce and Navigation*, Vol. I, p. 511. For related estimates, cf. Timothy Pitkin, *Political and Civil History*, 1817, pp. 41, 44.

[127] 1860 Census, *Statistics of the United States* ..., 1866, p. 550.

We can interpolate between these totals by the documented tonnage in the cod and mackerel fisheries, if there is some basis for believing that the ratio of men to tons changed linearly. It is possible to test this ratio for Barnstaple, which had a substantial portion of the Massachusetts total. In 1832 this county averaged 145 men per 1,000 tons in the cod fisheries; in 1845, 152, and in 1855, 135.[128] This stability suggests that we are reasonably safe in interpolating the 1800 and 1860 ratios, and then applying these to the tonnage data.[129]

Whaling

For whaling we interpolate between the 1860 Census total and an 1800 level of 1,000, by documented tonnage in whaling, given the constancy in both the crew-tonnage cleared ratios for New Bedford in 1835–60 and in various estimates of that ratio apparent in the sources.[130]

The New Bedford trend is an obvious source, to be used with some circumspection since perhaps 10 per cent of tonnage clearing the port was in coasting and foreign trade. More reliable, because more comprehensive, are figures on clearances for the South Seas or the Pacific Ocean—the port classification varying through the years. The trend in tons per man was the following:

Per 1,000 Tons of Shipping in New Bedford

	Men Entered	Cleared	Cleared for South Seas or Pacific Ocean
1835	71	76	81
1839	75	62	78
1850	78	81	(81)
1860	74	85	90

[128] For 1832 we use data in U.S. Congress, Serial Set 222, *Documents on the Manufactures of the United States*, p. 94; for 1845, John G. Palfrey, *Statistics of . . . Industry in Massachusetts for . . . 1845*, 1846, p. 366; for 1855, *Statistical Information . . . Industry in Massachusetts . . . 1855*, 1856, p. 612. The last two are the state industrial censuses. We compute 1800 and 1860 ratios of employment to documented tonnage (as reported in *Historical Statistics*, 1960, p. 445), interpolate the ratios, and then apply these to the tonnage series.

[129] Tonnage data from *Historical Statistics*, 1960, p. 445. We may compare the implicit ratio of 134 men per 1,000 tons in Adam Seybert, *Statistical Annals*, Phila., 1818, pp. 340, 341, with Jeremy Belknap's estimate (in his *The History of New Hampshire*, Dover, N.H., 1812, Vol. III, p. 158) of six to seven men and one to two boys per twenty- to forty-ton ship, and Samuel Davis's ("Notes on Plymouth, Mass." in *Collections of the Massachusetts Historical Society*, Second Series, Vol. III, 1815, pp. 167, 168) report for Plymouth in 1770 of seven to eight men in ships averaging thirty to forty-five tons.

[130] 1860 Census, *Statistics of the United States.* . . . For 1800 we take an arbitrary minimum of 1,000 in the light of Seybert's estimates (*Statistical Annals*, pp. 338, 341) that only 1,332 were employed in 1818 in New Bedford and Nantucket, and that the total tonnage employed for the U.S. in 1800 was only 651.

Data for whaling reported by the Massachusetts Censuses of 1837 and 1845 lead to similar results, variations between the entering, clearing data and other sources reflecting deaths, desertions, etc.[131]

The absolute level for whalers can be set at 87 men per 1,000 tons on the basis of various reports. For Nantucket in 1807 the ratio was approximately 86.[132] For 1832, Thomas Greene reported that the whalers out of New Bedford aggregated 39,623 tons, and had 3,105 men, giving a ratio of 86; *Niles' Register* indicates similar results.[133] A report by the Committee on Public Lands for 1846 and data on ships cleared for the whale fisheries for 1860 both lead to a figure of 88.[134] It is suggestive that no increase in labor productivity appears in this portion of the industry. At the same time, the increasing theme of all reports on the industry is the development of crimping, the increased hiring of foreigners, and the search for ever lower-cost labor.[135] It is in particularly sharp contrast with the decline that appears to have cut manpower requirements per ton in half from 1794 to 1832.[136] (There is some indication that employment aboard ship actually increased somewhat over the long term.[137])

---

[131] John P. Bigelow, *Statistical Tables Exhibiting the Conditions and Products of Certain Branches of Industry in Massachusetts*, 1838, p. 182; John G. Palfrey, *Statistics of . . . Industry in Massachusetts.*

[132] James Freeman, "Notes on Nantucket. August 1st, 1807" in *Collections of the Massachusetts Historical Society*, Second Series, Vol. III, p. 29 reports 46 ships with 10,525 tons belonging to Nantucket on July 27th, 1807, that forty of the forty-six ships were engaged in whaling, the larger ships with twenty-one men, the smaller with sixteen. We assume an average of 18.5 men per ship.

[133] U.S. Congress, Serial Set 222, *Documents on the Manufactures of the United States*, pp. 182–183. *Niles' Register*, Vol. XLI, Nov. 19, 1831, p. 218, gives data indicating 76 for the right whale fisheries and 81 for the sperm fisheries.

[134] 29th Congress, First Sess., Doc. 46, *Report of Committee on Public Lands*, Serial Set 478, p. 44; *Report of the Secretary of the Treasury . . . Commerce and Navigation, 1860*, p. 527. The preceding figures, based on clearances, are comparable indications of full-time equivalent employment requirements. A total employment figure of 12,301 for whaling (as of June 1, 1860) appears in the 1860 Census, *Statistics of the United States*, p. 550. No tonnage data, unfortunately, are given. But comparison of a January 1864 tonnage figure shown in the same source (p. 549) with that in *Historical Statistics*, 1960, p. 445 suggests that the latter's figure of 167,000 tons for 1860 might be reasonably comparable. If so, a ratio of 72 is indicated on a total basis—to be contrasted with a figure of 88 when allowing for the fact that the average vessel made more than one trip. Since whaling voyages averaged 2½ years, however, this reconciliation is hardly adequate.

[135] Cf. Samuel Eliot Morison, *The Maritime History of Massachusetts, 1783-1860*, Boston and New York, 1921, *passim*.

[136] U.S. Congress, *American State Papers, Commerce and Navigation*, I, p. 511, estimates 4,139 tons and 600–700 men for 1794 in a report made by a House Committee in February 1803.

[137] *Annals of Congress*, December 1822, p. 402. Representative Floyd of Virginia gives ship clearances from Nantucket as 18,765 tons and 1,315 seaman—or 70 men per 1,000 tons.

Coasting Trade

The trend of employment in coastal shipping cannot be derived directly from the port data, for no such classification is shown. However, as an indication we may take clearances for Canada and for Cuba. Voyages to the former were mostly short lake runs across from New York; the Cuban voyages originated in various ports, tending to be somewhat longer in duration:[138]

	Men Per 1,000 Tons, Clearances for	
	Canada	Cuba
1835	66	50
1839	57	50
1850	43	43
1860	(41)	37

Coastal voyages would have been of somewhat shorter duration than those to Cuba but hardly as short as the brief trips from New York to Canada that dominate the clearances for Canada. The Cuban rates were therefore adopted for coasting voyages, with the 50 rate carried unchanged to earlier decades on the analogy of the very stable data for foreign trade. (See below.) Confirmation of the rate is suggested by one of 48 for coasting voyages out of New York City in 1831.[139]

Foreign Trade

As an indication of the trends for foreign trade we have U.S. clearances for foreign countries, which include a heavy volume of shipping to Cuba and Canada (noted above) as well as to other foreign countries:[140]

Men Per 1,000 Tons
(clearances for foreign countries
from all U.S. ports)

1830	45
1835	52
1839	48
1850	41
1860	29
1870	29

It must be emphasized that this trend necessarily reflects the varying proportions of shipping headed for the separate ports. The variation depending on destination was marked. Thus for 1839 the 57 for Canada

---

[138] *Report on Commerce and Navigation:* 1835, pp. 277–278; 1839, pp. 283–284; 1850, pp. 319–320; 1860, p. 557. From *ibid.*, 1960, pp. 526 and 594, one can derive a ratio of 26 for Canada as a whole, and of 41 for Canada excluding an enormous volume of shipments out of Buffalo Creek (which accounted for nearly 17,000 crew members). For extrapolating man-tonnage ratios, the exclusion seemed desirable.

[139] Based on a report of the *New-York American Advocate*, summarized in the *New England Magazine*, 1831, Vol. I, p. 529.

contrasts with 37 for trips to England, while for virtually the same year we have a report by the French attaché in San Franciso indicating a 64 rate for American vessels entering that port.[141]

Given the above figures for coasting and foreign trade we adopt a two-step procedure for estimating employment in these trades. The clearance data, of course, cannot be used directly since a given ship and its crew could make many clearances during the year. We therefore use the documented tonnage data, on the assumption that, by applying the ratios derived above to the tonnage in foreign trade and in coasting, we can compute a series for the trend in ocean navigation employment.[142] We deduct tonnage on western rivers from the total for "coasting plus internal." The ratios for Cuba were used for the years shown (50 for the earlier years), while a similar procedure for foreign trade utilized 45 for 1800–20.

The resultant series yields an 1830 figure within a few percentage points of one estimated for 1834 by Secretary of the Treasury Woodbury, and an 1840 estimate close to that estimated for 1839 by the Chairman of the House Committee on Naval Affairs (see following section). As a result, we adopt the series as the final one for ocean navigation. The results are as follows:

Employment in Ocean Navigation and Fishing
(in thousands)

	Total	Ocean Trade	Fishing and Whaling
1800	50	45	5
1810	69	64	5
1820	68	56	12
1830	87	72	15
1840	114	90	24
1850	153	124	29
1860	191	161	30

---

[140] *Report on Commerce and Navigation:* 1835, p. 256; 1839, p. 262; 1850, p. 286; 1860, p. 530. 21st Congress, Second Sess., Senate Doc. 76, *Report of the Secretary of the Treasury . . . Commerce and Navigation*, p. 286, shows tons entered, tons departed, and number of seamen employed. I have averaged tonnage entered and departed in computing tons per man. See *Report on Commerce and Navigation:* 1835, p. 278; 1839, p. 268; 1850, p. 284; 1860, p. 527. For 1870, we used U.S. Bureau of Statistics, *Immigration and Navigation Statements, being Parts II and III of the Annual Report . . . on Commerce and Navigation*, 1870, p. 59.

[141] Duflot de Mafros, *Travels on the Pacific Coast*, 1844 (reprint 1937, p. 266).

[142] Tonnage data from *Historical Statistics*, 1960, pp. 445–446. For 1830 the documented tonnage figures appear too low. An enormous 1828-30 decline is reported, whereas clearances and entrances continued to grow over those years. (Presumably this decline reflects the 1829-30 clearance of ghost tonnage from the totals.) See *ibid.*, p. 439. We therefore use the 1820-30 change shown by the average of entrances and clearances of American vessels to extrapolate the 1820 tonnage rate. This 37 per cent rise is virtually identical with the 1820-28 rise in documented tonnage.

Comparisons with Census of Population

A continuing source of data, beginning with 1840, is the decennial Census of Population.

That Census was probably the most inadequate in our history and the navigation figures are peculiarly affected. There are several indications that the employment figure of 89,000 was inadequate. First, the industrial portion of the Census finds fewer people in such industries as fishing, construction, lumbering, etc., than does the population enumeration. Nevertheless, the industrial portion of this Census reports only 36,000 persons employed in fisheries and whaling—compared with a total of 56,000 in navigation of the ocean.[143] This, in turn, implies that well under 20,000 persons were engaged in coastal and foreign trade. However, the 8,312 U.S. ships clearing from the country in the year ending September 30, 1839, had crews of nearly 71,000.[144] It is conceivable, but hardly likely, that the explanation lies in the difference being a measure of the employment of foreign sailors by American ships. However, hospital money deductions are too large for this to be likely.

Another indication of the limitations of the 1840 Census is to be found in comparison of the detailed local data. Barnstable reported more people gainfully occupied in ocean voyages than New York State, while Boston reported twice as many. This is wholly improbable in the light of clearance data, hospital money receipts, and later Censuses. Thus, the 1839 Commerce and Navigation Report shows documented tonnage in Barnstable at a tenth of that for New York. It is characteristic of inadequate Census practice for there to be overallocation to some categories in the midst of general underenumeration. The 1840 Census offers a classic example.

For 1850, the Census was restricted to free males aged fifteen and over. The omission of slaves in this industry would not have affected the totals much. On the other hand, the 1850 Census reports a decrease in navigation employment for the leading states of Massachusetts and Virginia,[145] although their tonnage had increased significantly from 1840 to 1850. Conceivably the institution of the catch-all category for "laborer not agricultural," a doubtful improvement instituted by De Bow, tended to reduce the navigation as well as other industrial categories. In any event the 1840–50 trend of a decline from an incomplete 1840 Census makes the 1850 report dubious.

---

[143] 1840 Census, *Compendium* pp. 103, 361 (1841, T. Allen edition). That the industrial portion was too low can be seen by examining the data on Boston fishing shipments.
[144] 1839, *Report on Commerce and Navigation*, p. 288.
[145] As can be seen by comparing the 1840 Census with *The Seventh Census of the United States: 1850*, p. lxxx.

LABOR FORCE AND EMPLOYMENT, 1800–1960

The 1860 Census, as the last prewar Census, should be invaluable as a benchmark for 1800–60 trends in navigation employment. It appears unsatisfactory, however, for two main reasons. The 1860 Census actually reports less employment in the major navigation occupations than does the 1850 Census: 96,000 for mariners, fishermen, sailors, and boatmen as against 116,000. Yet the documented tonnage in 1860 was fully 55 per cent greater than in 1850.[146] Tonnage of American vessels cleared was 135 per cent greater, with ships making more passages during the course of the year.[147]

It is quite unlikely, under the circumstances, that employment in 1860 should have been below that in 1850, and in addition lower than that for 1870, when the decline in the merchant marine that began during the war years was well under way. The explanation of the Census count may simply have been that so many voyages were under way in midsummer 1860 that the sailors were not at home to be counted by the enumerators—and, mostly being unmarried, were not reported to the Census enumerators by wives or relatives.

What about the third category of information on employment in navigation—namely, contemporary reports, surmises, and wishful hypotheses? We can begin with the year 1810.

In protesting an increase on import duties in 1824, a committee of Boston merchants—their number including not merely William Gray and Nathaniel Thorndike, but Daniel Webster, Nathan Appleton, and Abbott Lawrence[148]—was clearly concerned with emphasizing the importance of the export trade. We may, therefore, take as a maximum or reasonable estimate their assertion that "one seaman is required on an average, for every twenty tons" of shipping—yielding their figure of more than 71,000 employed in 1810.

In 1815, Josiah Quincy estimated that the nation possessed 120,000 seamen each worth $500 a year to the country.[149] Study of his language suggests the likelihood that he was speaking of all those competent as fishermen, and of the fishing interest he was then engaged in magnifying and lauding. One must assume that he included the Haverhill cobbler or the Barnstable artisan who fished part of the year. This is the only explanation of a figure far above that for 1850, when our documented tonnage

---

[146] *Historical Statistics*, 1949, p. 208. The data for 1849 and 1859 were used. The tonnage data was as of December 31, while the Population Census was taken as of June 1, 1860.

[147] *Report on Commerce and Navigation: 1850*, p. 284; 1860, p. 532.

[148] *American State Papers, Finance*, IV, p. 470.

[149] *Speech of Hon. Josiah Quincy in the House, January 25, 1812*, Alexandria, 1812, p. 10. Quincy quotes Congressman Reed as to the valuation of our tonnage, adds his own surmise as to seamen.

was enormously greater. (Of course in 1815 a further factor was at work: the overloading of vessels during wartime in order to provide a crew to handle prizes.[150] But numerically this can hardly have been a major factor.

For 1816, however, we have the much more reasonable estimate of 70,000 men, made by the Secretary of the Treasury. As the officer responsible for the registration of all ships, the publication of our figures on foreign trade, and the collection of hospital money from mariners, he must be considered a preferable source.[151]

For the early 1830's we have a number of estimates. As of 1829, the Secretary of the Treasury reported for each state on the number of vessels and "the seamen usually employed in navigating the same which belonged to each state or territory." His figure for December 31, 1829, was 61,672.[152] As his entries of 22 men for Ohio and 262 for South Carolina indicate, however, it would appear that employment on rivers and canals (and seagoing employment in vessels not large enough to be licensed by the Treasury) was probably omitted. For December 1834, Secretary Woodbury, of the Treasury, made another report, finding "our whole number of seamen of every kind, exclusive of about five thousand in the navy is computed to be seventy-five thousand."[153] It will be noted that unlike the earlier report, whose estimate is tied to the listed vessels, this purports to include all seamen "of every kind"—the difference presumably explaining the difference of 13,000, rather than any marked gain in employment.

For January 1833, we also have an estimate by Chairman Reed of 70,000.[154] A week later, however, another Congressman, Pearce of Rhode Island, found 80,000 seamen working at $10 a month—the same figures as appear in a memorial from his constituents, published elsewhere.[155]

---

[150] In 1813 the Grand Turk had a complement of 120 men in a boat half the size of one launched in 1916—for a crew of 7. Cf. Robert R. Peabody, *The Log of the Grand Turk*, 1926, p. 223.

[151] Adam Seybert, *Statistical Annals*, p. 315. Presumably it is this same estimate that is quoted in Franklin D. Scott, editor, *Baron Klinkowstrom's America, 1818-1820*, 1952, p. 169. The Baron (a lieutenant colonel of the Swedish fleet) toured the U.S. in 1818-20, eyes open and pen in hand.

[152] 21st Congress, Second Sess., Sen. Doc. 76, *Report of the Secretary of the Treasury, Commerce and Navigation of the U.S.*, pp. 284-285.

[153] 23rd Congress, Second Sess. Sen. Doc. 7, Serial Set 266, *Report from the Secretary of the Treasury on . . . Marine Hospitals in the United States*, December 8, 1834, p. 2.

[154] Gales and Seaton, *Congressional Globe*, January 22, 1833, p. 1189. Reed estimates more than 300 ships, over 100,000 tons, with an annual fitting out cost of $1.9 million exclusive of labor, a value of ships and outfits of $7 million.

[155] *Ibid.*, January 30, 1833, p. 1519. Pearce estimates 2 million tons of registered and enrolled shipping, slightly above the official figure, worth $30 per ton. The memorial appears in 22nd Congress, Second Sess., 1832-33, Doc. 101, Serial Set 234, p. 4.

For 1839 Chairman Reed estimated the total then at 109,000 men.[156] However, according to the same source, the *New York Herald* of January 1845 estimated 7,000 seamen in the navy, 17,000 on whalers, and 44,000 merchant seamen—or 68,000 exclusive of fishermen.[157]

In the same year, Captain Marryat adopted Henry Carey's figure of $4\frac{1}{3}$ seamen per 100 tons, and applied this to the documented tonnage estimate of 86,000—or well below Chairman Reed's estimate.[158]

As a final sample of *ad hoc* estimation we may refer to the estimate of a Navy surgeon, who found that "the number of persons employed as seamen and boatmen, including those engaged in the cod, whale, and other fisheries, is not less than 160,000."[159]

The decisive movement in the preceding figures is given by the components for coasting and foreign trade. Our implicit ratios for these categories may be compared with various contemporary reports. As of 1818, Seybert estimated six men per 100 tons in foreign and coasting trade.[160] In 1821, the Committee on Manufactures quoted and accepted a statement of the Mercantile Society of New York, giving ratios of 6.6 in coasting, 5.0 in the West India trade and 4.8 in the European trade.[161] In 1820 and 1824, committees of Boston merchants found five seamen "required on an average" for every hundred tons, both then and in 1810.[162] More than a decade later, in 1836, the same ratio continued to be used.[163] And the Andrews report, as of 1853, also implicitly adopted this ratio.[164] This over-all ratio of fifty men for every 1,000 tons in the early years compares favorably with our over-all foreign trade ratios of forty-five men in 1830 and forty-eight in 1839. Our 1870 ratio of twenty-nine may be compared with the later figure of thirty-five estimated by David A. Wells in 1890 (for sailing vessels of 200–300 tons).[165]

---

[156] Quoted in *Remarks on the Scarcity of American Seamen and the Remedy, by a Gentleman Connected With the New York Press*, New York, 1845, p. 17. Possibly he was misquoted.

[157] *Ibid.*, p. 24.

[158] Captain Frederick Marryat, *A Diary in America* (New York, 1839), I, p. 183. Marryat estimated the British ratio at 5, from his own knowledge.

[159] William S. W. Ruschenberger, *Remarks on the Condition of the Marine Hospital Fund . . . by a Surgeon U.S. Navy* (Philadelphia, 1848), p. 9.

[160] Adam Seybert, *Statistical Annals*, Philadelphia, 1818, p. 35.

[161] *American State Papers, Finance*, III, Report of the Committee of Manufactures, January 15, 1821, p. 643.

[162] *Report of the Committee of Merchants and Others of Boston on the Tariff*, Boston, 1820, p. 13. *American State Papers, Finance*, IV, p. 470.

[163] Daniel Drake, "Report on Ohio Hospitals," in 24th Congress, First Sess., House Doc. 195, 1836.

[164] 32nd Congress, First Sess., Senate Exec. Doc. 112, *Andrews Report on Trade and Commerce*, 1853, p. 834.

[165] David A. Wells, *Recent Economic Changes*, 1890.

Peering circumspectly into the morass of data noted above, what can we conclude as to the numbers employed in navigation? For 1816 we have a reported figure by the Secretary of the Treasury that is within 2 per cent of the present estimate based on clearance data, as is the 1829 estimate of the then incumbent Secretary. For 1834, Secretary Woodbury's figure falls between that estimated here for 1830 and 1840, as does Chairman Reed's figure. For 1839 Chairman Reed estimated a figure within 5 per cent of the figure given here for 1840. And finally, Navy Surgeon Ruschenberger's figure of 160,000, published in 1848, is within 2 per cent of that estimated here for 1850. We regretfully differ with an unnamed gentleman from the *New York Herald* and with the good Captain Marryat, the latter using much the same procedure as ours, but relying not on clearance data for his ratio but on Henry Carey's figure of 43. We note, finally, that the present figures implicitly assume the equivalent of full time activity in our interpolating series—tending to overstate marine employment. In the other direction, however, we omit employment in nondocumented vessels —whether small fishing vessels, broadhorn arks, or sloops navigated by men whose primary work was farming or cobbling shoes. The assumption that these contrary biases more or less cancelled is hopefully demonstrated by the great similarity between the present estimates, derived from clearance data, and the *ad hoc* estimates noted above.

NAVIGATION EMPLOYMENT: 1870–1960

For 1870–90, employment was estimated as the sum of employment aboard sailing vessels, steam vessels, and unrigged vessels.

For sailing vessels the ratio of men per 1,000 tons in 1870, 1880, and 1890 is computed in connection with the productivity estimates and discussed in the section on that subject. These figures applied to the U.S. documented tonnage under sail give sailing vessel employment.

For steam vessels employment in 1880 and 1890 is reported by the Census of Water Transportation. For 1870 we compute a ratio of 57 men per 1,000 tons, interpolating between the 70 figure available for 1851 and the 51 figure implicit in the 1880 Census data. Applying this to the documented tonnage total gives steam vessel employment in 1870.

For unrigged vessels we extrapolate to 1870–90 the 7,129 total employment on such vessels in 1906 by changes in documented tonnage of canal boats and barges.

The resultant 1870 estimate of 137,000 appears consistent with a contemporary estimate by Joseph Nimmo, Chief of the Division of

LABOR FORCE AND EMPLOYMENT, 1800–1960 173

Tonnage of the Treasury's Register's Office.[166] Nimmo estimated 36,300 men in American vessels engaged in foreign trade—or 36 per cent of the present total, which may be compared with the 35 per cent that foreign trade vessels constituted of total foreign plus coastwise and internal tonnage.[167]

For 1906, 1916, and 1926, we adopt the Census totals, merely adjusting the 1906 Census to include fishing and yacht employment.[168]

For 1900, 1910, 1920, and 1930, we interpolate between the above data by documented tonnage totals, exclusive of canal boats and barges—the exclusion reflecting the trivial amount of employment associated with the fairly substantial tonnage of schooner barges, tugs, etc.

The 1930 total is extrapolated to 1940–60 by the full time employment estimates of the national income accounts.[169]

GAINFUL WORKERS IN MINING: 1800–1900

We extrapolate a 1900 benchmark to earlier decades by a series reflecting the trend of employment in coal, iron, precious metals, lead, stone, and salt mining.

1. For the trend of employment in coal mining we adopt the Census of Mining totals for 1840–70, extrapolating to earlier years by the 1840–60 regression of employment against a weighted series for coal production.[170]

2. For iron mining we take the total for the 1850–70 Census of employment in mines, extrapolating back by the trend of pig iron shipments.[171]

3. For precious metals the 1870 Census of Mines is clearly too low and we use instead the Population Census reports for 1850–70, giving the

---

[166] Nimmo is quoted in 41st Congress, Third Sess., House Executive Document 76, *Letter from the Secretary of the Treasury: Foreign Commerce and the Practical Workings of Maritime Reciprocity*, February 1870, p. 30.

[167] An estimate of 100,000 in coasting and internal navigation for 1870 may be a reasonable figure, but it is not clear from Nimmo's estimate of 36,300 men aboard American vessels, and 45,372 aboard foreign vessels engaged in the U.S. foreign trade, just what the basis or purpose of his estimate was.

[168] 1926 Census, p. 21. The yacht employment total is given on page 5 of the 1906 *Census of Transport*.

[169] U.S. Office of Business Economics, *National Income*, 1954 Edition, Table 26; *U.S. Income and Output*, Table VI-13; and *Survey of Current Business*, July 1962, Table 53.

[170] Employment and coal production, 1840-70: *Historical Statistics*, 1960, pp. 349, 357, 360. We reduce bituminous tonnage by one-third before adding it to anthracite to reflect the difference in tons produced per man for each type as indicated by data in the 1870 Census, *Industry and Wealth*, p. 760.

[171] *Ibid.*, pp. 401, 407, 768. *Historical Statistics*, 1960, p. 366.

total count of miners in California and Colorado.[172] For those years there was no precious metal mining of numerical consequence outside those states, while virtually all mining in the two states was for precious metals. For 1840, gold production figures and the (inadequate) Census of Mines alike suggest a trivial number in mining—under 1,000. For 1830, however, we must reckon with a widely reported, if short-lived, mining boom from 1829 to 1831—when substantial gold deposits were discovered in Georgia and the Carolinas. The major historical consequence is associated with the displacement of the Indians from Georgia, but the discoveries did briefly affect the gainful worker totals. Her Brittanic Majesty's attaché in Washington in 1832 reported that men "are flocking to the mines from all parts and find ready employment," and he estimated 20,000 hands so employed.[173] A contemporary Congressional report found 6,000 to 7,000 engaged at times in mining gold in Georgia, but not more than 100 at other times.[174] We take contemporary figures of $2.50 worth of ore raised per day per hand, divide into the mint receipts for the year to get less than 1,000 full-time employees.[175]

4. For lead mining the 1840–70 Censuses report about 1,000 employees, possibly missing some considered as self-employed contractors.

From contemporary reports we find less than 500 in 1811, about 1,200 in 1819 and about 2,000 persons in all branches—mining, smelting, and transporting—in 1826, though "part of their time is devoted to farming."[176]

5. For stone we sum Census reports on employment in granite, marble, and other stone mining for 1840, 1850, and 1870.[177] For 1860 employment was estimated as 79 per cent of the number employed in granite and marble manufacturing—the same ratio as for 1870.[178] This procedure gives an 1860 total about halfway between 1850 and 1870, or much the same

---

[172] 1850 Census, *Statistics of Progress*, p. 976, for California. Gold was not discovered in Colorado until later years and we assume 0 in 1850. 1860 Census, *Population*, p. 668. 1870 Census, *Population and Social Statistics*, p. 820.

[173] William G. Ouseley, *Remarks on the Statistics and Political Institutions of United States*, Philadelphia, 1832, p. 172. Ouseley estimates 20,000 hands and $100,000 weekly value.

[174] U.S. Congress, Serial Set 210, Report 82, *Assay Offices, Gold Districts, North Carolina and Georgia*, p. 26.

[175] Yield per hand from *ibid.*, pp. 24, 29. Mint data from *1850 Report of the Secretary of the Treasury*, p. 138.

[176] For 1811, H. M. Brackenridge, *Views of Louisiana*, Baltimore, 1817, pp. 125, 267. For 1819, Henry Schoolcraft, *A View of the Lead Mines of Missouri*, New York, 1819, p. 127; and *American State Papers, Public Lands*, Vol. III, p. 663. For 1826, *ibid.*, Vol. IV, p. 558. A 1,200 estimate appears in *Steeles Western Guide and Emigrants Directory*, Buffalo, 1836, p. 96, making 3 million pounds annually.

[177] 1840 Census, *Compendium*, p. 359. 1870 Census, *Statistics of Industry and Wealth*, Vol. III, pp. 468, 769.

[178] 1860 Census, *Manufactures*, p. 738. 1870 Census, *Statistics of Industry and Wealth*, pp. 396, 769.

result as if the Population Census count of stone and marble cutters plus quarrymen had been used for interpolation.[179]

6. Rough estimates for the minor category of salt mining were made.[180]

We sum the above six series and use the total to extrapolate our 1870 benchmark for mining employment. We take our 1900 benchmark for employees in mining (derived in connection with the employment estimates for 1900 ff.) and extrapolate to 1870 by the Decennial Census of Population totals for miners and quarrymen.[181] We then extrapolate the 1870 figure by the sum of the six series obtained above. For 1800 we arbitrarily assume 10,000 as compared to the 11,000 in 1810 derived by the above procedure.

CONSTRUCTION: 1840–1900 (DECENNIAL)

For 1850–1900, the logical starting point for extrapolating the 1900 construction employment total is the set of Population Census reports on the number of gainful workers in the major construction trades for the decennial years.[182] The important group of construction laborers must be omitted because it was never separately reported by the Census, being mixed in with totals for laborers in manufacturing, service, transportation, etc.[183]

How reasonable is the trend shown by these figures? Since it is possible to rationalize virtually any trend, we must seek an external check on the data. One check we use is the value of construction materials (constant dollars) per construction worker; by using Shaw's estimates we find a reasonable trend in the 1870–1910 data.[184] A second check is indicated

---

[179] The *Seventh Census*, pp. lxxv, lxxvii. 1860 Census, *Population*, pp. 673, 677. 1870 Census, *Manufactures*, p. 821.

[180] Data for 1840 from the 1840 *Compendium*, p. 359; 1850 *Compendium*, p. 183; 1870 Census, III, p. 622. Running between 2,000 and 3,000 for these years, the total was taken as 2,000 for earlier dates.

[181] 1900 Census, *Occupations*, pp. xlii, xliii, lviii, lix. The Census of Mines and Quarries gives broadly similar results, but its incomplete coverage of precious metal mining and small-scale operations in the earlier years makes the Population source a preferable one.

[182] We use the data for carpenters, masons, plasterers, painters, electricians, plumbers, paper hangers, roofers. For 1870 ff. we use the summary in Edwards, p. 105. For 1850 and 1860, that in the 1900 Census, *Occupations*, p. vii.

[183] Edwards attempts a figure for laborers in the "building, general and not specified category," but even for this broad group his estimates are labeled as "largely estimates" (cf. *Comparative Occupation Statistics*, p. 105, n. 44 and p. 144). "Building contractors" are likewise intermingled with "manufacturers" in the reports. Carson's esimates for laborers in construction have no external basis of support and he apparently does not use them in creating his construction estimate.

[184] William H. Shaw, *Value of Commodity Output since 1869*, 1947, p. 76. These data are essentially summaries of Census of Manufactures reports, with allocation of mixed commodities and deflation by Shaw's price index. The materials per mechanic beginning with 1869 are estimated (in thousand dollars) as follows: 6.0, 8.1, 12.7, 11.4, 12.3.

because of the difficulty of properly deflating the value of product data. For this check we relate employment of carpenters to employment in the factories producing carpentry materials, of masons to that of brick factories, etc.[185] The trend in each set of ratios, while less reliable because such an approach implies parallel productivity trends in both industries, is stable enough to confirm some major shifts of absolute level in the separate gainful worker series that are questionable a priori. Construction employment was therefore estimated by extrapolating 1900 employment by the trend in the specified trades, taken as a group.

For 1840, the relationship between construction employment for 1850–1910 and the number of white families was computed, and the figures proved to have a narrow range.[186] We use the 1850 ratio for estimating 1840 from the figure for white families. The resultant estimate for 1840 is much the same as using an average of 1850–1910 ratios. It is likewise within a few percentage points of an estimate arrived at simply by fitting a least squares trend to the absolute number of construction employees 1850–1910, and extrapolating to 1840.[187]

Since we are dealing with the 1840 figure, irregular trends arising from the construction of railroads are no problem. For roads and turnpikes we assume, in the absence of knowledge, a closeness of trend so that the growth of this volume of social capital is assumed to parallel the growth in white families. The canal frenzy of the 1830's was ended by 1840, nor had the volume involved been great. At the peak of construction in Ohio, Pennsylvania, and Maryland in the early 1830's, less than 30,000 persons were at work—and allowing for the lack of work during the winter months, the full time equivalent would have been substantially less.

Finally we may compare our estimate of 210,000 employed in construction in 1840 with the 85,501 employed in house construction in 1840 according to the Census of that year.[188] The implication that 41 per cent of

[185] Manufacturing employment data are from the 1900 Census of Manufactures, I, pp. 11 and *passim*; 1870 Census, pp. 396, 402, 403, 407.

[186] White families were estimated in connection with the present series for domestic servants. The figures rest ultimately on Population Census count of families, as adjusted. The number of white families per person in the trades used above was as follows for Census years 1850–1910: 11.7, 13.3, 11.4, 12.9, 9.9, 11.2, 10.6.

[187] Since an increasing volume of construction reflected streets and industrial buildings in the cities, the ratio of construction mechanics to urban population was likewise computed, but the results have a wider range over the decades, hence are inferior for extrapolation. Allowances for variations in canal, road, and railroad construction are obviously not worth making. If we had sound data for such activity and building construction, a direct estimate would have been made in the first place. The limited coverage of the building permit series makes their use equally pointless.

[188] *Compendium of the Enumeration of the Inhabitants and Statistics of the United States*, Thomas Allen, ed., 1841, p. 364.

all construction employees were engaged in house construction may be compared with the 37 per cent ratio of nonfarm residential to gross construction in the 1869–78 period, the earliest shown by the Kuznets estimates.[189] The encouraging similarity simply tells us that the basic Kuznets data for this later period do not suggest any obvious inadequacy in our estimate.

The pathbreaking work by Daniel Carson for 1870 and after constitutes the major previous estimate for this industry, his figures having been used with almost no change by Fabricant, and his trend having been used with little change by John Kendrick.[190] Carson uses much the same group of mechanics' occupations for interpolations and secures similar results for trend.[191] (His absolute figures differ markedly, because he relies on Population Census levels, which are drastically different from those in the establishment-benchmarked series of the BLS beginning in 1929, as well as those in the National Income Division estimates for 1929 and after.)[192] Since the present estimates were tested against Shaw's materials output, they will also be reasonably consistent with Kuznets' output figures for the period, the movement of the latter series being essentially that of materials output.[193]

MANUFACTURING EMPLOYMENT

Wage Earners

Estimates of manufacturing employment presented below are intended to reflect trends in factory employment. The growth of the factory system as an employer of manpower can be seen in its most useful single dimension by such a measurement. We do not, therefore, rely on the Population Census materials, which comprehend, at all times, employment in small

---

[189] Kuznets, *Capital in the American Economy*, Part C. We use gross construction in 1929 prices from his Table 15 and nonfarm residential data for 1871–78, averaging the latter 5-year moving averages.

[190] Carson, p. 47, and Fabricant, p. 42, in *Studies in Income and Wealth, 11*; John W. Kendrick, *Productivity Trends in the United States*, Appendix A, Table A-VII and Appendix E. Kendrick relied on an earlier version of the present estimates for deriving his intercensal figures for 1900 ff. and for a revised 1920 benchmark, but uses Carson for 1870–1910 decennial trends.

[191] Carson, in *Studies in Income and Wealth, 11*, pp. 115, 117. He also includes "builders and contractors," but since Census reports confused that group with "manufacturers" there is no basis for making a reliable estimate of their number, nor for using it to give added illumination here.

[192] His 1930 figure of 3.0 million compares with an NID figure of 2.2. Hence, all his figures are much above the present ones.

[193] Kuznets' data in his *Capital in the American Economy* move largely as do the construction figures in his *National Product Since 1869*, New York, NBER, 1946. These reflect a constant percentage margin added on to Shaw's output data, plus minor inventory adjustment.

handicraft shops. While there is certainly distinct value in a really comprehensive measurement, it is difficult to separate out of the Population Census materials a moderately clean-cut category, so that we can say—this is the definition of manufacturing employment, or factory employment, or productive activity that is comprehended under the category for which we present figures.

Even in the measurement of factory employment that derives from a factory Census we have serious limitations. In the early decades the Gallatin report, or the early Censuses, intermingled factory employment with employment in homes under the putting out system and under contract. This, of course, was the way the factory system did develop. But it is difficult to see how one could usefully add up employment of persons working twelve hours a day in a factory with that of women binding shoes at home for a few hours each day. In any event, the course we attempt here is simply to measure employment in factories exclusive of hand trade activities, such as saddlers, photographers, mechanical dentists, blacksmiths, and others included at various times. The resultant series will show a sharper transition from handicraft to factory activity than would be measured if we had some means of measuring hours of input and units of output under home and factory conditions.

1840. The Census sought to measure comprehensively all factory production, as well as output in fishing and other industries that would now be considered under other headings. We estimate employment under two headings: that reported by the Census and that not reported.

Employment is reported by the Census for about 80 per cent of total factory production. We sum employment as reported for cotton, wool, iron, tanneries, furniture, mills, breweries, and many lesser industries.[194] For some 221,000 employees thus covered (excluding the 72,000 in cotton textiles), the average value of product per employee is approximately $1,000. Dividing this $1,000 average into the value of product for which no employment data are given, we derive employment in this category.[195] The combination of both categories gives an employment total of 471,000. We round to 500,000 as a means of allowing for remaining omissions—primarily of small proprietors—and to emphasize by the even half million the arbitrariness of any figure for this industry at this period.

1810. We estimate employment in this year as the total of the six major components. (1) Cotton textiles and (2) iron manufacturing are both

---

[194] 1840 Census, *Compendium*, pp. 358 ff. By including all iron employment we undoubtedly include some iron mining and woodcutting employment. We exclude, however, employment in house construction, though shown under manufacturing.

[195] The value of product data are taken from the *Report of the Secretary of the Treasury on Finances for the Year Ending 1855*, p. 93.

estimated as described in the sections below. (3) For the largest single employment category—mills—we make separate estimates for grain mills and other. For grain mills we compute the number of mills and bushels of grain ground in both 1810 and 1840, the result suggesting approximately 5,000 bushels ground per mill per year at both dates.[196] From the estimates of the consumption of wheat at both dates, as made by Towne and Rasmussen,[197] we then derive the number of mills in 1810 more accurately than the incomplete 1810 Census would indicate. We then extrapolate the 1840 Census count of mill employment by the percentage change in mills between the two Censuses. We then apply the same percentage change to the 1840 employment in saw and oil mills, implicitly making the same assumption for these products as Towne and Rasmussen did for wheat, i.e., that population change accounts for change in final takings. (4) For breweries the 1810 product total is somewhat above that for the 1840 Census, but the price index of spirits was also greater, so that we take the trivial change in production to reflect a constant employment total.[198] (5) For woolens even Gallatin could state no more enthusiastic an evaluation of woolen manufacturing in 1810 than that "there are yet but few establishments for the manufacture of woolen cloths."[199] Taking the partial data listed, and allowing for various incompletenesses in reporting, one can estimate 400 factory employees, approximately 1,000, when home workers are included.[200] (6) The product in the balance of manufacturing in 1810 is estimated at $62 million.[201] For this group we assume the same value of product per person engaged as that for mills, the two being similar in 1840.[202] The summation of the above groups comes to 61,000, which we round to 65,000 to include

[196] For 1810 data appear in *American State Papers, Finance*, II, p. 761. The fullest reports are those for the counties of Pennsylvania and we take those where both number of mills and bushels and barrels are reported. (We convert one barrel as equivalent to 3.4 bushels.) For 1840 we use Pennsylvania figures from the Census *Compendium*, pp. 140, 152.

[197] *Trends in the American Economy in the Nineteenth Century*, p. 294. Their data reflect primarily the change in the population.

[198] Output data from 1900 Census of Manufactures, Part 1, pp. li, lii. Price data from *Historical Statistics*, 1960, p. 115.

[199] *American State Papers, Finance*, II, p. 427.

[200] From the data Gallatin gives on page 434 one can compute yards made per worker and per establishment. Expanding the more comprehensive reports from the final Digest (p. 691) one can estimate total production of 240,000 yards. A later, but authoritative, reading of this source finds "nearly 200,000 yards produced." 50th Congress, First Session, H.D. 550, *Wool*, 1888.

[201] 1900 Census, *Manufactures*, I, p. li. The total for manufactures in this source is between the preliminary and final figures given in *American State Papers*, pp. 712, 713.

[202] The relevant 1840 groups are leather and tanning, which average $1,270 in 1840 compared with $1,250 in mills, and smaller amounts for building materials and furniture.

proprietors in minor omitted categories (prices for which changed little according to Warren and Pearson); building materials (prices for which were likewise much the same); and furniture (for which it was also assumed that prices were stable). We therefore adapt the 1840 product per employee figures and apply to 1810 product totals. Summing these categories gives us 75,000 employment.

1850–60. The Census of Manufactures provides totals that require an infinity of adjustments in principle, but little adjustment in practice. As the factory system developed, the shading between hand trades, service, and manufacturing became progressively more difficult to discern. The Census included an unknown number of shoe repairers together with shoe factory operatives, for example. Moreover, the definition of manufacturing was broad enough to encompass iron mining, fisheries, plumbing, and mechanical dentistry. Three adjustments can be singled out for notice.

1. Inclusion of nonmanufacturing categories such as iron mining, fishing, etc., can be handled by deducting these from Census totals.[203]

2. For major hand-trade categories, such as boot and shoe, bakery, blacksmith, wheelwrights, tobacconist, examination of the ratio of male to female operatives in these and later years, and comparison with the Population Census totals (for 1850–70) for related occupation titles suggests that the Census covered few except bona fide factory operatives. A considerable number of males were reported for these industries in the Population Census, but not in the Census of Manufactures. Yet the ratio of shoemakers to population must, after all, have an upper limit.

3. It is well known that enumeration in recent years of industries with characteristically small establishments—e.g., sawmills—is a peculiarly difficult business. Even with vast improvements in roads, means of enumeration, lists, and skills, we still fail to enumerate such categories at all well. How much less comprehensively the Censuses of a century ago must have covered such categories as small shoe repair shops. This consideration gains particular force from the practice of excluding all establishments with under $500 in manufacturing activity—an exclusion that would have kept out a far greater proportion of these essentially service activities than one assumes at first blush. To the extent that there was coverage of such establishments, it helps to compensate for the undoubted failure of the *ad hoc* Census organizations of the time to do an adequate job of covering what they purported to cover.

[203] The following categories are subtracted in one or more of the years under review: carpentering, coal mining, fishing, gold mining, stone quarrying, copper mining, iron ore, millinery, dentistry, masonry, painting, photography, plastering, plumbing. Totals, and all adjustment items, from 1870 Census, *The Statistics of the Wealth and Industry of the United States*, III , Table VIII B.

We therefore accept the Census reports, excluding only categories clearly to be omitted from the purview of manufacturing, and assume that the small amount of overcoverage of cobblers and hand trades helped offset the failure to get a comprehensive coverage of those properly included.

1870–1900. A recent very careful study by Richard Easterlin provides the basis of the wage-earner estimates.[204] We adjust his results only for 1880.[205]

The estimates made above for 1850–1900 were all adjusted in the same ratio as the 1900–29 figures were (in the manner described in the section on annual estimates of employment, 1900 and following). Essentially this adjustment is one for undercoverage in the Census of Manufactures enumerations. It reflects the percentage difference between total employment as enumerated by the Census of Manufactures and that reported by the Social Security system in the late 1930's and 1940's before the Census utilized the Social Security lists of employers in checking the completeness of its coverage. We have little basis for varying the ratio to earlier years. On the one hand, Census enumerators were less well trained than nowadays. On the other hand, factories were then much more noteworthy phenomena, particularly since the Census of Population and Manufactures enumerations were then combined.

Persons Engaged

The wage-earner totals for 1810 and 1840 having already been rounded upward no separate addition is made for proprietors or salaried workers in 1810–40.

For 1850–90, persons engaged were estimated by adding to the wage-earner totals 150 per cent of the count of establishments for the same industries. This ratio is that prevailing in 1890. Since it is somewhat below that for 1900–10, it is more appropriate for use in earlier years,

---

[204] In Lee *et al.*, *Population Redistribution*, p. 636. Data on page 623 of this study will differ from ours, being derived from the Population Census. The totals, presumably inclusive of self-employed and salaried workers, are nonetheless below the wage-earner count of the Manufactures Census as adjusted by Easterlin.

[205] For 1880 we increase the total to allow for employment in the periodical press, omitted by the Census in that year. The reported newspaper employee group as a per cent of all printing employees ran to about 44 per cent in 1870, and to 65 per cent in both 1890 and 1900. (Computed from data in 1870 Census, III, p. 397; 1890 Census, VI, Part 1, p. 105; 1900 Census, VII, Part 1, p. 368.) We assume 50 per cent in 1880, so that the omitted category was assumed equal to the reported printing group of 58,506 (1900 Census, VII, Part 1, p. 13). The 1870 Census explicitly covered the press, reporting 13,130 employees. We make no adjustment for the still smaller group prior to that date. Adjustments for 1850-80 are made by Robert E. Gallman in *Trends in the American Economy*, 1960, p. 58.

when most other persons were proprietors, and the salaried group was of limited size.

For 1900, the ratio of proprietors and firm members to establishments was extrapolated from 1909 and 1904, while the count of salaried personnel was available from the Census.[206]

The 1850–1900 data thus derived were increased in the same percentage for undercoverage as the wage-earner data for these decades—on the assumption that the percentage undercoverage of establishments in the days of smaller establishments was similar to that for wage earners.

COTTON TEXTILE FACTORY EMPLOYMENT

We estimate cotton textile employment for 1800 at 1,000, for the convenience of having a round number—although the total in the industry was almost certainly less than this sum. The estimate was arrived at by using data for later years, as follows.[207]

	Operatives Per 1,000 Spindles	Bale of Cotton	Spindles Per Bale of Cotton
1805	—	—	5
1815	100	1	4.5
1831	48	3	4.8
1860	23	12	—

We have an estimate by John Whipple (owner of the Hope Textile Mill in Providence) of 5,000 spindles in the industry in 1806.[208] His total is much the same as the 4,000 estimated for 1807 by Zechariah Allen (and quoted by Samuel Batchelder, one of the earliest manufacturers in the industry), and the 4,500 for 1805 as estimated by Secretary of the Treasury Woodbury.[209]

For 1805 we then compute spindles per bale of cotton using Whipple's 5,000 spindle figure and the 1,000-bale cotton consumption estimate made by the Committee on Commerce and Manufactures.[210] This gives a ratio much the same as that for 1831. Applying the ratio to the 500 bales consumed in 1800 gives 2,500 spindles. This figure is confirmed by Clark's itemization of the eight operating mills in 1800, working—he

---

[206] Proprietors equaled 103 per cent of establishments in 1909, 105.5 per cent in 1904, and were taken as 105 per cent in 1899.

[207] Computed from sources noted below in connection with estimates for each year.

[208] Gales and Seaton, *Register of Congressional Debates*, January 30, 1833, col. 1511.

[209] Samuel Batchelder, *Introduction and Early Progress of the Cotton Manufacture in the United States*, 1863, p. 53. Woodbury Report, 24th Congress, 1st Session, House Doc. No. 146, *Cotton*, p. 51.

[210] Quoted in *Addresses of the Philadelphia Society for the Preservation of National Industry*, Philadelphia, 1819, p. 165. Also, *Niles' Register*, Vol. 9, p. 448.

guesses—"less than 2,000 spindles."[211] Assuming that the 1805–15 constancy in the spindle-bale ratio suggests that technological change during this preliminary period of the industry was limited, we may use the 1815 ratio of operatives to spindles to estimate 500 operatives in 1800. For convenience, and to emphasize the limitations of our knowledge, the figure is rounded to an even 1,000.

For 1810, Gallatin's summary of the 1810 Census gives a total for mills of 80,000 spindles and 4,000 employees.[212] We double that figure to include home weavers, giving a full-time equivalent figure of 100 employees in cotton textile production per 1,000 spindles for 1810 and earlier years—half in the factories and half weaving at home. This addition is computed as follows: (1) Coxe's *Digest* has an implicit ratio of 50 factory workers per 1,000 spindles, plus 80 additional women engaged in home weaving at an average of half a day each.[213] (2) Data for one Providence factory in Gallatin's report imply 53 factory employees and 178 total employees, including neighborhood workers.[214] (3) One of the most advanced of the early factories, and apparently the first to do any substantial amount of weaving in the factory, averaged 100 employees (inclusive of weavers) some five years after the Census.[215] On the other hand, one of the most modern factories in 1815, but one without power looms, averaged 55 operatives per spindle in December 1815.[216]

For 1815, a memorial presented to the Congress in that year by cotton manufacturers stated that within thirty miles of Providence, then the center of the textile industry, there were 130,000 spindles at work,

---

[211] Victor S. Clark, *History of Manufactures in the United States*, 1929, Vol. 1, p. 535.

[212] *American State Papers, Finance*, II, p. 427. A comparison of the state spindle detail given on page 432 with Tench Coxe's Digest (printed in *ibid.*, p. 694) shows unbelievable increases in spindles for New Jersey, Maryland, South Carolina, etc., with only small increases in quantities spun in mills (p. 692). Coxe obviously included home spindles. Gallatin's data also appear in *Annals of Congress*, 11th Congress, Part 2, Appendix, col. 2227, 2228, distinguishing the actual 31,000 spindles in operation at the end of 1809 from the 80,000 projected for 1810. His figures are repeated in the Woodbury Report, p. 51.

[213] *American State Papers, Finance*, II, p. 669. Coxe asserts that 1,160,000 spindles and 58,000 persons would suffice to work up into yarn all the cotton produced in the United States, the weaving of which would take "100,000 women with the fly shuttle during one-half of each working day in the year..."

[214] *Ibid.*, p. 434.

[215] Samuel Ripley, "A Topographical and Historical Description of Waltham... 1815" in *Collections of the Massachusetts Historical Society*, 2d Series, Vol. III, 1815 (reprinted 1896), pp. 263–264.

[216] Data for the Fall River Manufacturing Company appear in William R. Bagnall, "Sketches of Manufacturing and Textile Establishments" (edited by V. S. Clark; typescript in the Baker Library of the Harvard Business School), 1908, pp. 1915–1916.

consuming 29,000 bales of cotton, with 26,000 persons steadily employed.[217]

Because of the urgent nature of the request for tariff protection we may assume that the largest possible total for persons affected by such protection would be reported. Was there significant employment beyond the thirty-mile radius? At this point none of the Lowell mills had yet been erected, and only limited production took place in Massachusetts. For the important textile center of Philadelphia, employment was about 2,000, according to estimates of a protectionist group similarly desirous of maximizing the count.[218] For the dominant production center in New York State, no more than 2,000 could be estimated.[219] We can thus hardly estimate more than 30,000 to 35,000 for the United States. This figure is a third of that estimated in 1816 by the Committee on Commerce and Manufactures which estimated the U.S. consumption of cotton in manufacturing at three times that of the area around Providence, and the employment at nearly four times as much.[220] The Committee's estimate of 100,000 is almost double the number counted fifteen years later, after the rise of Lowell, by a vigorous and alert protectionist group.[221] We are therefore unable to rely on these figures, either for employment or cotton consumption, as a basis for setting the 1815 level. We take 30,000 as the U.S. total in that year, adding to the 26,000 for the Providence area, an arbitrary 2,000 each for Pennsylvania and New York.

For 1820, we estimate 12,000 employed, the remarkable decline reflecting not an absence of growth in the industry but the aftermath of the

---

[217] *American State Papers, Finance*, Vol. III, p. 52. *Niles' Register*, IX, p. 190. Cf. the *Transactions of the Rhode Island Society for the Encouragement of Domestic Industry, in the Year 1863*, pp. 21 and 73, for a list of mills and spindle count in November 1815. Batchelder, *Cotton Manufacture*, p. 59, gives a slightly smaller count.

[218] *Circular and Address of the National Institution for Promoting Industry in the United States, to Their Fellow Citizens*, New York, 1820, p. 21. In this circular an estimate is made for 1814, 1816, and for 1819. The intent being to show the great decline to 1819, the earlier figures would, if anything, be biased upwards. Data on page 27 show that Rhode Island employment in 1816 was 15,253—hence 11,000 of the above 26,000 were outside that state.

[219] *Ibid.*, p. 25, gives spindles for Oneida County, which were assumed to require manpower at 200 per 1,000 spindles.

For the area around Wilmington, a promanufacturing group implicitly estimated 600 children (for less than a full year) in cotton and wool manufacturing (*Niles' Register*, Vol. 9, p. 96).

[220] *American State Papers, Finance*, Vol. III, p. 82. Also, quoted in *Addresses of the Philadelphia Society*, p. 166. See also, Isaac Briggs in *Niles' Register*, Vol. 9, p. 391. Niles himself takes the same figures as the Committee uses but rates the 66,000 women and children at equal to "16,000 hands fit for agricultural services." From his adjusted components one then derives a total of 54,000. *Niles' Register*, Vol. 9, p. 277.

[221] *General Convention of the Friends of Domestic Industry*, New York, 1831.

1819 recession. A total of 11,423 employees is reported in the 1820 *Digest of Manufactures*, to which we add an additional complement (for firms omitted by the first Census report), yielding a total of 12,000.[222] Despite the variety of doubts cast on the 1820 Census, its results are not unreasonable. One arrives at an 1819 figure of 5,000 in the major production centers of Rhode Island, Pennsylvania, and New York by using data from the key protectionist group that attacked the Census results.[223] Hence, allowing for omitted areas, a revival in 1820 would hardly bring the total above the amended *Digest* figure of 12,000.[224]

For 1831, a comprehensive listing of mills, output, and employment made for the New York Convention of the Friends of the Manufacturing Interest provides us with total employment.[225] Our only adjustment in these totals is to exclude hand-loom weavers, as not being part of the factory employment.[226]

For 1840–1940, the regular U.S. Census reports were used after minor adjustment.[227]

[222] *American State Papers, Finance*, IV, pp. 28 ff. These data are summarized by Edith Abbott, "The History of Industrial Employment of Women in the United States," *Journal of Political Economy*, October 1906, p. 482. We add to Miss Abbott's total, individual figures from Adams' *Letter*, 17th Congress, 2d Session, Exec. Doc. 90.

[223] *Circular and Address of the National Institution for Promoting Industry in the United States, to Their Fellow Citizens*, pp. 21, 25, 27, enables us to make an 1819 estimate of 5,000 for the main production centers.

[224] As an indication of the extreme order of magnitude that might have been involved in the 1820 Census, we can take a statement by Niles in 1823, when recovery from 1818–20 was well advanced. He states that the 250,572 spindles reported in the Census reflect "an imperfect return" but that even if the Census were not imperfect there has since been a marked rise to "not less than 330,000" (*Niles' Register*, Vol. 24, p. 34). Computing the implicit ratio for Rhode Island, from the Census, of employees per 1,000 spindles gives a figure just under 50—implying less than 13,000 employees as late as 1823.

[225] Printed in *Niles' Register*, Addendum to Vol. 42, March–August 1832, p. 7. These data are reprinted in a number of sources indicating their authoritative character. See, for example, Gales and Seaton, *Register of Congressional Debates*, January 23, 1833, col. 1327; 1860 Census, *Manufactures*, p. xix; 1880 Census, II, *Manufactures*, pp. 10–11. Presumably the same source lies behind figures quoted by Carroll Wright in the 1880 Census, *Manufactures* (pp. 542 ff.) although minor differences appear.

[226] Hand-loom weavers in Pennsylvania and New Jersey, the major remaining centers of such work, are reported by the New York *General Convention of the Friends of Domestic Industry*.

[227] 1840–90 data: 1890 Census, *Manufactures*, Part III, p. 186. 1919 Census, *Manufactures*, X, p. 177. 1929 *Census of Manufactures*, Part II, pp. 247, 261. 1939 *Census of Manufactures*, Vol. II, Part 1, p. 287. The 1909 Census figure was adjusted to a 1910 level by using data on the 1909–10 reduction in cotton consumption and the decline in finished textile production. Production data were from Shaw, *Value of Commodity Output*, p. 72. For 1920 and 1930 we use Kuznets' adaptation of BLS data (Kuznets, *National Income and Its Composition*, p. 598). For 1940 we extrapolate the Census total for cotton textiles by the 1939-40 change for textile mill employment (Office of Business Economics, *Business Statistics*, 1951, p. 57).

By 1950, different raw materials were used in textile production, and the same mill would work on cotton at one time, synthetics at another. As a result, the Census gives us no separate figures for cotton textile employment. We make a rough estimate, however, in order to complete the long time series, by relating (a) adjusted bales of cotton consumed in manufacturing to (b) cotton textile employment in 1909, 1919, 1929, and 1937—the regression relationship implying 350,000 employed in 1950, and 300,000 in 1960.[228] We divide bales consumed by an index of output per person employed in the production of textile mill products.[229] The adjusted figures thus derived provide for productivity trends and yield a very simple and close relationship between bales and employment.

IRON MANUFACTURING: WAGE EARNERS

For 1830 we have the results of a comprehensive survey by a convention of iron manufacturers.[230] The general reasonableness of their findings may perhaps be validated by their acceptance by Henry Clay.[231] But somewhat reminiscent of Russian statistics, though they appear to purvey the truth, it is not the truth we assume them to offer. Their level, 29,524, is virtually the same as that of the 1840 Census figure of 30,497.[232] Now the 1840 figure may reflect the doldrums of that period, but is it likely to have shown no advance in employment? We suggest that the Convention figure included, as was more or less commonly done at the time in protariff statements, all employment connected with iron production, inclusive not merely of iron mining, but also coal mining, woodcutting, carting, etc.[233]

As one basis for adjustment we take detailed 1832 data for 100 New

[228] Consumption data from *Historical Statistics*, 1960, pp. 414–415, extrapolated to 1960 by *Business Statistics*, 1961 edition, p. 189.
[229] The output-per-person index is from Kendrick, *Productivity Trends*, Table D-IV.
[230] *General Convention of the Friends of Domestic Industry*, assembled at New York, October 26, 1831, *Reports of Committees, on the Product and Manufacture of Iron and Steel*, pp. 20, 32.
[231] Gales and Seaton, *Register of Debates*, February 2, 1832, col. 260.
[232] 1840 Census, *Compendium*, p. 358.
[233] The convention report rests on an implicit figure of five tons of bar iron and castings produced per man in Center and Huntingdon Counties, an extremely low figure for the most efficient area in the United States if only forgemen were included, but quite reasonable if all personnel were in.
In 1837 a New York State report noted that iron ore furnaces near Amenia made 10,000 tons of iron a year and employed 1,000 men as "ore diggers, coal men, teamsters, smelters, limestone diggers, etc." This implies the same ratio of ten tons per man. New York State, Geological Survey, *Annual Report, 1837*, p. 181, State of New York, in Assembly, February 20, 1838.
Manufacturers' reports in the McLane Report, the Harrisburgh Convention in 1827, and the Iron Masters Convention in 1849 similarly appear to include nonfactory employees; and the line between manufacturing and mining continues to be difficult to distinguish even in our own day.

Jersey furnaces, from which we can estimate that 30 per cent of all wage payments per ton of bar iron went for work in the forge.[234]

We compute an alternative ratio from the 1880 Census. In that Census the reports included mining employees as well as those in iron manufacture, while they did not in 1890.[235] We take the data for furnaces working with charcoal, as most comparable with the 1840 data, and find that in Tennessee and Virginia, where the charcoal iron industry was most important, the tons of ore per hand in 1850, 1860, 1870, and 1880 were much the same.[236] Rising from this level (of about 60) at all four Censuses, it suddenly jumped to 368 in 1890—suggesting that this was the first Census to exclude mining employees. We make three estimates of the overcount in 1880, i.e., of the number of mining and related employees other than manufacturing workers. (a) If we assume that employees per establishment, reported at 144 in 1880 and 43 in 1890, should have been the same at both dates (and one could make an even stronger argument, since the Census reported that the 1890 establishments were larger and better equipped),[237] we derive a figure of 4,988 employees. (b) If we assume that capital per employee in charcoal furnaces in 1880 was 20 per cent below that in 1890—for mineral furnaces it was 24 per cent lower—the figure comes to 4,812. (c) If we assume that the employment per ton of pig iron produced was cut in half from 1880 to 1890 in charcoal furnaces as in mineral (instead of the reported 80 per cent charcoal decline), we estimate 5,224. All three procedures suggest that manufacturing employment was perhaps one-third of the reported 16,670 in the 1880 charcoal iron industry. Given the New Jersey ratio of 30 per cent and the 1880 Census one of 33 per cent, we stipulate a 30 per cent ratio for adjusting the 1831 Convention total down to a round 20,000 employment in 1830.

The Census for 1840 reports under iron, "number of men employed including mining operations," at 30,497.[238] Contemporary evidence indicates that the only nonmanufacturing activities included in that total were those of iron mining.[239] We adjust the Census total down by the detailed New Jersey data used above.

---

[234] U.S. Congress, Serial Set 223, *Documents Relative to the Manufactures in the United States*, 1833, p. 183. A ton finished at the forge was valued at $59.25, $18.00 of which was for making the iron at the forge and repairs of the forge—the rest for mining and carting ore and coal, plus woodcutting.

[235] 1890 Census, *Manufactures*, VI, Part 3, p. 385.

[236] 1880, 1890: 1890 Census, *Manufactures*, VI, Part 3, pp. 407–09. Charcoal iron industry only. 1850–70: 1870 Census, *Wealth and Industry*, pp. 602–603, all iron manufacturing in specified states.

[237] 1890 Census, *Report on Manufacturing Industries*, Part III, p. 406.

[238] 1840 Census, *Compendium* (Allen edition, 1841), p. 358.

[239] Okeley's 1842 report, reprinted in 62d Congress, 1st Session, Senate Doc. 21, p. 499.

Pig iron production in 1810 is estimated at 20 per cent of that in 1840, and bar iron output at about 23 per cent.[240] With little change in the locus of the industry, or the sources of fuel and ore, in this period, we arbitrarily assume only a modest productivity gain, and stipulate 1810 employment as 25 per cent of that in 1831.

For 1800–20, we estimate, from the 5,000 level in 1810, an arbitrary 1,000 in 1800. The 1810 level is kept for 1820; although the industry did grow significantly over the decade, production fell by over one-half to the recession year of 1820.

For 1850–1900, we use Census data, with an adjustment for the Census overcount in 1880 based on considerations discussed above with respect to 1840.[241]

For 1910–60, because the coverage of the Census shifts over this period, and because most Census enumerations relate to the last year of the decade (e.g., 1909 rather than 1910), we have to make a number of adjustments in reported Census data. We utilize Fabricant's presentation of the Census materials for 1910–39, plus data from BLS and Census to arrive at a final series most comparable with the Census iron and steel figures for the nineteenth century.[242]

TRADE: 1840–1900

For 1870–1900, we extrapolate the 1900 employment total by the number reported in selected occupations in the Census of Population.[243] The

---

[240] Pig iron: *Historical Statistics*, 1960, p. 366. Bar iron: data from *General Convention of the Friends of Domestic Industry*, p. 2.

[241] 1850 and 1860: 1870 Census, *Wealth and Industry*, III, pp. 407, 401. 1870: 1890 Census, *Manufactures*, VI, Part 3, p. 383. This last source adjusts the 1870 Census for the duplicate inclusion of establishments under more than one detailed industry. 1890: 1900 Census, VII, Part 1, p. 9. This source adjusts the 1890 Census, presumably because it reported peak rather than average employment. 1880: We use data from the last source, but exclude an estimated 12,000 overcount. (For this estimate, see the above discussion with respect to the 1840 estimate.)

[242] Solomon Fabricant (*Employment in Manufacturing, 1899-1939: An Analysis of Its Relation to the Volume of Production*, New York, NBER, 1942, p. 199) gives an 1899 figure for blast furnaces and steel mill products that is identical with the 1900 Census total for "iron and steel." We therefore use his comparable 1909, 1919, 1929, and 1939 figures. These are moved forward one year by, respectively: 1909-10 production of steel ingots, rails, shapes (*Historical Statistics*, 1960, p. 416); 1919-20, monthly FRB indexes of employment in the industry (*Federal Reserve Bulletin*, December 1923, p. 273); 1929-30, BLS indexes (Lewis Talbert and Alic Olenin, *Revised Indexes of Factory Payrolls*, BLS Bul. 610, 1935, p. 27); and 1939-40, 1950 and 1960 employment in SIC 331 (BLS, *Employment and Earnings Statistics, 1909–1960*, Bul. 1312, p. 95).

[243] We use Edwards' summary for trade (*Comparative Occupation Statistics*, p. 110) exclusive of his "all other occupations" category. We find his estimates for this latter group dubious because they imply a ratio of laborers to dealers 1870–1900 that is wholly inconsistent with the level from 1910 on, where we do have direct data. It is also necessary to exclude bankers, brokers, real estate agents, and others in the group that were not in trade.

procedure is similar to that of Carson and the trend is therefore similar to his. (Because they are consistent with the figures for 1929 and after, however, our figures are at a much different level than is his Population Census-benchmarked series.)[244]

For 1850–60, we extrapolate our 1870 figure by the trend in selected trade occupations.[245]

For 1840, the Population Census reported the number of persons engaged in commerce, as they did for 1820. We find it necessary to ignore these totals for four main reasons.

1. The procedures used in the 1840 Census of Manufactures clearly include persons engaged in transportation under the heading of "commerce," and we assume the same procedures were used for the enumeration of the population, both activities being under the charge of William Weaver and his small staff.[246]

2. The distinction between manufactures and trade was particularly difficult to make in a period when the "hand trades" were as important as they were in 1820 and 1840. The allocation between the two published categories cannot, therefore, be assumed a priori to be a reliable one. The cobbler made and sold his product, as did the wheelwright, candle-maker, carriage maker, etc. Hence, there was equal warrant for including any given person either under the hand trade group (assigned for many decades to "manufactures") or commerce.

3. For three states (with nearly half of trade employment in these years) we compute the number of persons served per person engaged in trade:[247]

	1820	1840	1850	1860	1870	1880	1890	1900
Massachusetts	38.8	90.4	42.5	36.5	43.7			
New York	146.1	83.5	45.5	39.0	26.4			
Pennsylvania	143.5	109.2	53.7	59.9	40.7			
U.S.	108.5	120.7	34.3	30.3	25.7	22.5	18.6	16.4

[244] Kendrick adopts Carson's benchmarks, except for using the present writer's preliminary 1920 benchmark, and interpolates by an early version of the present series. His figures differ from Carson not merely in level but in trend, particularly 1870–1900, possibly because he "adjusted to an employment basis" (*Productivity Trends*, p. F-7).

[245] Data on agents, clerks, livery stable operators, peddlers, merchants, and dealers as shown in the 1900 Census, *Occupations*, pp. lv–lvi. Because female clerks were not enumerated in 1850, we extrapolate our 1860 estimate to 1850 by the above group exclusive of clerks.

[246] 1840 *Compendium*, p. 360, shows "internal transportation" as part of commerce.

[247] We use published Census state trade totals, U.S. totals for 1820 and 1840, and our U.S. estimates for 1850 ff., and relate these to the count of the white population. Purchasing by and for nonwhites prior to 1863 required almost no added distribution network, while their money income level afterwards was so low as to make it more useful to exclude them from the ratio computation. For 1850 we sum figures for thirteen occupations—e.g., clerks, dealers, etc.; for 1860 the itemized dealer and clerk figures; and for 1870, the trade and transport total minus transport occupations, to give trade totals.

Is it possible to accept that, in Massachusetts, for example, 1820, 1850, and 1860 are at similar levels, but 1840 is so much higher? Or should we simply assume underenumeration of trade employment in 1840? Can we believe that the distribution network in New York and Pennsylvania expanded so rapidly from 1820 to 1850 and so little from 1850 to 1860?

4. The internal relationships in the 1840 Census reports suggest gross undercoverage—leading to the high ratios noted above. For example, two countries with almost identical population in Vermont respectively reported 4 and 136 persons in commerce. Other pairs, in New York, reported 51 and 238, 3 and 376; in Virginia, 0 and 70; in North Carolina, 7 and 245; in South Carolina, 7 and 121. The Census also reported no persons in commerce in various New England towns, in Jersey City, Raleigh, Macon, Fredericktown, Portsmouth—and only one person for the nine thousand inhabitants of Detroit.[248]

A state Census of Michigan in 1837 reported 795 merchants, whereas the 1840 U.S. Census reported only 728 in commerce, including merchants and employees.[249]

Our procedure for 1840 was to extrapolate the trend in ratios of white population to trade employment shown in the table above. By doing so we assume that the growth of cities, the development of the distribution system in new areas and the proliferation in old ones, the changing productivity in distribution were all subsumed in a trend closely correlated with the trend in ratios of white population to trade employment.

Without some absolute benchmark on the figures for an earlier date, this is hardly a very satisfactory method; a great deal turns on the question of the inflection of the curve of ratios. In this respect our estimates for this category, and construction, differ significantly from our other early figures which do, in general, have some absolute control. (The link to the population series does, of course, provide one constraint.) It may be helpful, however, to note that a linear extrapolation—surely the outside estimate, since it implies no improvement in the distribution system—will increase our 1840 figures by less than 10 per cent. The result of the foregoing adjustment is a more than doubling of the 1840 Census figure. Hence the level, albeit more consistent with later figures (the population data and so on), is well above that of previous writers who have adopted the Census results.[250]

[248] All data from the 1840 *Compendium*. Vermont: Caledonia, Chittenden; New York: Orleans and Cortland, Greene and Suffolk; Virginia: Augusta and Harrison; North Carolina: Ashe and Bladen; South Carolina: Pickens and Richmond.

[249] James H. Lanman, *History of Michigan*, New York, 1839, p. 294.

[250] In *Trends in the American Economy*, Theodore Marburg adjusted the Census upward to a level of 135,000 (p. 319). Other estimates, relying on Seaman's figures for

RAILROADS

Our estimates for railroads are the sum of ones made for each of the six individual regions into which the 1880 Census divided the country (Table 4).

Region I: New England

We estimate the trend of employment for two categories—Massachusetts railroads, and all others in the region. Data on the Massachusetts roads are readily available for 1870 and 1880.[251] For 1850, stockholder investigating committee reports give us figures for seven major roads,

TABLE 4

RAILROAD EMPLOYMENT, 1840-80

	U.S.	Region[a]					
		I	II	III	IV	V	VI
1840							
Mileage	2,265	356	1,399	488		21	
Empl. per mile	(3.24)						
Employment	7,000						
1850							
Mileage	7,310	2,256	3,207	1,706	97	46	
Empl. per mile	(2.88)	3.5	2.22	3.3	3.43	2.74	
Employment	21,080	7,900	7,120	5,600	330	130	
1860							
Mileage	27,420	3,639	11,687	7,130	4,700	147	115
Empl. per mile	(2.98)	3.7	2.42	3.3	3.43	2.74	2.74
Employment	81,790	13,460	28,300	23,200	16,120	400	310
1870							
Mileage	43,510	4,115	16,552	10,007	9,110	331	3,394
Empl. per mile	(3.66)	5.12	3.95	3.3	3.61	2.74	1.72
Employment	159,120	21,070	65,400	33,000	32,900	910	5,840
1880							
Mileage	87,070	5,894	28,647	13,869	23,586	897	14,174
Empl. per mile	(4.78)	5.37	7.33	3.1	4.06	2.49	2.36
Employment	415,967	31,634	209,725	43,154	95,776	2,238	33,440

[a] I: New England; II: Middle Atlantic, Michigan, and Ohio; III: Southeast; IV: Illinois, Iowa, Minnesota, Missouri, Wisconsin; V: Louisiana, Arkansas; VI: Mountain and Pacific.

---

income produced (*Essays on the Progress of Nations*, 1868, p. 459) will reflect either his adoption of the Census employment figures or his 10 per cent increase in reported Census capital totals for trade.

[251] Massachusetts, Board of Railroad Commissioners, *Fourth Annual Report*, January 1873, pp. 178, 188, gives data for 1872, which we use for the 1870 ratio. *Ibid.*, January 1881, pp. 111 ff., gives data for 1880.

which account for a third of all employment in 1872.[252] For 1860, selected annual reports for 1856–61 plus the findings of a Congressional committee are available.[253] However, the ratio of 7.6 men per mile, to which the 1860 data lead, comes to well above a reasonable 1860 figure, being at just about the 1870 level of 8.0 for all roads. Presumably this is so because they were primarily the larger roads. For 1860, therefore, we compute instead the ratio of the Connecticut rates for 1860 to 1870, and apply that ratio to the Massachusetts 1870 ratio.[254] Connecticut rates for 1860, 1870, and 1880 are readily computed from the reports of the Railroad Commissioners.[255]

We then weight the Connecticut and Massachusetts rates together for 1860 and 1870. Extrapolating the region's 1880 average gives a regional figure for 1870.[256] For 1860 we assume the weighted ratio represents that for the region. The resultant average of 3.7 for the region in 1860 seems consistent with the 1850 Massachusetts average of 3.72. We therefore adopt the latter (rounded down to 3.5), for the region's average in 1850.

Region II: Middle Atlantic, Michigan, and Ohio

Because of the diversity of experience in this region, estimates of employee per mile in each of the major states in the region were made. The summation of the employment figures thus derived were then used to extrapolate the region's 1880 total to earlier years.

[252] Onslow Stearns, Superintendent of the Northern Railroad, provides data for the Boston and Maine, Boston and Providence, Old Colony, Fitchburg, and Northern (of New Hampshire) in *Report of the Committee of Investigation of the Northern Railroad, to the Stockholders, May 1850*, Concord, Appendix. *Report of the Committee of Investigations of the Nashua and Lowell Railroad... May 29, 1850*, Boston, 1851, pp. 77, 89, gives data for this railroad. Boston and Concord data are in *Annual Report, 1850*, pp. 14, 32.

We adopt the weighted average for these roads as being representative for the state in 1850. By later years, of course, more marginal roads had entered, with lighter traffic, hence employment, per mile. Therefore, one cannot use the seven road-state relationship in these later years to adjust the 1850 data.

[253] Annual reports, and averages: Boston and Maine, 1858 (6.5); Boston, Concord and Montreal, September 1859 (1.1); Western, January 1861 (9.2). For 1856, data on the Boston and Worcester (8.0), Boston and Maine (8.0), and Western (9.0) appear in 34th Congress, 1st Session, House Report 358, *Pacific Railroads and Telegraphs*, pp. 48, 60.

[254] The Connecticut rate, 54 per cent, is virtually identical with that for the Boston and Maine.

[255] Connecticut, General Railroad Commissioners, *Seventh Report*, 1861, pp. 28, 33; *Seventeenth Report*, 1870, pp. 54, 58; *Twenty-Eighth Report*, 1881, pp. 80, 95. Employment data were not used for a few small roads that did not report mileage data.

[256] 1880 regional data from: 1880 Census, *Report on the Agencies of Transportation*, 1883, pp. 259, 377. For weights we use mileage as of 1880 from the 1880 Census (*ibid.*, p. 507) and as of 1860 and 1870 from H. V. Poor, *Manual of the Railroads of the United States for 1872-73*, 1872, p. xxxiii.

LABOR FORCE AND EMPLOYMENT, 1800–1960 193

For New York State the *Annual Report* of the State Engineer provides us with data on miles laid in the state.[257] For 1880 we derive a state employment total from 1880 Census data for individual roads.[258] For 1860 we begin from employment as reported for the New York and Erie, the Hudson River, and the New York Central railroads in 1856.[259] For these same roads, and for all other New York State roads in 1857, we then computed total expenditures on the major employment activities (conductors and baggagemen, engineers and firemen, freight labor, porters, watchmen, and switchmen).[260] The three roads mentioned above accounted for 68.1 per cent of expenditures in the sum of these categories—and about 70 per cent of freight cars as well. We concluded that their employment accounted for 68 per cent of all state employment in 1856 as well, thus implicity assuming that the average expenditure per employee was the same for all roads as for the average of these three. For 1860 we then interpolate between the 1856 and 1870 employee-per-mile estimates.[261] For 1850 and 1870 we utilize the State Engineer's count of N.Y. freight cars, as follows. Ratios of employees per freight car were computed for all state roads in 1860 and 1880; they averaged 1.23 and 1.22 respectively.[262] Given this stability we therefore assumed a 1.23 ratio for 1850.[263]

As for New York, so for Pennsylvania we have comprehensive figures on freight cars and miles of main line.[264] We estimate employees per

[257] *Annual Report of the New York State Engineer and Surveyor, January 1851*, Table A; *September 1860*, Table C; *1870*, p. 211, and *1880*, pp. 97–98.

[258] 1880 Census, *Report on the Agencies of Transportation*, Tables 6 and 9, using the reports for the New York Central, Erie, Syracuse and Binghampton, New York and Ontario, Rome and Watertown, Albany and Susquehanna.

[259] Erie: Edward H. Mott, *Between the Ocean and the Lakes, The Story of the Erie*, 1899, p. 483; Hudson River, NYC: 34th Congress, 1st Session, H.R. 274, *Pacific Railroads and Telegraphs*, pp. 47, 48, 64.

[260] *Report of the State Engineer and Surveyor on the Railroads of the State of New York for . . . 1857*, 1858, pp. 295 ff. We exclude the 3rd, 6th and 8th Avenue railroads.

[261] The rate of gain appears reasonable in the light of tonnage increases for the Erie and New York Central, as reported in 52d Congress, 2d Session, Senate Report No. 1394, Part I, p. 618, and the stability in mileage as shown in Poor, *Manual of the Railroads*.

[262] *Report of the State Engineer and Surveyor*, 1857.

[263] Data for employment on a few individual roads are available for 1850, but appear inconsistent with these results, possibly because construction employees were included in the early reports. The *Annual Report* of the Albany and Schenectady Railroad, January 31, 1852 (p. 13) leads to an average of 9.5; that of the Syracuse and Utica for 1848 (pp. 1, 9) to an average of 10.0. For the Erie (Mott, *Between the Ocean and the Lakes*) an average of 2.9 per mile operated is indicated, much closer to the state average we use.

[264] Pennsylvania, *Reports of the Several Railroad Companies of Pennsylvania Communicated by the Auditor*, 1863, pp. 328, 331; *Annual Report of the Auditor General of State of Pennsylvania . . . Railroad, Canal and Telegraph Companies*, 1871, pp. xxv, xxix; *Annual Report of the Secretary of Internal Affairs*, Part IV for 1880, pp. xxiii.

freight car on the Pennsylvania Railroad in 1856, 1866, and 1880;[265] then use these ratios to extrapolate the 1880 state ratio of employees per freight car.[266]

Employment data are available for Ohio in 1870 and 1880, while for 1860, the unsatisfactory expedient of applying the average employees-per-mile figure for 1868 to the 1860 mileage total was adopted.[267] For Maryland, data on employment and freight cars in 1856 and 1880 are available, with interpolated ratios then computed for 1860 and 1870.[268] Since the B & O and Maryland ratios in 1880 were very similar—30.0 and 33.6—the interpolated ratios were then applied to mileage totals computed for the state.[269]

The regional employment total for 1880 was then extrapolated to 1870 and 1860 by the summation of the above state employment estimates.[270] The 1850 data available for a few roads in the region, and computed above, are hardly a very satisfactory basis for a direct estimate. We have something of an upper limit in the form of the 2.9 employees per mile on the Erie Railroad, one of the two largest in New York and one of the four or five largest in the region. With lighter loads on the other roads, the ratios for them should be still smaller. We assumed instead that the New England ratio was a solid reference point, and estimated the Region II ratio as much below the New England 1850 ratios as the two were apart in 1850, thus giving an 1850 estimate of 2.22.

Region III: Southeast

We estimate that employment per mile of railroad for the entire 1850–70 period averaged 3.3, as compared with an 1880 Census ratio of 3.1. Our

[265] 1880: 1880 Census, *Transportation*, pp. 266–267. 1866: *Guide for the Pennsylvania Railroad, with an extensive map . . . and information useful to travellers*, Philadelphia, 1866, p. 40, indicates that the road when completed will have "not less than 4,050 employees" and 4,800 eight-wheel cars. The booklet originated in the Pennsylvania Railroad offices. Data in *Pacific Railroad and Telegraph* (34th Congress, pp. 48, 62) lead to an average of 2.79 as compared with our 2.89 figure for New York State.

[266] The 1880 ratio for the state was computed from the 1880 Census, *Transportation*, Table VI, using data for the Allegheny, Delaware and Hudson, Lehigh, New York and Pennsylvania PRR, Philadelphia and Reading, Pittsburgh and Titusville.

[267] 1870, 1880: Ohio, *Annual Report of the Commissioner of Railroads and Telegraphs, 1880*, pp. 45, 67. 1868: *Ohio Annual Report . . . 1868*, p. 228. 1860: mileage estimated by subtracting from the 1868 total the 1860-67 construction on all lines listed as having mileage in Ohio. 1880 Census, *Transportation*, Table VIII p. 228.

[268] 1856: U.S. Congress, *Pacific Railroads and Telegraphs;* 1880 Census, *Transportation*, p. 261.

[269] For 1880 the Maryland total was computed by totaling employment and mileage for ten major roads in the state, then applying the average to the state mileage total. Data from 1880 Census, *Transportation*, pp. 260 ff. and 509 ff. Mileage for earlier years computed from *ibid.*, pp. 321 ff., using data for selected roads.

[270] 1880 Census, *Transportation*, p. 257, gives the regional total.

rate is the average of rates computed from twenty-one reports for various southern railroads in the years from 1840 through 1867.[271] These roads were scattered throughout the old South, from the Central of Georgia to the Richmond and Danville, Louisville and Nashville, etc. No temporal trend is obvious in the ratios; hence, the superior estimate based on a pooling of all the averages was used for all dates.

Region IV: Illinois, Iowa, Minnesota, Missouri, and Wisconsin

For employment in 1870 we (1) estimate employees per mile of road built in the region by 1870, 1874, and 1880; (2) apply these averages to the Census data on miles built. For ten individual roads we have data on employment in 1874 as given in a special Treasury study.[272] From the 1880 Census we can compile data for employment on these same roads in 1880, as well as mileage built as of both dates.[273] The ratio of one average to the other, when applied to the implicit 1880 region average (of 4.26 employees per mile built), gives a 3.79 figure for 1874. This figure was then extrapolated to an 1870 level (of 3.70) on the basis of data for seven major roads showing tons (of freight carried one mile) per mile of road built.[274] These roads account for most of the mileage in the region.

[271] Annual reports with year and computed men per mile indicated as follows: 1840, Charleston and South Carolina, 3.9; 1842, Central of Georgia, 0.8; 1848, Charleston and S.C., 3.0; 1848, Louisa, 7.0; 1853, Richmond and Petersburg, 4.1; 1854, R and P, 8.5; 1855, R and P, 4.7; 1854, Virginia and Tennessee, 3.3; 1856, Virginia and Tennessee, 2.8; 1857, Virginia and Tennessee, 2.5; 1857, Memphis and Charleston, 2.4; 1855, Richmond and Danville, 3.5; 1856, Richmond and Danville, 4.5; 1857, Richmond and Danville, 3.2; 1858, Richmond and Danville, 3.2; 1859, Richmond and Danville, 3.5; 1857, North Carolina, 1.7; 1858, North Carolina, 1.9; 1859, Norfolk and Petersburg, 1.7; 1861, Louisville and Nashville, 2.2; 1862, Louisville and Nashville, 3.4; 1867, Louisville, Cincinnati, and Lexington, 3.2. Reports for these railroads for other years, and for other railroads, in the collection of the Bureau of Railway Economics were examined. Out of several hundred reports, however, these and half a dozen reports for railroads in Massachusetts and New York were the only ones with employment figures. The sample is unquestionably not a random one; fortunately its variability is not great.

[272] Data for employment on individual roads appear in Edward Young, *Labor in Europe and America*, 1875, pp. 787–788. Young was Chief of the Treasury's Bureau of Statistics.

[273] 1880 Census, *Transportation*. Employment data from Table VI, mileage data from Table VIII. The roads were: Cairo and St. Louis, Cairo and Vincennes, Chicago and Alton, Chicago and Iowa, Chicago and Pekin, Chicago and Rock Island, Indiana and Bloomington, Quincy and Alton, St. Louis and Alton, Evansville and Terre Haute, Indianapolis and St. Louis.

[274] Data on tons carried one mile by the Chicago and Milwaukee, Chicago and Northwestern, C B and Q, Chicago and Rock Island, Illinois Central, Pittsburgh and Fort Wayne, Ohio and Mississippi were taken to reflect trends in Region IV. (A few, however, were actually classed in Region II.) Data from 52d Congress, 2d Session, Senate Report 1394, Part I, *Wholesale Prices, Wages and Transportation*, 1893, pp. 618 ff. Mileage data from 1880 Census, *Transportation*.

	1870	*1874*	*1880*
Employment per mile built (ten-road sample)		5.27	5.92
Index of employment in Region IV (based on ten-road sample)		38	100
Index of tons of freight (one mile equivalent) per mile built (seven-road sample)	45	52	100

It will be seen that the tonnage data suggest only a mild change in the ratio from 1870 to 1874, and it is likely that the rise in employment per mile built was even milder.

As a rough indication of a check on this estimate, the following may be helpful. We know that the number of freight cars in the region in 1880—and indeed in all regions but the South—approximately equaled the number of employees. For three major roads (C B and Q; Milwaukee and St. Paul; Chicago and Northwestern), which together had about a third of all mileage in the region in 1870, we have data from annual reports.[275] These can be condensed to indicate that freight cars per mile in 1860 were 89.3 per cent of their 1870 average. This compares quite favorably with our implicit estimate that employees per mile in 1860 averaged 87 per cent of the figure for 1870.

We estimate that employment per mile built changed in 1860–70 in the same proportion that freight (tons carried one mile) per mile built did over this decade. The 1860 Census reports total freight tonnage and mileage for nine major roads in the region.[276] We infer the number of tons carried per mile from the tonnage total by applying a ratio computed from data in the Aldrich report. That report provides (for the C B and Q, the Chicago and Rock Island, the Illinois Central, and the Pittsburgh and Fort Wayne) figures on both total tonnage and tonnage carried one mile.[277] The 12.4 per cent ratio thus derived is virtually identical with the 11.8 per cent one can compute for the same year for all railroads in New York State.[278]

Region V: Lousiana and Arkansas

For the numerically trivial mileage in Census Region V, we assume that the 1850–70 averages per mile were 10 per cent above that implicit in the 1880 Census—as the averages were for Region III.

[275] Chicago, Burlington and Quincy, *Annual Report, 1870*, pp. 13, 16, 45; *1880*, p. 15. Milwaukee and St. Paul, *Annual Report, 1870*, pp. 34–35; *1880*, pp. 50, 272. Chicago and Northwestern, *Annual Report, 1870*, pp. 15, 34, 36; *1880*, pp. 50, 272–73.
[276] 1860 Census, *Population*, p. 324.
[277] Senate Report 1394, Part I, *Wholesale Prices*, 1893. For some roads the ratio was computed from data for 1863 rather than 1860.
[278] *Annual Report of the New York State Engineer and Surveyor*, September 30, 1860, p. 362, and 1860 Census, *Population*.

LABOR FORCE AND EMPLOYMENT, 1800–1960 197

Region VI: Mountain and Pacific

Employment in this region was nonexistent in 1850. In 1860 the bulk of mileage was in two roads, the Houston and the Texas. We assume the same ratio as that estimated for Region V, a region of similar physical and traffic density conditions.

As of 1870 the bulk of the mileage operated was in the Central Pacific and Union Pacific. For each road we estimate freight cars per mile operated, assume a constant 1870–80 employment per freight car, and compute employment.[279]

U.S. Total: 1840

For 1840, we estimate U.S. railroad employment at 9,000 by one procedure, and give some rough checks by others. Our basic estimate begins from an 1838 Treasury Department survey which reports 350 locomotives in use in the United States.[280] Given the limited number of roads at the time, and publication of figures for individual locomotives, the result should be reasonably solid. For 1840 the number of locomotives as estimated in Hunts (bringing up to date figures from Von Gerstner) is consistent with growth from this Census count, at 465.[281] We estimate twenty employees per locomotive at this date. The average for the Erie railroad in 1841 is somewhat greater, that for five New England roads in 1850 somewhat lower.[282] The average for the New York Central, Erie, Pennsylvania, B & O, Boston and Worcester in 1856 was twenty-one.[283]

If we take the independent data on mileage operated and apply estimates of men employed per mile in the same roads noted above, we derive estimates of from 6,700 to 9,500.[284] Because railroad mileage was

[279] 1880 Census, *Transportation*, pp. 276, 278, 413, gives data for 1880. For 1869: 44th Congress, 1st Session, House Report 440, *Pacific Railroads*, pp. 248, 136. Tons of freight carried one mile rose sixfold on the Union Pacific whereas freight cars increased only 50 per cent. Tonnage data from Senate Report 1394, Part I, *Wholesale Prices, 1893*, p. 620. The increase in freight handled per freight car is not unreasonable.

[280] 25th Congress, 3d Session, House Doc. 21, U.S. Treasury Department, *Steam Engines*, December 13, 1838.

[281] *Hunts' Merchants' Magazine*, Vol. IV, p. 481.

[282] Erie Official report data in Mott (*Between the Ocean and the Lakes*, p. 483). New England Roads: *Report of the Committee of Investigation of the Northern Railroad to the Stockholders*, May 1850, Concord, Appendix. Data from the 1880 Census, *Transportation*, Part IV, p. 257, also indicate about twenty men per locomotive in the Northeast, where 1840 railroads were concentrated, and twenty to twenty-five in other regions.

[283] All data from House Report 358, *Pacific Railroads and Telegraphs*, 1856, pp. 47, 48, 60, 62, except Erie employment from Mott, *Between the Ocean and the Lakes*.

[284] Mileage data from *Historical Statistics*, 1960, p. 427. Employees per mile on the Erie averaged 2.4. The mean for seven New England roads in 1851 (see 1850 estimates below) was 3.4.

concentrated in the Northeast the bounding of our 9,000 estimate by these figures is some support. Since roads with low traffic tended to have relatively high employment per mile, it is useful confirmation that the average for the Charleston and South Carolina in 1840 ran no greater than 3.9 men per mile.[285]

Other Estimates

The Population Census count of "railroad employees" is well below the official ICC and BLS figures for 1890 and after, and is therefore below our estimates, linked as they are to the BLS level.[286] This is so because the large number of trackmen, carpenters, clerks, and others employed by railroads in occupations not peculiar to railroading, are included in other Census occupation rubrics.

The above estimates are similar to the results in the impressive study by Fishlow in the present volume, except for the year 1870.

Two factors account for the latter difference. His estimate was derived by multiplying the total earnings for that year (of $385 million) by .592, the pooled coefficient relating employment to earnings on Massachusetts and Ohio railroads.

His total earnings for 1870 appear to be too low because the sample of five railroads he used to extrapolate the 1880 ton-mile total back to 1870 reflects the smaller gain in Census Region II than actually characterized the U.S. in general.[287]

Using data for individual railroads in Massachusetts and Ohio, Fishlow computes a regression coefficient to link earnings and employment. When he used this procedure in the check year of 1880 it led to an underestimate of the actual U.S. total by 12 per cent.

If one combines a 10 per cent overestimate (as for the 1870 earnings figure) and a 12 per cent error in the regression coefficient (as for 1880), Fishlow's procedure could yield an 1870 total at about the level of the present estimates.

The above discussion has all rested on the assumption that the 1860 level of railroad earnings, from which Fishlow departs to estimate the

---

[285] *Annual Report* for 1840. The 1842 *Report* for the Central of Georgia reported 126 employees and 147 miles of road, but did not report any agents. For 1848 the Charleston and South Carolina average was 3.0.

[286] Population Census data appear in Edwards, *Comparative Occupation Statistics*, p. 109, and 1890 Census, *Population*, Part I, p. cvii.

[287] Fishlow indicates his source as the Aldrich report. Ninety per cent of his ton-mile total was activity on four railroads in Census Region II. Frickey's data—used by Fishlow for estimating passenger miles—indicate, however, that the ton-mile gain for Region II was about 10 per cent less than that for the U.S. See Edwin Frickey, *Production Trends*, 1947, pp. 84, 87.

1870 level, is satisfactory. However the 1860 figure itself was no reliable total. It derives from Fishlow's adjustment of an 1855-56 Treasury survey riddled with errors. Substantial corrections of the survey results were properly made by Fishlow, but it is doubtful if at this remove one could be confident of estimating within, say, 10 per cent accuracy.

The use of Fishlow's regression procedure is, as he puts it, acceptable on the assumption that the product of the operating ratio, the share of wages in expenses, and the reciprocal of the wage rate are constant throughout the country. It would be a happy conclusion if the average of the cross product of three such ratios in Massachusetts and Ohio yielded a U.S. average that was accurate within 10 per cent.

Although Fishlow makes no regional estimates, by comparing 1860 and 1880 regional data we can consider where the additional 68,000 employment in Fishlow's 1870 U.S. estimate might be. Surely not in Regions I, III, and V, given the 1880 total and assuming some growth from 1860 to 1880. For Region VI our 1870 estimate is based on data both for the Union Pacific and the Central Pacific (plus allowance for minor roads)—which account for virtually all mileage in the region at this date. Hence Regions II and IV account for most of the difference.

For Region II our component estimates for Pennsylvania rest on the data for the entire state showing that freight cars were almost five times as numerous in 1880 as 1870; New York and Maryland data showing roughly a tripling in each; and Ohio, less than doubling of employees. The combination of weighting, allowance for other states, and adjustments to an employment level leads to our tripling for the region. For Region IV our component estimates rest on data for employment of ten major roads in the region in both 1874 and 1880. We extrapolated from 1874 to 1870, using the ton-mile data for all railroads in the region as reported in the same Aldrich report from which Fishlow drew the five-road sample he used to extrapolate total U.S. ton miles. With these fairly specific, localized, and comprehensive indications we find it difficult to believe that the rise for these two regions was only two-thirds of what we estimate. It seems preferable to leave open the question of the validity of extrapolating for the U.S. from ton mileage on five railroads, and the accuracy of an implicit cross product of three ratios, to yield results—via extrapolation from two states to the U.S.—that are precise within 10 per cent.

Daniel Carson, working from the Population Census data, provides a higher estimate for 1870, and Fishlow derives a similar figure in confirmation of his estimate for that year.[288] Implicitly both analysts assume that

---

[288] Daniel Carson, "Changes in the Industrial Composition of Manpower since the Civil War," in *Studies in Income and Wealth*, *11*, 1949, p. 127. See also Fishlow in this volume.

the ratio of all railroad employees, classified as such, to the number not so classified was virtually identical in 1880 and 1870.

But a look at the details of the Census indicates that the undercoverage of this group by the Census "railroad men" category was not in fact constant. Thus, the Census reports a 36 per cent decline of "employees of railroad companies" in Connecticut from 1870 to 1880 (despite statewide data showing a marked rise in tonnage hauled), and about the same decline in Alabama; and though it reports a U.S. gain of 70 per cent from 1870 to 1880, that gain is net of a substantial (and unreasonable) decline in the absolute number of railroad employees of Irish origin.[289] It would appear that variations in the rate of railroad construction affected the reported series on employees, making them useless as a basis for estimating operating employment, or checking such estimates at all precisely.[290] If we exclude natives of Ireland from the total count of railroad men in both 1870 and 1880, on the assumption that the count for this group was distorted by the inclusion of construction activity, we arrive at a Population Census figure that rises 140 per cent from 1870 to 1880. This gain compares with 160 per cent implicit in the present estimates and 80 per cent in Fishlow's (see Table 4).

TEACHERS: 1830–1960

For 1900–60 we adopt the Population Census count of teachers, in preference to the figures reported to the Office of Education by cities, county boards of education, and other units.[291] The Office of Education is concerned with public education, providing data on private education only at intervals. Moreover, the usual possibilities of duplication in an establishment source (for example, an art teacher working in two different schools during the payroll period and being reported by each) are further complicated by the inevitable difficulty in securing reports from the host of private schools. Both factors make it desirable to rely on an unduplicated source.

[289] 1870 Census, *Population*, pp. 676, 707. 1880 Census, *Population*, pp. 762, 763.

[290] In Georgia, for example, substantial construction work in 1869 was done on the Marietta and North Georgia, on the Chester and Lenoir, Savannah, Florida and Western. Instead of 196 miles completed in 1869, the state had only 24 miles completed for these roads in 1879. For Alabama, the Alabama Great Southern, Montgomery and Eufalfa, Selma, Rome and Dalton completed 159 miles in the earlier year and none in the later. (1880 Census, *Railroads*, p. 349.) The decline in the natives of Ireland clearly suggests a decline in construction employees, not in operating ones.

[291] 1900 to 1950: Population Census data are summarized in Kaplan and Casey (*Occupational Trends*, p. 22). 1960: U.S. Census of Population, 1960, Report PC(1)1 C, pp. 1–216, Table 87. Office of Education, *Biennial Survey of Education in the United States*, 1954–56, Chapter 1, Table 13.

For the years prior to 1890 the Population Census totals, though including employment in many small, private schools, prove to be smaller than those from the early Censuses of educational institutions.[292] This gap reflects the greater tendency in those early decades for teaching to have been a part-time job. A fair number of teachers presumably reported another occupation in the population count, because they spent more time in it than teaching, because they were engaged in it at the time of the Census, etc. (For 1860, 1880, and 1890 the Population Census total runs about 80 per cent of the establishment Census count.[293] For 1870 the ratio is down to 57 per cent, but this discrepancy apparently reflects an inadequate Population Census: it gained by a mere 10 per cent over 1860, as compared with 30 per cent and greater gains in other decades, and a near 50 per cent gain in the school Census.[294]) We, therefore, use the 75 per cent ratio prevailing in 1860 as a basis for extrapolating the 1860 Population Census count back to 1850.[295] We have a rough confirmation of this figure from the count of male teachers reported by the Census of that year.[296]

The Census for 1840 reported the number of scholars and schools, but not the number of teachers.[297] We approximate the number of teachers by the close relationship between the number of schools and teachers in 1850 and 1860, extrapolating that ratio.[298] The implicit size of class per teacher—thirty-five—is slightly above the 1850 figure of thirty-three. As such it compares quite favorably with what contemporary sources indicate. The school returns for Massachusetts in 1836 have an implicit average of 29.5 students per teacher.[299] In the late 1830's James Hall described

[292] For example, 1870 Census, *The Statistics of the Population of the United States*, Vol. I, p. 458.

[293] Population Census count: 1880 Census, *Population*, pp. 760, 792; 1870 Census, *Population*, I, pp. 676, 688; 1860 Census, *Population*, pp. 673, 677. 1890 Census, *Population*, II, pp. 354–355. Establishment count: 1870 Census, *Population*, I, p. 458; 1880 Census, *Population*, p. 918; 1890 Census, *Population*, II, *Report on Education*, p. 51.

[294] The school Census data gain itself may be understated: for example, it reports teachers in New York and Illinois decreasing markedly from 1870 to 1880, whereas the Population Census data more reasonably show a marked rise. The 1880 school Census specified "whole number employed at one time," whereas the 1870 Census makes no such qualification, and apparently thus arrived at a high, duplicated, count.

[295] 1850 teachers count from *The Seventh Census of the United States: 1850*, p. lx.

[296] Our estimate of 79,300 teachers times the 66 per cent ratio of female teachers to all teachers (shown in 1870) implies 27,000 male teachers and professors. The reported number was 30,530.

[297] Summarized in *The Seventh Census of the United States: 1850*, p. lxi.

[298] 1850: *ibid.*, p. lx, lxi. 1860: *Statistics of the United States (including mortality, property, etc.)*, 1866, pp. 505–506. The ratio (130 per cent in 1860 and 122 per cent in 1850) was taken as 115 per cent in 1840.

[299] Summarized in *North American Review*, Vol. 44, April 1837, p. 503. The data cover reports from 289 towns and cities.

a typical Midwestern "school of 30 scholars."[300] (Moreover, the report of the New York State superintendent of common schools has an implicit average for 1840 that is virtually identical with that for the state in the 1850 Census.)[301]

For 1830 we compute the number of teachers from the ratio to pupils. A survey by the U.S. Secretary of State provides figures on the number of pupils in free schools in Connecticut, Maine, Rhode Island, and Indiana in 1830.[302] To this number we can add that for New York State, from the superintendent's reports.[303] These data are then used to estimate the ratio of pupils to total white population in the five states in 1830. They are then compared with a ratio similarly computed for 1840.[304] The lack of change in the ratio for these states is taken to indicate that the U.S. 1840 ratio of pupils to white population also was unchanged from the 1830 ratio.

Applying the 1840 ratio, therefore, to the 1830 white population gives a figure of 1,503,000 pupils. Dividing this by a class of thirty-eight pupils per average teacher, we obtain 40,000 teachers associated with the school system.[305] Adjusting this to the Population Census level, we then get a figure of 30,000 for persons whose primary activity was teaching, whether in public or private schools.[306]

For 1790 information collected with the Census for Philadelphia gives the number of professors and teachers in middle and south Philadelphia and in Southwark.[307] Applying the ratio of teachers to population in middle and south Philadelphia (then including the bulk of the city's population) to the urban population total in 1790, we obtain a total of 570. (Philadelphia then had about 12 per cent of that group.[308]) For the rural population we rely on the evidence of the 1840 Census, which reported

---

[300] James Hall, *Notes on the Western States*, 1838, p. 204.

[301] S. S. Randall, *A Digest of the Common School System of the State of New York*, Albany, 1844, p. 83. *The Seventh Census of the United States: 1850.*

[302] 23d Congress, 1st Session, *Statistical View of the Population of the United States from 1790 to 1830 Inclusive*, 1835, pp. 190, 209, 214.

[303] New York State data for 1830 and 1840 from Randall, *Digest of Common School System.*

[304] Data for the first four states from 1850 Census, p. lxi. For New York we use Randall, *Digest of Common School System*, for both dates, in order to obtain maximum comparability.

[305] The ratio of 27.4 pupils per teacher in 1860, 30.6 in 1850, and 35.6 in 1840 was assumed at 38.0 in 1830.

[306] In 1860 the Population Census count of teachers was 75 per cent of the total derived from the school Census, and that ratio was adopted for earlier years as well. (The ratio for 1870 was 57 per cent; for 1880, 81 per cent; and for 1890, 82 per cent.)

[307] Census, *A Century of Population Growth*, p. 142.

[308] Population data from *Historical Statistics*, 1960, p. 9.

very few "scholars" outside the states with major cities. A fortiori, there would have been still fewer in 1790. We therefore increase our urban teacher total of 570 to an arbitrary 1,000 to give a national total for 1790.

For 1800, 1810, 1820, we apply ratios of teachers to the white population. Ratios were computed from the above data for 1790, 1830, 1840, 1850, and 1860. The curvilinear trend, fitting these figures rather easily, was used to interpolate the 1800–20 ratios.

EMPLOYEES IN DOMESTIC SERVICE

Procedures used for 1900–60 are described elsewhere.[309] This series related to all persons in the domestic service industry—a group that includes a small number of the servant aristocracy (coachmen, private policemen, and so on) in addition to servants *per se*, and will therefore exceed the Census counts for domestic servants.

For 1870–90, the 1900 total was extrapolated by the Census total for servants and laundresses (except laundry) as adjusted by Edwards for these years.[310]

For 1860, the Census of Population count for servants and allied occupations was used, with minor adjustment.[311]

The Census count of male domestic servants in 1850 was inflated on the assumption that the ratio of female domestics to total domestics showed a mild trend over time, declining to the 87 per cent in the reported Census figures for 1870, and the 84 per cent in 1880, 1890, and 1900.[312]

For 1790 we estimate one servant per family in northern and Middle Atlantic cities, to give just under 25,000.[313] This ratio is based on an 1830 statement by a Philadelphia society for placing domestic servants to the effect that there were "at least the same number of domestics as houses in the city."[314] (Note that our series applies to free servants

---

[309] Cf. the author's *Manpower in Economic Growth*.

[310] Edwards, *Comparative Occupation Statistics*, p. 112.

[311] 1860 Census, *Population*, pp. 662 ff. Total for servants, domestics, housekeepers, laundresses, matrons, cooks and stewards, porters. This 634,000 total was reduced 34,000 on the assumption that the 40,000 hotelkeepers, restaurant and saloon keepers, and bartenders in that year had that many employees.

[312] The proportion for 1850 was estimated at 93.9 per cent, and for 1860 at 90.6 per cent, on the assumption of a 3.3 per cent change from one decade to the next—the same as that occurring between 1870 and 1880.

The resultant 365,000 was reduced to 350,000 to allow for servants in hotels, saloons, etc.

[313] All cities of any consequence in 1790 are listed in Census, *A Century of Population Growth*, pp. 11, 78. We exclude southern cities since the bulk of servants in private households were slaves. Dividing this population total by the average size of free family (*ibid.*, p. 96) gives an estimate of 22,000 families in the cities having free servants.

[314] Address to the Public of the Society for the Encouragement of Faithful Domestics, Philadelphia, July 20, 1830, p. 1.

only, hence excludes the bulk of those in the South, and does not include servants in taverns, hotels, etc., who are classified under trade.)

As an approximate check we can make an estimate from data for 1860. The 1860 ratios of servants per family in each region applied to the number of free families in 1790 gives a figure of about 50,000.[315] This figure is clearly on the high side, reflecting (1) the considerable advance in the supply of domestic servants (indicated by the stability and weakness of wage rates for the group after the 1847–50 migrations), and (2) the general increase in the standard of living. Both forces would tend to increase the number in domestic service, the former much more strongly.

For 1800–40, if we chart the relationship between servants as estimated above for 1790 and 1850–1950 against the number of white families at those twelve decadal dates, a simple curvilinear relationship appears.[316]

This relationship was interpolated for the 1800–40 period to give servant-family ratios which, applied to estimates of families, gives an interpolating series.[317]

---

[315] 1860 data from Census, *Population in the United States in 1860*, 1864, pp. 674–675. 1870 data from 1870 Census, *Population*, 1872, I, p. 595. We compute ratios for New York, the New England States, New Jersey, Pennsylvania, and Delaware as a group.

[316] Simple for purposes of extrapolation. The 1870 figure is above trend, reflecting the results of abolition and the tremendous increment to the free labor force. 1920 and 1950 are below trend, reflecting the effects of prosperity attached to both wars. These years were excluded in determining the trend line.

[317] Estimates of white families were derived by the writer in *Demographic and Economic Change in Developed Countries*, Special Conference 11, Princeton for NBER, 1960, pp. 413–414.

General Comment

BRINLEY THOMAS

UNIVERSITY COLLEGE, WALES

Like its predecessor, this volume presents important additions to our stock of statistical time series and throws new light on old problems in the interpretation of nineteenth century development. This general comment draws together the parts of the papers which bear on the comparative study of American and British economic growth and on the mechanism of long swings.

I shall take first the interesting section of Lebergott's paper dealing with the role of agriculture in the process of growth in the United States. "Occupying nearly 75 per cent of our labor force in 1800, farming occupied over half the labor force until some time between 1880 and 1890 . . . . The Kuznets and Gallman estimates of national income are consistent with this conclusion." Lebergott goes on to say that "an industry that used from 50 to 90 per cent of the nation's labor input over the first century of our national existence was surely more likely to make greater demands on most supplying industries than one that merely accounted for 5 or 10 per cent."

Before the above figures are discussed, it will be instructive to look at the corresponding data for the United Kingdom. Agriculture occupied 36 per cent of the British labor force in 1801, 22 per cent in 1851, and 11 per cent in 1891. By 1871 there were more persons in domestic service than in farming. Phyllis Deane and W. A. Cole have concluded that in the eighteenth century in the United Kingdom ". . . the growth of the home market for industrial goods was closely bound up with the fortunes of the agricultural community, in much the same way as the growth of the export trade depended on the prosperity of the primary producers overseas."[1] The part which agriculture played in the United States during the nineteenth century (which has been underestimated) was very different from the increasingly passive and subsidiary role that British agriculture played in that century. Lebergott cites Leontieff's estimates which show that a dollar's worth of agricultural output requires far more iron and

---

[1] *British Economic Growth, 1688–1959*, Cambridge, Eng., 1962, p. 92.

steel output and machinery production than does a dollar's worth of output generated by textiles or miscellaneous manufacturing or even by the iron and steel industry itself; and he rightly emphasizes the part of agricultural investment devoted to land clearing, which entailed copious demands for breaking plows, axes, scythes, and hoes. Added to this were the innovations which made farming more capital-intensive and the system of land ownership which gave owner-occupiers an incentive to invest liberally in the land.

To make sense of the differing roles of agriculture in the economic growth of the United States and the United Kingdom in the nineteenth century, it is essential to contrast the status of the two countries in the international economy. Instead of seeking an interpretation along these lines, Lebergott jumped to the conclusion that his data "... indicate only one thing: U.S. experience reveals no higher law at work forcing a decline of the share of the labor force in agriculture during economic growth." This assertion is unwarranted. It can be shown to be wrong both theoretically and empirically.

The deeper questions raised by these data could have been more fruitfully dealt with if the time series had been handled according to a different chronology. This criticism applies to other papers as well. Instead of presenting the data mechanically for periods such as 1840–60, 1860–80, 1880–1910, etc., would it not have been more instructive to arrange them according to the chronology of the long swing? Several papers in this volume, e.g., those by Gallman, Lebergott, and Gottlieb, have confirmed the statistical findings of Kuznets, Abramovitz, and others in this field. It would have been an advantage to relate these to similar findings for the United Kingdom and other countries.

For example, Gallman shows that during periods of rising or peak long-swing rates of advance for construction, the share of construction in gross domestic capital formation remained roughly constant, whereas during periods of declining or trough rates of advance, the share of construction fell sharply (Tables 4 and 6). Lebergott, in his comparative analysis of the United States and the United Kingdom in the period 1880–1910, makes the point that the "... contrasting migration flows plus variations in the rate of natural increase generated differing manpower requirements in residential construction. The induced effects on highway and public building construction, on plant for making steel, brick, and lumber can be surmised, although not measured at present." Gottlieb demonstrates an oscillation in the ratios of residential to total building —a rise in residential booms and a fall in residential contractions. This confirms Kuznets' results for the years 1870–1920, but Gottlieb shows that the process was at work over a longer period.

I would like to suggest that an approach centering on the interaction between the United States and the United Kingdom would give a more satisfactory explanation of some of the phenomena. Let me summarize some of the statistical findings on this process of interaction between 1870 to 1913.

1. There were long swings in additions to population and internal migration in the United States which were inverse to those in the United Kingdom.

2. In the United States there was an inverse relation between long swings in population-sensitive capital formation (i.e., nonfarm residential construction and durable capital expenditure by railroads) and long swings in other capital formation (i.e., net changes in inventories, producer durable equipment, and foreign claims) and in the flow of goods to consumers.

3. The long swings in British domestic capital formation were positively related to those in British additions to population and inverse to those in population-sensitive capital formation in the United States.

4. Long swings in British domestic capital formation were synchronized with long swings in American "other" capital formation and exports.

5. The long swings in British and European emigration and British capital exports were synchronized with the long swings in population-sensitive capital formation and internal migration in the United States.

6. There were long swings in American imports at constant prices.

7. The gross volume of population-sensitive capital formation in the United States was over 40 per cent of total capital formation in the 1870's and even in the first decade of the twentieth century it was still 25 per cent of the total.

There is a strong disposition among American economists to seek an explanation of every feature of American long swings exclusively in terms of forces operating within the United States itself.[2] The essence of the causal sequence suggested by Kuznets can be put as follows. Starting with long swings in the flow of goods to consumers in the United States, we have—after some lag—swings in immigration, additions to population, and population-sensitive capital formation; this causes inverted swings in other capital formation and additions to flow of goods to consumers, and the latter then induces a new inflow of population.[3] The weakness of

---

[2] Richard A. Easterlin's attempt to show that the inverse relation between long swings in the United States and the United Kingdom in the period 1870-1913 does not exist can hardly be said to succeed ("Influences in European Overseas Emigration before World War I," *Economic Development and Cultural Change*, April 1961, pp. 347–348). He himself admits that "... to speak of conclusions on the basis of this preliminary reconnaissance would be presumptuous" (p. 348).

[3] Kuznets, *Capital in the American Economy: Its Formation and Financing*, Princeton for NBER, 1961, pp. 346–349.

this model is that the whole thing turns on a suspiciously long "lag" between additions to the flow of goods to consumers and additions to population. Kuznets has to ask us " . . . to allow for a long lag that would, in a sense, turn negative into positive association."[4] Surely this is going too far. We are dealing here with *long* swings with a span of about twenty years. There is something arbitrary about singling out one case of negative association among many and treating it as a long lag. If it is right to do this in one case, why not in others too?

Much needs to be done before the mechanism of long swings can be satisfactorily explained. I am still convinced that these problems can best be resolved on the basis of a two-country model.[5] This kind of analysis may be briefly summarized as follows.

Let there be two countries: A, an industrialized creditor, and B, a country underpopulated but rich in natural resources. Each country is divided into two sectors—domestic capital construction and export. There is free mobility of factors between the sectors and between the countries. Let the word "period" refer to either the upswing or the downswing phase of the construction cycle, i.e., half the span of a long swing. The following relationships are assumed:

a. The level of activity in the export sector of one country depends on the marginal efficiency of capital in the capital construction sector of the other country in the same period.

b. The export capacity of each country is a function of the rate of expansion achieved in the capital construction sector of that country in the previous period.

c. A major fraction of total capital formation is population-sensitive.

d. Changes in population growth are dominated by changes in the net external migration balance.

Let us suppose that an outflow of population takes place from A to B. This stimulates activity in population-sensitive capital formation in B and retards it in A. The upswing in B attracts a flow of capital funds from A and this in turn stimulates A's export sector. Thus in A there is a downswing in population-sensitive capital formation and an upswing in exports. Meanwhile in B the construction boom triggered off by the inflowing population gives rise to an internal suction of factors. There are two phases in this upswing in B. First, there is a rapid build-up of population-sensitive capital formation as an *effect* of population increase; in this phase, the boom is supply-determined and there is a widening of the capital structure. In the second phase, limitational factors become

---

[4] *Ibid.*, p. 347.
[5] See my *Migration and Economic Growth*, Cambridge, Eng., 1954, p. 189.

important and the wage-price spiral starts. The rise in wages will now lead to renewed immigration of labor, and the inflow is now demand-determined.

Through the interaction of the multiplier and the accelerator (with lags), the construction boom in B reaches its downturn. It is a crucial matter whether the timing of this downturn in B is dependent on what is simultaneously happening in A. Statistical analysis of British long swings suggests that upturns in home construction preceded downturns in the export sector, and upturns in the export sector preceded downturns in home construction.[6] If this is so, the downturn of the boom in B is determined by the revival in population-sensitive capital formation in A. Factors and loanable funds in A now move into the reviving construction sector; migration into towns in A takes the place of emigration from A to B. The upswing in construction in A is accompanied by a downswing in population-sensitive capital formation and a revival in the export sector in B. This goes on until the upturn of the construction cycle is reached in B, which is a signal for a revival in exports and then a downturn in home construction in A. Space does not allow an account of the price and balance-of-payments implications of these inverse cycles. The model can demonstrate that international migration, through its influence on changes in population growth and population-sensitive capital formation, is the most plausible explanation of inverse long swings.

Within such a framework one can see more clearly the external determinants of the components and sources of output growth in the United States. Let me now return to the role of agriculture which I mentioned earlier. I would interpret the evolution as follows. Each upswing phase of the long cycle in the United States saw a great extension of railways, construction, roads, land clearing, etc.; the capacity thus built up fulfilled itself in large additions to agricultural output in the next period, i.e., the downswing phase of the long cycle. It was in these latter phases, coinciding with the capital formation booms in the United Kingdom, that the United States had its upswings in additions to the flow of goods to consumers and to net claims on foreign countries. In this connection, it is significant that crude and manufactured foodstuffs and crude materials comprised as much as 80 per cent of total United States exports even as late as 1890; the corresponding figures for 1856 and 1875 were 84 and 80 per cent, respectively. The major market was the United Kingdom. There was a very high industrial input (drawing heavily on the metal industries) into agriculture and new settlement; the United States was meeting not only the rapidly increasing demand of her fast-growing

[6] See *ibid.*, p. 186.

population but was also undermining the obsolete agrarian economies of Europe. In the 1880's European agriculture felt the boomerang effects of earlier railroad and steamship investment; although wages on the wheat farms of northwestern United States in 1887 were four times as high as in Rhenish Prussia, nevertheless the cost of production in Prussia was double the American (80 cents a bushel against 40 cents). William N. Parker's illuminating paper on the increase in productivity in the production of small grains is very relevant in this context.

The economic interplay between the New World and the Old entailed a vast expansion of remarkably efficient agriculture in the United States, and its upsurges of output and exports were most prominent during the downswings of the long cycle. The impact on the rural economies of Europe meant an intensification of population pressure and renewed outflows of migrants across the Atlantic. The mass rural exodus of "new" immigrants from southern and eastern Europe between 1900 and 1913 must be attributed partly to the delayed force of New World innovations reaching back to the Old World with disruptive power. In Schumpeter's words, ". . . the story of the way in which civilized humanity got and fought cheap bread ... is the story of American railroads and American machinery."[7]

In seeking to explain the complex process of output growth over the last century, we must as economists and economic historians pay due attention to the complex interaction between major economies from one period to the next. Analysis which tends to be inward-looking is apt to miss the interesting clues.

[7] Joseph A. Schumpeter, *Business Cycles*, Vol. 1, New York, 1939, p. 319.

Output of Final Products

The New England Textile Industry, 1825–60: Trends and Fluctuations

LANCE E. DAVIS

PURDUE UNIVERSITY

H. LOUIS STETTLER III

THE JOHNS HOPKINS UNIVERSITY

I. Introduction

Many studies of the growth of the American economy during the first three-quarters of the nineteenth century have suffered both from the diffusion of their focus and from the weakness of their statistical foundations. Reliance on the federal Census returns necessitates estimating the underreportings that plague those volumes and, perhaps more important, limits the study to benchmark years—a limitation that excludes any discussion of fluctuations between those years. Moreover, these studies have frequently attempted to draw economy-wide conclusions from data indicative of conditions in only a local area or, conversely, to describe each area in terms of a national average. Thus, there have been frequent attempts to apply technical coefficients or economic relations derived from one area to another region characterized by an entirely different technology or subject to very different market forces.

For most of the period before 1870, no national market existed for many industries. In this period it appears almost useless to study the economy as an integrated whole rather than as a sum of various quite heterogeneous parts. Any national figures, almost by definition, blur the local characteristics; and, all too frequently, marginal changes (changes that are small, but possibly of crucial importance for growth) in local figures get lost in the national totals. Ideally, it would be best to study every industry in every region; however, because of the scarcity of source material, it is improbable that this goal can be achieved. What is more possible, however, is a study of the important industries in each region during those periods when the industry was changing rapidly (either expanding or contracting). Since almost all industries that play a significant role in the process of economic growth could be included in this more restrictive study, such an approach seems fruitful despite the narrowing of focus. At the same time, insofar as it is possible to estimate output from all industries in all regions, studies of aggregate income will not be hampered. In fact, even these broader studies may be improved, since a comparison of local firm records with Census reports may provide a better measure of underreporting than anything previously available.[1]

Furthermore, from the point of view of economic theory, a certain economy is achieved in studying those sectors that are undergoing rapid change. Moreover, since rapidly changing industries or those heavily concentrated (geographically) tend to leave business records and other economic artifacts which may substitute for the Census as a source of statistical data, there is an additional incentive to adopt this regional methodology.

II. The Estimates

THE DATA

The cotton textile industry provides a fine example of what can be done with a regional approach (in fact, the example may be too good). The industry was heavily concentrated in New England and, within New England, the growth of the industry after 1820 was associated with the rise of a particular group of firms. Fortunately there exist today, at various museums and libraries on the East Coast, the original business records of a significant proportion of these firms.[2]

[1] The records of some of the textile mills, for example, include copies of the Census enumerators' reports as well as their mill records—which show quite different totals. The following table shows the discrepancies for six mills reporting to the 1859 Census (in millions of yards).

	Total	Merrimack	Hamilton	Suffolk	Tremont	Lawrence	Massachusetts Cotton
Reported	96.0	20.8	12.2	8.0	11.9	18.7	24.4
Actual	96.4	22.1	11.6	8.5	11.0	18.6	24.6

These six firms are a sample; all errors exceed 5 per cent, but they appear random with mean equal to zero. There is, of course, no guarantee of the randomness of the errors. More important objections to use of the federal Census after 1820 for the textile industry flow from (1) constantly changing definitions (e.g., establishment); (2) the paucity of information gathered and reported vis-à-vis the Massachusetts Census (especially the failure to report yards of output in 1839); (3) the lack of a local breakdown in 1850; and (4) the number of different "official" versions.

[2] The materials for this paper were gathered from various textile collections deposited at the Baker Library, Harvard, the Merrimack Valley Textile Museum, and the Manchester Historical Society Museum. There are to the knowledge of the authors other collections in the possession of local historical and academic units; in fact, the problem

There were two principal types of textile firms in New England.[3] The firms of the Massachusetts type, modeled after the Boston Manufacturing Company, were located on the major rivers of northern New England. They were typically multifacility operations (sometimes with bleacheries or printworks) and generally capitalized in excess of $500,000. From the beginning, these firms were integrated producing units heavily concentrated in the low-count goods. Existing alongside the Massachusetts-type operations were a much larger number of small proprietary single-mill firms located on the streams of lower New England. These so-called Rhode Island-type mills were small, often specialized, and tended to produce medium grades of cloth (particularly printing cloths).[4] During the mid-1840's the structure of the industry in lower New England changed, and thereafter the Rhode Island mills began to resemble their northern counterparts more closely.

Table 1 lists the firms whose records were used in this study. The distribution of firms is not wholly geographical. Firms of the Rhode Island type were located in Massachusetts as well as in Rhode Island and Connecticut. Moreover, although Metacomet was located in New Bedford and resembled the larger firms, it was really representative of the post-1845 mills springing up in lower New England, and was therefore included as a Rhode Island-type mill.

The company records for at least sixteen of the Massachusetts-type mills exist for at least some part of the period prior to the Civil War. The journals, semiannual accounts, and treasurer's reports give sufficient information to ascertain the amount of cloth produced by each firm.[5] Generally, the output by six-month periods is available, but for certain years the output of Nashua and Jackson was only given on a yearly basis.

---

of data is not its insufficiency but rather its overabundance. In choosing the particular data used in this paper, preference was given to those records that were complete and continuous for long periods, and reasonably intelligible.

[3] The relative contributions of the two families of firms to the growth of the New England textile industry might be seen from the county returns in Massachusetts. Middlesex and Hampden counties included the large mill centers of Chicopee and Lowell; Bristol and Worcester counties were always centers of the smaller enterprises. The number of spindles (in thousands) follows:

	1820	1837	1855
Middlesex-Hampden	10	233	635
Bristol-Worcester	18	240	508

[4] The best discussion of the two types of enterprise is found in Caroline Ware's *The Early New England Cotton Manufacture*, Cambridge, Mass., 1931.

[5] In some cases it was necessary to add yards sold to the change in inventory for a period, since output figures were not always present whereas sales and inventory values were more common.

TABLE 1

SAMPLE MILLS, 1815-60

Firm Name	Years for Which Records Were Examined
MASSACHUSETTS TYPE	
Boston	1815-60
Merrimack	1824-60
Hamilton	1828-60
Suffolk	1832-60
Tremont	1832-60
Lawrence	1833-60
Massachusetts Cotton	1840-60
Naumkeag	1848-60
Lancaster	1847-60
Dwight[a]	1834-60[b]
Lyman[a]	1850-60
Nashua	1826-60
Jackson	1832-60
Amoskeag[a]	1837-60
Laconia	1845-60
Pepperill	1851-60
RHODE ISLAND TYPE	
Rampo	1821-33
Slater	1826-34
Slater and Tiffany	1827-37
Sutton	1830-59
Metacomet	1848-60

[a] Records are available for different groups of mills operated by these companies.

[b] Includes the records of the Cabot and Perkins Companies which were merged into Dwight in 1856.

For the Rhode Island-type firms, output could be estimated from either weaving ledgers or consignment books.[6] Typically, these data were available on a monthly basis.

ADJUSTMENTS TO A UNIFORM ACCOUNTING PERIOD

Reporting techniques differed greatly from mill to mill. Every month represented the closing of at least one company's accounts. Moreover, companies changed their final closing date from time to time. To make the data comparable, each mill's output was allocated uniformly over the months covered by the accounting period. Uniform allocation was

[6] The use of the consignment books introduces a sequence of lags, since there is no way of dating consignments in terms of the time elapsed between production and shipment. In the case of Slater and Tiffany, pounds of cloth produced were known but not yards.

TABLE 2

QUARTERLY OUTPUT OF SELECTED FIRMS OF THE
MASSACHUSETTS TYPE, 1821-60
(thousand yards)

Year	Jan. 1–Mar. 31	April 1–June 30	July 1–Sept. 30	Oct. 1–Dec. 31
1821	337	363	370	384
1822	403	439	420	380
1823	408	465	454	433
1824	451	487	483	865
1825	1,044	886	813	774
1826	783	988	1,138	1,168
1827	1,193	1,442	1,659	1,658
1828	2,414	2,688	2,858	2,746
1829	2,569	2,681	2,702	2,711
1830	2,711	3,032	3,354	3,282
1831	3,701	3,899	4,036	3,956
1832	4,094	4,604	5,799	6,091
1833	6,956	7,481	7,975	8,052
1834	8,570	9,050	9,001	9,038
1835	9,971	10,920	11,282	11,196
1836	11,534	12,328	12,642	12,186
1837	12,690	12,620	12,063	11,746
1838	12,610	13,884	14,241	14,028
1839	14,461	15,080	14,961	14,540
1840	14,471	13,877	13,406	14,083
1841	14,734	15,088	16,020	16,234
1842	16,888	16,913	16,404	16,047
1843	15,727	15,868	16,697	16,958
1844	17,878	17,916	18,783	18,596
1845	19,622	19,994	21,163	21,296
1846	21,747	21,863	21,718	21,882
1847	22,271	22,474	22,593	22,923
1848	24,073	25,680	27,260	26,786
1849	28,030	28,229	28,994	29,827
1850	27,782	27,604	29,129	28,022
1851	28,249	28,799	31,452	32,832
1852	34,427	35,246	37,102	37,272
1853	34,580	31,172	35,938	36,509
1854	35,198	35,241	35,993	36,446
1855	35,976	36,018	35,396	35,956
1856	37,316	37,258	38,132	38,570
1857	34,638	33,929	31,451	29,474
1858	33,561	35,984	39,373	40,840
1859	40,962	41,350	42,889	43,430
1860	43,084	43,229	40,881	40,447

Source: Company records of Boston, Merrimack, Hamilton, Suffolk, Tremont, Lawrence, Naumkeag, Lancaster, Dwight, Lyman, Nashua, Jackson, Amoskeag, Laconia, and Pepperill.

selected for a number of reasons. First, it has been generally inferred that the short-run marginal cost curves were steep about some "capacity" level.[7] Second, the data required to construct a monthly deflating index were not available. What data exist are generally of poor quality because they represent only the history of the Rhode Island-type mills, whose output was often curtailed in response to a freshet or to install new equipment. These conditions were not characteristic of the rest of the industry. Third, a monthly index constructed from Rhode Island data, Nashua reports, and consumption of cotton in Cabot mill number 1 (where monthly data were available), with adjustments made for floods, replacement of equipment, and other local events, is not significantly different from the results achieved by uniform allocation. The monthly figures were summed to standard quarters and calendar years. Annual output by firm may be found in Table A-1, while the quarterly output of selected Massachusetts firms will be found summed in Table 2.

CHOICE OF OUTPUT MEASURE

The output reported in these tables is in terms of yards. The company books described output variously—in pieces, pounds, and yards. Yards were, however, the most common; they were available in all but a single case. Pieces, a widely varying metric, were the least common. Pounds of output, when not reported, could usually be estimated from the weight of cotton inputs corrected for the mill's waste factor or from the yards of output, if the weight per yard ratios were available for the full range of product. Yards were used as the measure for this study because they represented a more meaningful unit of demand, because they were more generally available, and because they appear to be a more useful tool in comparing input and output ratios.[8] Despite these advantages, yards are not a perfect measure. Because of the lack of homogeneity arising from changing width and count patterns, they too may be criticized. However, tests suggest that for this sample at least the bias injected by nonhomogeneity may not be too significant.[9]

[7] See R. C. O. Mathews, *A Study in Trade-Cycle History*, Cambridge, Eng., p. 130.
[8] Robert Layer (in *Earnings of Cotton Mill Operatives, 1825-1914*, Cambridge, Mass., 1955) feels that pounds are a superior measure. Nominally, pounds ignore width differentials but not count differences. Since yards and standard yards appear to have a relatively stable relationship on the aggregate level, the question of unit recedes in significance. Output in yards or pounds would differ by a scale factor. There are, however, certain marginal considerations which make yards preferable. In estimating production functions, the marginal products of labor, cotton, and capital can be ascertained by using yards as a measure of output; if, on the other hand, pounds were used, the rather constant proportion between cotton input and cloth output (regardless of the quality of output) does not allow the other parameters to have significant coefficients.

ADJUSTMENT TO REGIONAL LEVELS

Table 3 shows the relationship between sample output and the output of various regions as they appeared in the Census. The relationships are rather stable except for the initial change in the thirties. This jump may

TABLE 3

RATIO OF OUTPUTS OF SAMPLE FIRMS TO CENSUS OUTPUT TOTALS, SELECTED YEARS, 1831-59

Year	Rhode Island, Connecticut	Massachusetts	Maine, New Hampshire	New England
1831	.04	.16	.10	.11
1837		.29		
1839[a]	.01	.30	.40	.22
1845		.28		
1849	.04	.23	.32	.19
1855		.29		
1859	.03	.26	.29	.20

Source: 1840, 1850, 1860, federal Censuses; 1831, from Survey of Friends of Domestic Industry; 1837, 1845, 1855, Massachusetts Censuses.

[a] Ratio based on estimate of textile output.

be explained in terms of the development of Lowell and the later textile centers in Manchester and Springfield. In Massachusetts, the sample shows a gain of three firms between 1832 and 1834 (Suffolk, Tremont, and Lawrence), while eighteen were actually incorporated. During the period 1828-30, however, only Hamilton enters the sample while twenty-four firms were incorporated. A portion of the fall in relative output recorded in 1860 is a result of firms leaving the sample in that year.

The movements in the sample output series may be used to provide an index of cyclical movements in textile output. Moreover, when inflated to approximate total production they furnish a measure of total output between benchmark years. Their usefulness, however, depends on the

[9] For those mills that reported both yardage and pounds, it was possible to compensate for quality and width differences by converting their output into standard yards. Therefore, the thirty-nine mills with data on both yardage and pounds were divided into two classes, those producing mostly low-count (under 16) goods and those producing finer quality (over 18) cottons. The output of the former class were then converted into standard 14s × 14s, 48 × 48, 36" wide brown sheeting; and the latter's output into 22s × 22s, 40, 36" wide sheeting. Apart from the late twenties, when the sample size was small (only thirteen mills), the relations between standard and recorded yards was extremely stable (see Table A-2); and both series tend to move together. In the thirties, for example, there was first a movement toward heavier goods and then a movement toward lighter ones. If, however, total standard yards (the sum of the standard columns) is compared with total recorded yards, a slight trend toward the coarser cloth is evident.

validity of two assumptions: that the sample of firms is representative of the entire industry, and that the Census provides adequate estimates of total output in the benchmark years. As to the first, the evidence suggests that the sample of Massachusetts-type firms was quite representative of the entire universe, but that the sample of Rhode Island firms was probably less so. As to the second, the Massachusetts inflations based on the state Census probably provide a more trustworthy measure of activity in the benchmark years than the New Hampshire-Maine and New England inflations based on the federal Census do.[10]

The results of such inflations are presented in Table 4. The Massachusetts mills were inflated on a straight-line basis between the Census years 1831, 1837, 1845, and 1855. The series was extrapolated to 1860 according to the 1845–55 increments; and backward extrapolation from 1831 to 1825 was based on an incremental change of zero. This latter extrapolation probably overstates output in the earlier years (since it appears that the sample size increased relative to the universe in the period) but, in the absence of any benchmark, it remains the best alternative.

The Maine and New Hampshire inflation was based on the federal Census and the survey of 1831. The same basic procedure used in the Massachusetts inflation was followed, but the 1825–31 period was treated in the same manner as the 1831–39 interval. Rhode Island and Connecticut were treated in the same manner as Maine and New Hampshire; however, the changing sample size during the intercensual period created additional problems. Even if these could be solved, however, the smallness of the Rhode Island and Connecticut sample, particularly in the 1839–47 period, argues against attaching great significance to the inflated series.

Two different estimates of total New England output suggest themselves. First, it is possible to sum up the inflated series for the three state groups; second, one could inflate, in a manner similar to the treatment of New Hampshire and Maine, the output figures from all mills for which the data were of good quality. The difference between the two series is much greater than can be explained by the exclusion of Vermont from the first variant. Unfortunately, the former series is also subject to wide fluctuations induced by the inclusion of the poor-quality series from Rhode Island. As a result, the second technique is almost certainly a better indicator of textile activity in New England, and it is a series so constructed that is included on Table 4.

[10] Because of omissions in the 1840 Census, it was necessary to estimate textile output and this procedure injects another element of bias into the inflation based on the federal Census.

TABLE 4

OUTPUT BY REGION, 1826-60
(thousand yards)

	Sample			Inflated			
Year	Rhode Island, Connecticut	Massachusetts	Maine, New Hampshire	Rhode Island, Connecticut	Massachusetts	Maine, New Hampshire	New England
1826	770	3,757	321	27,018	23,475	3,211	37,072
1827	1,337	4,920	1,031	37,100	30,753	10,313	54,107
1828	1,427	8,768	1,939	35,731	54,797	19,386	97,327
1829	1,050	8,313	2,350	32,194	51,957	23,501	96,937
1830	1,536	9,895	2,484	42,715	61,845	24,838	112,536
1831	2,035	12,637	2,955	47,212	79,231	30,810	141,727
1832	1,813	14,730	5,858	44,167	81,832	41,841	175,961
1833	1,605	22,434	8,029	45,650	106,828	50,183	245,672
1834	1,240	27,558	8,092	47,391	111,482	44,954	272,129
1835	1,125	35,262	8,107	44,642	130,598	40,533	314,262
1836	1,489	38,531	10,159	58,421	132,866	46,178	335,797
1837	1,014	35,453	13,667	69,926	118,175	56,944	323,151
1838	1,190	39,675	15,088	99,083	134,449	58,030	359,868
1839	893	42,714	16,328	85,769	147,288	58,316	369,014
1840	372	40,900	14,938	80,870	143,509	53,349	371,012
1841	710	44,876	17,201	151,064	160,272	61,431	432,476
1842	235	45,526	20,726	48,750	165,548	74,023	437,005
1843	993	44,936	20,520	198,400	175,756	73,303	426,125
1844	741	48,332	24,842	145,098	179,000	88,721	465,593
1845	778	48,434	33,643	149,231	179,958	120,152	497,640
1846	711	51,752	35,461	133,962	190,097	126,649	527,120
1847	834	54,849	35,417	151,455	203,150	126,487	532,422
1848	5,157	65,790	38,642	117,740	240,988	138,007	589,217
1849	6,346	68,233	46,848	148,271	242,902	145,989	600,318
1850	6,143	72,210	40,320	146,962	262,583	127,032	585,179
1851	4,245	73,513	48,741	103,790	266,350	155,276	634,098
1852	5,612	86,440	57,609	140,652	312,058	185,596	745,210
1853	6,168	85,195	55,417	158,154	306,452	180,570	725,552
1854	5,989	88,627	54,255	157,605	316,521	178,823	735,368
1855	5,731	91,647	51,603	154,474	328,482	172,067	735,370
1856	6,142	97,437	53,840	170,139	347,990	181,769	774,588
1857	5,812	86,445	43,048	165,114	308,732	147,072	661,353
1858	7,464	93,840	56,510	218,246	347,555	195,401	765,920
1859	7,456	105,192	63,440	225,939	404,580	222,051	856,362
1860	6,130	104,570	63,172	245,200	402,192	224,332	850,188

III. Analysis of Fluctuations in Textile Output

Recently there has been some controversy over business fluctuations in the antebellum decades, and the textile series cast some light on the dating of the so-called cycles.[11]

[11] The question of specific cycles was recently raised by J. R. T. Hughes and N. Rosenberg in "The United States Business Cycle before 1860: Some Problems of Interpretation," *Economic History Review*, 1963; and some questions of long swings were discussed by L. E. Davis in "The New England Textile Mills and the Capital Markets: A Study of Industrial Borrowing, 1840-1860," *Journal of Economic History*, March 1960.

SPECIFIC CYCLES

The strong trend element and the discrete entry of firms into the sample somewhat complicate the analysis of the fluctuations of textile activity; however, Table 5 presents two series that throw some light on the subject; and Table 6 contrasts the textile series with the National Bureau turning points. Column 1 of Table 5 shows changes in the output of the sample firms of the Massachusetts type after correction for lumpiness of entry. To correct for lumpiness, change in output for any pair of years was calculated by summing the differences in production for only those firms which operated throughout both years. This corrected series indicates four years (1825, 1832, 1842, and 1850) when total sample output increased, but the level of production by established firms contracted. Column 2 presents the first differences for the inflated New England series previously presented in Table 4.

The first cycle apparently reached its peak during the last quarter of 1824 and the first quarter of 1825. Output increased until that date for both sample firms and fell for the next three quarters. Inventories were accumulated during the second half of 1824 and remained at substantial levels until mid-1826. Real sales fell during 1824, recovered during the first three quarters of 1825, subsided again until the fourth period of 1826 and the first of 1827. These fluctuations are somewhat at variance with the standard cycle that shows contraction to 1824, recovery during 1824, and contraction again in 1825-26.

All indexes show an increase in textile activity until 1828; however, output increases stopped in the third quarter of that year and did not reach the same level until mid-1830. Inventories increased in the first quarter of 1829 and fell only in the second half of 1830, but real sales fell after mid-1828 and rose again in late 1829. These movements seem to be in accord with measures of general business activity.

The recovery continued until mid-1831, and at that time output peaked for each of the operating companies. Large inventories were, however, accumulated during the next year, and sales fell for the entire sample in 1832. Complete recovery was evident in 1833. It is an open question whether this dip was a product of the smoothing process (since it is not noticeable in the second series) or actually represented a decrease in industry activity. However, additional evidence appears to support the latter conclusions. The standard works on business fluctuations show a general contraction during 1831–34; the Cole and Smith volume-of-trade index shows a drop in 1832; and, perhaps most important, the Rhode Island sample (containing a constant number of firms

TABLE 5

ANNUAL CHANGES IN TEXTILE OUTPUT, 1821-60

Year	Massachusetts, Corrected (thous. yards) (1)	New England (mill. yards) (2)
1821	194	
1822	187	
1823	119	
1824	137	
1825	−188	
1826	236	
1827	1,164	17
1828	1,037	43
1829	−454	0
1830	1,582	16
1831	2,742	29
1832	−305	34
1833	1,439	70
1834	55	27
1835	7,704	42
1836	3,270	21
1837	−3,079	−13
1838	4,223	37
1839	3,038	9
1840	−1,813	2
1841	3,524	61
1842	−3,039	5
1843	−589	−11
1844	3,396	39
1845	102	32
1846	3,317	29
1847	2,961	5
1848	4,957	57
1849	2,544	11
1850	−855	−15
1851	977	49
1852	10,292	11
1853	−1,247	−20
1854	3,433	10
1855	3,020	0
1856	5,791	39
1857	−10,992	−113
1858	7,394	105
1859	11,353	90
1860	−723	−6

Source: Col. 1 from Table A-1; col. 2 from Table 4.

TABLE 6

OUTPUT TURNING POINTS, TEXTILE SERIES AND
NBER ESTIMATES, 1825-58

Trough		Peak	
Textile	NBER	Textile	NBER
1825	1826	1828	1828
1829	1829	1831	
1832		1833	1833
1834	1834	1836	1836
1837	1838	1839	1839
1840		1842	
1843	1843	1844	1845
1845		1846	
1847	1846	1848	1847
1850	1848	1852	1853
1853	1855	1856	1856
1857	1858	1858	1860

Source: Textile series from Table 5; NBER series from W.C. Mitchell, *Business Cycles, The Problem and Its Setting*, New York, NBER, 1927, pp. 425-427.

during this period) shows a decrease in output and a substantial fall in profits.[12]

Textile output increased from 1833 to 1836; however, the rate of growth of output and sales fell substantially during middle and late 1834. After the peak of the third quarter of 1836, output stagnated at a lower level for a year. In the last half of 1837, sales fell sharply, output less rapidly, and inventories accumulated. The severity of the contraction was limited to the third and fourth quarters of 1837. Recovery was immediate and continued sporadically through 1838 and 1839, but the rate of increase was lower in the latter year. Late 1839 and 1840 witnessed a fall in output and sales, and inventories tended to accumulate during the first and third quarters of 1840. The year 1841 marked a moderate recovery in sales and output but only a slight reduction in inventories. The peak of this moderate recovery was reached in the second quarter of 1842, and was followed by three quarters of contracting output. Sales fell during the same period but not as rapidly as production, and as a result inventories were reduced. The trough was reached by the second quarter of 1843. These movements do not coincide precisely with the accepted series. The National Bureau marks 1838 as the trough, but textile activity increased

[12] Arthur H. Cole and Walter B. Smith, *Fluctuations in American Business, 1790-1860*, Cambridge, Mass., 1935.

throughout most of that year. Similarly, the NBER figures do not show the recovery of 1841 and fall in 1842. The troughs of 1843, however, do coincide. The Cole and Smith series also shows no recovery in 1838, but it does reflect the contraction-expansion-contraction phases of 1840, 1841, and 1842.

The middle and late 1840's witnessed a period of sustained increase in both output and sales of cotton cloth. Inventories were reduced to virtually nothing in 1846, but accumulated during the later forties. The industry's growth rate accelerated rapidly during this period through the expansion of existing mills, the entry of new firms at existing sites (the sample added nine new mills built by the old firms and four by new firms), and the development of new sites (for example, Lawrence and New Bedford). The expansion slackened after mid-1848 and ended in the first quarter of 1850, when sales and output fell. Inventories peaked in the fourth quarter, and the trough was reached by the first quarter of 1851. After that date all indexes showed continued improvement until 1853 when output in the first half fell but immediately recovered. This brief drop in production and the concomitant fall in sales were borne entirely by the firms producing finer goods. Output continued to grow during the years 1854, 1855, and 1856. This pattern differs substantially from the general cycle pattern of the National Bureau. Although the Massachusetts data suggest a mild contraction in 1845 and the New Hampshire mills cut back production in late 1847, the Bureau's dates are noticeably out of phase with the fluctuations of textile activity during the 1845–48 period. The textile depression of 1850 seemingly occurred at a time of relative prosperity. Thereafter, however, throughout the mid-1850's textiles appear to have led the economy.

The dominating economic event of the late 1850's was the panic of 1857. Sales and output began to erode early in that year but dropped swiftly during the last half. The year-end inventories were almost twice the level of 1856. Recovery was evident with the new year; predepression levels of sales and output were attained in the third quarter of 1858. Inventories were parred by over 35 per cent but were still substantially above the 1856 level. Output and sales reached a prewar peak in 1859; inventories dropped below the 1856 levels; and profits attained their highest level since 1846. Again, the textile industry apparently led the economy.

LONG SWINGS

The data also appear to cast some light on the existence of long swings in the American economy during the antebellum decades. Cole has observed an apparent long swing during the 1840's and 1850's. In his examination

TABLE 7

THREE-YEAR MOVING AVERAGES OF CHANGE IN TEXTILE OUTPUT, 1822-59

Year	Massachusetts (thous. yards) (1)	New England (mill. yards) (2)
1822	167	
1823	148	
1824	23	
1825	62	
1826	404	
1827	812	
1828	582	20
1829	722	20
1830	1,292	15
1831	1,341	26
1832	1,294	44
1833	398	44
1834	3,068	46
1835	3,678	30
1836	2,633	17
1837	1,473	15
1838	1,396	11
1839	1,818	16
1840	1,585	24
1841	441	23
1842	33	18
1843	76	11
1844	971	20
1845	2,273	33
1846	2,128	22
1847	3,747	30
1848	3,489	24
1849	2,217	18
1850	890	15
1851	3,473	15
1852	3,342	13
1853	4,161	0
1854	1,737	-3
1855	4,083	16
1856	725	-28
1857	733	10
1858	2,587	27
1859	6,010	-66

Source: Table 5.

of the period, he found a trough in 1843 and another in 1858 and concluded "there evidently were forces which, acting slowly, gave decade-long sustaining power to values."[13] A similar swing was found in interest rates of the period.[14] The textile series also show such a swing and, in addition, suggest that the 1843–58 swing was not unique. Table 7 presents three-year moving averages of the changes in textile output. These series indicate a trough in 1843-44, an intervening peak in 1847–49, and a trough in 1856-57. Moreover, they show an earlier swing with an apparent trough in 1824-25 and a peak in 1834-35.

IV. Productivity Change

Table 8, constructed from the spindleage, payroll, and output records of the Tremont, Suffolk, Hamilton, Lawrence, Boston, and Nashua mills, provides some measure of the increase in productivity realized by the Massachusetts-type firms during the antebellum period.[15] Column 1 shows average output per man-day and column 4 average output per spindle in the six mills. Columns 2 and 5 present a three-year moving average of the raw data, and columns 3 and 6 five-year moving averages chosen to smooth the effect of short-term fluctuations.

The series presented in columns 4, 5, and 6 show an increase in output per spindle of almost 50 per cent during the 1830's. Thereafter, the series decline slightly, but the fall is merely a reflection of a reduction in the hours of work and implies no decrease in productivity. (When the series are adjusted to a common workday, they display no significant change after 1840.) The labor figures (columns 1, 2, and 3) also indicate a substantial increase in productivity during a part of the period. These latter series display a monotonic increase until about 1850; thereafter, however, output per worker remained relatively constant.

A comparison of the two series with each other and with the usual indexes of business activity produces only tenuous results. Despite what one might think a priori, the two do not move together in the short run. In the long run, the upward trend of labor productivity and the change in the direction of the capital index about 1840 might suggest that innovation before 1840 was capital-saving as well as laborsaving, while thereafter the laborsaving, capital-using innovations appear to have been more prevalent. The short-cycle behavior also displays no consistent patterns,

[13] Cole and Smith, *Fluctuations*, pp. 126–127.
[14] Davis, "The New England Textile Mills," p. 13.
[15] The hourly and daily data for the last four mills was originally gathered by Robert Layer. He generously allowed us to use his worksheets.

TABLE 8

OUTPUT PER MAN-DAY AND OUTPUT PER SPINDLE-YEAR, 1821-60

Year	Yards Per Man-Day			Yards Per Spindle-Year		
	Raw (1)	3-Year Moving Average (2)	5-Year Moving Average (3)	Raw (4)	3-Year Moving Average (5)	5-Year Moving Average (6)
1821						
1822						
1823						
1824						
1825						
1826						
1827				130		
1828				161	166	
1829				206	195	191
1830				219	222	213
1831	40.07			241	229	243
1832	35.56	35.15		228	264	249
1833	30.02	34.66	37.23	323	262	254
1834	38.59	36.93	37.25	235	267	258
1835	42.19	40.62	39.45	244	246	262
1836	40.07	42.88	42.64	260	251	255
1837	46.39	44.13	44.40	249	265	273
1838	45.94	46.91	45.13	287	287	287
1839	48.39	46.77	46.39	326	309	294
1840	45.88	46.88	47.26	314	311	301
1841	46.38	47.66	48.18	294	297	299
1842	50.72	50.04	50.52	283	285	292
1843	53.03	53.46	52.32	277	284	285
1844	56.62	54.82	53.14	293	283	284
1845	54.82	53.97	53.22	278	287	287
1846	50.47	52.13	54.48	290	288	293
1847	51.09	53.63	55.39	295	298	291
1848	59.33	57.20	57.61	309	296	293
1849	61.19	62.17	60.73	284	294	286
1850	66.00	64.41	63.94	289	276	280
1851	66.05	66.22	65.48	255	269	270
1852	66.62	66.40	66.20	263	260	279
1853	66.53	65.98	65.78	261	280	274
1854	64.80	65.07	64.79	316	285	276
1855	63.88	63.27	62.78	277	285	272
1856	61.14	63.28	65.69	262	261	270
1857	63.83	66.60	67.74	245	252	265
1858	74.82	71.22	66.98	250	262	272
1859	75.01	69.97		292	285	
1860	60.12			313		

Source: Company records of Tremont, Suffolk, Hamilton, Lawrence, Boston, and Nashua.

although, as Layer has indicated, the labor productivity fell during the boom of 1846 and increased significantly during the depression of the early 1840's. Moreover, the long swings in productivity usually noted in studies of the long cycles of the post-Civil War economy are notably absent from the textile data.[16]

It has been frequently argued that the American textile industry underwent a revolutionary transformation in the decade between 1814 and 1824, but that thereafter technical progress was relatively slow. Certainly the industry during the decade following the War of 1812 was marked by many obvious changes. In terms of industrial organization, the integrated mill, so successfully pioneered by the Boston Manufacturing Company, was widely introduced. Moreover, there were also important developments in machine technology. The decade saw the invention and widespread innovation of the power loom, the Waltham dresser, the double speeder and filling frame, the self-acting loom temple, and a number of pickers and openers.[17]

These developments, admittedly of an almost revolutionary character, antedate this study. The study does, however, cast some light on the developments after 1824. The case for relative stagnation in the latter period rests, not on any quantitative evidence, but only on the observation that the changes that did occur do not appear particularly important when contrasted with the developments in the previous period.[18] Gibb, working from the records of the textile machinery firm, has argued that techniques continued to improve rapidly after 1824, and the productivity figures from the Massachusetts mills bear out his conclusions. True, there were no revolutionary changes in industrial organization, but there were developments in textile machinery. The twenty years after 1824 saw the development of the cap spinner and an imperfect ring spinner, the self-acting mule, and improvements in the roving frames. These developments almost certainly contributed to a steady increase in per worker productivity during the 1830's and 1840's. Moreover, new machines do not tell the

[16] The relationships between output and wages are discussed in Robert Layer, *Earnings*, pp. 24–27.
[17] For a detailed account of these developments, see George S. Gibb, *The Saco Lowell Shops*, Cambridge, Mass., 1950.
[18] In discussing the technology, most works on American history describe the early contributions of Slater and the Waltham system, but fail to emphasize the long-run revolutionary impact of the succession of gradual changes which characterized the period between 1825 and 1850.

At times the authors emphasize innovations which were not widely adopted before the 1860's. (See, for example, H. Faulkner, *American Economic History*, New York, 1960, p. 248.) In other cases, they single out the period after 1850 and gloss over the middle period. (See R. Russel, *A History of the American Economic System*, New York, 1964, p. 179).

entire story. More important, throughout the period new refinements were worked out in the machine shop and then were incorporated in the latest models of the old machine.

Also contributing to the rise in per worker productivity and certainly the prime cause of the increasing output per spindle, reflected in the figures in Table 7, was the great rise in spindle speed associated with the innovation of the belt drive. Even if machine technology had not changed, the output of the existing capital stock would have risen significantly because of the much higher operating speeds which could be attained with the new drive system. Before 1828 mechanical problems inherent in the English gear drive (problems that multiplied rapidly in the imperfect American copies of the drive) severely limited operating speed. In that year, however, Paul Moody in building the Appleton mills introduced the belt drive principle to textile production. Thereafter, the new principle was widely imitated and American mills greatly increased their speed of operation.[19]

The increase in speed meant that output per spindle rose by almost 50 per cent. The number of spindles has been used typically as an index of textile capital, but the increase in output per spindle implies that such a measure cannot be reliably employed in any study spanning the period before and after the innovation of the belt drive.[20]

Finally, Table 9 compares the productivity of the six firms with the average productivity of the entire industry. The United States figures are drawn from the federal Censuses and, therefore, may be distorted. However, since even the Massachusetts figures based on the fairly reliable state Census show lower productivity, it seems unlikely that the weakness of the Census fully explains the difference. The sample firms, although not on the average producing a coarser cloth, show remarkably higher output-to-labor and output-to-spindle ratios even though the figures on output per man-year for the sample firms have been deliberately underestimated (the data in Table 9 are based on an average 265-day work

[19] Gibb, *Saco Lowell Shops*, pp. 76–80.
[20] This is also true in the period after the introduction of belt drive, even though Taussig claims that spindles are "the best single indication of the extent and growth of such an industry as cotton manufacture" (Taussig, *Some Aspects of the Tariff Question*, Cambridge, Mass., 1915, p. 265).

It is easy to cite examples in which the use of the number of spindles as a measure of output leads to error:

Example 1. With use of the spindle figures, the estimated output for the U.S. in 1839 would be 412 million. If the value of output in 1839 were deflated by a price index and then converted into yards (on the basis of 1840), the estimate would be 357.

Example 2. Between 1890 and 1910, spindle speed increased by 11 per cent, spindles by 93 per cent, and consequently expected output by 112 per cent. Actual output increased by 108 per cent.

year).[21] An examination of the six sample firms indicates that they were substantially larger than the average.[22] Therefore, insofar as economies of scale exist, they probably account for part of the differences between the sample and the average.

TABLE 9

MEASURES OF TECHNICAL EFFICIENCY IN COTTON TEXTILE PRODUCTION FOR SAMPLE, UNITED STATES, AND MASSACHUSETTS
(yards)

	Output Per Spindle Per Year			Output Per Man Per Year		
Year	U.S.	Massa-chusetts	Sample	U.S.	Massa-chusetts	Sample
1820[a]	142		234	2,000		
1831	185	233	241	3,707	5,938	10,618
1837		224	260		6,395	12,823
1845		215	278		8,483	14,437
1849	217	232	284	7,796	10,399	16,215
1855		207	277		9,055	16,928
1859	219	248	292	9,410	10,800	19,878

Source: 1820, 1850, and 1860 for Mass. and U.S., from Censuses; 1831, from survey of that year; 1837, 1845, and 1855 Mass. figures, from state censuses.

[a] Based on a sample of firms which reported yardage produced.

Moreover, as Anne Grosse has shown, there are usually significant differences between the technically "best practice" and the normal mill practices in the textile industry. Management is not omnipotent and

[21] A further bias may be introduced against the tentative conclusion by assuming that the proportion of inputs used in the production of intermediate products (batting, thread, yarn, etc.) was the same as the proportion of cotton used in each. Also it may be assumed that at least 10 per cent of the employees were not in the four major departments. The Census data ratios would then be increased by 14 per cent for the U.S and 12 per cent for Massachusetts. The only case in which the relative values would be changed is 1831 for which the Massachusetts capital ratio would exceed the sample ratio.

[22] The number of spindles per mill was as follows:

	Sample	Mass.	U.S.
1831	6,073	1,359	1,567
1837	7,164	2,011	
1839	6,794	2,145	1,842
1845	6,842	2,707	
1849	8,370	6,048[a]	3,284[a]
1855	8,403	5,168	
1859	9,912	7,711[a]	4,799[a]

[a] The Census reported only firms. Many firms were multimill operations.

even if it were, it must utilize both old and new equipment.[23] Since the sample mills were the industry leaders or large mills with machine shops, which actually developed new machines and improved existing ones, it seems reasonable to assume that their technology was closer to "best practice" than was that of the average mill. The histories of the mills seem to indicate that the equipment of the leading firms was continually modified by their shop crews, while firms without shop connections had to buy complete new machines. These latter purchases could not be made until the entire new machine was available and, under any conditions, would not have been made until the total cost of the new machines' output was below the variable cost of the old machines' product. In fact, the average of 21 per cent, by which spindle productivity in the sample mills exceeded productivity in all Massachusetts during 1831–55, is not very far from Mrs. Grosse's 25 per cent estimate for the excess of best over average practice in the United States in 1941–46.[24]

A comparison of the figures presented in Table 9 and Mrs. Grosse's work shows at least one marked difference. In the more recent period, Mrs. Grosse has shown that average practice tends to approach best practice (i.e., the divergence between average output per spindle and best practice output tends to fall over time). No such convergence is apparent in Table 9. Although it is not possible to explain this difference with certainty, one partial answer suggests itself. In the period covered by the Grosse study, best-practice techniques were fairly constant. (Her best-practice estimates were constant from 1926 to 1935 and from 1936 to 1949.) In the earlier period, on the other hand, there was a continual improvement in best-practice techniques until the late 1840's. During this period, therefore, it is possible that average mills continually adopted newer practices but were still unable to close the gap between themselves and the leading firms. Even this answer, however, fails to explain the lack of convergence in the 1850's, when best-practice techniques showed little improvement. As for best practice relative to spindles (as opposed to labor), however, the early textile data do show the Massachusetts figures closing on the sample data.

---

[23] Anne Grosse, "The Technological Structure of the Cotton Textile Industry," in *Studies in the Structure of the American Economy*, W. Leontief, ed., New York, 1953, pp. 360–420.

One should also mention T. Y. Shen's thesis, "Technological Change in the Cotton Textile Industry," a best-practice study. This work has been reported in T. Y. Shen, "Job Analysis and Historical Productivities in the American Cotton Textile Industry," *Review of Economics and Statistics*, May 1958.

[24] Grosse, "Technological Structure," p. 410.

V. Summary

The records of business firms represent a relatively untouched resource for studies of the American economy in the nineteenth century. In this study, based on the records of a number of cotton textile firms, we have attempted to provide some information on the fluctuations in industrial output between the Census benchmark years. Firm records can also serve as a basis for studies of fluctuations in inventories, sales, costs, profits, and other variables for which we have only benchmark data. If studied regionally, firm records may be able to yield quantitative evidence regarding interregional differences in relative costs and techniques. In addition, an analysis based on firm records can yield some estimates of productivity changes and their causes and can, perhaps, suggest some ways by which these changes are transmitted through the economy.

TABLE A-1

ACTUAL OUTPUT STANDARDIZED WITH TABLE 3 TREATMENT, BY MILLS, 1815-60
(thousand yards)

Date	Boston	Merrimack	Hamilton	Suffolk	Tremont	Lawrence
1815	76					
1816	189					
1817	261					
1818	530					
1819	996					
1820	1,262					
1821	1,456					
1822	1,643					
1823	1,762					
1824	1,899	389				
1825	1,711	1,807				
1826	1,785	1,971				
1827	1,862	3,057				
1828	1,967	3,988	2,810			
1829	1,813	4,357	2,141			
1830	1,879	5,778	2,236			
1831	2,003	6,358	4,275			
1832	1,982	6,244	3,490	2,396	615	
1833	1,674	6,673	3,526	4,294	5,161	1,104
1834	2,090	6,948	3,926	3,839	4,578	6,173
1835	2,420	9,026	4,171	4,329	6,182	9,130
1836	2,450	9,961	4,612	4,579	6,661	10,265
1837	2,703	9,047	4,061	3,566	6,848	9,224
1838	2,820	9,998	5,104	4,520	6,713	10,518
1839	2,862	11,711	5,528	4,857	6,735	11,018
1840	2,580	9,922	5,606	5,118	7,123	10,549
1841	2,705	12,488	5,447	5,132	7,089	11,561
1842	2,488	12,625	5,212	4,807	6,314	10,858
1843	2,190	12,612	5,215	4,532	6,083	10,339
1844	2,683	13,299	5,828	5,061	6,624	10,480
1845	2,559	13,320	5,445	5,553	6,642	10,604
1846	2,477	13,052	6,343	5,689	6,584	12,816
1847	2,936	13,193	7,074	5,843	6,748	13,476
1848	3,172	14,729	8,953	5,732	7,401	14,199
1849	3,416	14,635	9,024	6,155	7,515	13,607
1850	3,472	17,663	10,208	4,656	5,666	11,128
1851	2,788	19,602	10,338	4,695	4,235	10,859
1852	2,724	19,989	11,696	6,420	7,715	13,696
1853	2,968	17,862	12,058	6,935	8,706	14,743
1854	3,154	17,354	11,658	6,709	10,021	13,726
1855	3,310	18,553	11,724	7,109	8,755	14,187
1856	3,698	19,765	11,218	8,049	10,797	15,927
1857	3,538	19,029	10,508	6,523	9,602	14,696
1858	3,497	19,680	10,336	7,361	8,552	16,371
1859	4,166	22,103	11,592	8,507	11,003	18,627
1860	4,573	22,447	12,917	8,455	11,169	19,183

(continued)

TABLE A-1 (continued)

Date	Naumkeag	Lancaster	Dwight I	Dwight II	Dwight III	Lyman I
1815						
1816						
1817						
1818						
1819						
1820						
1821						
1822						
1823						
1824						
1825						
1826						
1827						
1828						
1829						
1830						
1831						
1832						
1833						
1834						
1835						
1836						
1837						
1838						
1839						
1840						
1841			407	44		
1842			1,654	1,565		
1843			1,403	2,558		
1844			1,563	2,791		
1845			1,555	2,515	236	
1846			1,445	2,386	956	
1847		327	1,553	2,607	1,087	
1848	4,877	1,422	1,510	2,558	1,231	
1849	5,394	3,192	1,523	2,373	1,393	
1850	5,458	4,058	1,461	2,185	1,418	4,831
1851	5,614	4,102	1,528	1,787	1,384	6,574
1852	5,701	4,369	1,543	2,356	1,306	8,595
1853	5,294	4,026	1,572	2,444	1,440	4,172
1854	5,112	3,869	1,486	2,521	1,361	8,202
1855	5,460	4,678	1,679	2,486	1,403	8,289
1856	5,392	5,039	1,447	2,546	1,519	8,110
1857	4,925	4,612	1,284	2,885	1,154	4,240
1858	4,868	4,636	1,789	2,139	1,204	8,955
1859	5,366	5,340	1,930	2,485	1,342	8,553
1860	5,539	5,309	1,776	3,503	1,449	4,309

(continued)

TABLE A-1 (continued)

Date	Lyman II	Nashua	Jackson	Laconia	Pepperill	Amoskeag I
1815						
1816						
1817						
1818						
1819						
1820						
1821						
1822						
1823						
1824						
1825						
1826		321				
1827		1,031				
1828		1,938				
1829		2,350				
1830		2,483				
1831		2,954				
1832		3,347	2,510			
1833		3,721	4,308			
1834		3,706	4,385			
1835		3,646	4,460			
1836		5,591	4,567			
1837		6,560	4,975			1,272
1838		7,755	4,997			1,407
1839		9,033	4,980			1,453
1840		7,690	5,173			1,126
1841		8,203	5,321			1,077
1842		7,582	4,803			1,102
1843		7,814	4,867			1,148
1844		8,485	5,368			1,160
1845		11,041	5,473	1,974		1,189
1846		11,133	5,250	3,093		1,190
1847		10,479	4,980	5,509		540
1848		11,767	5,219	8,355		
1849		12,445	5,115	10,787		
1850		12,548	4,339	8,111	755	
1851		9,535	3,713	10,937	7,174	
1852	324	10,529	4,162	12,296	10,624	
1853	2,969	11,005	4,818	10,569	9,919	
1854	3,446	11,975	5,000	9,571	9,794	
1855	4,006	12,840	5,098	3,811	10,707	
1856	3,923	11,689	6,474		13,291	
1857	3,440	10,430	6,103		10,946	
1858	4,446	11,834	6,843		14,768	
1859	4,172	13,881	8,441		14,976	
1860	3,834	15,265	8,568		15,792	

(continued)

TABLE A-1 (concluded)

Date	Amoskeag II	Amoskeag III	Rampo	Slater	Slater and Tiffany	Sutton	Metacomet
1815							
1816							
1817							
1818							
1819							
1820							
1821			149				
1822			181				
1823			317				
1824			350				
1825			245				
1826			281	488			
1827			320	678	338		
1828			261	643	604		
1829			305	458	546		
1830			336	608	589	1	
1831			271	696	662	405	
1832			247	619	610	335	
1833			119	461	572	452	
1834				393	365	479	
1835				247	374	503	
1836				415	600	473	
1837	857			620	61	331	
1838	927			506		683	
1839	860			509		383	
1840	946					372	
1841	884	1,714				710	
1842	914	6,322				234	
1843	507	6,182				992	
1844	1,050	8,777				740	
1845	1,242	12,719				776	
1846	1,549	13,242				710	
1847	1,537	12,369				833	
1848		13,299				403	5,020
1849	498	17,999				845	5,500
1850	1,568	12,996				842	5,303
1851	1,564	15,816				875	3,370
1852	1,700	18,295				955	4,656
1853	1,458	17,646				998	5,170
1854	1,452	16,460				757	5,232
1855	1,423	17,721				1,286	4,445
1856	1,411	20,972				1,000	5,141
1857	544	15,024				511	5,300
1858	504	22,559				1,645	5,820
1859	1,900	24,240				1,800	5,656
1860	1,681	21,863					6,130

TABLE A-2

COMPARISON OF STANDARD AND RECORDED YARDS, BY COUNT, 1825-60
(thousand yards)

Year	Producers of Low-Count Goods[a]			Producers of Medium-Count Goods[b]		
	Number of Mills	Estimated Standard Output	Recorded Output	Number of Mills	Estimated Standard Output	Recorded Output
1825	3	1,586	1,711	3	2,349	1,801
1826	5	1,629	2,106	3	2,562	1,971
1827	5	2,669	2,893	3	3,975	3,057
1828	7	6,787	6,717	3	5,185	3,988
1829	7	6,282	6,305	3	5,664	4,357
1830	8	6,588	6,600	5	7,512	5,778
1831	8	9,492	9,233	6	8,266	6,358
1832	11	14,361	14,343	6	8,117	6,244
1833	15	24,224	23,790	6	8,674	6,673
1834	15	31,127	28,700	6	9,032	6,948
1835	17	36,094	34,341	7	12,118	9,323
1836	17	39,271	38,729	7	14,980	11,657
1837	17	37,759	37,940	7	12,554	9,944
1838	18	42,823	42,429	8	16,050	12,350
1839	18	45,921	45,016	8	18,713	14,393
1840	18	45,022	43,842	8	19,504	13,016
1841	18	47,218	45,460	9	20,218	15,952
1842	18	43,948	42,066	9	21,645	18,264
1843	18	42,782	41,043	9	20,240	17,227
1844	18	47,263	44,531	9	23,094	19,205
1845	21	52,362	49,295	10	25,249	19,435
1846	22	57,126	53,388	10	26,936	20,598
1847	24	60,901	57,048	10	25,797	19,937
1848	24	73,018	69,679	10	27,755	21,757
1849	24	77,505	73,463	10	27,008	21,755
1850	24	68,311	66,346	10	31,373	24,829
1851	24	73,055	69,893	10	32,481	26,245
1852	24	91,907	85,568	10	34,105	27,555
1853	24	90,909	87,019	9	30,330	24,317
1854	24	88,982	86,725	9	26,567	23,824
1855	24	87,353	83,006	9	31,573	25,411
1856	24	91,798	86,540	9	34,671	26,428
1857	24	81,018	77,276	9	31,372	24,896
1858	24	89,255	84,433	9	32,168	26,914
1859	24	102,670	96,562	9	37,801	30,119
1860	24	107,829	101,466	9	36,060	29,604

[a] Boston, Hamilton, Suffolk, Tremont, Lawrence, Naumkeag, Jackson, Nashua, Pepperill.
[b] Merrimack, Dwight.

COMMENT

Paul F. McGouldrick, Board of Governors, Federal Reserve System

In estimating the annual output of the New England textile industry between 1825 and 1860, Davis and Stettler assumed that their sample was representative of established companies with respect to percentage changes in output and that additions to output by new companies occurred evenly between benchmark dates. The first of these assumptions is supported broadly by my analysis of much the same group of companies which shows a very low variance of company output changes around mean output changes.[1] My comparison of the output of these companies with the raw cotton consumed annually in the United States after the Civil War showed a steeper upward trend, as would be expected from the entry of new companies, particularly those in the Fall River-New Bedford area and the South. But the contours of both series around their respective trends do not disprove the hypothesis that the companies in the Baker Library manuscript collection represented the established component of the industry in their output decisions.

But the second assumption made by Davis and Stettler—that output of new companies can be distributed evenly between benchmark dates—is purely arbitrary. Specifically, plant and equipment spending of my own sample of companies shows very pronounced cycles from 1835 to the Civil War and thereafter. It would be astonishing if similar or even more pronounced cycles did not occur in the formation of new companies.[2] An index reflecting the number of new firms beginning operations year by year would certainly improve the estimates of this component of the output series. Such an index could be derived from the annual Massachusetts data on the incorporation of new companies by identifying textile companies among these from listings in trade directories available for the early 1870's. Lags between incorporation and output could be estimated from manuscripts in the Baker Library at Harvard University and applied

---

[1] After adjusting for entry of new companies but *not* for gross or net investment. The analysis is in my unpublished Ph.D. dissertation for Harvard University.

[2] All contemporary and present-day literature on the industry shows that orders for textile machinery fluctuated sharply in a cyclical fashion (George Gibb's *The Saco Lowell Shops* and Thomas Navin's *The Whitin Machine Works* might be referred to). Erastus Bigelow complained bitterly about the very uneven pace of textile company formation, asserting that what we would call information lags and desires of promoters to keep projects going, even in the face of declining markets, caused large amounts of excess capacity to appear when prices and profits were falling (Harris-Gastrell report on American cotton manufacturers, H.M. Stationers Office, London, 1873).

to textile company incorporations. The index constructed in this way could then be used to raise the percentage changes of the established companies to industry output between benchmark dates. It would also serve as a check on the reliability of the benchmarks.

In my opinion, the authors are correct in using yards instead of pounds of cloth as a measure of output. My own Laspeyres and Paasche index tests among companies over time show that the differences between the unweighted and the weighted output estimates were small enough to make weighting unnecessary for nearly every purpose. But the procedure for converting actual yards to standard yards is not clearly specified in the paper, nor are reasons given for not converting all output to the standard-yard equivalent.

In the last section on capital and labor productivity, the authors use the spindle as an index of capital. My work shows a large decline in the constant-dollar cost of structures and equipment per spindle for a standard mill producing a homogeneous output (in constant dollars) between the 1820's and the 1860's. Hence, the spindle has a strong upward bias over time as a capital index. Furthermore, the productivity of spindles and that of other textile capital goods changed at different times and at different rates. Thus these two biases affect not only long-run but also decade-to-decade comparisons.

The authors' estimates of labor productivity agree with mine. Labor productivity not only rose after 1840 but rose as rapidly as before, despite the substitution of unskilled and largely illiterate immigrant labor for the celebrated New England farm girls.

Data on inventories and imports would help the interpretation of the output series. Changes in cloth inventories can be obtained, in value and in physical terms, for a subsample of the companies in Davis' and Stettler's sample. Inventory cycles could illuminate changes in general economic conditions as well as those in output and prices of this industry. While the authors concentrated on the regional industry, imports competed strongly with home output of standard cloth types before the Civil War.

REPLY *by Davis and Stettler*

Paul McGouldrick calls attention to two assumptions in our paper and takes exception to the second. Unfortunately, he has slightly misunderstood the argument. We have not assumed that additions to output resulting from the entry of new firms occur evenly over the intercensual period; rather, we have assumed that the change in the sample proportion occurs evenly between benchmarks. This change occurs for three reasons: existing sample mills expand output more or less rapidly than nonsample mills; existing sample firms add new facilities more or less rapidly than nonsample firms; and firms enter the sample at a rate other than the entry rate for the industry.

We well realize the difficulties associated with the even allocation technique and considered a number of alternatives; however, the more complicated hypotheses are underpinned by some equally tenuous assumptions.[1] McGouldrick suggests that we construct an index based on the differences in the rate of entry into the cotton industry. Such an index could simply assume that the yearly change in the sample proportion equals the difference between the benchmark ratios multiplied by the ratio of entry since the first benchmark to entry between benchmarks. Entry could be defined as a lagged function of incorporations in Massachusetts. If one wishes to take into account entries into the sample, the ratio of incorporations could be deflated by a similar ratio using yearly sample entries.

There are a number of unstated assumptions underlying such indexes. First, it is assumed that the number of investment projects consummated by production is proportional to the number of incorporations. This assumption is hardly justified. The ratios of net entry to incorporations appropriately lagged (given below) vary widely for six- and eight-year periods. Second, the proposed index assumes a lag invariant with respect

Period	Lag of 0 Years	Lag of 1 Year	Lag of 2 Years
1831–37	.36	.46	.67
1837–45	.43	.32	.22

to the business cycle and other disturbances. For the textile firms listed on the Boston Stock Exchange in the pre-Civil War era, the average lag between the year of incorporation and the commencement of operations is 1.25 years; however, the variance of the distribution is 1.58 years. In

---

[1] Changes in the weighting system induce considerable changes in the intercensual estimates. Massachusetts output estimates for 1832–36 using various inflation techniques are presented below:

Year	Sample	Even All.	Cumulative Inc. Lagged		Cumulative Inc. Corrected and Lagged	
			1 Year	2 Years	1 Year	2 Years
1832	14	82	98	105	113	105
1833	22	107	131	149	173	125
1834	28	111	153	138	212	125
1835	35	131	176	160	176	141
1836	39	133	183	160	183	160
1837	35	118	118	118	118	118

A particular difficulty arises in a number of instances with the last two indexes on an arithmetic count—division by zero.

addition to the lag between charter and start up, there is a shake-down period before normal levels of output are achieved. For the sample firms, the average is 6.0 quarters; for individual mills the figure is 2.9; the variances are 6.44 and 1.36 quarters, respectively. Use of a constant-lag index is questionable in the light of the high variances associated with the distributions. McGouldrick has observed that variability of the investment-output lag dampens the cycle in output. Third, the indexes based on entry or investment tacitly assume that changes in the productivity occur in proportion to entry. During the middle 1830's, the sample firms' output per spindle rose 8 per cent, while the Massachusetts figure fell 4 per cent; output per sample worker increased 20 per cent, while the Massachusetts workers' output rose only 8 per cent. It is of interest that the largest increases in sample output per worker occurred in 1834-35, the end of a three-year period of low entry.

Building in Ohio Between 1837 and 1914

MANUEL GOTTLIEB

UNIVERSITY OF WISCONSIN-MILWAUKEE

In a preceding work an effort was made to develop a new nationwide series of the number of dwelling units erected annually in the United States since 1840.[1] This series was made possible only by utilization of the newly discovered statistical record of residential building in the state of Ohio during the second half of the nineteenth century. Of that record only the data on the number of dwelling units erected were used in the earlier study. This study is a preliminary report on the value of building by class of structure derived from the Ohio building statistics and on the adjustments required to make raw data usable for analytic purposes.

The original agencies that collected building statistics in Ohio were local personal property tax assessors who were following a program of statistical reporting inaugurated by state law in 1857 and maintained through 1915. The local assessors were required as a matter of official duty to make an annual return for personal property and for alterations in the roster of real property caused by demolitions, destruction, or improvements. Local assessor reports for towns or townships were submitted to county auditors who consolidated them and forwarded them

NOTE: The present paper grows out of intensive research during the past four years into all phases of urban building carried on with the aid of the National Bureau (see progress reports in the Annual Reports of the National Bureau for 1962, pp. 48–51, and 1963, pp. 46–47) and particularly with the aid of Moses Abramovitz. In its technical form, the study owes much to my research helpers, first of all to Mary D'Amico of the University of Wisconsin-Milwaukee. Paul Sampson, formerly of the Numerical Analysis Laboratory of the University of Wisconsin, and Asa Maeshiro were responsible for carrying through many of the regression calculations. The processing and tabulation of Ohio data were made possible by a special grant in 1961 from the Rockefeller Foundation. A grant from the Wisconsin Urban Program (Ford Foundation) provided timely help in the summer of 1961. The graduate school of the University of Wisconsin provided the research assistance which made possible the regression tabulations and analysis.

[1] Manuel Gottlieb, *Estimates of Residential Building, United States, 1840–1939*, NBER Technical Paper 17, New York, 1964 (hereinafter referred to as Gottlieb, *Estimates*).

to the Commissioner of Statistics. After 1868 this office was filled *ex officio* by the Secretary of State. Between 1858 and 1868, the county returns were published with statewide totals and extended commentaries in the report of the Statistics Commissioner. After 1868, building returns were published in the statistical supplement to the annual report of the Secretary of State along with vital, election, production, court, and other statistics. For most of these years, the statistical reports and schedules were included in the annual compilation *Ohio Executive Documents.*

The reports on new building were presented with shifting detail and coverage. From the first report in 1858 until 1915, the number and value by counties of newly erected taxable structures were published annually (except for the omission of value figures in 1866). Detail on type of building was first attempted in 1862 for industrial buildings ("mills, factories, machine shops, foundries, furnaces"). In 1865 the number of "dwellings" and "barns and stables" was enumerated, and commercial-type buildings ("stores, warehouses, shops, and other places of business") were added in 1867. Although these categories were given in finer detail for some later years, by 1887 reporting under these categories had been stabilized. For 1873–79 returns for barns and stables and for 1887–90 returns for industrial buildings were included under a residual miscellaneous category.

No exact definitions of the building categories were ever spelled out in the published reports, nor were any archives or records for the statistical department ever located. Since the reported headings varied (e.g., industrial building finally turned into "manufacturing establishments" while commercial establishments became "stores and warehouses"), there is no certainty that classification of types of building was consistent from year to year or that there were not gradual drifts in reporting the aggregates or the types of building. In a few cases, misclassification of building was discovered chiefly by reference to inverse variation for a particular type of building and for the "miscellaneous" residual.

Reporting of tax-exempt construction was less comprehensive and was slower to get started. Enumeration of the number of newly erected churches and schools was attempted in 1859 and in 1860, but returns were manifestly incomplete. Only in the 1869 report were returns presented by counties for construction of schools and churches by number and value. In 1873 construction of "public halls" and "county building" was added. Returns for public halls were soon included in the miscellaneous heading and county building was unreported as a separate category for seven years between 1887 and 1894.

The data on residential building by number of "dwellings" were found internally consistent for degree of urbanization, real estate activity, and other variables.[2] Comparison of building permit data available for four central cities for 1900–12 and for one central city back to 1888 with corresponding assessor data for the counties involved showed the expected order of magnitude and parallel patterns. All comparisons showed divergences due partly to variations in coverage and definition. But these divergences did not impair broad comparability of level and pattern, even on the county level, and were reduced to minor proportions for county returns consolidated into group or statewide totals. A detailed presentation of this evidence is reserved for later treatment.[3] Confidence in the validity of the Ohio statistics was further buttressed by the professional skill and judgment indicated by the founder and designer of the Ohio statistical system and by the continued willingness of the state legislature to sponsor collections of data and to impose statistical reporting obligations on local government officers.

Adjustments in the building reports filed by these officers, as originally reported in our source documents, are made under five headings: (1) for clearly deficient returns or for occasional printing errors whereby digits were dropped or added; (2) for inadequate reporting of tax-exempt building; (3) for extrapolation to 1837; (4) to allow for alterations in standards of appraisal or for changing purchasing power; and (5) to convert a record of "completions" to a record of building activity. A positive program of data evaluation and analysis is then presented in the last three sections: (6) farm unit residential values, 1850–1912; (7) nonfarm unit residential values; and (8) adjusted time series of building in Ohio.

1. *Deficient Returns*

Deficient returns for the three categories of building—total, residential, and barns and stables—are summarized in Table 1. Coverage in the reporting system was best maintained between 1866 and 1905. There were many omissions in the reports before the Civil War and omissions rose ominously after 1910. In 1914 less than half the counties reported, a still smaller number in 1915, and thereafter reports ceased. Deficiencies for a particular year in either value or number were generally corrected by adjustment with the average per unit value of adjoining years. Where number and values were both lacking, we resorted to linear interpolation.

[2] See Gottlieb, *Estimates*, pp. 18–35.
[3] It is hoped that the complete results of the investigation into Ohio building, marriage, and conveyance statistics will be set forth in a special monograph, with appropriate detail including, of course, evidence related to evaluation.

Deficiencies for several consecutive years were generally corrected by reference to the index behavior of comparable counties. Our estimates for the last two reporting years were derived from the percentage change for the thirty-nine and thirty-three identical counties reporting both for 1913/14 and for 1914/15 and accounting for over half of all building. Somewhat more care was exercised in adjustment for deficiencies of the sample groups compared to the state aggregate. Generally, the missing counties accounted for a small fraction of statewide building so that crudities of adjustment could hardly exert an appreciable influence on the levels of the returns.

TABLE 1

COUNTIES NOT REPORTING IN ORIGINAL OHIO BUILDING SCHEDULES, 1858-1915

	1858-70		1871-1909		1910-14	
	Number	Dollars	Number	Dollars	Number	Dollars
Total building	31	38	70	19	112	98
Residential building			58	11	122	99
Barns and stables[a]	15	3	103	69	163	133

[a] Between 1873 and 1879, detail on barns and stables was grouped in with "miscellaneous" building.

On our statewide listings, it was more difficult to allow for deficiencies in the categories of building where construction was intermittent and ran a wider range of year-to-year change. Hence our statewide totals for industrial and commercial building were adjusted by different standards depending upon whether the deficient counties were highly urbanized or not. Between 1867 and 1910 the thirty-eight deficient industrial returns for the ten counties making up the first three sample groups were adjusted by the most appropriate method, the details of which must be left to the future monograph. Other deficiencies were only caught for counties that were deficient on total production. In these cases, the deficient counties were credited with their prorated share for the seventy-three nonurban counties and the totals were adjusted accordingly. The same method in principle was used to adjust commercial building for deficient returns. Our statewide totals for industrial and commercial building probably tend to understatement because of inadequate allowance for deficiencies. The behavior of the sample groups of counties under the indicated headings was subject to little or no bias on that account since adjustments for deficiencies were consistently made for all members of sample groups. Indeed, the rationale for utilizing sample groups partly grew out of the need to concentrate adjustments for deficiencies (and other errors in reporting) in a smaller group of returns.

2. Tax-Exempt Building

The inadequacies in reporting tax-exempt building were made apparent by the independent reports of school statistics collected by county auditors and later by county school superintendents; these statistics were obtained from local school boards and screened and republished with statewide totals by the state superintendent of schools. From 1837 to 1852, these reports were published in *Ohio Legislative Documents* with commentary and analysis. Between 1853 and 1902, school reports were included in the *Ohio Executive Documents*. After 1903, school data were published as Annual Reports of the Superintendent of Public Instruction. State superintendents frequently lamented inadequate and irregular reporting, particularly in the early years by local school officials. After eight years of annual reporting, the Secretary of State avowed "it is impossible even to conjecture what is the number or condition of the school houses in Ohio."[4]

In 1854 the reporting system was revamped, coverage was extended, and the school statistics for most items became reliable. After 1856 only one or two counties failed to report and coverage within reporting counties was more complete.[5] Blank register forms were prepared and furnished local school officials and statewide efforts were made to instruct local officials in how to keep records and make out reports.[6] Even under the new regime, items on average pupil enrollment or attendance were reported year after year with a floating margin of error.[7] Both internal and external checks indicate that items of school statistics with which we are concerned —number and cost of new school buildings erected—were reliably and consistently reported.[8]

For the aggregate period 1870–1910, school reports showed a building

---

[4] *Ohio Legislative Documents*, no. 33, 1845, p. 502.

[5] The number of deficient counties included in statewide school statistics are as follows: 1837, 13; 1845, 29; 1850, 8; 1853, 18; 1858, 7; 1856, 3; 1857, 1; 1858, 0; 1863–67, 0 (*Ohio Statistics*, 1880, p. 303).

[6] *Ohio Executive Documents*, 1868, Pt. I, p. 625.

[7] *Ibid.*, pp. 623 ff. Enumeration of "school houses" was made difficult by the diverse practices of school clerks in reporting school rooms or school structures (*ibid.*, 1865, p. 354).

[8] Report on these checks must be deferred to the monograph noted earlier. It is noteworthy that the time series of the cost of school building moved in close and plausible correspondence with an item drawn from local government budgets, namely, "total expenditures on sites and building."

value 2.6 times higher than assessor reports.[9] The school reports were believed to be more reliable since they were collected together with a broad variety of other school data, since the reported cost of school construction tallied well with school budgeted capital expenditures, and since both reporting and state compiling officers were professionals. A commentary found in an early assessor report indicated the grounds for incomplete assessor coverage. "School houses and churches not being taxable," it was noted, "are not often returned." This note was appended to the 1860 report at which time classification of new construction by type was not attempted.[10] The original reports on new building were collected by township or municipal assessors whose primary responsibility was to collect assessments on taxable personal property and to make an assessment allowance for the destruction of old taxable buildings and for the erection of new taxable buildings. Assessors could easily submit a statistical report on new building that would or could tally exactly with positive adjustments for realty assessment. Separate inquiry would need to be made for a report of tax-exempt building. It was hoped that, when separate breakdowns of exempt construction were added to assessor reports at various times between 1869 and 1872, the returns had become reasonably complete. But the school statistics demonstrated the reverse.

It was easy, of course, to substitute values for school building, as reported in school statistics, for assessor reports both before and after 1872. These values were deflated with a Riggleman index to adjust for changes in cost of building. The results are shown in column 1 of Table 2 from 1860 on.

Besides school and assessor-derived reports of building, there was available from 1856 on a well-audited series extracted from the annual financial reports of the state auditor on the property taxes raised by county governments for "building purposes." This was one of the breakdowns of property taxes classified by purpose or broad expenditure use. The deflated decade totals are shown in column 2 of Table 2. Unfortunately building expenditures could also be financed by borrowing; and taxes could be levied to accumulate in a building "fund." Thus neither the year-to-year movements nor trend of the series accurately represents local government building. The series does, however, accurately measure the willingness of public officials in Ohio to impose a levy on property owners to raise funds for building purposes.

[9] Our deflated (in 1913 dollars) aggregate of the cost of school building was $67.28 million against $26.06 million for assessor reports of school building for the 1870–1910 period. Even undeflated school building costs were approximately double the assessor reports ($53.6 million).

[10] *Annual Report*, Ohio Statistics Commissioner, 1860, p. 80.

TABLE 2

DECADE AGGREGATES OF TAX-EXEMPT OHIO BUILDING, 1850-1910
(THOUSAND 1913 DOLLARS)

Decade	School Building (1)	County Building Tax (2)	Estimated Total Exempt Building (3)
1860's	6,975	3,733	23,000
1870's	11,233	7,719	47,000
1880's	12,708	7,822	35,000
1890's	18,158	6,834	61,000
1900's	25,181	5,359	65,000
1870-1910	67,280	27,734	208,000

Source: NBER files, series nos. 0257,0263

With the aid of the school and county building data and periodic measures of the net change in stocks of exempt building, a set of decade estimates (listed in column 3 of Table 2) was prepared from the 1860's on of total estimated exempt construction. These estimates, when expressed as a percentage of total decade building, follow the countercyclical pattern shown by the increments to standing stock.[11]

3. Extrapolation to 1837

The collection of formal statistics on new building began in 1857. As noted in a previous publication, a close correspondence was found, as might be expected, between annual increments to total real estate assessments and new taxable building.[12] This correspondence derived from the state policy of freezing realty assessments between formal reappraisals except for property destruction or new building.[13] Any interim assessment

[11] Thus the ratio of exempt to taxable building as shown by growth increments of standing stock is as follows (in per cent):

1853–59	41.2
1859–70	16.9
1870–80	42.9
1880–90	9.7
1890–1900	27.3
1904–12	6.1

For 1890–1900 and 1904–12, on all real property including land values. The sources for these figures are *Wealth, Debt and Taxation*, Bureau of Census, 1907 and 1915; reports of Ohio State Board of Equalization for 1853, 1859, 1870, 1880, 1890, 1900.

[12] See Gottlieb, *Estimates*, pp. 76–79.

[13] I.e., for destruction or building exceeding $100 in value.

reductions for particular old properties were to be offset by assessment increases in other old properties. This policy was substantially, if not exactly, carried out. To allow for these irregularities, assessment increments were smoothed by computation from overlapping pairs of years. Chart 1 indicates that the correlation between new building and assessment increments was close for the period 1857–99. The coefficient of correlation was .938, using a set of linear estimation equations fitted separately to long expansion and contraction phases. These estimating equations were then applied to the smoothed assessment increments for 1837–56. These increments were doubled from 1837 to 1845 because of the low ratios of tax appraisals to market values prevailing at that time.[14] Unlike the tax assessment regression made to project nonfarm construction, this estimation had a more refined base and involved a projection of total building including farm building.[15]

Since exempt property was virtually not reported before the July 1869 report, it was necessary for our total building series, which by extrapolation with realty assessment increments is carried back to 1837, to be increased by an allowance for exempt construction. This allowance was devised by comparing movements of exempt and taxable building as disclosed in the periodic realty appraisals carried out in 1853, 1859, and thereafter decennially.[16] For this purpose, exempt building was represented

[14] Perhaps the tax increments before 1846 should have been raised two and a half times to put them on a level with post-1846 assessments, but, as in our earlier study, doubling seemed advisable in view of early assessment irregularities.

[15] Differences between the two regressions are summarily listed here (for earlier projections, see Gottlieb, *Estimates* pp. 75–80):

Area of Difference	Value Projection	Earlier Projection
1. Scope of building variable	Taxable building	All building incl.
2. Tax increment	All assessed realty	Assessed realty of towns and cities
3. Method of averaging tax increment	Average of overlapping pairs of years	Three-year moving average
4. Number of regressed observations	1857–1900	1857–1889

[16] School reports indicate that exempt property was appraised with the same standards as taxable property. Reported total value of school property shows the same degree in slippage from market value characteristic of appraisals of taxable property. The ratio of assessed to school statistics book values were 99.5, 74.7, 66.0, and 56.9 per cent for 1859, 1870, 1880, and 1890 respectively. See footnotes 19 and 20 for similar ratios for taxable property. On the other hand, the total value of church property as reported in the U.S. Census was 177, 141, and 179 per cent of the appraised total for 1860, 1870, and 1890. Since appraised values were taken for tax purposes and were screened where possible against sale values, I am inclined to accept their findings for present purposes.

Chart 1

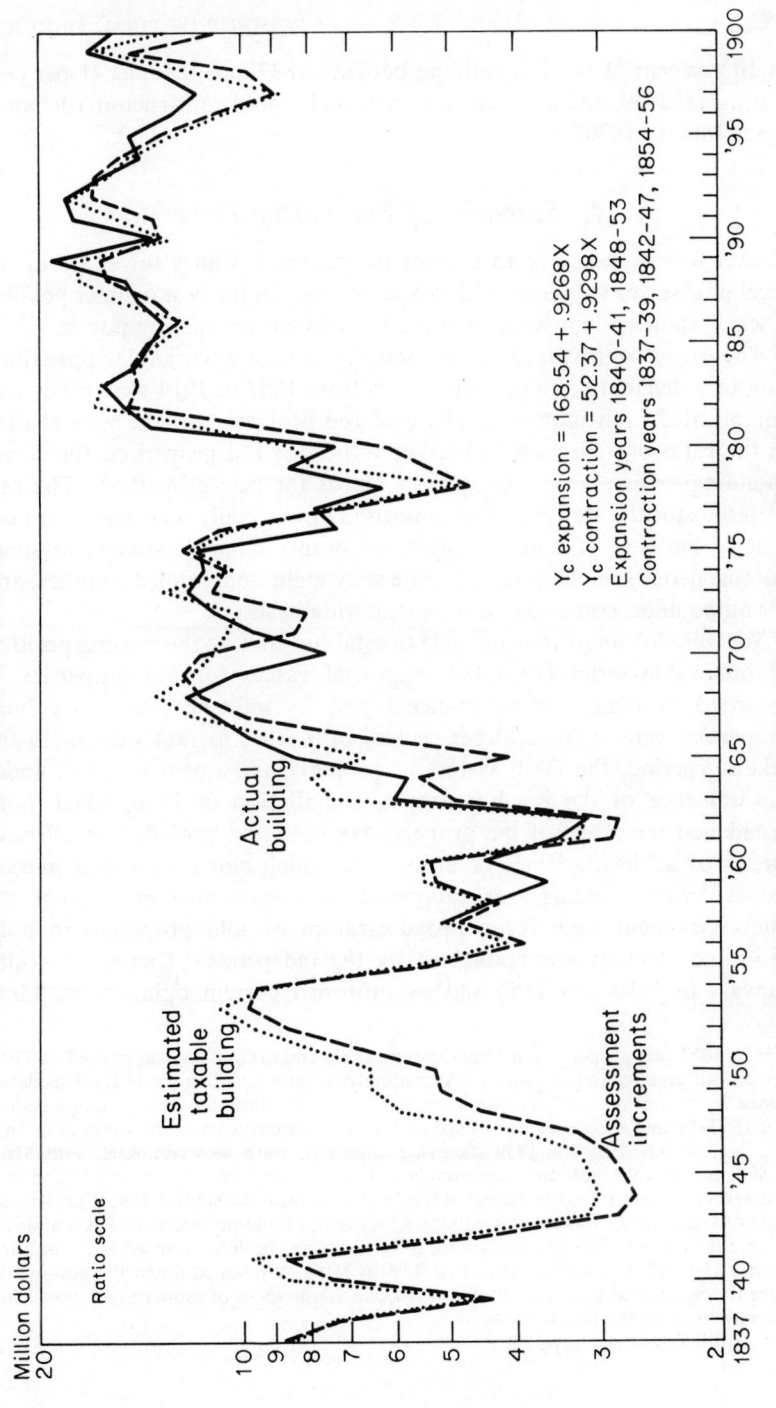

Taxable New Building, Actual and Estimated, and
Realty Assessment Increments, Ohio, 1837–1900

Y_c expansion = 168.54 + .9268X
Y_c contraction = 52.37 + .9298X
Expansion years 1840–41, 1848–53
Contraction years 1837–39, 1842–47, 1854–56

Source: Annual reports of the Auditor and Secretary of State of Ohio for 1835–1900.

at 10 per cent of taxable building between 1837 and 1853, at 41 per cent during 1853–59, and at three times reported school construction (deflated) from 1860 to 1870.[17]

4. Standard of Purchasing Power

It was found necessary to correct the reported values for shifts in the level of assessor valuation and to approximate in the best manner possible a set of values of constant or at least consistent purchasing power.

The problems involved in the establishment of a workable approximation to a deflated building value series from 1857 to 1914 were many and complicated. An important phase of the problem was the variant drift in the ratio of appraised to market values for old properties, the corresponding or divergent appraisal standards for new properties. The raw materials for the analysis were comprised of originally reported values per unit of building of different types for twenty counties studied in small homogeneous sample groups, for the sixty-eight nonsampled counties, and for all counties consolidated on a statewide basis.

We were fortunate in being able to establish that, at the starting position of our value series (1857–60), appraisal values for old properties as recorded in realty tax assessments and by inference for newly built properties were at least 90 per cent of prevailing market values. In the take-off period, the Ohio system of property assessment was still under the influence of the revolutionary tax legislation of 1846, which both broadened the scope of the property tax base and provided an effective means of achieving levels of assessment which closely followed market values.[18] This finding was supported by documented analysis of the sales-assessment ratio for a broad stratum of sold properties in both 1854 and 1859. It was reaffirmed by the independent Census of wealth canvass in 1850 and 1860 and by informed current opinion, including

---

[17] In 1853 real property in all buildings (taxable and exempt) was appraised at 110.6 per cent of taxable building alone. Accordingly, taxable building for 1837–56 as determined by regression analysis was increased by 1.1 to allow for exempt construction. For 1853–59 this allowance was raised to 1.41 since increments in the value of exempt building, disclosed by the 1859 statewide appraisal, were $6.8 compared with $16.5 million for taxable building. Apparently public building was shifted or adapted to a countercyclical basis. In the decade of the 1860's the ratio shifted to 1.169. Consistently higher for the 1860's was a set of estimates for exempt building determined as a multiple of school building. The ratio of school to total exempt building showed little variation from 4.85 in 1859 to 4.47 in 1870 and 4.68 in 1890. We scaled down the multiple to three times, since at this rate we obtained a consistent series of estimates for the 1860's midway between the two variants.

[18] Gottlieb, *Estimates*, pp. 76 ff.

that of the State Auditor, the independent Commissioner of Statistics, and a competent contemporary observer, Ezra Seaman.[19]

After the Civil War another consensus of evidence indicated that appraisal levels of valuation receded with reference to market values. Assessors regarded early post-Civil War market values as speculative and hence in the basic 1870 appraisals aimed at a level about 65 per cent of prevailing market values. At the succeeding decennial appraisals, conducted concurrently but independently by the agents of the Census Bureau and local and state appraisers, the evidence indicates a sliding tendency in the assessment-to-market ratio, which by 1900 was around 47 per cent.[20] The sustained price lift and boom in real estate values that marked the early 1900's lowered assessment-to-market ratios still farther.[21]

[19] The statistical surveys of bona fide deeds disclosed the following relationships between the aggregate of sale value and assessed value:

Time	Sales Proceeds (dollars)	Appraised and Equalized Value as Percentage of Sales Proceeds
April–October 1853	8,309,421	87.4
Year ending July 1, 1859	12,109,306	101.0
1869, agricultural	33,666,285	60.8
1870, agricultural	24,920,440	60.2
1869, urban	22,471,254	68.9
1870, urban	16,932,600	66.3
Jan. 1, 1879–Oct. 1, 1880, city and town transfers only	12,131,806	58.9

Statistical Report Secretary of State, Ohio, 1881, pp. 643–644; *Ohio Executive Documents*, 1859, Pt. I, pp. 857–60; *Abstract of Transfers on Sales of Lands and Lots 1869-70*, Auditor, State of Ohio, 1871; Board of Equalization *Proceedings 1853*, 1854. The ratios of total taxable (including personal) to true value was estimated by the assistant marshalls who took the Censuses of 1850 and 1860 at 85.96 and 80.39 per cent, respectively. The Ohio Statistics Commissioner concluded after a detailed review of the evidence that the 1859 appraisements "come very near the commercial value" (*Ohio Executive Documents*, 1859, Pt. I, p. 798).

[20] The U.S. Bureau of Census estimated that assessed values of taxable real estate were 64.03 per cent (1890), 47.6 per cent (1900), and 46.4 per cent (1904) of true value. The auditor's records for September 1892 indicated use by a local assessment officer of a 65 per cent ratio to get a "tax valuation" (Sept. 1892, letter books, p. 620, State Archive Building, Columbus, Ohio). Recognition was frequently expressed by tax officials of the practice of appraising real estate between 40 and 66 per cent of its true value. (*Proceedings of the First Annual Conference*, Tax Commission of Ohio, 1912, pp. 14, 23, 63.) Erosion of appraisals is clearly registered in farm lands assessed and equalized in 1870 and 1880 at $27.00 and at $24.06 per acre in 1900. The recorded sale of farm lands in bona fide deeds involved a mean acre price of $34.11 in 1880, $38.44 in 1890, and $34.18 in 1900 (Annual Report, Secretary of State, for conveyance statistics).

[21] Thus farm lands alone per acre sold in 1910 at nearly double 1900 values (1903, $37.34; 1907, $49.75; 1911, $60.42). By 1910 the ratio of assessed to market values had slipped to around 33 per cent. Average value of town lot deeds rose 20 per cent between 1895–1900 and 1908–12.

Dissatisfaction with assessments became widespread. Authority over assessment was shifted to a newly appointed state tax commission with a legislative mandate to raise appraisal standards up to market levels while ensuring that actual tax levies were not to be expanded.[22] The resulting decennial property assessment raised the assessed value of realty by some 250 per cent. Per unit values of newly erected buildings were

[22] Attached is an authoritative account of the process of lifting appraised values to nearly market levels:

"The work of the Commission in connection with the assessment of real property covered a period of more than four months, during which period the entire time of its members was devoted to the work. The county auditors' abstracts were analyzed, and the average values were ascertained and compared. Copies of the printed pamphlets issued by city boards of assessors and county auditors, showing the assessed value of the real estate in the cities, townships and villages, were secured and bound together by counties. County and State maps were procured, and the average value per acre in each township was marked on the county maps. These were combined to form a large wall map of the entire State. As the average values in every township in the State were shown upon this map, comparisons between townships in the same and other counties were readily made.

"The State was divided into convenient districts, containing from five to eight counties in each, and the county auditors from each district were called before the Commission on different days. These hearings, which extended over a period of three weeks, were informal in character, and the valuations in each county and all the counties in the district were inquired into and discussed by all the auditors present. The county auditors were requested to furnish a list of transfers extending over a period of three years, excluding therefrom so-called 'dollar sales' and forced sales. The auditors were requested to give the consideration stated in the deeds, or the consideration where it was known and not stated in the deeds, and the value at which each parcel was assessed. These transfers were analyzed and tabulated by the Commission, and after the county boards of equalization, which had been ordered reconvened, had completed their work, the county auditors were again called before the Commission by districts, the districts, however, being composed of a larger number of counties than in the first instance. The average selling value shown by the transfers, as compared with the average assessed value, was discussed with these county auditors, as were the relative values of the districts in each county. The relative values in all the counties were discussed and compared with all the county auditors present. Many persons familiar with values, members of boards of equalization and others appeared before the Commission, at its request, and others at their own instance.

"Members of city boards of equalization and city boards of land assessors were called before the Commission and inquired of as to the assessment in their respective cities. In determining the true aggregate value of the real property in the cities and villages, it also compared the per capita assessment in municipalities similar in character and population.

"Experts were sent into many of the counties, and, in some instances, persons familiar with local conditions and values were employed to make investigations for the Commission.

"Many boards of equalization made specific recommendations of changes thought by them necessary to properly equalize the districts with each other and to place the property on the duplicate at the true value. Other boards preferred to do this work themselves, notably in Adams and Franklin Counties and the larger cities, except Cincinnati. In these cases few changes were made by the Commission."
(*Report, 1911*, Tax Commission of Ohio, Columbus, pp. 66–68.)

increased by 198 per cent for residential and by 149–228 per cent for varying nonresidential classes. Contemporary beliefs, the express judgment of the U.S. Census Bureau after its independent survey of tax assessment in 1912, and our own evaluation of the appraised per unit values of new building all strongly indicate that assessor valuations in 1910–14 for old and new building were around 90 per cent of market values, a level of appraisal nearly identical with that prevailing in 1857–60.[23]

In order to align the whole time series of building value from 1857 to 1914, two special adjustments seemed called for. First, allowance had to be made for the Civil War inflation which over two years (1862-63) nearly doubled reported unit values of newly erected building. It was easy to compute this adjustment which restated 1857–62 values in post-Civil War appraisal standards (1867–70 base).[24] More of a problem was encountered thereafter. Did the current Kondratieff movement of building costs with its long decline to 1897 and its upward climb to 1909 affect reported assessment valuations? Did the erosion of appraisal standards,

---

[23] "As a result of carrying out the provisions of the law creating the tax commission ... the assessed valuation of real property ... was increased from about one third of the true value to the actual value." But the tables of value ascribed assessed values of taxable realty 90.0 per cent of estimated true value. (*Wealth, Debt and Taxation*, 1915, I, pp. 16, 619.) Our own evaluation of the level of the 1911–14 assessments concerned residential dwelling values reported by tax assessors for new building. We separately enumerated by two different methods counties which clearly undervalued. One method involved comparison of reported per unit dwelling values in the years immediately preceding revaluation with values in the years succeeding. We identified twenty and twenty-four deficient returns for 1911 and 1912. We established the deficiency by using the county's averaged relative standing of statewide per unit dwelling value for the standing nonfarm stock of 1890 and 1930, as disclosed by the Census returns. We found these relative standings of per unit values consistent with degree of urbanization and other relevant values. For 1910, 1911, and 1912, there were between 3000 to 5500 dwellings reported with clearly deficient value and the deficiency when applied to the reported totals resulted in per unit increases of 7.65, 8.07, and 11.03 per cent, respectively. We then computed for each county its estimated statewide average using its recorded per unit value and its mean relative standing of 1890–1930. With the reported statewide average for 1910–14 of $1790, we then arbitrarily established a trustworthy zone of estimates between $1500 and $2500. Utilizing only counties which between 1890 and 1930 had shifted in relative standing by less than 10 percentage points, we obtained 47 observations with a mean statewide per unit value of $1977. Using all 130 observations within the $1500–2500 range, the mean is $1909. From all this it seems clear that, either out of resistance to assessment at market levels or careless reporting, reported residential values for new building in the 1910–14 period understate true value by between 7 and 10 per cent.

[24] Our percentage adjustment derived from the ratio of building "pers" for 1867–70 to 1857–61 was 1.39. This is close to the boost in 1870 of 1860 property values by 43.1 per cent to correspond to the new level of postwar values. The price and cost-of-living indexes over the same years experienced a rise of a similar order of magnitude.

relative to market values, occur at a steady, or a varying pace and persistently through the period? The evidence was ambiguous and our judgments are tentative. One possible scheme of adjustment presupposed that assessor valuations reflected changes in current levels of building costs and that the erosion of appraisal standards was a steady and persistent process. This scheme of adjustment involved deflating by a convenient building cost index (Riggleman) and stepping up values from 1866 to 1909 at a constant rate to bring 1909 per unit values in line with the established 1910–14 level. The resulting series which reproduces the pattern of movement of the recorded original values is displayed in Chart 2 for unit values of residential building. This scheme of adjustment boosts considerably reported building values in the 1870's and again in the 1890's and scales down the marked rise in reported per unit values of over 50 per cent between 1897 and 1909. The aggregate cumulative level of production is raised considerably. It is felt that this scheme of adjustment embodies an outer limit of possible building values using input factors priced at a constant set of prices.

The other scheme of adjustment reflects another set of assumptions. These were that the process of assessment was carried out with the aid of valuation benchmarks which, after the Civil War inflation, were only slightly or irregularly influenced by shifts in building cost indexes of the magnitude encountered between 1870 and 1914. Since the level of building costs in 1910–14 was only 15 per cent higher than in 1868–72, the 1870–1914 period as a whole involved a nearly stable level of input factor prices. At the same time this scheme of adjustment presupposed that the erosion of appraisal values relative to market values did not persist through the whole period but varied by class of property with different cut-off dates. For some categories (e.g., residential), after 1897 assessment standards of new building seemed to pick up relative to assessment standards for old property. For other categories (e.g., industrial), there seemed to be a steady erosion of appraisal standards until 1909. Such disparate movements in assessment standards by class of property and region are commonly encountered in assessment history. Adjustments were made case by case largely by assuring continuity of level or trend of per unit values.

This scheme of adjustment for statewide residential per unit values is plotted in Chart 2 as accepted adjusted values. It reproduces all the movements of the original series which is merely given an upward bias scaled to link without a break the eroded 1907–09 to the accepted 1910–12 values. Comparable graphs for total residential building are shown in Chart 3. On the whole, this set of values, which from 1873 to 1911 runs

CHART 2

Residential Farm and Nonfarm Building Per Unit Values, Ohio, 1866–1914

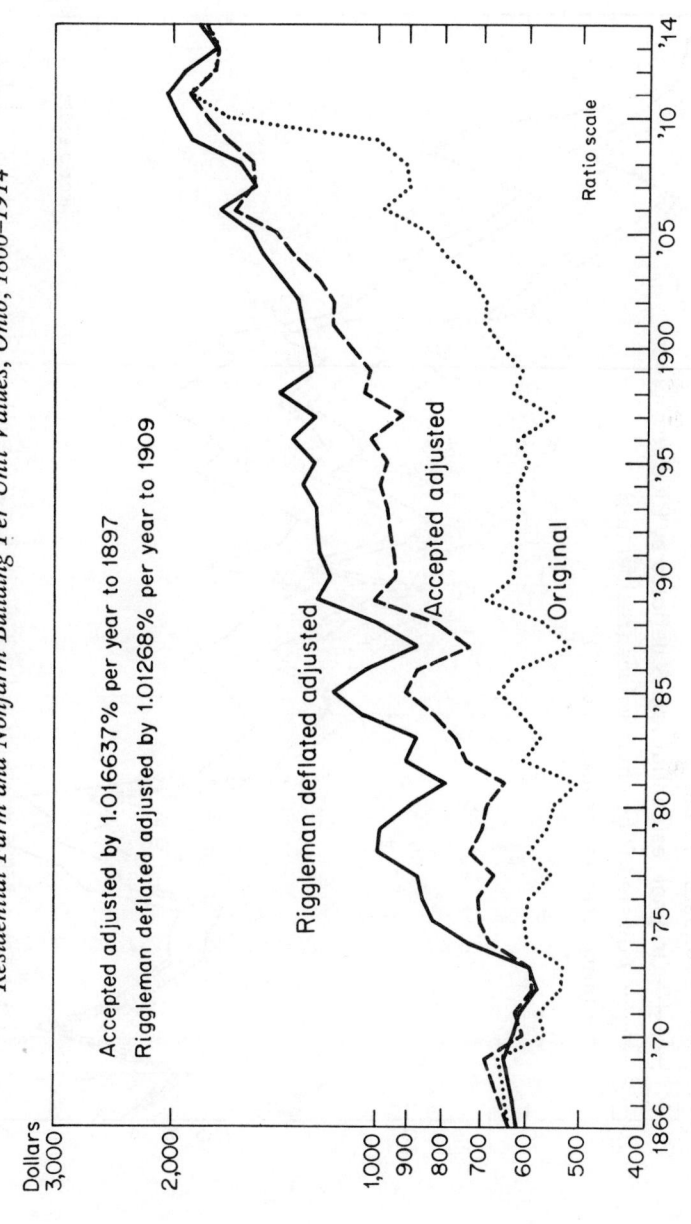

Source: NBER files.

CHART 3

Ohio Statewide Residential Building, Original and Adjusted Values, 1866–1914

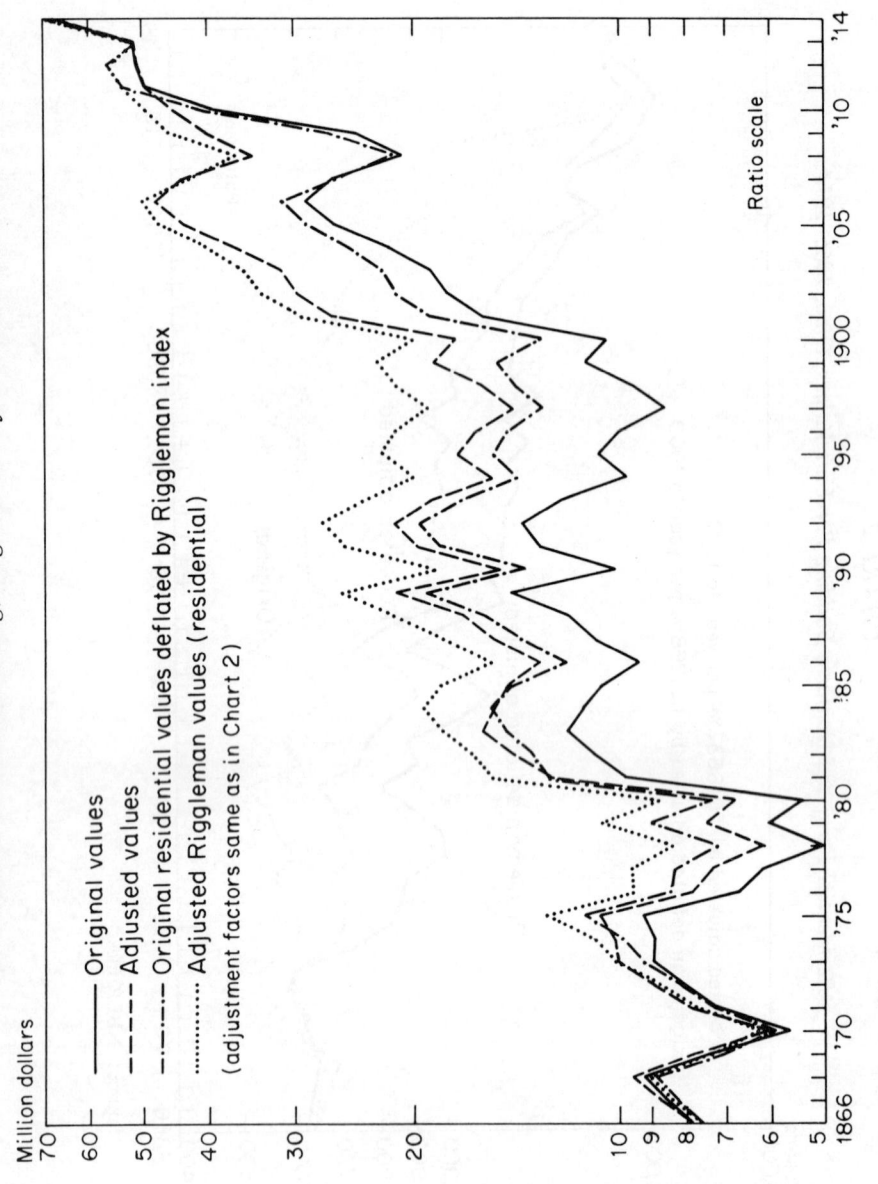

——— Original values
- - - - Adjusted values
—·—·— Original residential values deflated by Riggleman index
·········· Adjusted Riggleman values (residential)
(adjustment factors same as in Chart 2)

Source: NBER files.

midway between the originally recorded and the deflated per unit values, seemed more plausible and it was subjected to detailed statistical tabulation. In any case, it is felt that a trustworthy zone of annual movement for building activity was established for a period covering a vital growth phase of the American economy.

5. *Adjustments in Timing*

Building value indexes were also adjusted for timing to a calendar year record of building. The reported fiscal year data refer to "completions," unlike permit data which refer to "starts." The former (completions) needs to be shifted backward, the latter (permits) forward, to represent building activity.

The adjustment from a fiscal to a calendar year basis was hampered by shifting practice with regard to fiscal years.

	Period	*Fiscal Year Ending*
School Statistics	all	August 31
Assessor Reports	1857–70	July 1
	1871–72	May 1
	1873–79	April 12
	1880–81	July 13
	1882–93	April 12
	1894–	"for the year"

The assessor practice was in effect to make an early or late spring survey of construction undertaken in the preceding year(s) and completed—and hence put on the tax rolls—by the reporting date. Until 1870 but with difficulty thereafter, small structures could have been begun and completed within a reporting year. For most of our years most structures would have been begun in prior years and only completed in the reporting year. Hence a decision was made to regard building completions reported by assessors on a fiscal year basis as building activity carried on during the preceding calendar year. Thus the building report for completions during the year ending April 12 or May 1, 1874, would be credited to the calendar year 1873. School building was otherwise treated. It was felt that a significant proportion of school buildings could have been completed by August 31. Hence school reports were adjusted to a calendar year basis by shifting backward 35 per cent of building value and numbers to the prior year. This leaves open assessor reports which after 1894 bore no reference to fiscal year datings. Since no "break" in the reporting was indicated for 1893-94, the building reports "for the year" 1894 were interpreted as applying to the year ending April 12, 1894, or thereabouts. A report issued in 1894 would hardly cover the data of the reporting year; hence predating seemed justified.

Besides adjustment to a calendar year basis, further backward adjustment is needed for nonresidential buildings which run into large unit values and for which correspondingly longer building times could be presupposed. A scheme for displacement has been worked out by the BLS for current building permit data and has been roughly approximated by Chawner for total building (displacing 33.33 and 50 per cent) and Blank for residential building only (displacing 10 per cent). We used a sliding scale formula which varied in 1857–60 between 10 per cent, for buildings with a mean value of $1,000 and under, and 50 per cent, for buildings with a mean value of $30,000 and over. The scale of percentages was reduced slowly by decades to allow for improved building practices which accelerated building time.[25]

6. Farm Unit Residential Building Values, 1850–1912

The values for residential building analyzed thus far include both farm and nonfarm construction. For many purposes a segregation of the values of farm dwellings, which exerted a general but uneven influence over the course of statewide unit dwelling values, is required. We have

[25] The displacement schedule is as follows (in per cent):

Mean Value of Building (dollars)	1858–69	1870–79	1880–89	1890–1900	1900–09	1910–14
Under 1000	10.0	9.5	8.9	8.5	8.0	7.5
1000–1999	13.1	12.4	11.8	11.1	10.5	9.8
2000–2999	16.2	15.4	14.6	13.8	13.0	12.2
3000–3999	19.2	18.2	17.3	16.3	15.4	14.4
4000–4999	22.3	21.2	20.1	19.0	17.8	16.7
5000–6999	25.4	24.1	22.9	21.6	20.3	19.1
7000–8999	28.5	27.1	25.7	24.2	22.8	21.4
9000–11,999	31.6	30.0	28.4	26.9	25.3	23.7
12,000–14,999	34.6	32.9	31.1	29.4	27.7	26.0
15,000–18,999	37.7	35.8	33.9	32.0	30.2	28.3
19,000–22,999	40.1	38.1	36.1	34.1	32.1	30.0
23,000–25,999	43.9	41.7	39.5	37.3	35.1	32.9
26,000–29,999	46.7	43.4	42.0	39.7	37.4	35.0
30,000 and over	50.0	47.5	45.0	42.5	40.0	37.5

BUILDING IN OHIO BETWEEN 1837 AND 1914   261

already estimated by decade aggregates the approximate quantity of farm dwellings.[26] It was more difficult to arrive at a plausible set of values for farm dwellings.

We had available for each of the eighty-eight counties of the state average values of all dwellings erected in all locations—cities, villages, and farms—for the years between 1867 and 1910. The abundance of cross-section returns made it possible through multiple regression analysis to throw light on the strength of the influences which affected dwelling values. The object was to estimate the value of farm dwellings included in the totals or, conversely, to estimate the differential value of village and urban dwellings leaving farm dwelling value as a residual.

For this purpose, an annual estimate of urban and village population as a percentage of county population was extrapolated from decennial Census returns by individual counties. The two sets of population ratios would yield a biased and inaccurate version of the respective "weights," i.e., dwellings erected in urban and village centers as a percentage of the total. At the same time, the tendency to inverted correlation of the two ratios, if expressed in terms of total county population, might hamper measurement of the influence of urbanization and village shares on dwelling value. Hence mean dwelling values were regressed against urban population as a fraction of total and village population as a percentage of nonurban population. While the two sets of influences cannot be added mechanically, their separate contributions can be readily evaluated.

To test the influence of other variables, sample multiple regressions were run for selected individual years. The tested measures were average city size and farm land value per acre, and in both instances the separate influence of the two variables was inappreciable.[27] The tests, however, disclosed systematic and consistent relationships between mean dwelling

---

[26] See Gottlieb, *Estimates*, Table 11.

[27] Thus regression for thirty-seven nonurban counties in 1870-71 against village percentage and equalized dollar value per farm acre (mean $30.84) showed that at mean values acre values contributed only $14.56, or only some 4.5 per cent, to residual mean farm values, even though the range in per acre value by counties was relatively wide. For size of city, an index which combined urban percentage and average size of cities was devised for four Census years 1880–1910. Addition of the variable to the explanatory forces added practically nothing to the coefficient of multiple correlation adjusted for degrees of freedom. (Average coefficient for $R_{1.23}$, excluding size of city variable, was .485, and including the variable it was .489.) The direct simple correlation between dwelling value and city size was for the four years only .111, although the corresponding term for urban percentage alone was .474. As expected, the partial coefficients of city size, holding village and urban percentages constant, varied between $-.2997$ and $-.0626$ and averaged $-.06$, a statistically insignificant return.

values and urbanization and village percentages for both new housing and standing stock disclosed in the 1890 and 1930 Census returns.[28]

It was then decided to group the returns into the seven phases of Ohio statewide building cycles—1866–72, 1873–77, 1878–91, 1892–97, 1898–1904, 1905–09, 1910–14—and the period as a whole. The dwelling values used were unadjusted for the erosion of assessment standard between 1862 and 1909. Urban and nonurban counties were separately aggregated so that altogether there were sixteen separate multiple regression operations in which the full battery of correlation output was obtained. The principal output returns for the eight time periods are set forth for urban counties on a multiple regression basis in Table 3, for nonurban counties in Table 4. The results reached are significant in many contexts beyond the object of our present limited concern, estimation of value of newly erected farm dwellings in the state of Ohio.

It is necessary even for this limited concern to evaluate, however, the reliability of the regression returns. Our explanatory factors account for about 25 per cent of the dispersion of mean dwelling values around their own mean; the coefficient of correlation holds to a level of .47–.50, which is normalized for degrees of freedom. Both the regression coefficients and other output returns follow reasonably consistent time trends. The only exception here is provided by the 1910–14 period during which time the assessment and property valuation system of the state was recast. Counties did not move at a uniform pace to the newer levels of assessed value. Over a third of the county returns for 1911-12 showed clear-cut undervaluation which apparently was strong in rural communities. The conforming counties dominated the aggregates and generated relatively high urban and village differentials and correspondingly low residual values for farm dwellings. As expected, these low residual values in the constant term were associated with a relatively high standard error term and were disregarded since they grossly understate farm value dwellings.

[28] Although the results reached are superseded by the more inclusive regressions undertaken later, it is worth recording that a rise of 1 percentage point in the urbanization index involved the following percentage increase in average statewide dwelling value:

Year	Number of Counties	Urban or Nonurban	Regression Coefficient
A. Newly Erected Dwellings			
1870–71	51	urban	1.23
1911–14	65	urban	.86
B. Standing Stock Regressions			
1930	75	urban	.97
1890	70	urban	.75

The findings were more ambiguous for farm-village differentials, partly because of the small size of reporting nonurban counties in the later years.

TABLE 3

REGRESSION RETURNS (FORTRAN), MEAN DWELLING VALUE, OHIO,
SEVEN SUBPERIODS AND AGGREGATES, URBAN COUNTIES ONLY, 1866-1914

Item	1866-1914	1866-72	1873-77	1878-91	1892-97	1898-1904	1905-09	1910-14
Number of observations	3,158	221	325	968	413	486	318	251
Mean of variable								
X_1 (mean value, all newly erected dwellings)	614.5	549.9	509.1	511.1	513.7	594.1	687.4	1352.0
X_2 (urban population as per cent of total population)	30.32	21.94	21.66	27.31	33.32	35.57	38.50	40.20
X_3 (village population as per cent of nonurban population)	15.95	13.44	13.70	14.27	16.10	17.82	19.77	20.39
Standard deviation of variable								
X_1	385.5	255.7	230.2	241.2	253.9	321.2	336.8	619.8
X_2	19.31	16.37	16.33	17.38	18.37	19.57	20.46	20.82
X_3	7.40	7.60	6.13	6.42	6.84	7.38	7.48	8.03
Regression coefficients with standard error								
Constant	152.6	286.9	248.2	253.2	182.2	194.3	287.1	324.3
Standard error	±16.2	±28.6	±30.0	±19.6	±36.3	±42.3	±57.3	±99.8
b_2	8.61	7.09	5.37	5.84	6.10	7.96	7.12	16.10
Standard error	±.31	±.70	±.68	±.40	±.60	±.65	±.82	±1.49
b_3	12.58	7.98	10.56	6.89	7.94	6.55	6.39	18.66
Standard error	±.80	±1.51	±1.82	±1.07	±1.62	±1.73	±2.25	±3.87
Correlation coefficients								
$R_{12.3}$	.448	.4625	.3998	.4292	.4454	.4848	.4368	.5627
$R_{13.2}$	.27	.2627	.3063	.2027	.2347	.1696	.1573	.2911
$R_{1.23}$	.5125	.492	.4879	.4635	.4666	.4898	.4489	.6074
$R_{2.3}^2$ (adjusted)		-.08	.087	.037	-.100	-.100	-.006	.0886
Standard error of estimate	333.1		213.8	224.8	280.3	301.4	493.3	

TABLE 4

REGRESSION RETURNS (FORTRAN), MEAN DWELLING VALUE, OHIO,
SEVEN SUBPERIODS AND AGGREGATES, NONURBAN COUNTIES ONLY, 1866-1914

Item	1866-1914	1866-72	1873-77	1878-91	1892-97	1898-1904	1905-09	1910-14
Number of Observations	880	221	98	237	106	108	64	46
Mean of variable								
X_1 (mean value, all newly erected dwellings)	425.8	467.8	367.8	378.1	356.1	366.9	517.2	768.0
X_3 (village population as per cent of nonurban population)	18.46	15.97	14.61	16.53	20.99	23.23	23.94	23.68
Standard deviation of variable								
X_1	226.2	223.0	172.3	178.0	180.3	163.5	280.9	302.7
X_3	6.969	6.345	3.987	5.740	6.459	7.211	6.382	6.185
Regression coefficients with standard error								
Constant	371.6	275.1	219.3	458.9	503.5	406.6	623.9	650.8
Standard error	±21.5	±38.2	±64.9	±34.9	±57.6	±52.9	±137.8	±179.6
b_3	2.934	12.07	10.16	-4.89	-7.02	-1.71	-4.46	4.94
Standard error	±1.09	±2.22	±4.29	±1.99	±2.63	±2.18	±5.56	±7.34
Correlation coefficients								
$R_{1.23}$ (adjusted)	.084	.3374	.2131	-.1438	-.2332	0	0	0
Standard error of estimate	225.5	210.4	169.2	176.5	176.2	164.5	284.0	307.9

If the regression model is now tested by computation of estimated values for the Ohio sample groups, the movement of actual and calculated values are quite close for the intermediate groups (II, III, and IV) and are systematically deviant only for Groups I and V (see Chart 4). The regression, apparently because of its linear form, attributes too much influence to urbanization at the lower end of the urban spectrum and too little influence at the upper end. Expenses in programming for a nonlinear multiple correlation model unfortunately ruled out for the time being a wholesale nonlinear regression to all our data. For present purposes the bias is not too serious, although it indicates the danger of extrapolation beyond the main cluster of our observations.

The regression results indicate two different versions of the value of farm dwellings. In the first place since the regression clearly indicates that, for nonurban counties, the village differential after 1877 was nonexistent or negative, then the mean value of dwellings of nonurban counties represents for those areas a possible measure of farm dwelling values.[29] The trouble is that the measure applies to an increasingly smaller group of counties (only thirteen in the 1905–09 period) not at all representative of the farming life of the state. Moreover, the 1910–14 value needs to be revised upward because of utilization of older standards of appraisal in rural areas.

For urbanized areas there is of course a measure yielded by the constant terms of the multiple regression. The constant term represents those influences on dwelling value not reducible to or associated with urban or nonfarm building, and thus should reflect the position of newly erected farm dwellings. Again, the 1910–14 low figure illustrates that rural assessment standards were slow to change.

These two value estimates may be checked against two independent measures. One of these is the mean value of newly erected barns and stables, which should fluctuate in close correspondence with farm dwelling values. A second check is the mean value of newly erected dwellings in our Group V, consisting of five northwest primarily rural counties in which farm population greatly outnumbered nonfarm population and in which only one out of five resident persons in 1920 dwelt in cities over 2,500 persons in size. These dwelling values, which for 1910–14 have been adjusted for rural undervaluation, certainly constitute a limiting set of values.

The four series involved are plotted in Chart 5. There is a definite

[29] While the village regression coefficients are negative—interestingly enough until the 1910–14 period—the standard error of most of the coefficients is high and the correlation coefficient adjusted for degrees of freedom is zero or below standard significance levels.

CHART 4

Mean Values of Residential Dwellings, Ohio Sample Groups for Long-Cycle Phases, Actual and Calculated, 1866–1914

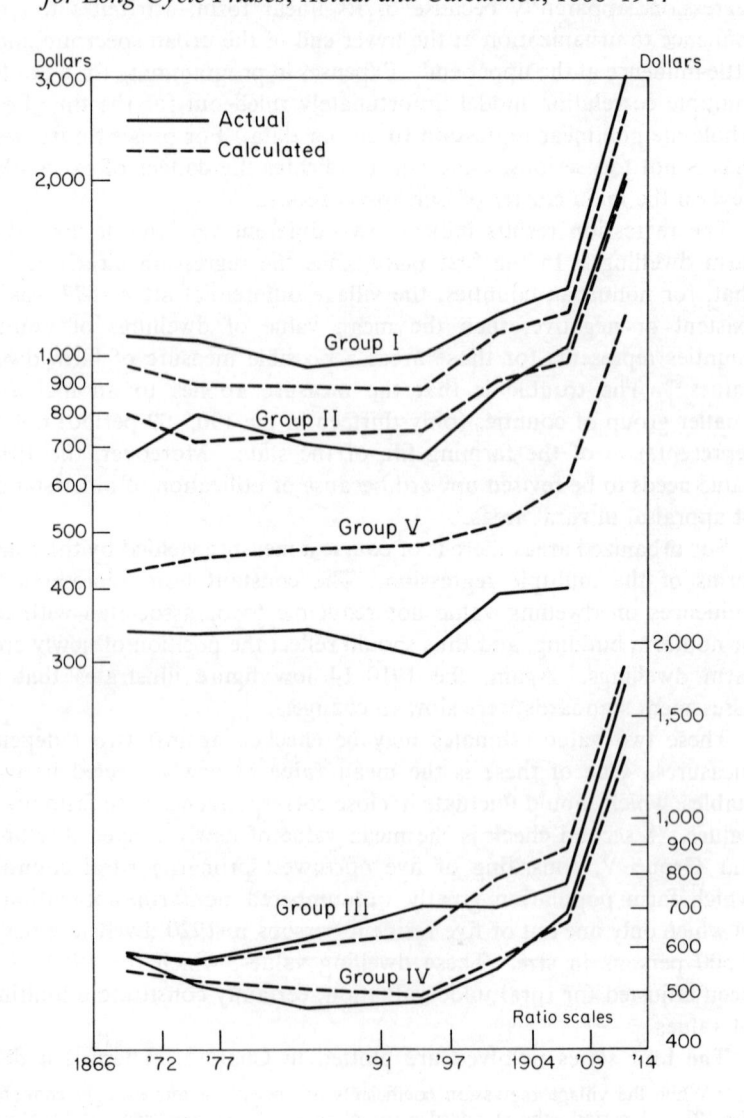

CHART 5

Variant Estimates of Mean Value of Newly Erected Farm Dwellings, Ohio, 1866–1914

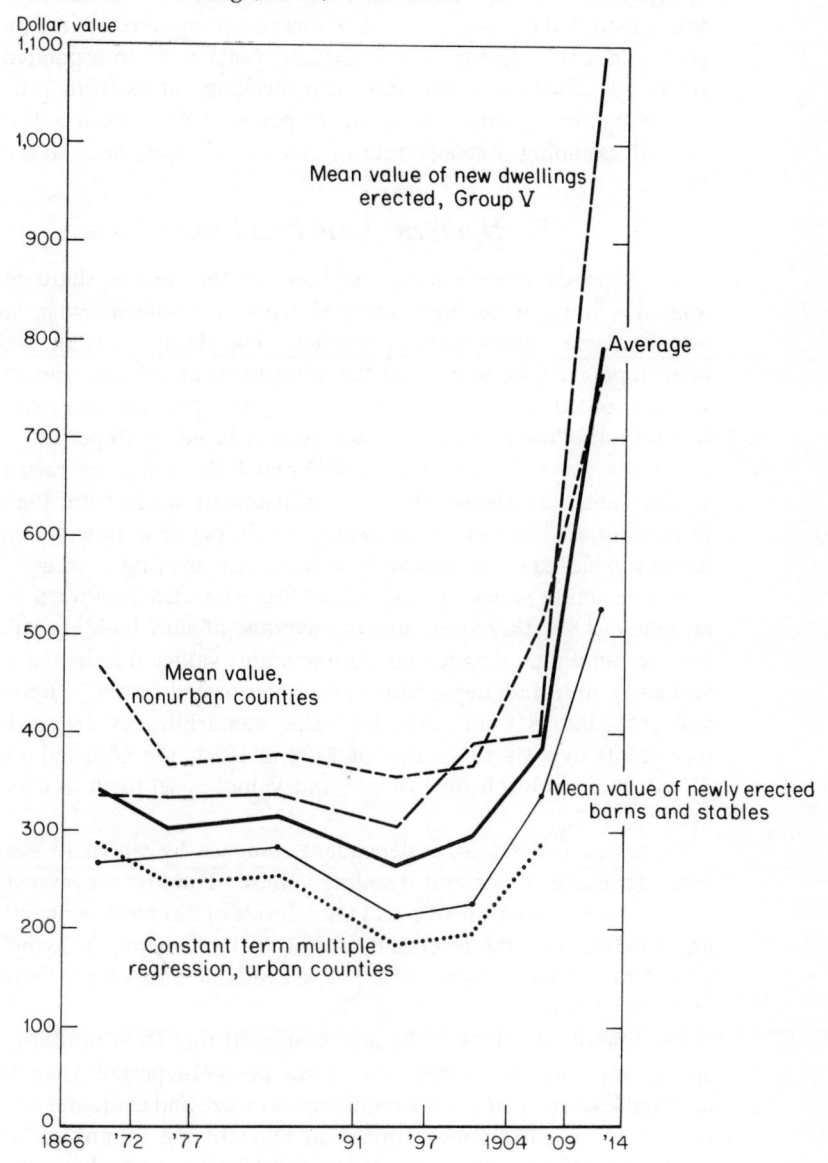

family likeness of movement. The true figure probably lies somewhere in between. Hence, we indulged our penchant for "divide and average" and constituted an average value of farm dwellings erected in Ohio between 1866 and 1914. Values for individual years were interpolated linearly, yielding a schedule of per unit farm dwelling values from 1867 to 1914. Accepting the terminal values at 90 per cent of market, a formula was devised assuming a steady rate of erosion of assessment standards until 1907.

7. *Nonfarm Unit Residential Values*

These farm dwelling values, weighted by the decade share of farm to total dwellings, were then removed from statewide average values and new nonfarm values were computed. The decade sets of these values fitted together with little need for adjustment at or near transition years. The computed unit values of the 1870's, with perhaps excessive estimates for farm dwelling production, had to be reduced by 10 per cent. And the values for two individual years, 1880 and 1881, had to be raised by some 15 per cent. Otherwise, the only adjustments made were the extension to residential values of the allowance for 10 per cent undervaluation from market values and smoothing by a three-year moving average.

The resulting series of unit values was extended backward to 1858 by an index of the three-year moving average of unit building values of all private building. Residential numbers and values dominated the totals, and total unit and residential values conformed closely. Between 1850 and 1858, the 1858 unit dwelling value was arbitrarily dropped by small increments to a starting value of $728 in 1850. As adjusted and extrapolated, the entire schedule of per unit values is set forth in Chart 6 from 1850 to 1912.

There are five sets of independent evidence by which to evaluate the entire schedule of per unit dwelling values. The first set involves a comparison between nationwide and Ohio levels of nonfarm per capita income and wealth. Did the level and pattern of Ohio nonfarm living conform to that of the nation as a whole? The available measures are summarized in Tables 5 and 6.

The line-up of values in the tables suggests that Ohio nonfarm property values and incomes during the entire pre-1914 period were below the national level, possibly due to differences in size and character of urbanization or to an early concentration in Ohio of the mining industry. Per capita living levels fell short of the national level which, particularly in the years before 1900, was dominated by the older Atlantic seaboard

CHART 6

Per Unit Nonfarm Dwelling Values, Ohio, 1850–1912

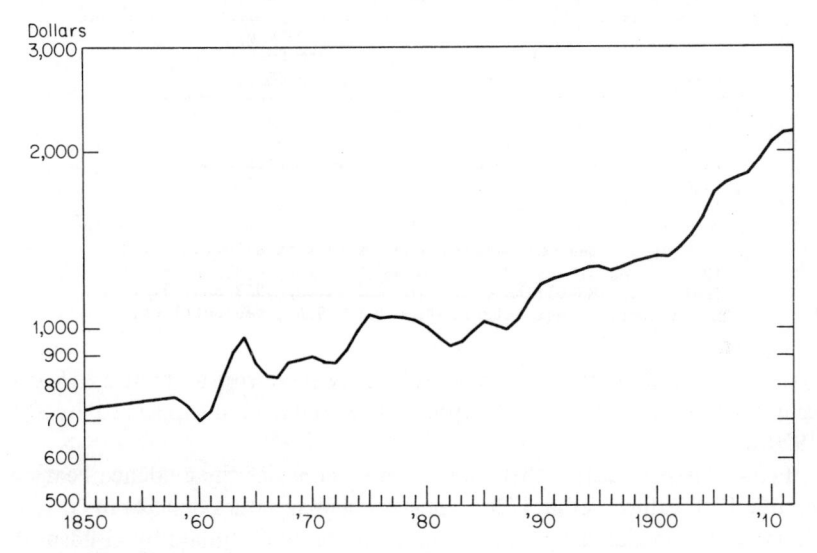

Source: Table 8, col. 6.

TABLE 5

PER CAPITA WEALTH AND INCOME, COMPARISON OF OHIO AND U.S.A., 1840-1930

Year	Item	Ohio (dollars)	Ohio as Per Cent of U.S.A.
1840	Total nonfarm income per nonfarm worker	356	75.3
1880	Total nonfarm income per nonfarm worker	551	96.3
1900	Total nonfarm income per nonfarm worker	609	97.9
1900	Nonfarm residential real property, true value per family	1,671	85.7
1930	Average value, all nonfarm dwellings	5,138	102.3

Source: Gottlieb, *Estimates*, Tables 4 and 6; *Wealth, Debt and Taxation*, 1907.

TABLE 6

NONFARM WEALTH PER NONFARM WORKER, COMPARISON OF OHIO AND U.S.A., 1850-1912

Year	Ohio (dollars)	Ohio as Per Cent of U.S.A.
1850	432.5	85.8
1860	644.8	88.5
1890	1,696.0	91.5
1900	1,664.0	86.6
1912	2,256.0	92.6

Source: See Ohio wealth appraisals listed footnote 19; *13th Census of Agriculture*, General Report, U.S. Bureau of Census, pp. 83-91; *Wealth, Debt and Taxes*, 1907 and 1915. For nonfarm worker data in Ohio and U.S.A., see Gottlieb, *Estimates*, Table 10.

states. The differential, which in 1850 may have run as high as 15 per cent, narrowed as the 1900's approached and had disappeared by the 1920's.

From this perspective then, we can now consider the evidence bearing on our 1850 values, set at $728 in post-Civil War values or $523 in current dollars. The evidence includes a highly inferential estimate by Goldsmith of average values of nonfarm standing dwellings in 1850,[30] which perhaps need not be given much weight. Nor need we be concerned about the Ohio Census returns for a newly erected Ohio dwelling in 1840 which,

---

[30] Raymond Goldsmith, "The Growth of Reproducible Wealth of the United States of America from 1805 to 1950," *Income and Wealth of the United States*, Income and Wealth Series II, Cambridge, Eng., 1952, pp. 318 f. His estimate of $800 is the average nonfarm residential value (allowing only for a quarter of realty value of land for 1890) reduced to the 1850 levels of building cost. Goldsmith found the $800 consistent with average dwelling values for standing stock of New York State in 1855 and with an estimate of $300 for the cost of building a standard room. The 1890 wealth estimate is itself questionable and a safer base would be the estimate of $1034 (see Leo Grebler, David Blank, and Louis Winnick, *Capital Formation in Residential Real Estate: Trends and Prospects*, Princeton for NBER, 1956, p. 365) derived from the Census of Homes returns. This probably is excessive in two regards, allocation of land value and allowance for use of structure value. Adjusting the $1034 accordingly and converting to post-Civil War values yields an equivalent in used values for 1850 of between $850 and $900. New dwelling values would on this base run to around $1170-$1700 depending on the used-to-new ratio prevailing. This is, of course, only an 1850 equivalent of an 1890 value assuming no upward drift of dwelling values because of urbanization and the rise of per capita incomes. Just that assumption is in question. The Chicago estimate of building costs needs to be interpreted in the light of central city-statewide differentials, which for Ohio were 84.6 per cent (1866-70), 47.6 per cent (1890), 51.5 per cent (1910-14), and 43 per cent (1930). Moreover, the average size rental unit—and the share of rental units—must be determined.

when adjusted for changes in purchasing power, was $721.[31] This includes farm dwellings with their presumably lower values, but it also includes an undetermined amount of site improvements and lot values.

A more important check on the soundness of the Ohio unit values consists of the Grebler-Blank-Winnick unit values derived from urban permit statistics overlapping with our Ohio returns for the years 1889–1912. The two annual series, together with the Chawner series for 1900–12 and benchmark estimates of the average values of the nonfarm dwelling stock, are set forth in Chart 7. The discrepancy between the annual estimates is widest at the start in 1889. The Ohio values gradually climb from 51.5 per cent of the permit-derived value in 1889–91 to 81.5 per cent in 1910–12. The Ohio values would seem to reflect more closely the shift of value experienced by the average unit value of standing stock.

This striking difference in level and behavior pattern is traceable in part to divergences of statistical method and composition. For the level in 1910–12, the gap is partly attributable to the tendency noted earlier for Ohio levels of per capita housing to understate national per capita levels by 7–10 per cent over the years in question. Then there were substantial adjustments for undervaluation, site, and preliminary costs in the permit-derived data. These adjustments involved a markup of 127.44 per cent over originally reported findings. Our returns are adjusted for undervaluation only by 10 per cent and no allowances for builders' profit or site improvements were made. Then nonfarm nonurban building was attributed a unit building value of 66 per cent of the mean urban value. Wickens attributed rural dwellings of the same class only 47.6 per cent of the unit value in the 1920's of all urban dwellings.[32] In the light of the 1890 Census differentials and the persistent emphasis on urbanization in our Ohio materials, the Wickens allowance for rural-urban differentials seems more appropriate. Using it for the years 1890–94 and 1905–09

---

[31] See Gottlieb, *Estimates*, pp. 51 f., for conversion of the Census returns of $1012 per house in Ohio to $639. Allowing for inflated building costs of 1840 compared to 1850 (Riggleman's 1840 index is 72.0, 1850 is 58.5) and converting to post-Civil War values, we derive the $721. It is anybody's guess how much to allow in this for farm dwellings, log cabins, and land or site values. Even in the late 1850's, the Ohio statistics commissioner could assert that "many" of the dwellings built were log cabins. In 1840 the proportion of log cabins was higher. Local Ohio historians, when asked about the validity of the 1840 Census value, cautioned the present author about the role of log cabins. Even in New York State, the 1858 Census described 6.3 per cent of all dwellings in the state as log cabins. (*Census of the State of New York for 1855*, Albany, 1857, p. 247. *Ohio Executive Documents*, 1858, Pt. 2, p. 587.)

[32] David L. Wickens, *Residential Real Estate*, New York, NBER, 1941, p. 73 (ratio of $2315 to $4865).

CHART 7

Per Unit Nonfarm Dwelling Values, Average Values of Standing Nonfarm Stock, Average Hourly Earnings in Manufacturing, 1889–1912

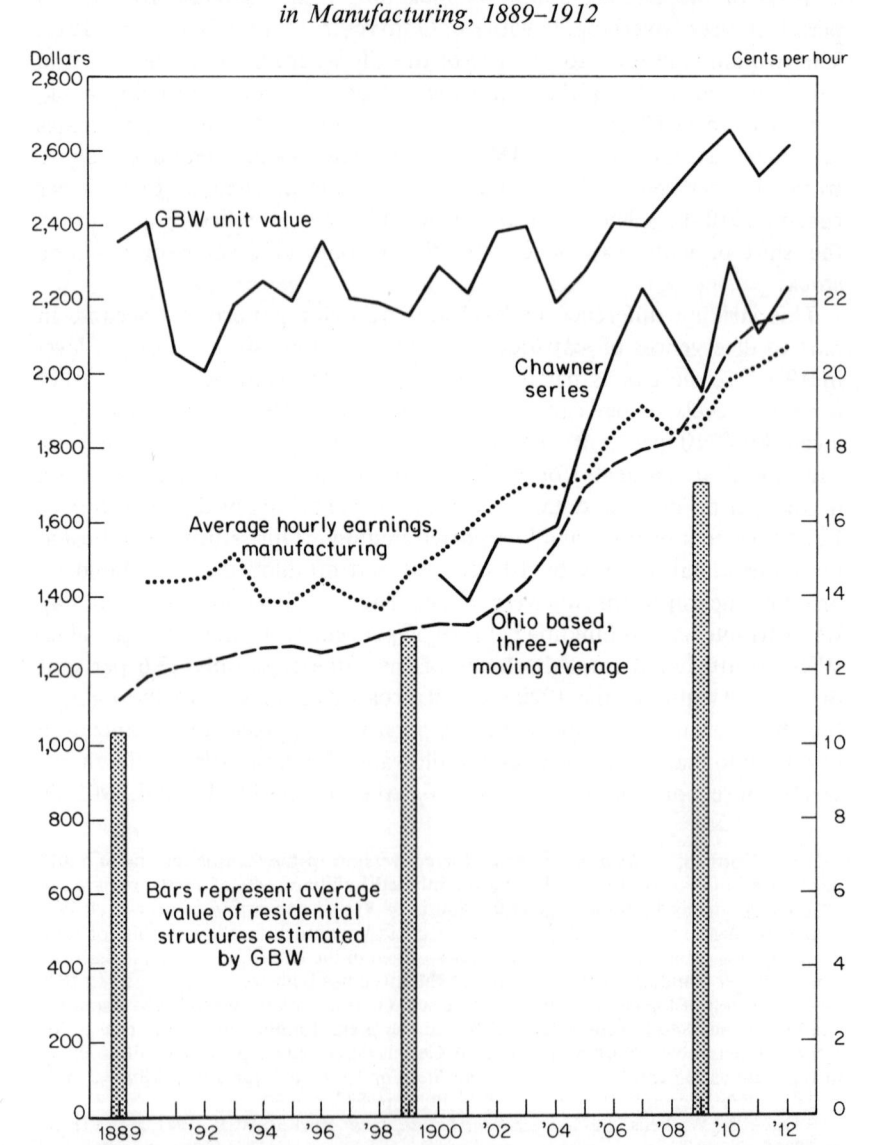

Source: L. J. Chawner, *The Residential Building Process*, 1939, Table 4.5; Grebler, Blank, and Winnick, *Residential Real Estate*, pp. 72 and 426; A. Rees, *Real Wages in Manufacturing, 1890–1914*, Princeton for NBER, 1961, p. 4; and Table 8, col. 6.

reduced the permit-derived values by 10 and 6.2 per cent, respectively.[33]

Finally, the sampled cities were grouped in relatively few size classes which would tend to be dominated, at least for size classes under 100,000, by the larger cities in the class. Thus the survey returns for the size class 2,500–24,999 was unduly influenced by cities at the upper end of the class because few of the smaller cities were caught in the reporting network. The probable bias was compounded since a technique used to make up for deficiencies arising from small sample size was projection of the experience of an adjoining size class of the same division.[34] Since sample deficiencies preponderated among lower size groups in the early years of the study, the building rates for these groups were biased upward when deficiences were remedied.

If we can thus explain the difference in the 1910–12 level between the Ohio and the permit-derived data, how can we explain the radically different time trend? The Ohio values double over the period while the permit values, ignoring apparently erratic returns for 1889 and 1890, rose only by 28 per cent by 1910–12. The standing stock per unit values rose by 65 per cent. The rise in the Ohio series accelerates in the 1900's, thus reflecting the real estate boom and lift of values that marked that boom. One possible explanation is the downward bias on unit values generated by the rapid expansion of the permit sampling base. For the four northeast and north central geographic regions, the mean percentage of population covered by the sample within the size class rose from 18.3 to 44.7 per cent for size class III (cities between 25,000 and 100,000) and from .5 to 5.3 per cent for size class IV (cities between 2,500 and 25,000). As the sample expanded, the weight of the smaller cities in the size classes increased. This would automatically tend to lower per unit values, thus offsetting the upward trend induced by higher incomes and the over-all price rise.

For these reasons both the level and pattern of movement of the Ohio per unit values for the 1889–1912 period could be considered more like the national averages than the Grebler-Blank-Winnick permit-derived unit values. Our analysis thus confirms the judgment of Margaret Reid

---

[33] We used for these purposes the original estimates of urban residential construction derived directly from the permit records and adjusted by Blank's 127.44 per cent markup for undervaluation and omissions (David Blank, *The Volume of Residential Construction, 1889–1950*, NBER Technical Paper 9, New York, 1954, p. 41). The breakdown disclosed that the nonurban component was estimated to account for 43.5 per cent and 34.5 per cent of all units during 1890–94 and 1905–09, respectively.

[34] D. Blank reports that three alternative techniques were used to make up for sample deficiencies. The first technique cited involved utilization of building experience of an adjoining size class in the region. Only when "data for the given class and adjacent class were not available for a closely preceding or following period" were other procedures employed (*ibid.*, pp. 36 f.).

that these values "substantially overstate the value of new dwellings except during the forties and early fifties."[35]

A third item of evidence concerns the 1890 benchmark appraisal by the Census Bureau of the value of owner-occupied encumbered nonfarm structures.[36] These made up one-tenth of all nonfarm dwellings. Including land, the mean declared value was $3,250. Grebler, Blank, and Winnick allow 40 per cent for value of land by extrapolation from benchmark estimates of the 1920's and 1907.[37] Considering the rising tide of land values that marked urban growth in the nineteenth century with horse and carriage transportation, this allowance seems inadequate. Ohio decennial appraisals, which allowed for little growth in land values after 1870, yielded a share of land value in the appraised value of realty in towns and cities that declined from 60.7 per cent in 1859 to 51.2 per cent in 1900.[38] The order of magnitude for Wisconsin in 1916 was somewhat lower.[39] At a 53 per cent share for land value, the average value of a dwelling falls to $1,527.5. This value however applies to the tenth of nonfarm dwellings in 1890 that were owner-occupied encumbered structures. These dwellings have higher mean values than all nonfarm dwellings the more home ownership is extended to lower income groups and the higher the ratio of structure to dwelling value. Between 1890 and 1940, the home-ownership ratio increased from 37 to 41 per cent; and Grebler, Blank, and Winnick accordingly allowed the 63 per cent ratio of encumbered owner-occupied dwellings to all dwellings in 1940 to be scaled down to 55 per cent. If this allowance is accepted, the mean value of a nonfarm standing dwelling in 1890 is $840. This is reduced by 5 per cent

---

[35] See Margaret G. Reid, "Capital Formation in Residential Real Estate," *Journal of Political Economy*, April 1958, p. 139. Margaret Reid was stimulated primarily by her conviction that the Grebler-Blank-Winnick thesis of a long-term decline in per unit real dwelling values was not sustainable. She expressed belief that "the evidence presented in this article indicates an upward rather than a downward trend in the quality of housing since 1890" (*ibid.*, p. 132).

[36] It has been specifically noted that "the high average value of the owned mortgaged homes may be greatly affected by the number and size of the apartment houses in which the owner himself resides. Such an apartment or flat building, if mortgaged, would be included in the average." *Mortgages on Homes*, U.S. Bureau of Census Monograph 11, Washington, 1923, p. 82.

[37] See Grebler, Blank, and Winnick, *Residential Real Estate*, p. 364.

[38] Equalization reports for the years in question.

[39] See H. M. Groves and J. Riem, "Statistical History of the Property Tax Base in Wisconsin, 1916–1958," *American Journal of Economics and Sociology*, January 1961, pp. 127–148. Assessed value of land in 1916 made up 59.7 per cent and 50.4 per cent of total assessed realty in Madison and Milwaukee, respectively, and around 33–35 per cent is indicated for the state. Wisconsin assessment data also indicate that the land improvement ratio for residential property exceeded the ratio for all nonfarm realty by only 1 or 2 percentage points. (See *ibid.*, p. 138, Table 4.)

to convert structures to dwelling unit values.[40] If we then apply to this mean value the high-low and mean ratio that prevailed in benchmark years after 1919 between the mean values of used dwellings and newly erected dwellings, we derive a high-low range for new 1890 mean dwelling values of $1,110–$1,520. Used values should be relatively high at building cycle peaks if new building is to be discouraged.[41] This would point to a level of new values nearer the lower end of the range, i.e., near the level predicated in our Ohio series ($1,186).

The lower end of the range is also indicated by qualitative information on costs of building in the pre-1914 period. In 1913, what was called an "ordinary house, 26 by 40 feet on the ground and 30 feet from the bottom of the cellar floor to the top of the roof," with three bedrooms, two and a half stories, and 31,200 cubic feet of total space was estimated for urban building in a standard building manual at $3,120.[42] If allowances are made for lower costs in small towns and for the many smaller rental units in multiunit structures, the range of values is not far from our Ohio 1913 value.[43] In 1879 three prizes were awarded for designs for a "cheap frame house" two stories high with three bedrooms which would be erected on detailed specification for $1,000. Work included excavation, foundation, framing, finishing, and all materials. No heating or plumbing equipment was included.[44] In 1890 these houses would have cost $1,089, or well in the neighborhood of our Ohio unit values.

Still another absolute comparison is yielded by purchase information

---

[40] We use the magnitude for overstatement estimated by Grebler, Blank, and Winnick (*Residential Real Estate*, pp. 369 ff.) in 1930 in connection with Wickens' use of 1930 Census data. They estimated that the shift from structure to unit values amounted to 10 per cent of the owner-occupied inventory or to roughly 5 per cent of over-all dwelling values. Blank (*Residential Construction*, p. 50) noted the use of structure values in the 1890 Census.

[41] Used realty prices for Paris, Manhattan, and Berlin show positive reference cycles with a lead in timing for Berlin, a lag in Manhattan, and virtually no reference decline in Paris. Experience in Los Angeles from 1900 onward suggests that used prices conform to price shifts of new building with greater amplitude. See R. M. Williams, "The Relationship of Housing Prices and Building Costs in Los Angeles, 1900–1953," *Journal of the American Statistical Association*, June 1955, pp. 370–376. See also Grebler, Blank, and Winnick, *Residential Real Estate*, Appendix C.

[42] William A. Radford, *Radford's Estimating and Contracting*, 1913, p. 767.

[43] Wickens estimated that in 1930 dwellings occupied by owners are in general worth about one-third more per unit than those which are rented (Wickens, *Residential Real Estate*, p. 3). Grebler, Blank, and Winnick (*Residential Real Estate*, pp. 112 f.) illustrate the same relationship with data indicating that dwelling units in varied multiunit structures were valued at between 70 and 80 per cent of single-family dwellings over the years.

[44] *Carpentry and Building*, September 1879, "Competition Design for Cheap Dwelling Houses."

obtained in the 1934 Financial Survey of Urban Housing.[45] In that survey acquisition prices were obtained in twenty-one large and medium-sized cities from 981 sampled owner-occupants who acquired their dwellings (mixed new and old) in the 1890's. The mean purchase price was $2,633. Applying thereto the Grebler-Blank-Winnick allowance for land value yields a structure cost of $1,627. Cities of that size and character were related to the statewide average, according to the pattern of our Ohio studies, at near 140 per cent, which yields a general average of nonfarm housing values (mostly new) of $1,162, a value that fits in with the Ohio-based series but not the permit series.

While comparison with the three independent sets of evidence indicates that our Ohio-based series fall within the range of acceptability, the comparisons also reinforce the suggestion disclosed by the income and wealth benchmark measures that our Ohio series was below the national level, particularly in the earlier years of the period.

8. Adjusted Time Series of Building in Ohio

The various adjusted series for the value of Ohio building cumulated by decades are shown in Table 7. The originally reported value of total building is presented in column 1 adjusted only for inclusion of estimates of tax-exempt building up to 1872 and of school building between 1857 and 1914, for deficiencies, for appraisal slippage, and for conversion from "completions" to current activity. The values in columns 2, 3, 4, and 5 represent the broad categories of taxable new building as separately reported in Ohio assessor building statistics and adjusted on the same scale as total value of building. The estimates of the value of newly erected farm dwellings, listed in column 6, were not derived from the assessor building statistics but were contrived on the basis of other information to form nonfarm estimates consistent with our other magnitudes. The final column adds together the estimates of industrial, commercial, and residential (columns 2, 3, 4, and 5) and the estimates of exempt construction previously reported (column 3 of Table 2) less only the estimated allowance for residential farm building (column 6) to make up estimates of total nonfarm building.

These estimates of nonfarm building could, of course, be converted into total building by adding back the farm components. When so converted, they differ considerably from the originally reported total value of building. Our derived total (from adding components) is higher

[45] David L. Wickens, *Financial Survey of Urban Housing, Statistics on Financial Aspects of Urban Housing*, Washington, 1937.

TABLE 7

DECADE AGGREGATES OF ADJUSTED VALUE OF NEW OHIO BUILDING, 1850-1910
(thousand dollars)

Decade	Reported Total Building (1)	Industrial Building[a] (2)	Commercial Building (3)	Residential Building (4)	Barns and Stables[b] (5)	Estimated Value		
						Farm Dwellings[c] (6)	Nonfarm Building[d] (7)	
1860's	99,599	5,147		63,835	5,519	7,291	94,691	
1870's	133,160	7,487	12,824	82,664	7,885	22,208	127,767	
1880's	220,868	21,517	24,648	146,308	12,889	9,743	217,730	
1890's	262,283	35,500	40,147	173,363	12,247	16,968	293,042	
1900's	550,379	116,000	79,580	353,908	22,682	15,011	599,477	
Total 1870-1910	1,166,690	180,504	157,199	756,243	55,703	63,930	1,238,016	
Adjustment factor for erosion slippage	1.579 (1897)	3.359 (1909)	2.585 (1899)	1.668 (1897)	1.50 (1907)	1.946 (1907)		

[a] Including only manufacturing establishments, mills, and foundries, but not railroads, canals, or public utilities. Return for 1900's is scaled down from our adjusted yearly series by some 10 per cent because of a probable overestimate in our industrial value correction factor.

[b] Decade aggregates for number of barns and stables were multiplied by adjusted unit values for the four decades after 1870: $270, $333, $292, and $409. Estimates for 1860's based upon building during 1865-69 and an assumed value of $250.

[c] Derived by applying to our independently derived decade estimates of newly erected farm dwellings (Gottlieb, *Estimates*, Table 11) per unit values derived from mean estimates of the value of newly erected farm dwellings (see Section 7) interpolated linearly for intervening years, adjusted for appraisal slippage with a correction factor of 1.95 (1907) and converted to decade average values from 1870 onward: $344.7, $400.3, $432.4, $642.7. (A $300 estimate was used for the 1860's.)

[d] Sum of columns 2, 3, and 4, minus column 6 plus column 3 of Table 2. For purposes of the 1860's estimate, $10 million was allowed for commercial building.

in all decades and for the forty years between 1870 and 1910 by 16 per cent than the originally reported aggregates as adjusted. In small part this was because industrial and commercial building were adjusted individually for deficiencies which were not transferred to the totals. A more important reason for divergence is the inadequate allowances after 1872, as noted previously, for exempt building other than school construction.[46]

Finally, our procedure of adjustment for erosion slippage involved disparate individual adjustments for the different categories of building. The coefficients of adjustment, expressed as a multiple of the recorded value for a stipulated cutoff year, are listed at the bottom of Table 7 for each adjusted series. The series was raised at a steadily growing rate by the cutoff year to the stipulated multiple of the reported value. The adjustment for slippage may have been slightly more favorable to the individual types of building than to the reported total. On the whole, the derived aggregate is believed to afford a more accurate measure of total building than the reported totals. However, it was not found feasible to distribute into annual returns these more accurate aggregates. The originally reported totals were, of course, available on a yearly basis and all adjustments were made to the yearly returns. Hence our yearly Ohio series of total building shows an order of magnitude and yearly variations but tends to understate as much as 15 per cent after 1872. The yearly totals for total, residential, commercial, and industrial building and unit values for farm and nonfarm residential building are listed in Table 8. A full publication of the tabular returns with the appropriate charts must await a more detailed monograph.

[46] There are indications that the assessors did a better job of reporting church and public building than school building. Thus, assessor reports of county building for the two sets of years 1873–80 and 1900–09 amounted to 88 and 79 per cent of the county tax levy for building purposes. Considering that appraised values were adjusted for appraisal slippage, the total holds up. Comparison of decennial appraisal reports with cumulated church building reports also indicates that assessor reports on church building were less lax than those on school building. Thus, comparison of assessor reports of new construction cumulated over the 1870–90 period compared with net increments of outstanding appraised value of building showed much more under-reporting of school than of church construction, as can be seen from the following figures (in thousand dollars):

	Schools	Churches
Net change in value of appraised building over 20 years	6,930	4,509
Cumulated assessor reports on new building	5,932	6,981

BUILDING IN OHIO BETWEEN 1837 AND 1914

TABLE 8

VALUE OF BUILDING, OHIO, BY TYPE, ANNUALLY, 1837-1914

Year	Aggregate Value of Building[a] (thousand dollars)				Unit Value of Residential Dwellings (dollars)	
	Total	Industrial	Commercial	Residential	All[a]	Nonfarm[b]
1837	13,165					
1838	11,177					
1839	7,002					
1840	13,104					
1841	14,892					
1842	7,415					
1843	4,816					
1844	4,618					
1845	4,786					
1846	4,954					
1847	5,978					
1848	8,975					
1849	9,938					
1850	10,015					728
1851	12,171					733
1852	14,740					737
1853	16,528					742
1854	14,403					746
1855	10,854					750
1856	8,106					755
1857	10,108					759
1858	8,557					764
1859	7,046	170				739
1860	7,892	115				695
1861	6,533	61		3,440		722
1862	5,486	238		3,598		817
1863	6,603	528		4,708		914
1864	6,968	586		4,304		964
1865	9,560	644		6,182		875
1866	11,800	702	925	7,926	640	828
1867	14,902	783	970	8,830	655	821
1868	16,291	756	1,359	9,376	673	875
1869	13,564	734	1,504	7,645	680	879
1870	10,987	619	897	6,202	607	892
1871	13,240	845	1,149	7,952	620	874
1872	14,762	1,208	1,456	9,031	588	872
1873	16,134	1,092	611	10,081	602	916
1874	15,980	771	1,503	10,245	683	993
1875	16,842	504	1,429	10,467	701	1,050
1876	13,191	499	987	7,797	700	1,040
1877	11,168	612	1,009	7,241	674	1,046
1878	9,119	455	1,191	6,260	726	1,041
1879	11,737	883	1,592	7,388	694	1,036
1880	11,764	1,463	1,293	7,289	685	1,001
1881	20,231	2,760	2,804	12,736	661	965
1882	23,404	2,528	2,218	14,423	787	935

(continued)

TABLE 8 (concluded)

Year	Aggregate Value of Building[a] (thousand dollars)				Unit Value of Residential Dwellings (dollars)	
	Total	Industrial	Commercial	Residential	All[a]	Nonfarm[b]
1883	23,837	2,021	2,025	15,788	769	940
1884	21,906	1,323	1,613	15,212	832	987
1885	20,887	1,811	2,358	14,365	902	1,024
1886	21,135	2,253	3,441	13,231	860	1,008
1887	22,843	2,108	2,851	15,430	743	992
1888	25,984	2,356	2,585	17,919	862	1,037
1889	28,877	2,894	3,460	19,915	967	1,120
1890	22,977	3,335	2,484	15,327	938	1,186
1891	29,811	3,714	3,883	19,830	944	1,207
1892	31,003	3,143	3,845	21,076	958	1,222
1893	28,260	2,946	4,067	18,648	968	1,242
1894	25,806	3,590	4,526	15,605	985	1,260
1895	26,599	2,688	4,226	17,570	999	1,265
1896	22,656	2,002	3,691	15,762	984	1,249
1897	21,432	3,173	3,745	14,899	960	1,266
1898	24,327	4,403	3,733	16,259	1,044	1,294
1899	29,412	6,506	5,947	18,397	1,024	1,309
1900	29,748	7,550	4,984	18,370	1,098	1,323
1901	41,102	11,062	5,868	26,702	1,156	1,321
1902	48,817	12,293	8,252	29,923	1,161	1,370
1903	49,200	8,342	9,868	31,797	1,227	1,437
1904	54,692	7,922	11,571	37,081	1,333	1,541
1905	65,250	15,080	8,859	44,015	1,429	1,693
1906	71,344	18,999	8,988	47,370	1,616	1,758
1907	65,201	15,793	8,610	42,675	1,504	1,795
1908	58,666	17,416	6,646	35,140	1,530	1,816
1909	66,359	15,185	5,934	40,835	1,674	1,926
1910	78,375	13,658	7,087	45,213	1,776	2,068
1911	91,131	14,543	7,950	49,893	1,883	2,139
1912	93,431	16,490	9,000	54,300	1,854	2,154
1913	89,192			52,913	1,735	
1914	107,047			67,799	1,792	

Source: NBER Series, No. 0148, 0150, 0163, 0164, 0223.

[a] At 90 per cent of market value.
[b] Adjusted to 100 per cent of market value, three-year moving average.

COMMENT

Paul A. David, Stanford University

As those of us who have had occasion to go prospecting for data are well aware, America's state and local documents contain virtually untapped mines of economic statistics. Manuel Gottlieb has not only struck a particularly rich lode buried in the *Reports of the Ohio Commissioner of Statistics* and kindred documents for that state, but has set up full-scale operations to exploit his find. In the present paper, he carries forward his earlier work on nineteenth-century building activity by utilizing assessment statistics in conjunction with other materials to fashion a series of estimates of new building in Ohio for the years 1837–1912. Specifically, Gottlieb provides us with a detailed account of the impressive amount of work that has gone into developing three major statistical series relating to the volume of construction in Ohio: (1) the "value" of total new building, (2) the "value" of new nonfarm residential building, and (3) the "value" of new commercial, industrial, and public (i.e., nonresidential) building. In each instance the final estimates are expressed in terms of 1870–1910 appraisal values, which Gottlieb considers tantamount to a constant price weighting of the various components of new building.

Use of this rather unorthodox standard of valuation appears to follow more directly from the nature of the original statistical deposit than from the particular techniques adopted in refining those materials. If the reliance upon property assessment data which this paper describes can be accepted, Gottlieb will have shown us a way of tapping a great vein of information. For, wherever property taxes were levied, at the state, the county, and municipal levels, there remain abundant deposits of assessed value statistics, many of which may be arranged in long time series disclosing, as is the case with the Ohio records, considerable historical detail.

Yet I remain rather more skeptical than Gottlieb as to the ultimate worth of nineteenth-century tax assessment records as bases for the estimation of stocks of real tangible wealth, or additions made thereto. It is, I suggest, generally symptomatic of the drawbacks inherent in this type of quantitative information that its use appears to require a great number of *ad hoc* adjustments, many of them guided only by adherence to the criterion of "assuring continuity of level or trend"—to cite Gottlieb's own description of the nature of the operation. Among the many difficulties posed by the character of the data with which Gottlieb has sought to cope, those concerning the meaning of assessed values are both particularly vexing and especially important to an undertaking of this kind.

It is apparently difficult to determine just what correspondence actually existed between the values entered by the Ohio assessors for *new* structures and the variables usually explained by economic analysis, i.e., market prices and quantities. Gottlieb's paper presents evidence regarding ratios of assessed values to current market values for conglomerations of Ohio real estate property at several dates during the 1837–1912 interval, and it is presumed that these observations, though they relate to land as well as to old structures, may be taken as applying specifically to the relationships that existed between assessed values and market prices of new structures. That, however, is not the end of the problem, for, as Gottlieb's discussion makes clear, assessors did not seek to maintain a consistent or indeed even a changing relationship between the current market values and the appraised values of "property." Over the long run, appraisal rates appear to have reflected the Ohio assessors' appraisals of the condition of the real estate market as well as the particular property being assessed. The principal source of slippage cited by Gottlieb was occasioned by the assessment of the level of prices prevailing in the Civil War and postwar property market—against the fictional standard of "normal," nonspeculative values so cherished in Marshallian long-run equilibrium analysis. On the other hand, it is also pointed out that, within shorter time intervals, the valuations established for buildings in existence on the occasion of the last equalization of assessments were likely to have been those that constituted the proximate reference point in the determination of assessed values entered for new structures.

Waiving any questions concerning the collective ability of Ohio assessors to accurately describe the structure of future prices while establishing property valuations congruent to those set down in the recent past, neither adherence to the prospective nor the retrospective standard of valuation is particularly helpful in the present connection. No simple mechanical adjustments would suffice to retrieve the history of annual current market values from assessment data derived by application of the former—despite the fact that such application might be thought of as an expectational transform of previous market prices. At the same time, it is only reasonable to entertain serious doubts that the retrospective standard of valuation led Ohio assessors to appraise new structures in accordance with some base-year cost of the resources currently required to erect the buildings under consideration. Yet that is precisely what we are obliged to assume if the estimates based on undeflated assessed values (corrected only for "secular slippage" in appraisal rates) are to be admitted into the company of conventional constant-dollar volume measures of gross additions to the stock of capital.

Beyond the difficulties and uncertainties surrounding efforts to adjust assessment statistics in order to eliminate the effects of shifts in assessment rates and appraisal standards over time, one encounters the problem of possible variations in assessment rates at given moments in time. Gottlieb's paper is rather ambivalent in its treatment of this issue. On the one hand, the general discussion seemingly proceeds on the tacit assumption that whatever actual assessment rates and appraisal standards happened to be, public awareness and pressure for fair treatment (although "just" values are not necessary market prices) tended to assure that they were uniformly applied to all taxable structures throughout the state. I do not know how reasonable a supposition about the assessment practices that prevailed in Ohio during the nineteenth century this may be, but, if Illinois' experience provides any guide, it would appear to bestow unwarranted praise upon the honesty and diligence of the average tax assessor. For example, land assessments in the Chicago central business district in 1896 were usually one-ninth of current market values; yet, in individual cases, assessments ranged from 1 per cent to 100 per cent of market value. Evidence at subsequent dates reveals similarly wide variations in assessment rates among different locations in Chicago and among different types of property and different owners, as well as among different urban places in Illinois.[1] If comparable variations existed in Ohio, they would not only jeopardize the validity of uniform adjustments of assessed values but, to the extent that the variations were of a systematic character, they would render disaggregations by type of property and location particularly treacherous.

On the other hand, Gottlieb reports having smoothed out the trends in assessed unit values for different categories of new structures individually, being prompted to do so because "assessment standards of new buildings *seemed* to pick up relative to assessment standards of old property" after 1897 in the case of dwellings, while, in the case of industrial buildings, "there *seemed* to be a steady erosion of appraisal standards until 1909." (The emphasis supplied is mine.) Assuming, of course, that the foregoing statements are not based solely on divergent movements in assessed unit values of different classes of property, it is a pity that Gottlieb's present paper does not make more use of his evidence of differential trends in ratios of assessed values to market prices for various categories of new

[1] Cf. *Report of the Illinois Tax Commission*, 1896, and Herbert D. Simpson, *Tax Racket and Tax Reform in Chicago*, Evanston, 1931, pp. 44–45; both cited in Homer Hoyt, *One Hundred Years of Land Values in Chicago*, Chicago, 1933, pp. 461–462.

buildings in Ohio. The latter would certainly seem to bear more directly on the question of the extent of the slippage in appraisal rates than do the ratios presented for Ohio real estate property *en masse*. In addition, the trends in the market value denominators of these ratios would throw further light on the secular decline in the prices of new industrial and commercial structures vis-à-vis prices for new houses, a relative price change that appears (from the recent work of Dorothy Brady) to have occurred in the U.S. during the decades that followed the Civil War. Moreover, the undisclosed magnitudes of cross-section variations in ratios of assessment value to market value are quite crucial; even if differential trend movements among the ratios implied by the assessment data have truly been rectified, a constant pattern of variation would still subject estimates of the aggregate volume of new construction to distortions caused by reliance on an inappropriate set of relative unit values in weighting the constitutent categories of new building.

The one major category of new construction for which assessed unit values could not be directly obtained from the Ohio data was farm residential building, and the specific technique employed by Gottlieb in closing the gap calls for some comment.

To obtain assessed unit values of farm dwellings Gottlieb sets up a model in which the mean unit cost of all kinds of dwellings in a given county is a linear function of (1) the proportion of the total county population living in urban places, (2) the proportion of the total nonurban population living in villages, and (3) a constant term. The idea underlying this approach is to employ the known locational distribution of the county population as proxy weights for the assessed unit values of farm, village, and urban dwellings, and, assuming the latter to be constants, estimate them by least squares regression analysis. The key assumption implicit in this approach—namely, that the annual ratio of the number of new dwellings to population is uniform across farm, village, and urban communities within a given county—may really be justifiable. But, if that is so, one rather wonders why investigators like Blank, Wickens, and Foster (let alone the Bureau of Labor Statistics, which also took up the work of estimating the total volume of residential construction from sample permit data) bothered to develop stratified sampling schema that classified communities according to size.[2]

[2] Cf. David M. Blank, *The Volume of Residential Construction 1889–1950*, National Bureau of Economic Research, Technical Paper 9, New York 1954; David L. Wickens and Ray R. Foster, *Non-Farm Residential Construction, 1920–1936*, NBER Bulletin 65, New York, 1937; *Construction Volume and Costs, 1915–1956*, a statistical supplement to *Construction Review*, U.S. Department of Commerce and Labor, 1958; *Techniques of Preparing Major BLS Statistical Series*, U.S. Department of Labor, Bureau of Labor Statistics, Bulletins 993, 1950, and 1168, 1954.

Suppose, however, that the implicit assumption is admissible. One could step up a regression model in which the mean unit value of all dwellings ($\bar{c}$) was a linear function of the proportion of the total population living in urban places (U/P), the proportion of the total population living in villages (V/P), plus a constant term. It would have the following form:

$$\bar{c} = a_0 + a_1(U/P) + a_2(V/P). \tag{1}$$

The constant term (a_0) in this function admits an unambiguous interpretation: it is the unit value of a farm dwelling. Similarly, the coefficient of the urban population share (a_1) is the urban-farm difference in the unit value of a dwelling; and the coefficient of the village population share in the total population (a_2) is the village-farm difference in unit value. All this is readily seen by invoking the assumption that new residential building is distributed exactly as is the population, writing the identity expressing the mean unit value of all dwellings as a weighted sum of the farm, village, and urban dwelling unit values, (c_f, c_u, c_v), thus,

$$\bar{c} = c_f[1 - (U/P) - (V/P)] + c_u(U/P) + c_v(V/P), \tag{2}$$

and then regrouping the terms to obtain the form of equation (1).

Note, however, that the regression model used in the paper differs from equation (1). Gottlieb reports that because the urban and village population shares in the total were found to be inversely correlated, to avoid multicollinearity he replaced the village share in total population with another variable, namely, the village share of the nonurban population. The revised model actually fitted was:

$$\bar{c} = a_0 + a_1(U/P) + a_2[V/(P - U)]. \tag{3}$$

Unfortunately, having done this, Gottlieb cannot legitimately then proceed, as he does, to interpret the estimate of the constant term in the revised regression model [equation (3)] as the unit value of a farm dwelling. There is, of course, one trivial case in which such an interpretation is legitimate—namely, when the proportion of the total population living in urban places is zero. This is especially significant in view of the vastly disparate estimates of the constant term that result when the model is applied to data for the urban counties in Ohio, on the one hand, and to data for the nonurban counties, on the other.

Finally, the estimate Gottlieb derived from this regression procedure was tossed into an average together with three other figures. The latter figures—being accorded three-fourths weight in determining the ultimate unit value for farm dwellings—would clearly concern us if we were particularly worried here about the precision of the final estimate obtained.

The issue raised above is, however, one that involves a principle of rather more general interest. The trouble would seem to have originated from a misplaced concern over the presence of multicollinearity. This econometric affliction is indeed troubling where one is primarily interested in securing a meaningful test of the statistical significance of regression coefficients; on the other hand, it does not lead to biased parameter estimates. In using a regression model such as that given by equation (1) to estimate unit values, it does not seem that one should be preoccupied with the question of whether or not a particular coefficient can really be taken to be significantly different from zero. Rather, one is attempting to decompose an aggregate into its *known* components; the general form of the appropriate regression model is derived from an identity, and cannot be in dispute. The two hypotheses that would, indirectly, be subjected to trial by the use of equation (1) are: the supposition that it is meaningful to talk about uniform unit values for each class of structures in the first place, and the more dubious assumption that the volume of new residential building was distributed among farms, villages, and urban places in precisely the same manner as the population.

Now it is one thing to call attention to the difficulties flowing from the character of assessed value statistics, or to note some particular technical defect in the way it has been handled, but it is quite another matter to venture assertions regarding the seriousness of such problems—that is, to appraise the over-all reliability of the final set of estimates that Gottlieb's paper presents. The latter issue can only arise legitimately once it is proposed that some particular use be made of these series: to evaluate the "reliability" of quantitative or any other sort of evidence *in vacuo* would be meaningless. After all, what considerations save those suggested by specific questions addressed to the data could serve to fix bounds within which inaccuracies—inevitably present in statistical reconstructions of the sort offered at this conference—are to be regarded as being of a tolerable order of magnitude? It is disappointing not to be able to find in Gottlieb's paper some definite indications of the context in which his estimates of three-quarters of a century of building activity in Ohio are meant to be evaluated. Being left, thus, at liberty to speculate on the question, it is entirely possible that I may unwittingly imagine these statistics will be used for purposes quite removed from those contemplated by their author. Yet, in view of the effort made in estimating them, we may at least be assured that this set of numbers was intended for ultimate service in capacities beyond that of providing corroborative detail for Margaret Reid's criticisms of the Grebler-Blank-Winnick data on unit real values of nonfarm dwellings.

Gottlieb's work will undoubtedly prove useful to inquiries into the role played by the instability of local construction activity in the generation and transmission of business cycles in Ohio's economy. Annual series relating to current expenditures on building (i.e., those derived from market prices) would, however, be rather more appropriate in that connection. Moreover, it would seem wise to hesitate before attaching particular significance to the precise pattern of turning points displayed by the annual series, if only in recognition of the rather arbitrary fashion in which the original observations relating to the completion of structures have been displaced backward in time. For each decade the percentage displacements are made to increase linearly with the absolute levels of assessed unit values, while over time the percentages displaced within each unit-value range decline from decade to decade in a strict arithmetic progression between 1858–69 and 1910–14. This is a convenient, but nontheless undocumented, specification of the course of technical change in the Ohio building trades.

The question of the extent of regional conformity with the long swing movements observed in new construction activity at the national level[3] obviously constitutes another context in which the Ohio building estimates are of immediate interest. Here, the common application of smoothing procedures would reduce the importance of imprecisions in annual observations. Yet, in attempting to ascertain the magnitudes of differences between relative amplitudes and durations of long-swing movements at the national and regional levels, it is presumably desirable to examine the movements of series that are in other respects closely comparable. Unfortunately, the mass of data available for the study of long swings in construction relate either to current values or to constant-dollar volumes obtained by the deflation of expenditure estimates; there is reason to suspect that, in comparison with the latter, the use of Gottlieb's assessment-derived estimates for Ohio might tend to promote an illusory reduction of the relative amplitude of regional fluctuations in the level of new building activity.

In a previous work, Gottlieb made use of annual data for the number of dwelling units erected in Ohio to fashion a corresponding aggregate series for the U.S. by projecting Ohio estimates to the national level. He has, I believe, shown commendable caution in not automatically following suit with the assessment value estimates described here. Can we trust that others will be equally judicious in their use of these statistics? Surely, their proper use would have been furthered had the present paper afforded

---

[3] Cf. Moses Abramovitz, *Evidences of Long Swings in Aggregate Construction Since the Civil War*, NBER, Occasional Paper 90, New York, 1964.

readers the benefit—even in most abridged form—of its author's insights into the specific differentiae of the nineteenth-century Ohio economy, particularly in their relationship to the history of building activity in that state.

REPLY *by Gottlieb*

The sense of apprehension with which Paul David evidently regards the use of property assessment statistics as a guide to building activity is, in my judgment, unfortunate. Property assessment figures are not a by-product of administrative operations, nor are they an information statistic gathered to appease curiosity seekers. They were used to apportion taxes among individuals and districts and hence were regarded seriously from the standpoint of accuracy and completeness of treatment of individual items, coverage, and summarization. Individual property assessments were listed in public records exposed to full public view and to pressures for equalized treatment; and consolidated assessment returns were meticulously examined at regular intervals by official bodies for disparities in assessment levels. In most communities, the basic facts on assessment process and outcome were matters of common knowledge among settled adult persons with property involvements. As a general rule it is true, as David asserts, that most assessment records, particularly of large and rapidly growing cities, will give erratic information on building because of the irregular revaluations of existing property and varying levels of assessment between assessment districts. But not always and surely not for all areas are property assessments so corrupt that a meaningful story cannot be extracted from them.

In Ohio, for reasons which I canvassed in the present paper and in an earlier study, assessment data have a high statistical potential. First, in Ohio the practice of revaluing existing properties was confined, substantially and by force of law and administrative practice, to the formal reassessment carried out every six years and after 1859 decenially by specially appointed appraisers. The shift of values arising out of revaluation can thus be largely isolated during most of our period from the shift in values reflecting new building net of demolitions. Secondly, at the opening (1846–59) and closing (1910–14) of our survey, a wide variety of evidence, including sales data, tax history, Census appraisals, and well-known facts, established nearly conclusively that realty assessments were adjusted to levels of market value. Contrary to David's impression, this approximation applied not only to land but also to improvements. And if old improvements or parcels of realty were valued for tax purposes at or near prevailing levels of market value, would not local pressure for equalization require that new improvements (so readily tagged with a current value) be appraised for tax purposes on the same basis? This pressure was not presumed to have worked, as David implies, without imperfection. I found evidence of resistance to assessment at market levels in many of the rural counties, at least for newly built residential property, and I estimated that for 1910, 1911, and 1912 between 3,000 to 5,500 dwellings were reported with "clearly deficient values." Doubtlessly similar imperfections existed in the earlier period. Hence it was presumed that average reported values approximated 90 per cent of market value rather than market value itself. This implies, of course, that the deviance from market values would be skewed on the underside.

In the half century after the Civil War, the level of assessment under the influence of wartime inflation and lax tax administration drifted away from market value levels. Both old property and newly erected buildings were valued at less than market value; and both sets of values jumped after the assessment revolution of 1910. But there was no simple way to correct for drift from market levels. As noted in my paper, I found indications that Ohio building values were derived not from sale transactions but appraisals by assessors accustomed to valuation by conventional benchmarks which disregarded year-to-year shifting in building costs. At any rate deflation of these appraisals by a standard building cost index presented an implausible picture (Chart 2). Hence I chose not to deflate my building values between the Civil War and World War I but adjusted the pre-Civil War and post-World War I values to the level of what I came to term 1870–1910 appraisal values. I agree with David that this limits the usefulness of the final estimates and that application of more conventional deflation techniques would have been desirable. After the detailed series have been released, other investigators will be able to experiment with other adjustment and deflation techniques. But, however adjusted, the Ohio building statistics provide us with a significant new measure of building reaching back into the 1830's and carrying up to World War I.

My estimates of residential building values were in part constructed from an annual series of Ohio unit dwelling values covering both farm and nonfarm building. An effort was made to extract the farm component in these series by a regression of countywide dwelling values against index measures of urban and village population ratios. David states that use of these ratios presupposes that "the annual ratio of the number of new dwellings to population is uniform across farm, village, and urban communities within a given county." On the contrary, I expressly asserted that the population ratios "would yield a biased and inaccurate version of the respective 'weights.'" These ratios were made wholly independent by computation on different bases, total population in one case and nonurban population in the other. This model produces a more reliable estimate of the respective parametric influence of urbanization and villages on dwelling values. David asserts that, with independent explanatory variables, the constant term in the regression equation no longer approximates farm values.

There is, I would contend, such an approximation if our regression coefficients of village and city influence can be reliably extrapolated to zero. In the test of the regression model for the Ohio sample groups, I pointed out that "the regression, apparently because of its linear form, attributes too much influence to urbanization at the lower end of the urban spectrum and too little influence at the upper end." Hence I would suggest the constant term in the urban county regression understates farm dwelling values less on account of the form of the village variable than on account of the linear form of the regression itself.

For the nonurban counties, our regression indicated that the village-farm differential was slight or nonexistent possibly because the suburban or satellite character of villages in urbanized areas was absent. Hence recorded mean values for these nonurban counties drawn chiefly from southeast Ohio were used to represent farm unit values. To facilitate evaluation of these two estimates of farm dwelling values, I added to the comparison another set of mean dwelling values for a slightly urbanized group of counties located chiefly in northwest Ohio. These three measures were then contrasted for balance with mean statewide values of barns and stables. Of the four measures plotted (see Chart 5), only one depended upon our regression. I found the four measures exhibited—and readers can judge for themselves—"a definite family likeness of movement." David can quarrel with my surmise that "the true figure probably lies somewhere in between" the series plotted. For the margins of accuracy needed for my projection and considering the variety of evidence brought to bear on the question, I would reaffirm that judgment. How one derives from there on some acceptable allowance for farm value, by averaging or some other method, is of secondary importance. What I averaged, in any case, was not two "disparate" regression estimates, one of them "meaningful," but one regression set of estimates and three sets of actual values believed to approximate farm dwelling values within some meaningful range.

Minerals and Fuels

Development of the Major Metal Mining Industries in the United States from 1839 to 1909

ORRIS C. HERFINDAHL

RESOURCES FOR THE FUTURE, INCORPORATED

The United States' rich endowment of mineral deposits made possible an enormous expansion in the output of metallic minerals from 1839 to 1909. The expansion of output was far from uniform in time or space, however. This paper develops the industry's statistical record of output and employment, by region and by mineral product. Some of the factors that produced the expansion are also briefly discussed.

Since this group of industries showed substantial changes in rate of output growth and large regional shifts, both among already producing regions and from them to new regions, the effects of some of the causal factors are shown more clearly here than in an already settled country with long-established mineral industries. In many cases, the record of output and employment gives some indication of the influence of the location of existing markets and economic activity, of changes in the cost of transportation, of technological change within the mining industries and, above all, of the quality of natural endowment and changes in it resulting from mining activity. A few comments are possible also on the bearing of the record of the mineral industries on some general issues of economic development.

The major metallic mining industries, which are the subject of this paper, constitute almost all of the industry—98 per cent or more if measured by value of output. This group is defined to include iron ore, copper, lead, zinc, gold, and silver. But major metal mining has been only a minor part of all the mineral industries, which include coal,

NOTE: I wish to acknowledge the valuable work of Selma Rein in investigating and evaluating a very extensive body of source material for output estimates and other data on the development of the mineral industries.

Jerome Milliman and Sam Schurr have given me the benefit of their helpful comments on an earlier draft of this paper.

petroleum, sand and gravel, and clay products, among others. And even this minor share was declining throughout much of the period under study—from one-third in 1870 to one-fifth in 1910, if measured by employment.

All mineral industries have been quantitatively unimportant for the United States as a whole—probably accounting for less than 3 per cent of total employment during the period under study. But in certain regions the mineral industries, and metal mining in particular, have been of far greater importance, and an examination of the record reveals their influence on the timing and pace of regional economic development and especially on the location of certain types of economic activity.

Until the beginning of metallic mining in the West, U.S. production of nonferrous ores was only a small part of the world total, except for lead. In the years that followed, however, the United States came to produce the sizable fractions of world output shown in Table 1.

The U.S. mine output of copper was sufficient to provide a sizable surplus over consumption of primary copper from about the 1860's on. Mine production of lead, however, was closer to apparent consumption over most of the period under study. Zinc mine output was roughly equal to consumption after 1879; in the 1850's none at all was produced.

In this study, the base year data are derived mainly from the various Censuses, supplemented in earlier years by Whitney's comprehensive

TABLE 1

U.S. MINE OUTPUT AS PERCENTAGE OF WORLD MINE OUTPUT: COPPER, LEAD, ZINC, GOLD, AND SILVER,[a] 1849, 1879, AND 1909

	1849[b]	1879[c]	1909[c]
Copper	1	15[d]	60
Lead	16	22	30
Zinc	0	12	35
Gold	24	30	17
Silver	0.3	44	27

[a] Metal content.

[b] World output is from J.D. Whitney, *The Metallic Wealth of the United States, Described and Compared with that of Other Countries*, Philadelphia, 1854. U.S. output is as estimated in the present paper.

[c] World output is from the various *Economic Papers* of the U.S. Bureau of Mines, except for copper in 1879. U.S. output is as estimated in the present paper.

[d] World output is Henry R. Merton's estimate in *Mines and Quarries, 1902*, Census Bureau, Washington, 1905, p. 491.

account of the mineral industries[1] and in later years by the data collected by the U.S. Geological Survey. It is perhaps not surprising that the Census data are defective in important ways. The 1870 Census was the first to make much effort to collect data for the mineral industries. The 1840 and 1850 Censuses are very doubtful, so much so that they seem unusable for our purposes. We have found it better to ignore them and to extrapolate employment back from 1860, by our estimate of output. Throughout the Censuses, the data on number of mines are particularly poor.[2]

The statistical record is not useless, however. It reveals faithfully the general movements of output, by commodity, and the fortunes of the different regions in the production of the various commodities. Less reliance can be placed on its employment data—though even here the general outlines of what happened are evident—and some reliance can be placed on even the more detailed quantitative aspects of the general picture.

Output Behavior, 1839-1909[3]

Within only seventy years, the mine output of the major metals grew to 117 times its 1839 level, a rate of growth averaging about 7 per cent per

[1] J. D. Whitney, *The Metallic Wealth of the United States, Described and Compared with That of Other Countries*, Philadelphia, 1854.

[2] It was difficult to get data and even to find mines in the areas of the Rocky Mountains, the Southwest, and the Pacific Coast. For example, the Census of 1880 explains that the collection of statistics was hampered by, among other things, "the assassination of Colonel Charles Potter, the expert in charge of this territory" (*Census of 1880, Precious Metals*, p. 100).

[3] In this section and in the rest of the paper it is necessary to speak of changes in the output of a "commodity" that in fact is made up of several commodities—iron in ore, copper in ore, etc. These must be combined by some weighting scheme. Since our interest is in mine output, it would be preferable to use as a weight the price of a "real" unit of value added, but this has not been feasible. Instead, 1879 market prices have been used as weights throughout.

However, the market prices used were the prices of metal for copper, lead, zinc, gold, and silver ores, but the price of ore for iron ore. While any weighting scheme other than the value-added one is arbitrary, it is true that the weight for one commodity, iron ore, is taken at the mine level but at the metal level for the others. The problem here is that the ratio of the price of metal to the price of the "ore" of that metal ("ore," because of the joint product problem at the ore level) was considerably higher in 1879 for iron than for copper, gold, silver, zinc, and probably lead.

If the price of iron—about 3.7 times the price of the same quantity of iron in ore in 1879—had been used to weight iron ore output in our tables, a number of statements in the paper would need extensive alteration. For example, all statements about the relative importance of iron ore compared with the other mine products would be liable to change. So also would all statements about movements of a composite that contained iron ore, provided the movement of iron ore differed substantially from the movements of the other members of the composite.

The weighting scheme that uses the price of pig iron instead of the price of ore as a weight will be called the "alternative weighting scheme." As we go along, some effort will be made to indicate the effect of using this scheme.

annum.[4] That growth, while exhibiting considerable steadiness in the aggregate, was punctuated by a number of large and sudden changes, both in the geographical location of production and also in the commodities produced. The two types of rapid change, clearly evident in Tables 2 and 3, are related, of course. We shall see that there have been two major initiators of change: (1) discovery of large mineralized areas with deposits far richer than those previously exploited; and (2) development of cheaper transportation which permitted the exploitation of extensive deposits where in many cases the grade of ore was only reasonably good.

The most obvious change in the location of mineral output over the period was the shift, just before the Civil War, from complete dominance by the East to a marked dominance by the West (including the Southwest, the Rocky Mountains, and the Pacific Coast), and to the West's continued but less imposing dominance at the end of the period (Table 3). In 1839 the share of the East in the total was 100 per cent, but by 1859 the precious metal discoveries in the West had reduced this share to 20 per cent. Thereafter, the share of the East increased to a level of 43 per cent by the end of our period in spite of the great development of mining in the West. Within the eastern region, the major shift was a steady increase in the share of the north central area (which in our classification includes, among other states, Michigan, Minnesota, and Missouri) from a share of one-third in 1839 to 84 per cent in 1909.

In 1839 iron ore and lead dominated major metal mining and continued to do so until the great precious metal discoveries gave gold and silver mining the leading position from 1849 to 1869. Since that time, the relative importance of the other four metals—especially of copper and iron ore—has increased considerably.

Changes in the fortunes of the different regions and the different commodities are so closely intertwined that they must be examined together in order to be understood. Table 3 indicates that the northeast region was of minor importance in 1839 and thereafter dwindled to practically nothing as far as major metal minerals were concerned. The middle Atlantic region was the dominant iron ore producer in 1839 and accounted for a little over one-third of the total major metal mineral output. The output of iron ore in this region increased steadily until 1879, after which it fell. After 1879, the region's output of zinc increased substantially, but neither these changes nor those in the output of iron ore were sufficient to prevent the region's decline to a comparatively low level until by the end of the

---

[4] All relative rates of growth in this paper are continuously compounded.

TABLE 2

INDEXES OF VALUE OF MINE OUTPUT OF MAJOR METALS, BY REGION
AND COMMODITY, 1839-1909
(in 1879 prices, 1909=100)

Region[a] and Commodity	1839	1849	1859	1869	1879	1889	1902	1909	Relative Percentage Share, 1909
Region									
New England	--	--	--	--	--	--	--	--	
Middle Atlantic	12	20	35	54	80	76	67	100	3
South Atlantic	21	34	28	43	61	81	132	100	1
North central	0.8	1.3	2.2	4.6	11	25	69	100	36
South central	2.5	2.6	3.9	2.3	6	35	80	100	4
East	2.0	3.1	4.8	8.0	16	30	71	100	44
Southwest	--	--	0.6	13.0	20	20	35	100	19
Rocky Mountain	--	--	1.0	5.9	18	47	92	100	32
Pacific	--	29.0	130.0	56.0	57	43	71	100	6
West	--	2.9	15.0	14.0	23	38	71	100	57
United States	0.9	3.2	11.0	11.0	20	34	71	100	101
Commodity									
Iron ore	1.9	3.2	4.7	7.4	14	28	69	100	26
Copper	--	0.1	1.4	2.6	4.7	20	58	100	37
Lead	4.5	6.1	4.3	4.5	24	40	71	100	6
Zinc	--	--	1.7	4.1	9.4	24	71	100	5
Gold	0.6	14.0	60.0	43.0	43	41	85	100	14
Silver	b	0.2	0.7	16.0	61	90	98	100	11
									99

Note: All tables with no source given are derived from one of the basic tables in the appendix.

If the alternative weighting scheme had been used (see text footnote 3), the following indexes would have resulted:

	1839	1879	1909
United States	1.3	16	100
Middle Atlantic	20.0	128	100
North central	0.3	8	100

[a]The following states (or predecessor territory) are included in the regions. New England: Me., N.H., Vt., R.I., Mass., Conn.; Middle Atlantic: N.Y., N.J., Pa.; South Atlantic: Del., Md., Va., W. Va., N.C., S.C., Ga., Fla.; West north central: Minn., Iowa, Mo., N.D., S.D., Neb., Kan.; East north central: Ohio, Ind., Ill., Mich., Wis.; South central: Ky., Tenn., Ala., Miss., La., Ark., Okla., Tex.; Southwest: Ariz., N.M., Nev.; Rocky Mountain: Mont., Idaho, Wyo., Colo., Utah; Pacific: Calif., Ore., Wash.; West: Rocky Mountain, Southwest, Pacific; East: all regions not in the West.

[b]Less than 0.5 per cent.

TABLE 3

PERCENTAGE VALUE OF MINE OUTPUT OF MAJOR METALS, BY REGION AND COMMODITY, 1839-1909
(in 1879 prices)

Region and Commodity	1839	1849	1859	1869	1879	1889	1902	1909
Region								
New England	4	1	a	a	1	a	--	--
Middle Atlantic	36	17	9	13	10	6	2	3
South Atlantic	15	6	2	2	2	1	1	1
North central	34	15	8	15	20	26	35	36
South central	10	3	1	1	1	4	4	4
East	100	42	20	31	34	37	43	43
Southwest		a	1	22	19	11	9	20
Rocky Mountain			3	17	29	44	41	32
Pacific		57	76	30	18	8	6	6
West	0	58	80	69	66	63	57	57
United States	100	100	100	100	100	100	100	100
Commodity								
Iron ore	60	27	12	17	18	22	26	26
Copper		2	5	8	9	22	30	37
Lead	30	11	2	2	7	7	6	6
Zinc	0	0	1	2	3	4	5	5
Gold	9	60	79	53	30	17	17	14
Silver	a	1	1	16	34	30	16	11
Total	100	100	100	100	100	100	100	100
Iron ore, copper lead, zinc	90	39	20	30	36	54	68	75
Gold and silver	10	61	80	70	64	46	32	25
Total	100	100	100	100	100	100	100	100

Note: If the alternative weighting had been used, the following percentage distributions would have resulted.

	1839	1879	1909
New England	6	1	0
Middle Atlantic	53	24	3
South Atlantic	12	3	1
North central	15	25	55
South central	14	2	6
Southwest	0	13	11
Rocky Mountain	0	19	19
Pacific	0	12	4
Total, U.S.	100	99	99
Iron ore	85	45	57
Copper	--	6	22
Lead	12	4	3
Zinc	--	2	3
Gold	4	20	8
Silver	--	23	7
Total	101	101	100

a Less than 0.5 per cent.

whole period it was decidedly a region of little importance for the major metallic minerals.

The south Atlantic region, which started off in 1839 with 15 per cent of the total metal ore output—a total made up of iron ore and gold in roughly equal parts—enjoyed a small spurt in iron ore output after 1869, but in every decade after 1839 it must be reckoned a region of practically no importance for major metallic mineral production.

TABLE 4

GROWTH OF VALUE OF MINE OUTPUT OF MAJOR METALS, BY REGION AND COMMODITY, 1839-1909
(per cent per year in 1879 prices)

Region and Commodity	1839-49	1849-59	1859-69	1869-79	1879-89	1889-1902	1902-09
Region							
New England	1.3	0	-7.2*	19.0*	-11.0*		
Middle Atlantic	5.5	5.5	4.3	3.8	-0.5	-1.0	5.8
South Atlantic	4.6	-2.1	4.3	3.7	2.8	3.8	-4.0
North central	5.0	4.9	7.5*	8.8*	8.0*	7.9*	5.2
South central	0.4	4.4	-5.6	9.5*	18.0*	6.4*	3.2
East	4.6	4.2	5.2	6.8*	6.3*	6.7*	4.9
Southwest	--	--	31.0*	4.4	-0.4	4.1	15.0
Rocky Mountain	--	--	17.0*	11.0*	9.5*	5.2	1.2
Pacific	--	15.0*	-8.5*	0.3	-2.8	3.9	4.8
West		15.0*	-0.8	5.3	4.8	4.9	5.0
United States	13.0*	12.0*	0.7	5.8	5.3	5.6	4.9
Commodity							
Iron ore	5.1	3.7	4.7	6.2*	7.1*	6.9*	5.3
Copper	--	--	6.3*	6.0*	15.0*	8.2*	7.7*
Lead	3.0	-3.6	0.6	16.0*	5.2	4.5	4.9
Zinc	--	--	8.8*	8.4*	9.3*	8.4*	5.0
Gold	--	15.0*	-3.3	0	-0.6	5.7	2.3
Silver	--	15.0*	32.0*	13.0*	4.0	0.6	0.4
Gold and silver		15.0*	-0.7	4.9	2.1	2.8	1.5

*Indicates rate of growth over 6 per cent per year.

The south central region, whose output of iron ore gave it about 10 per cent of the total in 1839, also enjoyed a rather greater increase in this output after 1869 but, sizable as it was, it was far from sufficient to give the region any more than minor importance in the total.

The remaining regions—east north central, west north central, and the West (the Southwest, Rocky Mountains, and Pacific Coast)—are ones in which spectacular development in metal ore output took place. This can be seen in the top panel of Table 4 since most of the asterisks, indicating an annual rate of growth over 6 per cent per year, are found in these regions. The rates of growth for New England are of no significance since they are based on very small outputs. The north central area enjoyed an early specialization in lead. Although its initial share of output declined because of the tremendous growth of output in the West, output in the north central area increased steadily and sizably after Michigan began to produce copper. Iron ore also enjoyed a steady and an even larger growth, first in Michigan and later in Minnesota. Added to these were the smaller but still significant increases in the outputs of lead and zinc after 1869 and of gold in South Dakota. The result of the growth in output over the whole range of ores was to make the north central region the leading metal ore producer by 1909.

The discovery of gold in California in 1849 opened the great metal mining era of the West. The gold deposits of California were so rich that in 1859 the Pacific region was producing three-quarters of the country's total metal ore output. After the peak Census year of 1859, however,

TABLE 5

VALUE OF GOLD AND SILVER AS PERCENTAGE OF TOTAL METAL ORE OUTPUT OF THE WESTERN REGIONS, 1869-1909

	Rocky Mountain		Southwest		Pacific	
	Gold	Silver	Gold	Silver	Gold	Silver
1869	90	10	36	64	98	2
1879	19	70	24	65	93	6
1889	11	55	24	41	91	8
1902	22	32	13	18	72	7
1909	17	25	19	15	61	7

the region declined in relative importance (6 per cent of the U.S. total in 1909), although its 1909 output was only 23 per cent below that of 1859. After 1890, copper increased to account for about one-third of that share, the remainder being made up of gold and a small quantity of silver.

Metal mining in the other two regions of the West—the Rocky Mountains and the Southwest—began on a significant scale a decade later than it did in California with the 1859 discovery of gold and silver in Nevada and Colorado. Initially, the output was made up almost entirely of gold and silver, since only their values could support the very high cost of moving concentrate or metal out of the producing areas. As time went on, transportation improved with the steady spread of railroads, and it became profitable to mine for products associated with the gold and silver, that is, copper and lead and, to a lesser extent, zinc. This is reflected in a steady decline in the importance of gold and silver in the outputs of those regions, as shown by Table 5.

The bulk of the absolute growth of metal mine output is shown by Table 2 to have taken place in the last two or three decades of the period under review. Although the annual rates of growth were very high for the country as a whole from 1839 to 1859—mainly because of gold—the absolute quantities involved were quite small. After a pause during the Civil War decade, metal mine output grew steadily for the country as a whole but with considerable variability among commodities and regions. For example, total metal mine output in 1889 was only one-third of the 1909 level. Southwest output was only one-fifth of the 1909 level. So also was copper output. Even gold in 1889 was only 41 per cent of the 1909 level, although it had been 60 per cent some twenty years earlier.

In summary, the period 1839–1909 began with all ore produced in the East—iron ore and a little gold on the eastern seaboard and lead in the upper Mississippi valley. In the East, iron ore output increased in the middle Atlantic region until the north central area (Michigan) began to displace it. The spectacular bursts of precious metal output in the West began in 1849, first in California and a decade later in Nevada and Colorado. As transportation improved in the West, the relative importance there of gold and silver declined and that of lead, zinc, and particularly copper increased. The north central region, an important early producer of lead, became the country's leading mineral producing region with the tremendous development of copper in Michigan after 1850 and of iron ore in the Great Lakes states after 1875. The period began with the middle Atlantic and north central regions as the main mineral producers and ended in 1909 with the north central, the Rocky Mountain, and the southwestern regions as the main producers. In 1839, iron ore and lead accounted for most of mineral output (60 and 30 per cent, respectively). In 1859, four-fifths of the country's metal ore output was in the form of gold and silver, practically all of which was gold. By 1909, copper was the leading mineral, accounting for a little over one-third; iron ore accounted for one-quarter and gold and silver together for one-quarter of the major metal ore output.

Employment

Our estimates of employment are based mainly on Census data. There have been special mineral censuses of widely varying worth beginning with the year 1879. In 1869, minerals were given a separate section in the *Census of Industry and Wealth*, but before that time minerals received no special attention. The treatment of minerals in the 1840 and 1850 Censuses is so poor that no useful estimates of employment—or output, for that matter—can be made from them.

Census employment data are for establishments, that is, for industries. While mining industries are identified by the names of commodities, a commodity with a particular name is not necessarily produced entirely within the industry of the same name, nor is an industry with a particular name restricted to production of the commodity of that name. For this reason, it was necessary to consolidate lead and zinc into a single industry and to do the same for gold and silver. In the latter part of the nineteenth century the discrepancy between commodity and industry became wider. In the Rocky Mountain and southwestern regions copper, lead, and zinc began to appear in considerable quantities although employment in these industries was often recorded by the Census in the gold and silver mining industry.

One of the major Census mysteries, especially in 1839, 1849, and 1859, is the definition of mining. A number of nonmining activities closely associated with mining appear to be included in Census tabulations for mining. The iron mining figures seem to be fairly comparable after the 1860 Census in which iron mines were classified with blast furnaces when owned by the same firm, and the noncaptive mines were tabulated separately. For gold and silver, the Census employment estimates definitely contain more than mining operations, but the nonmining operations included have always been closely associated with mining itself. The data for Michigan copper almost certainly include a considerable amount of smelting. This may also be true for some parts of the West in the later decades of the century, although the Census employment data probably exclude smelting more thoroughly in the West than in Michigan. Employment data for lead and zinc are obscure; a substantial number of smelter workers is probably included in the nominal mining employment.

With time, a somewhat clearer line has developed between mining and smelting operations, both in actuality and in the successive Censuses. This has caused "mining employment" to be lower in the later years than it would have been otherwise. Hence an observed decline in the ratio of employment to output is probably somewhat larger than it ought to be.

DISTRIBUTION OF EMPLOYMENT BY REGION AND INDUSTRY

There is naturally a rough correspondence between the distribution of employment among regions and among products and the distribution of output, but a comparison of Tables 6 and 3 reveals numerous departures from this conformity. The differences are all reflected in the ratio of employment to output, examined in detail later.

Because of the similarity between the distribution of employment and output, the general movements over time are much the same. Major metal employment was entirely in the East at the beginning of the period, with the middle Atlantic region dominating. During the next few decades, the initial distribution was radically changed because of the influx of workers into gold and silver mining, first in the Pacific Coast area and

TABLE 6

PERCENTAGE DISTRIBUTION OF EMPLOYMENT IN MAJOR METAL MINING, BY REGION AND COMMODITY, 1839-1909

Region and Commodity	1839	1849	1859	1869	1879	1889	1902	1909
Region								
New England	3.6	1.3	0.5	1.0	1.3	0.4	--	--
Middle Atlantic	43.0	22.0	11.0	23.0	19.0	9.0	4.1	3.0
South Atlantic	24.0	12.0	3.1	3.4	4.3	5.3	5.2	2.7
East north central					13.0	20.0	24.0	21.0
West north central					9.6	7.9	16.0	21.0
North central	19.0	11.0	8.8	17.0	23.0	28.0	40.0	43.0
South central	10.0	4.2	1.9	1.7	1.8	5.1	7.0	5.2
East	100	50	25	46	50	48	55	54
Southwest	0	0.2	1.4	7.3	12.0	8.6	6.3	11.0
Rocky Mountain	0	0.0	1.6	15.0	15.0	28.0	26.0	23.0
Pacific	0	50.0	72.0	32.0	23.0	16.0	13.0	13.0
West	0	50	75	54	50	52	45	46
United States	100	100	100	100	100	100	100	100
Commodity								
Iron ore	59.0	31.0	14.0	32.0	33.0	35.0	35.0	32.0
Copper	0	3.3	8.5	9.2	6.4	5.8	19.0	28.0
Lead and zinc	25.0	8.9	2.1	2.9	7.8	5.8	7.0	11.0
Gold and silver	16.0	57.0	75.0	56.0	53.0	53.0	38.0	29.0
Total	100	100	100	100	100	100	100	100

then around the beginning of the Civil War into the Rocky Mountain and southwestern regions. By 1902 Michigan copper and Lake Superior iron ore had brought the north central region to a leading position. In the West, development of the base metals limited the decline in the region's relative employment position.

MAJOR METAL MINING EMPLOYMENT COMPARED WITH ALL MINERAL EMPLOYMENT

The major metal mining industries do not, of course, constitute the whole mineral industry for, as Table 7 shows, they have accounted for less than one-third of all U.S. mineral employment since 1870. In addition to the major metals, the mineral industries include the comparatively unimportant minor metals, the so-called nonmetallics (e.g., sand and gravel, clay), and the very important category of mineral fuels, coal and petroleum.

While the regional percentages differ considerably even in the earlier years, there clearly has been an increasing regional specialization on major metallics since 1870. In 1870 major metal employment accounted

TABLE 7

COMPARISONS OF EMPLOYMENT IN MAJOR METAL MINING, ALL MINERAL INDUSTRIES, AND TOTAL EMPLOYMENT, BY REGION, 1870-1910

Region	1870	1880	1890	1900	1910
MAJOR METAL MINING AS PERCENTAGE OF ALL MINERAL INDUSTRIES					
New England	11.0	17.0	4.5	0	0
Middle Atlantic	23.0	19.0	6.3	2.2	1.7
South Atlantic	27.0	32.0	21.0	16.0	5.5
East north central	28.0	33.0	26.0	26.0	22.0
West north central	20.0	48.0	22.0	37.0	51.0
South central	28.0	27.0	22.0	15.0	11.0
Southwest	42.0	125.0[a]	71.0	52.0	68.0
Rocky Mountain	48.0	29.0	59.0	48.0	62.0
Pacific	41.0	45.0	54.0	36.0	54.0
United States	32.0	32.0	24.0	21.0	20.0
MAJOR METAL MINING AS PERCENTAGE OF TOTAL EMPLOYMENT[b]					
New England	0.04	0.07	0.02	0	0
Middle Atlantic	0.48	0.49	0.19	0.08	0.07
South Atlantic	0.10	0.15	0.18	0.19	0.10
East north central	0.31	0.35	0.53	0.60	0.58
West north central	0.09	0.46	0.28	0.64	0.93
South central	0.04	0.05	0.15	0.17	0.14
Southwest	6.9	20.0	9.0	6.5	8.4
Rocky Mountain	12.0	7.6	7.5	7.1	5.2
Pacific	6.7	4.8	2.1	1.8	1.3
United States	0.47	0.56	0.47	0.50	0.51
ALL MINERAL INDUSTRIES AS PERCENTAGE OF TOTAL EMPLOYMENT					
New England	0.40	0.46	0.46	0.34	0.31
Middle Atlantic	2.1	2.6	3.1	3.8	4.2
South Atlantic	0.37	0.48	0.86	1.2	1.8
East north central	1.1	1.1	2.0	2.3	2.6
West north central	0.49	0.95	1.3	1.7	1.8
South central	0.17	0.21	0.69	1.1	1.3
Southwest	17.0	16.0	13.0	13.0	12.0
Rocky Mountain	26.0	26.0	13.0	15.0	8.5
Pacific	16.0	11.0	3.9	5.1	2.4
United States	1.5	1.7	2.0	2.4	2.5

Source: Major metal employment is our estimate. 1870-90 are from Table A-4, and 1900 and 1910 (1902 and 1909) are from Table A-5. All mineral industries and total employment are from H.S. Perloff, *et al.*, *Regions, Resources and Economic Growth*, Baltimore, 1960.

[a] Obviously incorrect. The estimate for employment in all mineral industries is probably too low.

[b] Employment data from Table A-4 are used below to calculate major metal mining employment as a percentage of total employment, with the latter assumed equal to all males (including slaves) 15-60 years of age as recorded in the population Censuses.

NOTES TO TABLE 7 (concluded)

	1840	1850	1860	1870
New England	.04	.03	.03	.06
Middle Atlantic	.22	.25	.31	.56
South Atlantic	.15	.19	.13	.14
East and west north central	.14	.15	.20	.28
South central	.08	.07	.07	.06
Southwest	--	.26	2.9	7.2
Rocky Mountain	--	--	2.4	13.0
Pacific	--	11.00	18.00	6.6
United States	.14	.30	.67	.57

for no more than 48 per cent of all mineral employment in any region, but in 1910 four regions had over half of their mineral employment in the major metal mining industries. In 1870 no region had less than 11 per cent of its mineral employment in the metal mining industries, but in 1910 four regions were below that level. In the western regions major metal employment is a more important part of all mineral employment than it is in the East. Indeed, in one year a ratio of major metal employment to all mineral employment of 1.25 was obtained for the Southwest. No attempt has been made to correct this, since the more defective estimate is probably the estimate for all mineral employment, a series which is not the main concern of this paper. Still, taking the whole group of percentages for the western region, there can be little doubt of the dominant position held by major metals in the whole field of mineral employment.

MINERAL EMPLOYMENT COMPARED WITH ALL EMPLOYMENT

Mineral employment has always been a small part of total employment in the United States, although there has been a substantial increase from the 1.5 per cent level of 1870, shown in Table 7.[5] Regional differences are very great, reflecting the rich deposits of minerals in some regions and the other opportunities for economic activity in each of the regions. The regions best endowed with metallic minerals are perhaps less well endowed in other respects. The theory of ore genesis indicates that this is not entirely a matter of chance. The consequence is that in the southwestern and Rocky Mountain areas metallic mineral employment was 5 per cent or more of total employment in each of the five Census years

[5] It should be borne in mind that our estimates of employment are based upon the mineral Census and refer, in most cases, to average employment during the year of operations covered by the mineral Census. I believe the year of operations actually covered by most reports to the Census was the year preceding the nominal year of the Census. Thus the 1870 mineral Census, for example, very likely collected reports on calendar year 1869 from most of the reporting units. The occupational data on which the total mining employment estimates and the total employment estimates are based refer in all cases to the spring of the Census year.

from 1870 to 1910, and even in the Pacific Coast region was 5 per cent or more in two of the five years. In the other regions, major metal employment is a much smaller part of total employment, being under 1 per cent in all cases, even in the west north central region, whose major metal components are lead and zinc, Minnesota iron ore, Michigan copper and iron ore, and South Dakota gold.

The middle Atlantic region emerges as the third most specialized mineral region when all mineral employment is compared with total employment. In 1910 the middle Atlantic region had over 4 per cent of its employment in mineral industries compared with about 2 per cent for the north central region, which is specialized in metallic minerals rather than the nonmetallic minerals that are important in the middle Atlantic area.

In the Rocky Mountain and southwestern regions, mineral employment as compared with all employment was still very high in 1910, 12 and 8 per cent, respectively. In the earlier years, as many as one out of every four employed persons was working directly in a mining industry in the Rocky Mountain area.

These percentages are high for a developed region. Nevertheless, the picture of the West as completely dominated by metallic mineral mining is so strong that many may be surprised that they were not higher there. The circumstances under which major metallic employment could be close to 100 per cent would be most unusual. If a wholly empty region is entered for the first time by prospectors, all of whose activities—including shooting, preparation, and cooking of game—are regarded as part of mining, then employment in the major metallic industries in that area would equal total employment. (The probability would be, of course, that the Census taker would not be able to find the prospector.) But as soon as the region under scrutiny is enlarged, there are necessarily many kinds of activity other than metal mining, even though the mining industries may be the main or even the sole reason for them. Mining must be supported by industries that supply materials and equipment. Equipment, mineral products, food, and other consumer commodities must be transported. Men and their families must be fed, housed, and clothed. And in many mineralized areas there are other bases for economic activity, some of which would be carried on even in the absence of any mineral industries. The consequence is that throughout the whole of the period measured here metallic mineral employment and all mineral employment have been greatly outweighed by employment classified in other industries. Nevertheless, the existence of ghost towns and ghost areas proves that in many smaller areas almost all employment was derived from the major metallic industries. With the development of activity around other economic opportunities, later data show that mineral employment has declined relative to total employment. In 1950, for example, all mineral employment was 3.4 and 3.5 per cent of total employment in the Rocky Mountain and southwestern regions, respectively, compared with 26 and 17 per cent in 1870.

Ratio of Employment to Output

The behavior of employment in relation to output is summarized in Table 8. The ratio of employment to output E/O "eliminates" the size of the industry and can thereby show clearly one important aspect of the production structure. The ratio E/O for minerals is likely to vary more among industries, among regions, and over time than might be expected in the nonmineral industries. The factors that tend to produce this result are discussed below. Their effects are obscured by the fact that our data are sometimes grossly inadequate measures of the quantities we would like to measure. The reasons for this are explained below before we turn to the "real" factors making for variability in E/O. It should be borne in mind that E/O is not the inverse of total factor productivity in any meaningful sense but is simply the ratio of employment to output. In particular, it would be quite possible, though perhaps not likely, for E/O to increase from one period to the next even though a proper measure of the total productivity of an industry would show an increase. It would be desirable to discuss productivity, its variation among industries and regions, and its variation over time, but estimates of inputs other than labor have not been possible for these industries.[6]

We should like also to measure long-term changes, but our data refer to particular years and sometimes to particular dates. The level of output or of employment may be distorted in any one year by forces that are temporary and will therefore prevent the data from revealing in full clarity the long-run changes that are taking place.

In the mineral industries, the measurement of output is complicated by the fact that the labor force is engaged in producing two types of goods—a "current" good, which comes out in the form of concentrate or ore, and a capital good, which is visible as a developed mineral deposit. It would be desirable in the study of E/O to separate the two types of product and to separate the amount of labor used to produce each or— this not being very feasible even in principle—to make certain that the

[6] For comments on the problem of extracting measures of capital used in mineral industries from the censuses, see my review (*Journal of the American Statistical Association*, March 1957, p. 119) of Israel Borenstein's *Capital and Output Trends in Mining Industries, 1870–1948*, Occasional Paper 45, New York, NBER, 1954.

TABLE 8

INDEXES OF EMPLOYMENT DIVIDED BY OUTPUT, BY REGION AND COMMODITY,
1859-1909
(1889 = 100)

Region and Commodity	1859	1869	1879	1889	1902[a]	1909[a]
New England	--	--	--	--	--	--
Middle Atlantic	144	193	182	100	60	46
Iron ore	133	173	164	100	94	88
Lead and zinc	275	618	411	100	20	17
South Atlantic	95	66	95	100	81	74
Iron ore	99	169	105	100	91	102
North central	181	159	148	100	63	63
Iron ore	57	178	155	100	48	40
Copper	484	205	151	100	104	108
Lead and zinc	112	112	234	100	63	93
Gold and silver			55	100	70	86
South central	174	278	186	100	69	64
Iron ore	183	402	202	100	80	75
Southwest	314	69	121	100	58	46
Copper			112	100	112	117
Gold and silver		51	100	100	68	38
Rocky Mountain	147	241	123	100	63	70
Copper				100	131	213
Gold and silver	107	175	99	100	64	65
Pacific	81	83	98	100	65	61
Gold and silver	82	84	99	100	80	81
United States	175	161	148	100	64	60
Iron ore	131	189	168	100	55	46
Copper	752	427	267	100	96	111
Lead and zinc[b]	212	202	228	100	73	106
Gold and silver	149	116	111	100	67	64

Note: The employment estimates do not attempt to take into account changes in hours worked per year. If it had been possible to take account of the decline in the number of hours, the measures of E/O would have declined considerably more than they actually did.

1839 and 1849 are not included in the table, since employment in those years was estimated by extrapolating the 1859 quantity by output. Nor does Table 8 include all the possible E/O ratios that could be computed. It includes the regional E/O ratio (equivalent to a weighted average of the industry ratios within a region weighted by output in the given year divided by the same weighted average for the base year) for all regions having production in 1909, all, that is, except New England. The U.S. values of E/O are included, as are the U.S. industry averages, which may be viewed as equivalent to a weighted average of the regional industry ratios weighted by output in the given year divided by the same weighted average for the base year. Estimates have been included for individual industries within regions where these industries were of substantial size. Where outputs are very small, estimates of E/O are not to be relied upon, for the estimates of both output and employment for those states of minor importance in the industry are subject to sizable error, both because of undercount of employment and the independence of the regional distributions of output and employment. For example, E/O for Middle Atlantic lead and zinc in

NOTES TO TABLE 8 (continued)

the table exhibits a most unusual behavior. Another case of a very sharp change is that of the Southwest from 1859 to 1869 to 1879.

Each industry except iron ore is afflicted by a lack of correspondence between the definitions of output, which is measured on a commodity basis, and of employment, which is measured on an establishment basis. This lack of correspondence affects the estimates of E/O in various ways. The most important case is that of lead and zinc. The lead output of the West (produced from silver and lead ores) is included in the lead and zinc industry's output, but the number of mines classified in the lead and zinc industry in the West is very small throughout the whole period. Zinc adds to the problem to a lesser extent since the West's zinc output was a smaller part of the U.S. total. On the other hand, the distortion introduced into the gold and silver series is much less, for the lead omitted from the output of this industry was only a small part of the total output of the industry. Lead and zinc outputs of the West as a percentage of the U.S. total were as follows:

	Lead	Zinc
1859	0	0
1869	1	0
1879	70	0
1889	81	0
1902	72	11
1909	57	15

The change in the percentage of lead coming from the West caused E/O for the United States to fall more (or rise less) from 1859 to 1889 and to rise more (or fall less) from 1889 to 1909.

A less important problem is caused by the transfer of establishments from the gold and silver industry to the lead and zinc or copper industries. As copper, lead, and zinc became the major parts of the output of a number of mines in these regions, the Census Bureau began to recognize the mines as something other than gold and silver mines, even though gold and silver were contained in the ores. Hence the employment formerly attributed to the gold and silver industry was shifted, in part, to the other major metal industries.

In a number of cases, Census output of a commodity is greatly below our estimates of output based on other and presumably better information, and the Census ratio of employment to output has been applied to our estimate of output. There has been no attempt, however, to correct and make sense of every case of odd behavior in the employment-output ratio. Hence there are many anomalies in the behavior of employment in relation to output, especially where the quantities involved are small. These arise in part because the regional estimates of output are based in some cases on sources that may differ substantially from the regional distribution of output contained in the Census.

[a]1902 and 1909 Census employment data (see Table A-5) have been adjusted in an attempt to make them comparable with 1889 data. 1909 was first put on a 1902 basis by calculating average monthly employment (which was 11 per cent below the Dec. 15 figure used in the 1909 Census for major metallic minerals). 1902 and 1909 state figures were then multiplied by the following factor derived from 1889 data, to get to the 1889 basis:

$$\frac{\Sigma \text{ (employment)}_i}{\Sigma \text{ (employment)}_i \ \times \ \frac{\text{(days worked)}_i}{300}}$$

TABLE 8 (concluded)

where i is the skill level or occupational group. This method of adjustment is suggested in *Mines and Quarries, 1902*, p. 90. 1902 employment in all major metallic minerals for the United States on the 1889 basis is 22 per cent above 1902 employment reported in the Census.

I am indebted to Neal Potter for reminding me of this problem of consistency.

^bSee the second paragraph of the Note above.

output measured includes the capital good part of the output as well as the current product. We have not been able to do this but wish to call the difficulty to the reader's attention. In some years the capital good part of output can be very important. Some mines in any year are nonproducing, for instance, although not necessarily dead or even poor mines, for they may have a substantial labor force at work with little evidence of product visible above the ground. Hence, the labor force is engaged in developing the mine, that is, in producing a capital good. In the nonmining industries, on the other hand, there is usually a rather clear separation between operations on "current account" and operations on "capital account," although in agriculture and some industries the labor force does indeed engage in the production of capital goods to be used by the industry itself.

THE BEHAVIOR OF E/O

The interpretation of Table 8 presents great difficulty both because of the large number of factors affecting the behavior of the ratios and because of the largely unknown errors reflected in the estimates of output and employment. There are many puzzling features of the table that raise the possibility of systematic error in the estimates of employment or output, and also the possibility of fundamental changes in the conditions of production. Unfortunately, possible explanations abound.

1859–69

One of the puzzles is that E/O rose from 1859 to 1869 in two of the four eastern regions and in two of the three western regions. In the two eastern regions, a factor in this behavior is probably the estimate of employment in iron ore. Since the Census data on iron ore employment cover only a small portion of all the iron ore mined, we have been unable to find a satisfactory basis on which to construct an estimate. There are real factors that conceivably could explain the behavior of the ratio, such as that during the Civil War resort to lower grades of ore could have produced an increase in the ratio.

Rate of Change of E/O

The data for the United States as a whole show that E/O declined more rapidly from 1879 to 1902 than in the periods before or after (Table 9).

TABLE 9

ANNUAL RATE OF CHANGE IN EMPLOYMENT DIVIDED BY OUTPUT IN MAJOR METAL MINING, BY REGION AND COMMODITY, 1859-1909
(per cent per year)

Region and Commodity	1859 to 1869	1869 to 1879	1879 to 1889	1889 to 1902	1902 to 1909
Middle Atlantic	+2.9	-0.6	-6.0	-3.9	-3.8
Iron ore	+2.7	-0.5	-4.9	-0.5	-1.0
South Atlantic	-3.6	+3.5	+0.6	-2.1	-1.2
Iron ore	+5.4	-4.8	-0.5	-0.8	+1.7
North central	-1.3	-0.7	-4.0	-3.6	0
Iron ore	+11.5	-1.3	-4.4	-5.7	-2.4
Copper	-8.6	-3.1	-4.1	+0.3	+0.6
Lead and zinc	0	+7.3	-8.5	-3.5	+5.5
South central	+4.6	-4.0	-6.2	-2.9	-1.0
Iron ore	+7.9	-6.9	-7.1	-1.8	-0.8
Southwest	-15.2	+5.9	-1.9	-4.2	-3.2
Copper	--	--	-1.1	+0.9	+0.7
Gold and silver	--	+6.8	0	-2.9	-8.5
Rocky Mountain	+4.9	-6.8	-2.0	-3.6	+1.5
Copper	--	--	--	+2.1	+6.9
Gold and silver	+5.0	-5.7	0	-3.4	+0.3
Pacific	+0.2	+1.6	+0.2	-3.3	-0.8
Gold and silver	+0.2	+1.7	+0.1	-1.8	+0.2
United States (total)	-0.8	-0.9	-3.9	-3.5	-0.8
Iron ore	+3.7	-1.2	-5.2	-4.6	-2.5
Copper	-5.7	-4.7	-9.8	-0.3	+2.1
Lead and zinc	-0.5	+1.2	-8.3	-2.4	+5.3
Gold and silver	-2.5	-0.5	-1.1	-3.1	-0.7

Note: See note to Table 8.

This pattern is much less clear for industries within regions, however; in part, the movements of the weighted averages for the United States reflect shifts among regions and among industries. But the maximum annual rate of decline in E/O occurs in one or the other of the two "decades" from 1879 to 1902 in five of the seven regions shown in Table 9.

The annual rate of decline for the United States is less rapid from 1902 to 1909 than from 1889 to 1902. This is also true of each of the seven regions although for only eight out of the eleven industries within regions included in Table 9. Although the leveling off of the rate of decline in E/O may reflect in part the adjustment to the employment data for 1902 and 1909, these two years should be reasonably comparable if the Census adjustment for 1902 resulted in annual data equal to the average of monthly data. The adjustment of 1889 and 1902 to a common basis offers greater possibility for error.[7]

FACTORS AFFECTING E/O

In view of the uncertainties in the data, especially the levels and allocation of the employment data and the lack of correspondence between product and industry, any discussion of the factors affecting E/O must be tentative. However, some of the factors involved have left their traces even in our rough data.

Quality of Deposits

The decline in E/O for the period from 1879 to 1902 for the United States as a whole is explained in part by a shift of output to regions with lower ratios of E/O. For example, middle Atlantic E/O's for iron ore were considerably higher than those for the north central region, as can be seen below:

	1869	*1889*	*1909*
Middle Atlantic	1.76	1.02	0.89
North central	1.46	0.82	0.33

Similarly, the E/O ratio for copper in the West was considerably below that for copper produced in the north central region, as seen in the following figures:

	1869	*1889*	*1909*
North central	0.85	0.42	0.45
Rocky Mountain	—	0.10	0.22
Southwest	—	0.17	0.20

And the E/O ratio for gold and silver for the Rocky Mountain region was well below that for the Pacific Coast region. The main shift in output from 1879 to 1902 was between these two regions:

	1869	*1889*	*1909*
North central	—	0.60	0.52
Rocky Mountain	0.85	0.48	0.32
Southwest	0.30	0.59	0.22
Pacific Coast	0.97	1.16	0.94

Not all of the observed differences among the regional ratios are attributable to differences in the metal content of deposits, however, for

[7] See note b to Table 8.

the richness of the deposit is only one facet of quality. Perhaps even more important are size of the deposit and ease of working it.

Transportation and Exhaustion

For gold and silver it seems possible to see the effects on E/O of the discovery-exhaustion sequence and associated changes in transportation. When a new mining region is opened up, the heavy load of development work as new mines are brought to the point of production would tend to raise E/O. Independently of this factor, if the richest mines are discovered first we should expect a later rise in E/O. On the other hand, if the richer discoveries came later, E/O would decline, to rise at a still later point in time. What happened in the West was that both rich and poor deposits were discovered in these great mining regions. The factor that exerted the dominant influence on E/O probably was transportation. As transportation improved, it became profitable to work leaner and poorer deposits, and we should expect an increase in E/O, especially with improvements of the magnitude represented by the change from pack train and wagon to rail transport.

Something of this effect may be visible in Table 8. We do observe an increase in E/O in the gold and silver industry of the Southwest from 1869 to 1879. The change from 1859 to 1869 is unreliable because 1859 was the first big year for mining in the Southwest. Similarly, in the Rocky Mountain area, which opened up in 1859, we observe an increase in E/O for gold and silver. The Pacific Coast region had been producing since 1849, but here also an increase in E/O is shown from 1859 to 1869 and from 1869 to 1879.

The effects of improved transportation on the composition and level of output in the West are much clearer than the effect on E/O. When transportation is costly—as it was in the West until the development of the rail network—mining is restricted to ores that have a high value per unit weight of the material to be transported any distance. The products that met this requirement best were gold and silver, while the other metal products with lower value per unit weight were of no importance until 1879 and even then only small.[8] Ten years later, the "cheaper" commodities were considerably more important, and by 1902 they were more more important than gold and silver in the Southwest and nearly so in the Rocky Mountain region. The close association between the development of the rail network and the growth of the base metal commodities

---

[8] But gold and silver were not the only commodities to get over the transportation barrier. Furs succeeded in an earlier period as did cattle and sheep during the time when mining was spreading throughout the Rocky Mountain and southwestern regions.

is suggested very clearly by Table 10 which shows rail mileage by states at decadal points.[9] The association is much poorer for the Pacific Coast region than for the other two. Placer deposits (gold only) were far more important in California than in the states of the other two regions.

According to the 1880 Census, wagon haulage rates were seldom as low as 1 cent a pound for the trip from the mine and were as high as 6 to 8 cents a pound for the more distant mining camps.[10] In the same Census

TABLE 10

UNITED STATES RAILROAD MILEAGE IN THE WESTERN STATES, 1870-1910
(thousand miles)

	1870	1880	1890	1900	1910
Southwest	0.6	1.8	3.3	4.2	7.4
New Mexico	--	0.8	1.3	1.8	3.0
Arizona	--	0.3	1.1	1.5	2.1
Nevada	0.6	0.7	0.9	0.9	2.3
Rocky Mountain	0.9	3.2	9.3	11.6	15.5
Idaho	--	0.2	0.9	1.3	2.2
Wyoming	0.5	0.5	0.9	1.2	1.6
Utah	0.3	0.8	1.1	1.5	2.0
Montana	--	0.1	2.2	3.0	4.2
Colorado	0.2	1.6	4.2	4.6	5.5
Pacific region	1.1	3.0	7.6	10.4	14.9
Washington	--	0.3	1.8	2.9	4.9
Oregon	0.2	0.5	1.4	1.7	2.3
California	0.9	2.2	4.4	5.8	7.8

Source: *Statistical Abstract of the United States, 1920*, Census Bureau, Washington, 1921, Table 226, p. 333.

Note: Table 5 shows the decline of gold and silver relative to the total output of these regions.

it is observed that Arizona produces chiefly gold and silver, "though lead and copper, particularly the former, are rather abundant, and will, no doubt, be exploited on a large scale when the railroad system is further developed."[11] When mining first began in Arizona, some of the mines were as much as 300 miles from the nearest railroad. The only way to transport concentrate was by wagon or pack train. Even in 1880, no mine could be worked in Arizona unless its ore contained products worth at least $150 a ton. As late as 1885, some ores in Colorado bore freight charges ranging from $50 to $100 a ton before reaching a railroad

[9] See note to Table 10.
[10] *Census of Minerals, 1880*, p. ix.
[11] *Census of Precious Metals, Statistics and Technology, 1880*, 1885, p. 44.

but, by the turn of the century, all important mining camps were connected by rail with the main railroad lines.[12] Indeed, the Census report observed in 1880, "... now not only are there practically four great railway systems crossing the mountains from east to west, but a great number of short lines, generally narrow gauge, penetrate them in every direction, reaching mining towns which not many years since were only accessible by pack trains or saddle animals."[13]

Technological Change

To link major changes in technology with changes that have taken place in our data from 1859 to 1909 appears to be impossible, except, of course, for the persistent and very sizable decline in E/O for the United States as a whole and for each industry taken separately. It has not been possible, for example, to link definitely any particular change or set of changes with the accelerated declines in E/O in the two decades from 1879 to 1902, although some suggestions can be made. Fortunately, there are available two helpful examinations of technology in metal mining. One, by Lucien Eaton, is a straightforward account of changes in mining technology from 1871 to 1946.[14] The other is a survey of the long sweep of changes in technology in mining by C. E. Julihn.[15] It is on these two expert accounts of changes in mining technology that the following remarks are based.

Julihn's view is that over most of the period there were many small advances with significant cumulative effect but nothing that could be characterized as a major improvement. Toward the end of the century, however, a major change was in the making—abandonment of the selective, small-scale methods of mining where the miner had to make sure the ore he mined was not diluted or lost on the way to the smelter and development of nonselective, large-scale, mass production methods of mining. This shift in attitude and in method developed almost automatically as it became necessary to go to lower- and lower-grade ores during the latter part of the century. By using cheaper methods for breaking and handling large volumes of material, it was profitable to mine ores in which the desired mineral was diluted by large quantities of

---

[12] *Census of Mines and Quarries, 1902,* 1905, p. 577.
[13] *Census of Precious Metals, 1880,* p. xii.
[14] See his "75 Years of Progress in Metal Mining," in *75 Years of Progress in the Mineral Industry,* A. B. Parsons, ed., American Institute of Mining and Metallurgical Engineers, 1947, p. 40.
[15] See his "Copper: An Example of Advancing Technology and the Utilization of Low-Grade Ores" in *Mineral Economics,* F. G. Tryon and E. C. Eckle, eds., New York, 1932, p. 111.

waste material, which could by then be cheaply separated from the desired mineral.

There were, of course, many developments in technology during the last half of the nineteenth century, but not all constituted improvements. Many of the changes were associated with development of new regions which contained new types of deposits or deposits of sizes different from those in the older producing regions. Other changes represented not innovations but rather adaptations to improved transportation. All that can be done here is to enumerate some of the more significant changes and recall that no single one or set of them was powerful enough to leave traces in the data at our disposal.

One of the more significant changes was the introduction of dynamite around 1870. Drilling, formerly done by hand, gradually came to be carried on by compressed air. Steam power gradually became generally used for lifting—electric power not becoming a significant factor in metal mining until after the turn of the century. There were constant advances in the arts of breaking and grinding ore, one of the most important being improvements in the processes available for separating minerals from each other but, of course, the array of sink and float methods now in the mining engineer's repertoire came into use only after the turn of the century. Surveying and mapping became more accurate and helped to cut costs in developing and in working mines. The steam shovel came into general use in the last quarter of the century in open-pit mining. Loading became mechanized. And even in so simple a thing as the design of the hand shovel, abandonment of the old long-handled shovel was a significant improvement. Along with these narrower aspects of changing technology came a series of gradual advances in mining methods, such as the sequence of operations, the spacing of shafts and drifts, and so on. There were two innovations that had their origin in the United States: the square-set system developed on the Comstock lode and hydraulic mining.[16]

If this study had been carried beyond 1909, the major innovation involved in shifting to nonselective mass methods of mining would have been evident in copper.[17] In iron ore, methods of mass mining were applied in Michigan and Minnesota long before their use in the rather different problems of copper and other nonferrous ores. In mining districts in which lead and zinc deposits were sizable the impact of the change in method should also be apparent.

---

[16] *Census of Precious Metals, 1880*, p. vii.

[17] My study of the copper industry led to the conclusion that this development was clearly evident in the behavior of the price of copper (*Copper Costs and Prices: 1870–1957*, Baltimore, 1959).

NUMBER OF MINES AND OUTPUT PER MINE

The factors influencing E/O discussed above have had substantial effects on the number of producing mines and on the average product per mine. These effects have been concentrated on iron ore and copper and, to a smaller extent, on lead and zinc. While the number of gold and silver mines was quite different in 1909 from the number in 1869, the change in output per mine was much smaller than in the other three consolidated industries.[18]

TABLE 11

NUMBER OF IRON ORE MINES AND OUTPUT PER MINE, BY REGION, 1869 AND 1909

Region	Number of Mines		Annual Output Per Mine (thousand short tons)	
	1869	1909	1869	1909
Middle Atlantic	265	48	8	53
South Atlantic	51	89	2	24
East north central	89	88	12	164
West north central	3	144	60	221
South central	8	98	7	58
United States	420	483	8	76

Table 11 shows that the number of iron ore mines was about the same in 1909 as in 1869 for the country as a whole, although there were very large changes within regions. In the middle Atlantic region the number declined to less than one-fifth of its 1869 level, while in the west north central (Minnesota) and south central regions the numbers increased from a few in 1869 to over a hundred in 1909. Accompanying these shifts in the number of iron ore mines within regions were marked changes in output per mine. In the country as a whole, output increased about tenfold. Within each of the regions the increase was sizable, somewhat more than tenfold in the east north central region but decidedly less than tenfold in the west north central region. From the start iron ore mining was on a rather large scale in the west north central region.

In the lead and zinc industry, unlike the iron ore industry, there was a large increase in the number of mines from 1869 to 1909—from 127 to 1,213. As in the iron ore and copper industries, there appears also to

---

[18] All data on number of mines and output per mine are from the various Censuses. The data in Tables 11 and 12 are not consistent with the output data in Table A-2 but suffice to give a rough indication of change.

have been a substantial increase in the value of output per mine. In the west north central region (Missouri) the increase was of the order of sevenfold; in the east north central region (Wisconsin and Illinois) the increase was of the order of fourfold.

Gold and silver mines present a considerably more complex picture, partly because they consist of two different groups of mines, deep mines and a combination of placer and hydraulic operations. Between 1869 and 1909 the number of deep mines in operation increased very substantially, an increase which in absolute numbers took place mainly in the

TABLE 12

NUMBER OF COPPER MINES AND OUTPUT PER MINE, BY REGION, 1869 AND 1909

Region	Number of Mines		Annual Output Per Mine (thousand dollars)	
	1869	1909	1869[a]	1909
Northeast	2	--	97	--
Middle Atlantic	2	--	2	--
South Atlantic	4	--	24	--
East north central	27	21	86	1,430
South central	2	--	84	--
Southwest	3	120	6	308
Rocky Mountain	--	137	--	400
Pacific	--	23	--	232
United States	40	301	70	438

[a] Value of output per mine in 1869 was multiplied by the ratio of the price of copper in 1909 to the price of copper in 1869.

Rocky Mountain area, with the southwestern and the Pacific Coast regions also involved. The number of placer and hydraulic operations, on the other hand, declined to less than half its 1869 level, the decline taking place in absolute terms about equally in the Rocky Mountain and the Pacific Coast areas. Placer operations in the Southwest were never of any importance. Output per deep mine declined in every region which was in operation in both periods. With the placer and hydraulic operations, on the other hand, output per mine—which was only $5,000 in 1869—slightly more than doubled over the forty-year period, with almost all of this increase taking place in the Pacific Coast region.

Copper production in 1869 was an unimportant industry in every region with the exception of east north central (Michigan). As shown in Table 12, in this region, which had almost as many producing mines in 1909 as in 1870, output per mine in the later year was almost seventeen times that of the earlier year. In 1909, output per mine in the western areas was considerably lower than in the east north central region. Note, however, that we are dealing here with averages and with highly skewed distributions. An examination of the frequency distribution of mining enterprises in the copper industry by number of employees for Michigan, Arizona, and Utah in 1909 reveals that the largest mines were comparable in size to those in Michigan, but that in the western states there was a much larger number of very small mines than in Michigan.

Minerals and Regional Economic Development

The metallic mining industries played a major role in the westward spread of organized economic activity over the period of this study. Most important of all, they provided an export base for regional development and in some cases the only one. An export base is an essential part of the explanation of regional development where expansion of output and geographical expansion are associated, for, without it, there would be no reason to incur the locational disadvantage involved in moving away from established centers of activity. This economic opportunity often takes the form of an abundance of appropriable natural capital—forests, mineral deposits, or agricultural land. The richness of these pieces of natural capital permits higher costs for transportation to be incurred and thereby ensures the geographical expansion of production. The pace of this expansion then depends on the quality of the natural capital, the periods of gestation of man-made capital, the arrangements for spreading information about new economic opportunities, and the complicated mechanism governing the response to such opportunities.

The mere fact that minerals were produced, however, does not entail the conclusion that those deposits constituted valuable natural capital. The question is whether the amount of money spent in finding, developing, and producing the minerals was less than the cost of acquiring the products from the available alternative sources. The iron ore and copper deposits of the Lake Superior area and the coal deposits in all the areas where they were found were, in fact, valuable natural property, for the cost of finding and developing them was extremely low. As for gold and silver and the other metallic mineral deposits of the West, the size of the social surplus is more uncertain. If foresight were perfect, there would be no doubt about the answer, but the fact that the outcome of a mineral enterprise is uncertain opens the possibility that, from the point of view of society, the net rent earned was not large. This is quite different from saying that nobody made enormous profits on these deposits.

There is little possibility of estimating the net rent earned on the western deposits during the nineteenth century. The problem is very complicated from a conceptual point of view and involves expenditures not recorded in mining censuses. The impression, however, is almost universal that a substantial net rent was earned from the point of view of society as a whole. But even if net rents were negative, the fact that the activity was undertaken obviously exerted a profound influence on regional development. During the heyday of gold mining in California, gold was an important part of total economic activity, although there were other locational bases present before 1849. In parts of the Southwest, minerals furnished absolutely the only reason for settling there. In time, locational bases other than mining—agriculture, forestry, and lately "amenities"—developed, and mining diminished in relative importance both over the period under consideration and on down to the present.

Minerals were an export base with a peculiarity which aided regional development in another way. Since they had to be found, prospecting was an important means of accumulating knowledge of the different parts of the West. Assessment of the economic possibilities of a region—a necessity for all types of investment decisions—requires that a large stock of information of many kinds be at the command of many people, not just a few. The West was not unknown to a few white men who, before the Civil War, had roamed over most of the land as trappers. But the knowledge they had amassed was not comprehensive or systematic, and it provided only the bare essentials for the treks to Oregon and California. The extent of the ignorance of the West and the unreliability of what knowledge there was can be seen from the fact that even as late as 1867–79 the national government was induced to spend its money on four surveys of the West (King, Wheeler, Hayden, and Powell). Bartlett, in his account of those four surveys, sums up the situation as follows: "In 1867 men had asked, 'What lies out there?' By 1879, thanks to the work of the Great Surveys, their question had been answered. Now a new question was on men's lips: 'When shall we go there?'" [19]

IS THE PROCESS OF MINERAL DEVELOPMENT SYSTEMATIC?

A fundamental problem in the organization of the mineral industries is whether the finding and development of mineral deposits exhibits a systematic response to economic incentive. Our survey of the spread of metal mining across the continent is relevant to this important question, and I believe the evidence supports the conclusion that for fairly large

---

[19] Richard A. Bartlett, *Great Surveys of the American West*, Norman, Oklahoma, 1962, p. 376.

areas the search and development do respond in a systematic fashion to economic incentive, once it is widely understood that mineral deposits are probably present in the region. The actual pace of search and development and the actual growth of output do require other preconditions, the main one being transportation. For example, it was impossible for low-value ores of copper, lead, and zinc to be developed in the West until a rail network provided a cheap means of getting these mine products to market. It is true that, if we look at small areas or at the efforts of the lone prospector or even a single corporation, chance—from their point of view—plays an important part in success or failure. But if we step back and look at a larger area and a larger number of prospectors or corporations, the chance element begins to recede and a pattern of relentless expansion in response to economic incentive emerges in region after region.

The presence of copper in Michigan was known from very early times. The Smithsonian Institution has exhibited a large mass of native copper called the Ontonagon Boulder which had rested on the bank of the Ontonagon River in Michigan from prehistoric times until it was moved to Washington, D.C., in 1843. With knowledge of the presence of copper generally available, modern copper mining began in Michigan quite early, in about 1845. It expanded rapidly, and the search for additional deposits was conducted in an intensive manner. A similarly systematic and intense expansion of iron ore mining is observable in the Lake Superior deposits in Michigan and later in Wisconsin and Minnesota.

The presence of silver in what later was to be the southwestern United States was known for a long time. Indeed, the hope of finding large deposits of gold and silver had stimulated some of the earliest explorations in the Southwest—if they can be so dignified—but for a long time the presence of really rich deposits of gold or silver was hoped for rather than suspected. The well-known discovery which opened up the West for mineral exploitation was of California gold, a fortuitous discovery not to be ascribed to any economic activity in search of gold. But once the news of gold in large quantities in California was definite, the search was extended, covering more and more ground as time went on and alerting people to the possible presence of gold in areas far removed from California. The discovery of the Comstock lode in 1859 was an outgrowth of the California activity, and so probably, at a much greater distance, was the discovery of gold in Colorado the same year. After that, the process of combing the West for mineral deposits began. What prospectors and others were looking for depended naturally enough on what they could do with the mineral if they found it. This meant that in areas very far from the more settled ones attention was limited for a long time to ores containing large amounts of gold and silver. But with the development of the rail network, which was significant by 1880 but far more extensive by 1890, mining ores containing much smaller amounts of the precious metals began to pay.

Appendix

TABLE A-1

U.S. ANNUAL MINE PRODUCTION OF MAJOR METALLIC ORES, 1839-1909
(recoverable metallic content)

Year	Iron Ore (million long tons)	Copper (thousand short tons)	Lead	Zinc	Gold (million fine	Silver troy ounces)
1839	0.99		18		0.022	0.011[a]
1840			17		0.023	
1841			20		0.029	
1842			24		0.041	
1843			25		0.056	
1844			26		0.052	0.018
1845		0.112	30		0.054	
1846		0.168	28		0.060	
1847		0.336	28		0.047	
1848		0.560	25		0.145	
1849	1.67	0.784	24		0.517	0.088
1850	1.48	0.728	22		2.09	0.305
1851	1.43	1.01	18		3.84	0.544
1852	1.38	1.23	16	0.95	4.10	0.578
1853	1.63	2.24	17	1.6	3.40	0.490
1854	1.89	2.62	16	3.2	3.49	0.508
1855	2.06	5.09	16		2.79	0.419
1856	2.37	4.71	16		2.88	0.438
1857	2.19	6.39	16		2.18	0.349
1858	1.97	7.08	15		2.33	0.375
1859	2.40	7.72	16	5.1[a]	2.28	0.375
1860	2.55	8.66	16		2.09	0.761
1861	1.96	9.36	14		1.95	1.55
1862	2.05	11.3	14		1.76	3.48
1863	2.38	10.2	15		1.81	6.57
1864	2.76	11.1	15		2.07	8.51
1865	2.18	12.8	15		2.20	8.70
1866	3.05	12.9	16		1.79	7.73
1867	3.17	13.5	15		1.88	10.4
1868	3.34	14.7	16	12	1.69	9.28
1869	3.83	14.5	18	12	1.64	9.28
1870	3.66	14.5	18	13	1.64	11.1
1871	3.68	15.2	20	14	1.57	14.0
1872	5.38	15.0	26	14	1.70	15.5
1873	5.30	17.9	42	15	1.74	19.5
1874	4.87	19.8	51	18	1.71	19.7
1875	4.02	20.4	59	21	1.48	23.4
1876	4.00	24.0	63	22	1.94	29.6
1877	4.74	24.7	80	20	2.05	31.5
1878	5.62	24.6	89	24	1.74	37.2
1879	7.12	26.4	91	29	1.64	34.6
1880	9.13	30.5	96	33	1.51	33.0
1881	8.97	36.6	114	38	1.42	36.0
1882	9.00	45.9	130	42	1.34	39.0
1883	8.40	63.3	140	46	1.29	38.2
1884	8.20	84.8	136	49	1.14	40.1
1885	7.6	85.0	126	53	1.20	42.1
1886	10.0	78.9	128	57	1.38	41.4
1887	11.3	90.7	142	65	1.53	43.0
1888	12.1	113	148	72	1.39	47.4
1889	14.5	113	153	72	1.55	51.3

(continued)

TABLE A-1 (concluded)

Year	Iron Ore (million long tons)	Copper (thousand short tons)	Lead	Zinc	Gold (million fine troy ounces)	Silver
1890	16.0	130	141	80	1.42	51.6
1891	14.6	142	176	100	1.52	53.4
1892	16.3	172	170	111	1.51	53.6
1893	11.6	165	161	98	1.62	49.2
1894	11.9	177	156	91	1.91	47.6
1895	16.0	190	162	106	2.26	53.6
1896	16.0	230	183	98	2.45	50.9
1897	17.5	247	202	122	2.60	53.0
1898	19.4	263	206	145	2.84	52.6
1899	24.7	284	207	169	3.00	54.0
1900	27.6	303	267	176	3.21	57.6
1901	28.9	301	265	189	3.25	55.7
1902	35.6	328	274	216	3.24	55.8
1903	35.0	348	289	218	3.05	55.6
1904	27.6	405	307	248	3.34	56.4
1905	42.5	442	317	266	3.51	56.1
1906	47.8	456	347	266	3.64	57.2
1907	51.8	420	365	260	3.29	52.4
1908	36.0	476	330	234	3.50	50.7
1909	51.4	561	385	305	3.81	57.2

Note: These estimates were prepared by O.C. Herfindahl and Selma Rein. In Table A-2, the columns headed H.R. (Herfindahl-Rein) contain these estimates.

[a] Very rough estimate, see source.

NOTES TO TABLE A-1

Iron Ore

1839: Iron ore consumed in production of pig and wrought iron. Estimated by applying 1849 state ratios of iron ore consumed per ton of metal output (pig and wrought [bar] iron) to 1839 pig and wrought iron production. These were totaled for the national figure. The 1839 data are from *Compendium of the Sixth Census, 1840*, Washington, 1841, p. 358. The 1849 data are from *Compendium of the Seventh Census, 1850*, Washington, 1854, p. 181, and *Report of the Superintendent of the Census*, 32d Cong., 1st Sess., Serial 636, pp. 236–241.

1849: Iron ore consumed in production of pig and wrought iron. See above for sources of data.

1859: Iron ore consumed in production of pig and other iron. The Census figure of 2,310,000 long tons used in the manufacture of pig iron has been increased to 2,400,000 long tons to allow for ore consumed in other iron manufacturing (*Eighth Census, 1860, Manufactures*, Washington, 1865, p. clxxx).

Confusion exists about the unit of weight used for measuring iron ore production in this Census. The Census of 1902 states that the unit was the short ton. *Report on the Mineral Industries, 1892* (Vol. I, p. 271) considers it the long ton, as do James W. Swank (*Iron in All Ages*, Philadelphia, 1884, p. 380) and J. W. Foster (*The Geology and Metallurgy of the Iron Ores of Lake Superior*, New York, 1865, p. 70).

The *Preliminary Report on the Eighth Census* (Washington, 1862, p. 170) presents a figure of 2,514,000 tons for iron ore output, but the state data on which this figure is based are nowhere presented in the final report. See *Eighth Census, 1860, Manufactures*, 1865, pp. clxxiv ff. The iron ore output figure for 1860 generally carried in official U.S. statistics is 2,832,000 long tons or 3,210,000 short tons, obtained from the *Ninth Census, 1870* (pp. clxvii and clxxx) as follows: 2,310,000 tons consumed by furnaces and produced by captive mines, plus 900,000 tons produced by "regular mining" establishments.

An inspection of the Ninth Census for 1870 schedule (Carroll D. Wright and William C. Hunt, *History and Growth of the United States Census, 1790–1890,* Washington, 1900, p. 314) submitted to blast furnaces leads us to conclude that they would have entered all the pig iron they produced and not just pig iron from their own ores, and that they would have put down all the ore consumed and not just ore from their own mines. Hence, total ore consumed already includes that part of the 900,000 tons from noncaptive mines used in making pig iron.

The writer in *Mineral Industries, 1892* also held this view since he took the Census input figure of 2,310,000 and increased it by 100,000 tons for Michigan.

We believe Isaac Hourwich (writing in *Mines and Quarries, 1902*, 1905) and others accepted uncritically the figure of 3,218,000 tons. Since it appeared so high in relation to the input and preliminary mining figures, they assumed it must be in short tons and converted it to long tons to achieve some reduction.

1850–58: Estimated by interpolating by pig iron production series in Sam H. Schurr and Elizabeth K. Vogely, *Historical Statistics of Minerals in the United States*, Resources for the Future, 1960, series M 207.

1869: Iron ore mined (*Ninth Census, 1870*, Vol. III, *The Statistics of the Wealth and Industry of the United States*, 1872, pp. 749 and 768). Census data are in short tons.

1860–68: Estimated by interpolating by pig iron production series; see 1850–58.

1875: Schurr and Vogely, *Historical Statistics of Minerals*, series M 195.

1879: Iron ore mined (*Tenth Census, 1880*, Vol. XV, 1886, p. 19).

1870–74, 1876–78: Estimated by interpolating by pig iron production series; see 1850–58.

1882–1905: Iron ore mined except for 1885–88 which are consumption estimates (*Mineral Resources, 1913*, Pt. 1, p. 300).

1879–81: Estimated by interpolating by pig iron production series; see 1850–58.

1906–09: U.S. Geological Survey figures revised to include manganiferous iron ores (with manganese content higher than 5 per cent) to assure comparability. Before 1906, including 1902, Geological Survey estimates included these ores (see *Mineral Resources, 1913*, Pt. 1, p. 65, for data on production of manganiferous iron ores).

Copper

1845–53: Mine production from J. D. Whitney, *The Metallic Wealth of the United States*, p. 332 (some data used by *Mineral Resources* and Bureau of Mines, *Economic Paper 1*).

1854–1905: Smelter production from domestic ores plus copper content of U.S. ores imported into the United Kingdom.

Smelter production from *Mineral Resources, 1915*, Pt. 1, pp. 662–665 (excluding Alaska). *Mineral Resources* included in the year 1861 a total of 500 short tons produced in New Mexico during 1858–61; this amount has been equally distributed over this four-year period.

Copper ores imported into the United Kingdom in 1854–85 are from Great Britain Board of Trade Statistical Department, *Annual Statement of the Trade and Navigation of the United Kingdom*. Ore exports during 1854–85 went almost entirely to British smelters. After 1885, U.S. ore exports were negligible. British data are used, since U.S. exports do not distinguish between ore and regulus or matte (smelter products) Moreover, U.S. export data during that period are considered extremely unreliable (see F. E. Richter, *Quarterly Journal of Economics*, Vol. 41, pp. 260–262). Copper contents of the ore have been set at 30 per cent (see R. B. Pettengill, "The United States Foreign Trade in Copper by Classes and Countries, 1790–1932," unpublished Ph.D. dissertation, Stanford University, 1934, pp. 325–326). For 1883–85, actual copper contents are presented in *Mineral Resources* (*1884*, p. 360; *1887*, pp. 90–92). For those years, copper content of British ore imports from the United States was 46, 35, and 31 per cent, respectively. These were rich Montana ores, exports of which dropped off when the Anaconda Company started smelting operations (see Richter, *Quarterly Journal of Economics*, Vol. 41).

1906–09: Mine production from *Mineral Resources, 1913*, Pt. 1, p. 532 (excluding Alaska, *ibid.*, p. 215).

Lead

1839–85: Bureau of Mines, *Economic Paper 5*, Table 9, p. 13, "Annual smelter production of lead in the United States." Imports of lead ore did not become significant until 1886 (see W. R. Ingalls, *Lead and Zinc in the United States*, New York, 1908, p. 217). Hence smelter production represents domestic mine output.

1886–1906: U.S. Geological Survey, Bureau of Mines, series on refined pig lead from domestic ores and base bullion is apparently too low. Interpolated between 1885 and *Materials Survey* figure for 1907 by series on refined lead from domestic ores and base bullion (*Economic Paper 5*, p. 14).

1907–09: Bureau of Mines, *Materials Survey: Lead*, May 1951, Table IV-2.

Zinc

1852–1906: Smelter production from domestic ores, plus zinc content of zinc oxide from domestic ores, plus zinc content of ores exported (sources given below).

1907–09: Mine production as reported to the U.S. Geological Survey. Zinc oxide output dates from 1852 (Whitney, *Metallic Wealth*, p. 350, and Ingalls, *Lead and Zinc*, p. 281), while zinc metal was first produced commercially in 1858 (*Economic Paper 2*, p. 17). Zinc smelter product did not surpass zinc content of oxide until the early 1870's.

1. *Smelter production from domestic ores:* 1858–72, *Economic Paper 2*, p. 19; 1873–1909, *Mineral Resources, 1910*, Pt. 1, p. 263; data for 1904–05 include some zinc from foreign ores. Data for 1901–05 adjusted to exclude dross spelter (*Mineral Resources, 1913*, Pt. 1, p. 624).

2. *Zinc oxide production* (zinc content computed at 80 per cent, see Whitney, *Metallic Wealth*, p. 350; Ingalls, *Lead and Zinc*, p. 358; *Mines and Quarries, 1902*, p. 456): 1852–54, Whitney, *Metallic Wealth*, pp. 350–351; 1859, obtained by straight-line interpolation between 1854 and 1868 (smelter output was only 50 tons in 1859); 1868, 1871, 1873, 1874, 1878, Geological Survey of New Jersey, *Annual Report of the State Geologist for 1904*, Trenton, 1905, pp. 303–305; 1879, *Tenth Census, 1880*, Vol. XV, p. 822; 1880–1909, *Mineral Resources, 1890*; Ingalls, *Lead and Zinc*, p. 338; 1907–09, zinc contents of all zinc pigments from domestic ores, *Mineral Resources, 1909*, Pt. 2, p. 705.

In 1879, zinc oxide made in eastern works (*Tenth Census, 1880*, Vol. XV, p. 822) totaled about 10,107 short tons. From this we deducted an estimated 1,000 tons from Middle West ores smelted in the East (*Mineral Resources, 1882*, p. 367). The factor of 80 per cent was applied to the remaining zinc oxide production to yield approximate zinc content of eastern ores.

In the early period of zinc production, zinc oxide was produced largely from New Jersey ores, but between 1861 and 1876 (Ingalls, *Lead and Zinc*, p. 285) an unknown amount came from Pennsylvania. Since data are available only for New Jersey ores, the factor of 30 per cent (obtained by dividing 1879 zinc contents of eastern ores by New Jersey ore output) was applied to New Jersey ore production for 1868, 1871,

1873, 1874, and 1878 (New Jersey Geological Survey) to approximate zinc product of all eastern ores. The intervening years are straight-line interpolations.
3. *Zinc content of ores exported: Mineral Resources.* Zinc ore exports began in 1896, except for a single recorded shipment of Joplin ores in 1892. *Mineral Resources* records data for ore shipments without indicating metallic content. We have assumed that all ores shipped from New York City and Philadelphia were New Jersey ores, at a calculated 25 per cent zinc content (see *Mines and Quarries, 1902,* p. 454). Exports through all other ports were largely Colorado ores, with an estimated zinc content of 42 per cent (Ingalls, *Lead and Zinc,* pp. 264, 340). Actually some Virginia ores were shipped through East Coast ports; since these were richer than New Jersey ores, the estimates tend to minimize metallic content. Similarly, some ores coming from Western states were included in the Colorado totals since they were so small in relation to Colorado; these were also richer than Colorado ores.

Note on Accuracy of Method of Estimating Zinc Production for 1852–1906

The same method was applied to the years 1907–09 and the resulting estimates were checked with the mine production figures of the U.S. Geological Survey. Although there were significant differences between the two figures year by year, the totals for the three-year period were less than 3.5 per cent apart.

The present method, however, gives results very close to the U.S. Geological Survey smelter series for 1907–09, which includes zinc content of pigments, the yearly difference being the estimated zinc content of New Jersey ores exported.

In general, the major differences between a mine and smelter figure for any given year (excluding exports) are due to the various inventory changes and errors in calculating zinc content from assay and in estimating recovery.

Actually there are greater discrepancies between the U.S. Geological Survey smelter and mine figures than are indicated in *Mineral Resources, 1909* (Pt. 1, pp. 208–209) because no effort was made to add estimated zinc content of exported ores to the U.S. smelter data. In these years, New Jersey was shipping abroad a high-grade Willemite ore which assayed 48 per cent less 15 per cent loss for smelter recovery (Ingalls, *Lead and Zinc,* p. 270). For the three-year period, ore exports totaled 60,000 tons, with an estimated metal content of 24,000 tons, or 3 per cent of total mine output for the period.

Gold

Note: All values were converted to ounces at the coining rate of $20.67 which was constant throughout the series.

1839: The only official data of the period are gold deposits at the U.S. Mint, which are too low, since they do not take into account gold going directly to industry and exported.

Whitney increased these figures by 10 per cent to get his final U.S. figure (*Metallic Wealth,* p. 148), while the Bureau of Mines increased them by 15 per cent (*Economic Paper 6,* p. 14). We are taking Whitney's estimates because of the care with which he developed his data, and because he was alive at the time. Whitney says his figures are gold "of domestic origin." Gold deposits of domestic origin at U.S. Mint equal $404,200 times 1.10, i.e., $444,620.

1840: Whitney, *Metallic Wealth,* p. 148.
1841–44: Estimated as for 1839, from table of U.S. Mint deposits (*ibid.,* p. 146).
1845–47: *Ibid.,* p. 148.
1848: Set at 28 per cent of $10,694,000 (1849 revised Census figure) and converted into ounces. Ratio of 28 per cent obtained from *ibid,* p. 148, 1848/1849.
1849: California, U.S. Bureau of Mines, *Economic Paper 3,* estimate of C. G. Yale,

p. 20, Table 10 (value of gold in terms of recovered metal). Other states, value of product of gold mining (*Seventh Census, 1850*).

1859: Gold mining, annual value of product, *Eighth Census, 1860, Manufactures* p. 736. The 1870 Census disputes this figure (see pp. 751–753, *Ninth Census, 1870*, Vol. III), but Yale's figure for California, $45,846,000, is very close to the Census figure of $44,717,000.

1850–58: Interpolated by California production of gold, annual value of product (*Economic Paper 3*, Table 10, p. 20).

1869: Estimate of value of production based on data from *Ninth Census, 1870*, and other sources. See regional table for notes on sources and methods.

1860–68: Interpolated by Alexander Del Mar (*A History of the Precious Metals*, New York, 1902, p. 400). Del Mar was former Director of the U.S. Bureau of Statistics and Mining Commissioner to the U.S. Monetary Commission of 1876.

1870–78: Interpolated by Del Mar (*ibid.*, p. 400).

1879: Estimated from *Tenth Census, 1880*, Vol. XIII, pp. 354–357. State data built up to U.S. total as follows: deep mines, ore raised during Census year times average yield per ton of ore raised and treated, by state; placer gold, mint value of crude placer gold (see regional table).

1889: *Eleventh Census, 1890*, Vol. VII, p. 53, excluding Alaska. Census officials adjusted the U.S. figure to approximate U.S. Mint estimate (see *ibid.*, p. 51).

1880–88: Interpolated by Del Mar (see above).

1902: Mines and Quarries, 1902, p. 534, excluding Alaska.

1890–1901: Interpolated by mine production as gathered by Director of the Mint from reports of U.S. Mint operators and mine agents. (*Mines and Quarries, 1902*, p. 553; these data reduced by exclusion of Alaska, *Mineral Resources, 1913*, Pt. 1, p. 215.)

1905: Mine production of gold as reported to U.S. Geological Survey, excluding Alaska (*Mineral Resources*).

1903–04: Interpolated by mine production as reported by U.S. Mint operators and mine agents. Data for 1903 and 1904 from *Mineral Resources*.

1904–05: Interpolated by refinery production as reported by Director of the Mint, excluding Alaska. Indexes of refinery production linked to 1904 mine production to obtain 1902–05 mine series.

1905–09: Mine production as reported to U.S. Geological Survey, excluding Alaska, the Philippines, and Puerto Rico (*Mineral Resources*).

Silver

Note: All values were converted to ounces at the coining rate of $1.29 which was constant throughout this series.

1839: The only mine producing sizable amounts of silver was the Washington Mine, Davidson County, North Carolina. According to Whitney (*Metallic Wealth*, p. 399) this was a silver-lead mine discovered in 1836 and worked until 1852. In 1844, it produced 18,500 ounces and in 1851, 8,000 ounces of silver. We assume that 1837 was the first full year of production and that output in that year equaled output in the final full year of operation—8,000 ounces. If production rose linearly, output was 11,000 ounces in 1839 (*ibid.*, pp. 399–400).

Silver was also found in association with gold in the southern Appalachian Region, and with lead in the New England and middle Atlantic regions (see *ibid.*; Bureau of Mines, *Economic Paper 8*; Walter R. Crane, *Gold and Silver*, New York, 1908; Wm. P. Blake, *Report Upon the Precious Metals*, Washington, 1869; and Ingalls, *Lead and Zinc*).

1844: Whitney, *Metallic Wealth*.

1849: Silver was being mined at the Washington Mine, North Carolina, in New Mexico, associated with gold in California and the South, with copper in Michigan, with copper and lead in New England, and with lead in the middle Atlantic region.

	Ounces	
North Carolina (Washington Mine)	11,000	Output assumed to have declined linearly in 1844–51.
California	69,000	*Economic Paper 3*. Ratio of silver production (oz.) to value of gold production for 1848–59 (.0068) times 1849 value of gold production.
Lake Superior	300	Annual output of Cliff Mine for 1949 (Whitney, *Metallic Wealth*, p. 279). A number of authors state the major part of silver was stolen by the miners: Whitney; Blake, *Precious Metals*, p. 154; W. Gates, *Michigan Copper and Boston Dollars*, Cambridge, 1951, p. 13; A. P. Swineford, *Swineford's History of the Lake Superior Iron District*, Marquette, 1871, pp. 64–65. Silver was found at mines other than Cliff, e.g., Minnesota, Phoenix, Adventure Mines.
New Mexico	8,000	Whitney, Blake, and Ingalls (see above) refer to New Mexico silver output. See 1859 discussion for method of obtaining 1849 figure.
Total	88,300	

1859: Silver produced from California gold ores and silver quartz, Lake Superior copper ores, southern Appalachian gold and lead, New England and middle Atlantic argentiferous lead ores.

	Ounces	
California	312,000	.0068 times dollar value of gold production (*Economic Paper 3*); see also 1849 note.
Lake Superior	23,000	Blake, *Precious Metals*, p. 154.
New Mexico	40,000	1849 output calculated as follows: [40,000 (N.M. 1859) ÷ 312,000 (Cal. 1859)] × 69,000 (Cal. 1849) = 8,000 ounces.
Total	375,000	

1850–58: California plus New Mexico plus Lake Superior plus North Carolina equals United States. Straight-line interpolations for New Mexico, Lake Superior, and North Carolina (1849–51). Estimated annual silver production for California from *Economic Paper 3*, p. 20, using ratio of .0068 applied to annual value of gold.

1860: Growth assumed to be at constant rate of increase, i.e., 103.1 per cent per annum. The U.S. Mint series before 1861 are obviously highly arbitrary and appear less reasonable than the constant rate used here.

1861–68: U.S. Mint Series (*Economic Paper 8*, p. 18).

1869: Estimated from Census and other sources. See regional table for methodology and data. Estimated production came to $12 million, which was the U.S. Mint figure for 1869. Del Mar has a figure of $13 million for 1869 silver output.

1871–76: Del Mar, *Precious Metals*, p. 400. Del Mar made a study of silver production in Nevada for 1871–76 for the U.S. Monetary Commission of 1876 on the basis of returns from mining companies, and estimated the rest of U.S. production from current sources. See *Report of the Monetary Commission* (S. Rept. 703, 44th Cong. 2d Sess., 1877, Ser. 1738), for full discussion of precious metals data available during this period and for detailed data on Nevada production. Del Mar's data are in silver dollars at the coining rate ($1.29) and have been converted to ounces here. He appears to have revised his estimate for 1872 from $18,500,000 (*Report of the Monetary Commission*) to $20,000,000, the figure in his book. Except for 1872, data from the *Report*, which are millions and tenths, were used here instead of data from the book, which are in millions of dollars, for converting value to weight.

1870: Interpolated between 1869 and 1871 by U.S. Mint series.

1879: Tenth Census, 1880, Vol. XIII, pp. 354–357. State data built up to U.S. total as follows: deep mines, ore raised during Census year times average yield per ton of ore raised and treated; plus placer, mint value of silver in crude placer gold.

1876–78: Interpolated by U.S. Mint series.

1889: Eleventh Census, 1890, Vol. VII, pp. 52–53, excluding Alaska.

1880–88: Interpolated by U.S. Mint series.

1902: Mines and Quarries, 1902, p. 534, excluding Alaska.

1889–1901: Interpolated by mine production as gathered by Director of Mint from reports of U.S. Mint operators and mine agents (*Mines and Quarries, 1902*, p. 553). These data exclude Alaska (*Mineral Resources, 1913*, Pt. 1, p. 215).

1905–09: Mine production of silver as reported to the U.S. Geological Survey, excluding Alaska (*Mineral Resources*).

1903–04: Interpolated by mine production as reported to Director of Mint, excluding Alaska (see 1889–1902 above). Data for 1903–04 from *Mineral Resources*.

1904–05: Interpolated by refinery production as reported by Director of Mint, excluding Alaska. Indexes of refinery output linked to 1904 U.S. Mint mine production to obtain 1902–05 mine series.

Note: The mine agent series has been used for interpolating because it appears to represent better the movement of mine production. There is danger, however, that the early years especially are deficient.

TABLE A-2

U.S. MINE PRODUCTION OF MAJOR METALS, AS ESTIMATED BY VARIOUS SOURCES, 1839-1909

Year	Iron Ore (mill. long tons)[a]			Copper (thous. short tons)[b]				Lead (thous. short tons)[c]			
	H-R[d]	H.S.[e]	P-C[f]	H-R[d]	H.S.[e]	E.P.[g]	P-C[f]	Census Per P-C	H-R[d]	H.S.[e]	E.P.[h]	P-C[f]
1839	0.99											
1849	1.67											
1850	1.48											
1859	2.40	2.87		0.784	0.784	0.784			18	17.5	17.5	
1860	2.55			7.72	7.06	7.06			24	23.5	23.5	
1869	3.83	3.83		14.5	14.0	14.0			16	16.4	16.4	
1870	3.66											
1879	7.12		3.83	26.4	25.8	25.8	25.8	28.1	18	17.5	17.5	
1880	9.13	7.12	7.12									
1889	14.5	14.5	14.5	113.0	113.0	113.0	113.0	110.0	91	90.8	90.8	91
1890	16.0	16.0	16.0									
1899	24.7	24.6	24.7						153	178.0	178.0	152
1900	27.6	27.3	27.6									
1902	35.6	35.3	35.6	328.0	330.0	330.0	330.0	313.0	274	368.0	282.0	269
1903	35.0	34.8	35.0									
1909	51.4	51.3	51.3	561.0	563.0	546.0	546.0		385	447.0 / 385.0[i]	375.0	353

Note: The estimates of mine production in Table A-1 differ from previous series in a number of respects. It should be borne in mind that our goal is to estimate recoverable content of mine output (except for iron ore, for which the data refer to gross output), whereas the other sources may be estimating smelter output or, in some cases, a highly ambiguous quantity called U.S. "production." Needless to say, it has often been necessary to use smelter or refinery output as the basis for estimated mine output.

[a]The iron ore output figures in the sources cited in Table 1 appear to have been placed in the wrong year in 1859, 1869, and 1879. Hence the present 1859 estimate is comparable with the 1860 Census figure reproduced in Historical Statistics, 1960. We believe two errors have been made in the past in interpreting the 1860 Census: (1) Long tons have been called short tons, and (2) 900,000 tons have been double counted. The present 1869 and 1879 estimates agree with Historical Statistics, 1960, except for the displacement of one year.

[b]The present estimates for 1859 and 1869, e.g., differ from Historical Statistics, 1960, and the Economic Paper of the Bureau of Mines, because we have tried to include ore exports.

TABLE A-2 (concluded)

Year	Zinc (thous. short tons)[j]			Gold (mill. fine troy ounces)[k]				Silver (mill. fine troy ounces)[l]			
	H-R[d]	H.S.[e]	P-C[f]	H-R[d]	H.S.[e]	E.P.[h]	P-C[f]	H-R[d]	H.S.[e]	E.P.[h]	P-C[f]
1839				0.022	0.023	0.023		(0.011)	0.019	0.019	
1849				0.517	1.94	1.94		0.088	0.039	0.039	
1859	(5.15)			2.28	2.42	2.42		0.375	0.077	0.077	
1869	12.0		0.05	1.64	2.40	2.39		9.28	9.28	9.28	
1879	29.0	4.30	4.30	1.64	1.88	1.88	1.88	34.6	31.6	31.6	31.6
1889	72.0	21.3	28.1	1.55	1.60	1.59	1.60	51.3	50.1	50.0	50.1
1902	216.0	58.9	75.8	3.24	3.87	3.87	3.87	55.8	55.5	55.5	55.5
1909	305.0	157.0	208.0	3.81	4.80	4.81	4.80	57.2	57.3	54.7	57.3
		256.0	302.0								
		302.0									

[c]The difference between the present and the other lead series reflects lead smelted or refined from foreign ores. The present series covers only domestic mine output.

[d]Herfindahl-Rein, see note to Table A-1.

[e]*Historical Statistics*, 1960.

[f]Neal Potter and Francis T. Christy, Jr., *Trends in Natural Resource Commodities*, Baltimore, 1962.

[g]Bureau of Mines, various years.

[h]Smelter production.

[i]Production, mine (recoverable content).

[j]Inclusion of zinc oxide in the present series is the main reason for difference from the other series.

[k]The other series reproduced here all have their origin in what has come to be accepted as the "official" series, the series appearing in the reports of the Director of the Mint. Del Mar's devastating critique of this series (*Precious Metals*) and its deviation from other sources such as the *Economic Paper on California* (U.S. Bureau of Mines, *Historical Summary of Gold, Silver, Copper, Lead, and Zinc Produced in California, 1848 to 1926*, Washington, 1929) and the Censuses indicate that it should not be used to set levels of output. Levels in the present series are based on Censuses and the *Economic Paper on California*, the output of which was especially important in most of the nineteenth century. Note that the present series excludes Alaska and the Philippines, which accounts for the difference between the 1909 figures.

[l]The "official" series on silver appear to be very wide of the mark in some of the early ears. The present estimate for 1839 rests mainly on Whitney (*The Metallic Wealth*). For 1849, a difficult year to estimate since California's output was increasing rapidly, the present estimate makes use of the *Economic Paper on California* (*ibid.*) and Whitney. The 1859 estimate reflects mainly the *Economic Paper on California*. 1871-76 are directly from Del Mar who made a careful study of Nevada output in those years.

TABLE A-3

OUTPUT OF MAJOR METAL ORES IN QUANTITY AND VALUE, BY REGION AND COMMODITY, 1839-1909[a]

Region and Commodity	1839		1849		1859		1869	
	Quantity	Value	Quantity	Value	Quantity	Value	Quantity	Value
New England		203		232		232		113
Iron ore	0.070	203	0.08	232	0.070	203	0.03	87
Copper[b]	--	--	--	--	0.077	29	0.07	26
Lead[c]	--	--	--	--	--	--	--	--
Zinc	--	--	--	--	--	--	--	--
Gold	--	--	--	--	--	--	--	--
Silver	--	--	--	--	--	--	--	--
Middle Atlantic		1,743		3,007		5,202		7,993
Iron ore	0.591	1,714	1.03	2,987	1.61	4,669	2.57	7,453
Copper[b]	--	--	--	--	--	--	--	--
Lead[c]	0.35	29	0.24	20	0.164	14	0.18	15
Zinc	--	--	--	--	5.15	519	5.21	525
Gold	--	--	--	--	--	--	--	--
Silver	--	--	--	--	--	--	--	--
South Atlantic		703		1,110		903		1,393
Iron ore	0.088	255	0.18	522	0.11	319	0.11	319
Copper[b]	--	--	--	--	0.386	144	--	--
Lead[c]	0.52	43	0.47	39	0.328	27	0.52	43
Zinc	--	--	--	--	--	--	6.94	700
Gold	0.019	393	0.026	537	0.02	413	.016	331
Silver	0.011	12	0.011	12	--	--	--	--
North central		1,652		2,723		4,446		9,431
Iron ore	0.096	278	0.200	580	0.45	1,305	1.08	3,132
Copper[b]	--	--	0.753	280	4.864	1,811	13.2	4,916
Lead[c]	16.6	1,374	22.5	1,863	15.744	1,304	16.4	1,358
Zinc	--	--	--	--	--	--	0.25	25
Gold	--	--	--	--	--	--	--	--
Silver	--	--	--	--	0.023	26	--	--
South central		491		513		794		455
Iron ore	0.148	429	0.17	493	0.16	464	0.04	116
Copper[b]	--	--	--	--	0.849	316	0.87	324
Lead[c]	--	--	0.24	20	0.164	14	0.18	15
Zinc	--	--	--	--	--	--	--	--
Gold	0.003	62	--	--	--	--	--	--
Silver	--	--	--	--	--	--	--	--
Southwest				9		620		14,138
Iron ore	--	--	--	--	--	--	--	--
Copper[b]	--	--	--	--	1.544	575	--	--
Lead[c]	--	--	--	--	--	--	0.18	15
Zinc	--	--	--	--	--	--	--	--
Gold	--	--	--	--	--	--	0.246	5,085
Silver	--	--	0.008	9	0.040	45	8.07	9,038
Rocky Mountains						1,860		10,556
Iron ore	--	--	--	--	--	--	--	--
Copper[b]	--	--	--	--	--	--	0.07	26
Lead[c]	--	--	--	--	--	--	--	--
Zinc	--	--	--	--	--	--	--	--
Gold	--	--	--	--	0.09	1,860	0.459	9,488
Silver	--	--	--	--	--	--	0.93	1,042
Pacific				10,226		45,203		19,397
Iron ore	--	--	--	--	--	--	--	--
Copper[b]	--	--	--	--	--	--	0.29	108
Lead[c]	--	--	--	--	--	--	--	--
Zinc	--	--	--	--	--	--	--	--
Gold	--	--	0.491	10,149	2.17	44,854	0.918	18,975
Silver	--	--	0.069	77	0.312	349	0.28	314
United States		4,796		17,836		59,260		63,499
Iron ore	0.993	2,880	1.66	4,814	2.40	6,960	3.83	11,107
Copper[b]	--	--	0.784	292	7.72	2,875	14.5	5,400
Lead[c]	17.5	1,449	23.5	1,946	16.4	1,359	17.5	1,449
Zinc	--	--	--	--	5.15	519	12.4	1,250
Gold	0.022	455	0.517	10,686	2.28	47,128	1.64	33,879
Silver	0.011	12	0.088	98	0.375	420	9.28	10,394

(continued)

TABLE A-3 (concluded)

Region and Commodity	1879 Quantity	1879 Value	1889 Quantity	1889 Value	1902 Quantity	1902 Value	1909 Quantity	1909 Value
New England		753		261				
Iron ore	0.09	261	0.09	261	--	--	--	--
Copper[b]	1.32	492	--	--	--	--	--	--
Lead[c]	--	--	--	--	--	--	--	--
Zinc	--	--	--	--	--	--	--	--
Gold	--	--	--	--	--	--	--	--
Silver	--	--	--	--	--	--	--	--
Middle Atlantic		11,704		11,112		9,814		14,675
Iron ore	3.76	10,904	3.22	9,338	1.82	5,278	2.21	6,409
Copper[b]	0.13	48	--	--	--	--	--	--
Lead[c]	--	--	--	--	--	--	--	--
Zinc	7.46	752	17.6	1,774	0.45[d]	4,536	82.0	8,266
Gold	--	--	--	--	--	--	--	--
Silver	--	--	--	--	--	--	--	--
South Atlantic		2,011		2,651		4,333		3,277
Iron ore	0.43	1,247	0.80	2,320	1.38	4,002	1.13	3,277
Copper[b]	0.79	294	--	--	--	--	--	--
Lead[c]	--	--	--	--	--	--	--	--
Zinc	0.57	57	e	--	f	--	--	--
Gold	0.02	413	0.016	331	0.016	331	--	--
Silver	--	--	--	--	--	--	--	--
North central		22,791		50,544		141,181		203,063
Iron ore	2.51	7,279	8.07	23,403	27.2	78,880	42.2	122,380
Copper[b]	21.6	8,044	43.73	16,285	89.0	33,144	117.0	43,571
Lead[c]	27.2	2,252	28	2,318	77.0	6,376	162.0	13,414
Zinc	18.94	1,909	51.6	5,201	147.0	14,818	167.0	16,834
Gold	0.16	3,307	0.155	3,204	0.356	7,358	0.305	6,304
Silver	--	--	0.119	133	0.54	605	0.50	560
South central		1,178		6,861		15,824		19,750
Iron ore	0.33	957	2.13	6,177	4.52	13,108	5.09	14,761
Copper[b]	0.13	48	--	--	6.0	2,234	9.0	3,352
Lead[c]	--	--	--	--	--	--	3.0	248
Zinc	1.72	173	3.2[g]	322	--	--	9.0	907
Gold	--	--	--	--	--	--	--	--
Silver	--	--	0.323	362	0.43	482	0.43	482
Southwest		21,972		21,136		35,924		105,712
Iron ore	--	--	--	--	h	--	i	--
Copper[b]	1.58	588	17.65	6,573	66.0	24,578	184.0	68,522
Lead[c]	22.7	1,880	9.0	745	3.0	248	9.0	745
Zinc	--	--	--	--	--	--	11.0	1,109
Gold	0.25	5,168	0.248	5,126	0.227	4,692	0.952	19,678
Silver	12.80	14,336	7.761	8,692	5.72	6,406	13.98	15,658
Rocky Mountain		33,093		85,234		167,302		181,979
Iron ore	--	--	0.14	406	.67[j]	1,943	0.77[i]	2,233
Copper[b]	0.53	197	50.0	18,620	154.0	57,350	221.0	82,300
Lead[c]	39.1	3,237	115.0	9,522	192.0	15,898	211.0	17,471
Zinc	--	--	--	--	24.0	2,419	36.0	3,629
Gold	0.31	6,408	0.465	9,612	1.766	36,503	1.524	31,501
Silver	20.76	23,251	42.030	47,074	47.49	53,189	40.04	44,845
Pacific		20,016		15,123		24,967		34,905
Iron ore	--	--	0.03	87	--	--	k	--
Copper[b]	0.26	97	0.076	28	13.0	4,841	30.0	11,172
Lead[c]	1.82	151	--	--	3.0	248	--	--
Zinc	--	--	--	--	--	--	--	--
Gold	0.90	18,603	0.666	13,766	0.875	18,086	1.029	21,269
Silver	1.04	1,165	1.109	1,242	1.60	1,792	2.20	2,464
United States		113,541		193,577		399,285		563,415
Iron ore	7.12	20,648	14.5	41,992	35.6	103,211	51.4	149,060
Copper[b]	26.4	9,831	113.0	42,081	328.0	122,147	561.0	208,916
Lead[c]	90.8	7,518	153.0	12,668	274.0	22,687	385.0	31,878
Zinc	28.7	2,893	72.4	7,298	216.0	21,773	305.0	30,744
Gold	1.64	33,899	1.55	32,038	3.24	66,971	3.81	78,753
Silver	34.6	38,752	51.34	57,500	55.8	62,496	57.2	64,064

NOTES TO TABLE A-3

Source: Regional production tables unless otherwise specified. Where available, physical quantities from original sources have been used. Otherwise percentages have been applied to U.S. total from annual production series to estimate output. 1879 prices, as follows, have been used to measure value: iron ore, $2.90 a long ton; copper, 18.62 cents a pound ($372.40/S.T.); lead, 4.14 cents a pound ($82.80/S.T.); zinc, 5.04 cents a pound ($100.80/S.T.); gold, $20.67 a fine troy ounce; silver, $1.12 a fine troy ounce. All prices except following from *Historical Statistics*, 1960; iron ore, *1880 Census*, Vol. 15, p. 68 (value ÷ output); gold, coining rate.

^aQuantities are in the following units: iron ore--million long tons; copper, lead, and zinc--thousand short tons; gold and silver--million fine troy ounces. Values are in thousand dollars in 1879 prices.

^bCopper includes production undistributed to regions: 4 per cent in 1849 and 1.5 per cent in 1889. In 1889, this copper product was worth over $600,000 in 1879 prices. For 1859, undistributed copper totaled 34 per cent of smelter output. Although the geographical distribution of value of copper ore obtained from the 1860 Census is admittedly unsatisfactory, we have used this as a basis to estimate value of output. We have assumed north central and other eastern product to be distributed among regions according to relative census copper product of the regions. The remainder, 20 per cent, is allocated to the southwest region. *Mineral Resources*, 1915, Pt. 1, p. 662, gives north central as 63 per cent. The above procedure results in New England, 1 per cent; South Atlantic, 5 per cent; north central, 63 per cent; south central, 11 per cent; and Southwest, 20 per cent.

^cSince no data are available for 1859 regional output of lead, the relative regional distribution was arbitrarily held at 1849 for the purpose of computing value of output.

^dIncludes South Atlantic.

^eSee south central.

^fSee Middle Atlantic.

^gIncludes Virginia.

^hIncludes Rocky Mountains.

ⁱAll West.

^jSouthwest and Rocky Mountains.

^kSee Southwest.

TABLE A-3a

YIELD OF IRON ORE MINED, BY REGION, 1849-1909
(per cent)

Region	1849	1859	1869	1879	1889	1902	1909
New England	41			45	44		
Middle Atlantic	34			49	49		
South Atlantic	39	39		46	41		
North central	40	43		55	58		
South central	34			49	43		
United States	36	43	48	51	53		

TABLE A-3b

YIELD OF LAKE SUPERIOR AND SOUTHERN IRON ORES, 1879-1909
(per cent)

	1879	1889	1904	1909
Furnaces using:				
Lake Superior ores exclusively (Michigan, Minnesota, Wisconsin)	58.2	63.3	53.4	52.3
Southern ores exclusively (Alabama, Georgia, Tennessee)	43.6	44.1	41.4	40.6

NOTES TO REGIONAL DISTRIBUTION OF OUTPUT IN TABLE A-3 AND TO YIELD DATA IN TABLES A-3a AND A-3b

Iron ore

Table A-3
1839: Regional distribution is based on iron ore consumed in manufacture of pig and wrought iron. See notes to annual iron ore series for derivation.
1849, 1859: Regional distribution is based on iron ore used in manufacture of pig iron. See note to annual iron ore series for derivation.
1869-1909: Regional distribution is directly from data for iron ore mined. See notes to annual series. Totals for 1909 differ because of exclusion of high-grade manganiferous ores.

Tables A-3a and A-3b
1849: State and regional yield has been estimated by using pig iron production as though it were metallic yield of ores mined and consumed in the state in that year. This procedure is valid since there was little transportation of iron ores before 1850. "... Up to 1850 little iron ore was transported except for such distances as could be conveniently covered by wagons. The blast furnaces and forges, depending chiefly on charcoal for fuel, were located close to their supplies of raw materials" (*Eleventh Census, 1890*, Vol. VII, p. 13). See notes to annual series.
1859: Because of long-distance transportation of ores, state consumption figures, for the most part, no longer represent ores mined in the state. In the north central and

south Atlantic regions, however, pig iron was made almost entirely from local ores. Consequently, these regional yield estimates have been derived as in 1849. It should be noted that regional data conceal the richness of Michigan ores. They were in great demand by the iron works of Buffalo and Pittsburgh, as well as the iron centers of Ohio. In 1864, for example, only 12 per cent of Michigan ores were consumed in Michigan, 31 per cent going to Ohio and 57 per cent to New York and Pennsylvania (J. W. Foster, *The Geology and Metallurgy of the Iron Ores of Lake Superior*, New York, 1865, pp. 62–63). Michigan ores had yields as high as 65 per cent (*ibid.*, p. 51). The same source gives the following yield data for other ores: Lake Champlain, New York, 65 per cent; Clinton, New York, 45 per cent; and Tuscarawas, Ohio, 45 per cent. The national yield is a weighted average based on total U.S. ore consumption and pig metal output.

1869: Ratio of pig metal to iron ore consumed in manufacturing pig iron from *Compendium of the Ninth Census, 1870*, 1872, p. 909. Iron ore imports did not begin until the 1870's and were relatively unimportant until 1879.

1879: See notes to annual series.

1889: Eleventh Census, 1890, *Report on Mineral Industries in the United States*, 1895. Yield data on pp. 10–12 applied to state iron ore output. The resulting U.S. yield of 53.0 per cent is not significantly different from that calculated by the Census, 51.3 per cent.

1902: No yield data computed by Census.

Note to Table A-3b: Metallic yield for furnaces using exclusively southern or Lake Superior ores (*Thirteenth Census, 1910*, Vol. X, *Manufactures 1909*, 1913, p. 216). These yields would be generally higher than those in Table A-3a because of inclusion of states with lean ores in regional groupings, e.g., Ohio in north central, Kentucky in south central. Moreover, Georgia which produced relatively rich ores is part of the south Atlantic region.

Copper

Note: Although no official data on production of copper exist before 1845, copper was being mined in 1839 in the New England states (Maine, New Hampshire, Vermont, Massachusetts, and Connecticut), in the middle Atlantic region (New Jersey and Pennsylvania), in the South (Maryland and North Carolina), and in the north central region (Missouri). Copper ore was still being produced in these general regions in 1849, and in the south central (Tennessee) and Southwest (New Mexico) as well. All the above regions were still producing in 1859, but only one new state— Arizona—reported copper mine output. (It should be understood that the above states were not all producing in 1839, 1849, and 1859.) This information has been collected from Whitney, *Metallic Wealth*, and other sources.

1839: Whitney, *Metallic Wealth*.

1849: *Mineral Resources, 1913*, Pt. 1, pp. 662–665; also Whitney, *Metallic Wealth*.

1859: *Mineral Resources*, as above. For Vermont, see Edward Hitchcock *et al.*, *Report on the Geology of Vermont* (Claremont, N. H., 1861, Vol. II, pp. 850–859); for New Mexico, see note to copper annual series.

1869: *Mineral Resources*, as above. For New Mexico, Ross J. Browne, *Statistics of Mines and Mining West of the Rocky Mountains*, Vol. 2, *1869*, Washington, 1870, p. 403; for Vermont, estimated (see *Report of the State Geologist on the Mineral Resources of Vermont, 1899–1900*, Burlington, 1900); for Tennessee, J. B. Killebrew, *Introduction to the Resources of Tennessee*, Nashville, 1874, p. 249; *Ninth Census, 1870, Industry and Wealth*, p. 767.

1879: Tenth Census, 1880, Vol. XV, pp. 798–800.

1889: Eleventh Census, 1890, Vol. VII, *Report on Mineral Industries*, 1892, p. 155.

1902: *Mines and Quarries, 1902*, 1905, p. 486. For Tennessee, *Fourteenth Annual Report of the Mining Department*, Nashville, 1904, p. 184.

1909: *Mineral Resources, 1909*, Pt. 1, p. 159.

Lead

1839: Smelter production, *Compendium of the Sixth Census, 1840,* Washington, 1841, p. 358.
1849: Smelter production, annual value of product, *Message of the President of the United States Communicating a Digest of the Statistics of Manufactures according to the Returns of the Seventh Census,* 35th Cong., 2d Sess., S. Ex. Doc. 39, ser. 984, p. 42.
1859: Ingalls, *Lead and Zinc.* This was also the major source used to locate areas of production in 1849 and 1869.
1869: Mine output, annual value of product, *Ninth Census, 1870, Industry and Wealth,* p. 768.
1879: Refined lead from domestic ores, Bureau of Mines, *Economic Paper 5,* Table 11, p. 17.
1889: Production of lead as reported by mines. Except for Mississippi Valley output (29,258 short tons), these data do not take into account smelter losses. *Eleventh Census, 1890,* "Mineral Industries," pp. 163 and 168–169. See also *Mineral Resources, 1889.*
1902: Lead content of ores as reported by smelters. *Economic Paper 5,* Table 11, p. 17.
1909: Mine production of recoverable lead, *Materials Survey: Lead,* Table IV-2, p. IV-73.

Zinc

1859: See notes to annual series.
1869: Zinc mining, annual value of product, *Ninth Census, 1870,* "Industry and Wealth," p. 769. The extremely high figure for the south Atlantic states is misleading without explanation. Apparently, the mines of Davidson County, North Carolina, were worked for lead in a "desultory way" about this time and up to the early 1890's. According to Ingalls (*Lead and Zinc,* p. 89), those ores proved to be zinc rather than lead. A local smelter erected in 1887 was unable to smelt them, however, because of their complex character. There is no indication whether the zinc ores mined in 1869 or thereafter were ever smelted successfully.
1879: Mine production, *Tenth Census, 1880,* Vol. XV, *Directory of Lead and Zinc Mines,* pp. 978–981.
1889: Smelter output plus zinc contents of oxide, *Eleventh Census, 1890, Mineral Industries,* p. 174. Since these data do not trace ore back to state of origin, they tend to distort regional relationships. The Census states that only the New Jersey and Pennsylvania smelters used distant ores, but does not present data on the origin or proportion of such ores (see *ibid.,* p. 748). Other sources indicate that they were imported from Virginia and the Middle West.
1902: Zinc contents of ores mined (by assay), *Twelfth Census, 1902, Mines and Quarries, 1905,* pp. 454 and 456.
1909: Recoverable zinc contents of ores mined, *Mineral Resources, 1909,* Pt. 1, p. 208.

Gold

Note: All values were converted into ounces at the coining rate of $20.67 which was constant throughout these series.
1839: Value of gold produced by smelting houses, *Compendium of the Sixth Census, 1840,* p. 358.
1849: California, value of gold production; other states, value of product, gold mining. For California data, Bureau of Mines, *Economic Paper 3,* p. 20, Washington, 1929; for other states, *Message of the President, Seventh Census, 1850.*
1859: Value of product, gold mining. Although considerable doubt is cast on the California data by the next Census (1870), the 1859 Census figure for California is

very close to that given in *Economic Paper 3*. See *Eighth Census, 1860*, state tables, and U.S. total, p. 736.

1869: Value of gold product as follows: for California, *Economic Paper 3*, Table 10; for Colorado, C. W. Henderson, *Mining in Colorado*, Professional Paper 138, 1926, p. 69; for Utah, *Mineral Resources, 1913*, Pt. 1, p. 366; for Washington, *ibid.*, p. 790; for Wyoming, *ibid.*, p. 50; for other states, *Ninth Census, 1870, Industry and Wealth*, Table XIII, pp. 760–766, value of product at mine level. Gold quartz value figure inflated by 45 per cent to allow for value added by milling (see *ibid.*, p. 751); placer mined gold as given. This method appears to yield fairly good results. For example, New Mexico gold mining product was valued at $245,750 for quartz and $97,000 for placer gold. Inflating the quartz figure by 45 per cent, and totaling, gives a value of product of $454,000, very close to an independent estimate of $477,000 presented in Waldemar Lindgren et al., *The Ore Deposits of New Mexico*, Professional Paper 68, 1910, pp. 20–21. Adjustments were required in the case of Idaho and Nevada because separate values for gold and silver were not presented for gold- and silver-bearing ore. For Idaho, the value of gold and silver quartz was set at 57 per cent and 43 per cent silver. This is a rough estimate based on value relationships in Idaho deep mines as given in the *Tenth Census, 1880*, Vol. XV, p. 356. Gold value for Nevada was set at the 1869–70 relationship of gold and silver in the Comstock lode, 40 per cent gold and 60 per cent silver (*Mines and Quarries, 1902*, p. 255).

1879: Value of gold contents of ore raised during year computed as follows: Total of deep mines (tons of ore raised times average yield in dollars per ton of ore raised) plus placer mines (value of crude bullion). Source: *Tenth Census, 1880*, Vol. XIII, *Precious Metals*, 1885, pp. 354–355 for deep mines, p. 353 for crude placer gold.

1889: Eleventh Census, *1890*, Vol. VII, *Mineral Industries*, p. 53, excludes Alaska.

1902: Mines and Quarries, *1902*, p. 534, excludes Alaska.

1909: Mineral Resources, *1909*, Pt. 1, p. 130, excludes Alaska, the Philippines, and Puerto Rico.

Silver

All years except 1869: See notes to annual production series.

1869: See notes on regional distribution of gold production for method and sources. California silver output has been calculated by the method described in *Economic Paper 3*, notes to Table 6, p. 14. Gold and copper output are found in *Economic Paper 3*, Table 10, p. 20; lead output in Ingalls, *Lead and Zinc*, p. 145.

Source of Silver	Ounces
Gold, placer, and quartz	98,400
Copper	14,375
Lead	150,000
Total	262,775

All silver output values converted into ounces at the coining rate of $1.29.

TABLE A-4

EMPLOYMENT IN MAJOR METAL MINING, BY REGION AND COMMODITY, 1839-1909
(thousands)

Region and Commodity	1839[a]	1849[a]	1859	1869	1879	1889	1902	1909
New England								
Iron ore	.224	.256	.224	.218	.617	.426	0	0
Copper	0	.009[b]	.075	.355	.619	0	0	0
Lead and zinc	0	0	0	.005	0	0	0	0
Gold and silver	0	0	0	.008	0	0	0	0
Major metals	.224	.265	.299	.586	1.236	.426	0	0
All minerals				5.32	7.27	9.42	8.14	9.08
All employment				1,299	1,572	2,006	2,377	2,914
Middle Atlantic								
Iron ore	2.320	4.030	6.300[c]	13.100[d]	18.219	9.481	4.193	5.549
Copper	0	.007[b]	.065	.007	.010	0	0	0
Lead and zinc	.397	.270	.192	.437	.405	.233	.119[e]	.208[f]
Gold and silver	0	0	0	0	0	0	0	0
Major metals	2.717	4.307	6.557	13.544	18.634	9.714	4.312	5.757
All minerals				58.87	98.30	154.67	232.70	346.25
All employment				2,807	3,738	4,966	6,203	8,209
South Atlantic								
Iron ore	.380	.780	.475	.815	1.970	3.497	3.855	4.168
Copper	0	.034	.300	.189	.334	0	0	0
Lead and zinc	.064	.058	.040	.175[h]	.153	.810	.420[e]	0
Gold and silver	1.030	1.400	1.055[g]	.833	1.660	1.435	.836	.191
Major metals	1.474	2.272	1.870	2.012	4.117	5.742	5.111	4.359
All minerals				7.46	12.87	26.99	48.67	95.32
All employment				1,996	2,678	3,118	4,001	5,188
Eastern north central								
Iron ore					7.340	16.604	16.876	18.487
Copper					5.011	6.765	14.315	19.575
Lead and zinc					.530	1.145	.583	2.043
Gold and silver					0	.097	0	0
Major metals					12.881	24.611	31.774	40.105
All minerals					38.75	95.78	135.82	189.24
All employment					3,615	4,687	5,888	7,258

(continued)

TABLE A-4 (continued)

Region and Commodity	1839[a]	1849[a]	1859	1869	1879	1889	1902	1909
Western north central								
Iron ore						2.499	9.010	18.409
Copper					.041	0	0	0
Lead and zinc					6.253	4.100	7.661	19.590
Gold and silver					1.090	1.899	3.071	3.568
Major metals					9.277	8.498	19.742	41.567
All minerals					19.17	38.64	62.26	81.13
All employment					2,009	2,987	3,693	4,447
South central								
Iron ore	.650	.747	.706	.388	1.615	5.134	6.596	7.767
Copper	0	.046[b]	.405	.620	0	0	0	0
Lead and zinc	0	0	0	.005	.142	.357	0	.943
Gold and silver	0	0	0				.045	0
Major metals	.650	.793	1.111	1.013	1.757	5.491	6.641	8.710
All minerals				3.62	6.46	25.35	59.56	93.97
All employment				2,096	3,023	3,636	5,210	7,108
Southwest								
Iron ore	0	0	0	0	0	0	0	0
Copper	0	0	.580	.045	.110	1.097	4.395	13.608
Lead and zinc	0	0	0	.006[1]	0	0	0	.066
Gold and silver	0	.044	.277	4.250[1]	11.600	8.213	3.632	7.383
Major metals	0	.044	.857	4.301	11.710	9.310	8.027	21.057
All minerals				10.24	15.32	13.19	17.41	31.54
All employment				62	95	103	139	254
North central								
Iron ore	.128	.268	.603	4.556	9.233	19.103	25.886	36.896
Copper	0	.560	3.639	4.188	5.052	6.765	14.315	19.575
Lead and zinc	1.080	1.430	1.023	1.086	6.783	5.245	8.244	21.633
Gold and silver	0	0	0	0	1.090	1.996	3.071	3.568
Major metals	1.208	2.258	5.265	9.830	22.158	33.109	51.516	81.672
All minerals				36.60	57.92	134.42	198.08	270.37
All employment				3,896	5,624	7,674	9,581	11,705

(continued)

TABLE A-4 (concluded)

Region and Commodity	1839[a]	1849[a]	1859	1869	1879	1889	1902	1909
Rocky Mountain								
Iron ore	0	0	0	0	0	.545	1,224[j]	1,600[k]
Copper	0	0	0	0	0	1,958	7,407	17,947
Lead and zinc	0	0	0	0	0	0	.008	0
Gold and silver	0	0	.963	8,960[f]	14,300	27,450	23,100	23,542
Major metals			.963	8,960	14,300	29,953	31,739	43,089
All minerals				18.76	49.58	50.43	77.23	72.26
All employment				72	188	397	525	855
Pacific								
Iron ore	0	0	0	0	.014	.047	0	0
Copper	0	0	0	0	.020	0	.541	2,674
Lead and zinc	0	0	0	0	0	0	0	0
Gold and silver	0	9.800	42.762	18,700[f]	22,700	17,466	10,716	15,827
Major metals		9.800	42.762	18,700	22,734	17,513	11,257	18,501
All minerals				45.75	50.06	32.55	52.57	46.39
All employment				280	474	836	1,039	1,935
United States[l]								
Iron ore	3,702	6,081	8,308	19,077	31,668	38,233	41,754	55,980
Copper	0	.656	5,064	5,404	6,145	9,820	27,783	53,804
Lead and zinc	1,541	1,758	1,255	1,714	7,483	6,288	8,791	22,850
Gold and silver	1,030	11,244	45,057	32,751[i]	51,350	56,917	41,560	50,511
Major metals	6,273	19,739	59,684	58,946	96,646	111,258	119,888	183,145
All minerals				186.62	297.78	447.00	694.35	965.17
All employment				12,506	17,392	22,736	29,073	38,167

Note: 1859-1909 are Census data with adjustments as described. In the Censuses of 1850, 1860, 1870, 1880 (precious metals), and 1890, the Census forms asked for the "average number employed." In 1902 the form asked for the "average during the year," but this was reduced by the fraction of a 300-day year that the establishment was idle. This adjusted figure appears in this table. The 1909 figures are the number employed as of Dec. 15 or some date which would be reasonably representative thereof. In the 1840 and 1880 (nonprecious metals) Censuses, the form simply asked for the "number employed." See Table A-5 for 1902 and 1909 data adjusted to the 1889 concept of employment.

[a] All employment for 1839 and 1849 is estimated regionally on the assumption that employment moves proportionately with output from 1859. The employment figures in the Census for 1840 and 1850 include large amounts of manufacturing activity.

NOTES TO TABLE A-4 (continued)

[b] Estimated on basis of ratio for 1849/1859 of copper not allocated by state 0.31/2.7. The same states are in the "not allocated" group in both years.

[c] Iron ore employment per Census has been changed by the ratio of our estimate of regional output to the Census figure. The Census gives data on iron ore mining only for noncaptive mines. These accounted for 38 per cent of the output (see notes on 1859 output). If we take employment/output by region for the noncaptives and multiply by our estimated total output for each region (both captive and noncaptive), we get the figures for 1859 which are used in the tables. However, the employment/output ratios for noncaptives in the 1860 Census are far below the same ratios for 1869.

[d] Census figure of 9,036 multiplied by ratio of our estimate of output to Census output (2.57 L.T./1.77 L.T.). Census says Pennsylvania was undercounted. See output notes on 1869 iron ore.

[e] Unallocated employment is 539 in Census. This is distributed between New Jersey and Virginia on basis of 1889 employment in lead and zinc.

[f] Extrapolated from 1902 by output.

[g] Census figure of 573 increased by ratio of our estimate of output to Census output ($413t/224t).

[h] Census figure of 385 increased by output ratio of 331/153, as in footnote g.

[i] It appears that output of gold and silver was greatly underestimated in the 1870 Census. The table which follows compares Resources for the Future and Census value of gold and silver output in 1879 prices for Censuses of 1860 and later. Correspondence is good, except for 1869.

CORRECTION OF CENSUS, GOLD AND SILVER OUTPUT

	1859	1869	1879	1889	1902	1909
	CHANGE OF VALUE FIGURES FROM CURRENT PRICES TO 1879 PRICES (THOUS. DOLLARS)					
$_sp_i$, Census silver price	1.36	1.325	1.29	1.29	0.522	0.515
$r_i = {_sp_i}/{_sp_{79}}$	1.21	1.18	115.0	115.0	0.47	0.46
Rocky Mountain						
1. Value of gold	1,860	9,488	6,408	9,612	36,503	31,501
2. Value of silver	--	1,042	23,251	47,074	53,189	44,845
3. Index 1+2/1+2r_i	1.00	0.98	0.89	0.89	1.46	1.46
Southwest						
4. Value of gold	--	5,085	5,168	5,126	4,692	19,678
5. Value of silver	45	9,038	14,336	8,692	6,406	15,658
6. Index 4+5/4+5r_i	0.83	0.90	0.90	0.91	1.44	1.31
Pacific						
7. Value of gold	44,854	18,975	18,603	13,766	18,086	21,269
8. Value of silver	349	314	1,165	1,242	1,792	2,464
9. Index 7+8/7+8r_i	1.00	1.00	0.99	0.99	1.05	1.06
	CENSUS VALUE GOLD + SILVER IN 1897 PRICES = CENSUS VALUE IN CURRENT PRICES (MILL. DOLLARS) x INDEX					
Rocky Mountain						
10. Census	2.00	6.8	29.6	56.8	74.5	70.1
11. RFF	1.9	10.5	29.6	56.7	89.7	76.3
Southwest						
12. Census	0.20	10.4	19.4	13.9	9.9	27.0
13. RFF	0.04	14.1	19.5	13.8	11.1	35.3
Pacific						
14. Census	45.7	7.9	19.6	15.0	18.6	20.4
15. RFF	45.2	19.3	19.8	15.0	19.9	23.7
Total West						
16. Census	47.90	25.1	68.6	85.7	103.0	117.5
17. RFF	47.14	43.9	68.9	85.7	120.7	135.3

NOTES TO TABLE A-4 (concluded)

Census employment data for the western regions for 1869 have been increased by the ratio of our output value to Census value of output. Thus:

    Rocky Mountain   5,810 x 10.5/6.8 = 8,960
    Southwest   3,137 x 14.1/10.4 = 4,250
    Pacific   7,668 x 19.3/7.9 = 18,700

See notes on gold and silver output for details on output.

^{j}Includes Colorado, 451, and "all other" which I believe is mainly Rocky Mountain. Other states mentioned are Connecticut, Kentucky, Massachusetts, New Mexico, North Carolina, Texas, Utah, Vermont, and West Virginia.

^{k}Includes Utah, 81, and 1,519 "all other." I believe this is mainly Rocky Mountain, but other states mentioned in the Census (p. 259) are Connecticut, Kentucky, Massachusetts, Nevada, New Mexico, North Carolina, Texas, and West Virginia.

^{l}U.S. totals include unallocated.

TABLE A-5

EMPLOYMENT IN MAJOR METAL MINING, BY REGION AND COMMODITY, 1889, 1902, AND 1909 WITH 1902 AND 1909 ADJUSTED TO THE 1889 DEFINITION OF EMPLOYMENT

Region and Commodity	1889	1902[a]	1909[b]	1902[c] (1902 Definition Divided by 1889 Definition)	1909[d] (1902 Definition Divided by 1909 Definition)
New England					
Iron ore	426	0	0	.894	
Total	426				
Middle Atlantic					
Iron ore	9,481	5,040	5,720	.832	.859
Lead and zinc	233	121	182	.983	
Total	9,714	5,161	5,902		
South Atlantic					
Iron ore	3,497	5,490	5,040	.702	.851
Lead and zinc	810	697	0	.602	
Gold and silver	1,435	1,396	244	.598	.765
Total	5,742	7,583	5,284		
East north central					
Iron ore	16,604	19,840	20,150	.850	.928
Copper	6,765	14,315	19,575	1.00	1.00
Lead and zinc	1,145	877	2,100	.664	.683
Gold and silver	97	0	0	1.00	
Total	24,611	35,032	41,825		
West north central					
Iron ore	2,499	10,900	20,300	.827	.913
Lead and zinc	4,100	8,510	17,550	.900	.807
Gold and silver	1,899	3,420	3,540	.897	.890
Total	8,498	22,830	41,390		
North central					
Iron ore	19,103	30,740	40,450		
Copper	6,765	14,315	19,575		
Lead and zinc	5,245	9,387	19,650		
Gold and silver	1,996	3,420	3,540		
Total	33,109	57,862	83,215		
South central					
Iron ore	5,134	8,640	9,170	.763	.902
Lead and zinc	0	0	943		.563
Gold and silver	357	83	0	.540	
Total	5,491	8,723	10,113		
Southwest					
Copper	1,097	4,600	13,450	.954	.944
Lead and zinc	0	0	66		.515
Gold and silver	8,213	4,510	7,920	.806	.863
Total	9,310	9,110	21,436		
Rocky Mountain					
Iron ore	545	1,680	1,950	.730	.890
Copper	1,958	7,890	18,400	.939	.965
Lead and zinc	0	8	0		
Gold and silver	27,450	27,800	24,160	.832	.854
Total	29,953	37,378	44,510		
Pacific					
Iron ore	47	0	0	.810	
Copper	0	566	2,674		.954
Gold and silver	17,466	18,300	22,150	.584	.818
Total	17,513	18,866	24,824		
United States					
Iron ore	38,233	51,590	62,330		
Copper	9,820	27,371	54,099		
Lead and zinc	6,288	10,213	20,841		
Gold and silver	56,917	55,509	58,014		
Total	111,258	145,968[e]	195,284		

NOTES TO TABLE A-5

^aFactor described in note c applied to Table A-4.

^bFactors described in notes c and d applied to Table A-4.

$$c_{Equals} \frac{\Sigma \text{ (employment)}_i \text{ times } \frac{\text{(days worked)}_i}{300}}{\Sigma \text{ (employment)}_i}$$

where i is the skill level or occupational group. This correction is suggested in *Mines and Quarries, 1902*, p. 90.

^dEquals average monthly employment in 1909 per Census divided by 1909 Census employment on Dec. 15.

^eIncludes 1,125 for unallocated copper and 160 for unallocated gold and silver.

Note: Reported employment in 1902 was reduced by the fraction of a 300-day year that an establishment was not operating. This adjustment was thought to yield a result approximately equal to average monthly employment (see *Mines and Quarries, 1902*, p. 90).

Census employment in 1909 was the employment of Dec. 15 or for "the nearest representative date" if the mine was not in operation. Fortunately, data were collected also for the 15th of each month.

COMMENT

Paul W. McGann, U.S. Department of Commerce

The Herfindahl paper provides a survey of the Census and other sources of metal mining production, employment, prices, and numbers of mines in production over the period from 1839 to 1909, by region and commodity. There are problems of comparable industry definition in the early Censuses where smelting was included with mining and concentrating, but the current convention of treating smelting as a manufacturing activity was soon adopted after regional and company distribution of smelting became more concentrated. Data distinctions between current mineral production and mining development were not available during the period (and are still not available despite the fact that such distinctions are made annually for tax purposes and the data are regularly submitted to the Internal Revenue Service—which does not tabulate or publish them).

Dramatic shifts in regional concentrations of metal mining took place during the period and were extremely important in the economic history of the eleven western states. It is remarkable to note that employment in metal mining was greater during that period than it is today; and, therefore, the relative importance of these industries has shrunk drastically since then, while the population has increased tenfold.

Herfindahl examines the effect of these industries on regional development and notes that metal mining rather than agriculture was the primary source of export of major commodity production and employment for many of the western states for several decades. These statistics, therefore, justify the preoccupation of historians of these eras with the varying success of metal mining and remind us that this emphasis, despite the violence in these localities, is not based on nostalgic romanticism. For a number of states, such as the Rocky Mountain group between 1870 and 1880, metal mining provided over 25 per cent of all employment. This means that metal mining contributed more than one-half of the income-generating exports from these areas and thus was responsible for more than one-half the economic activity in these states.

Herfindahl discusses the relationship between transportation, markets, and location of mining activities. By the end of the Civil War with the expansion of railroads, transportation no longer had such a crucial bearing on regional distribution of metal mining, which had by that time decisively shifted to the eleven western states region for nonferrous mining, and to the Great Lakes states for iron. On the other hand, the shift from precious metals to base nonferrous metals was accomplished only as transportation resources were increased within these states, and such deposits could then be profitably exploited.

Growth in metal mining production roughly equals that of all manufacturing during the period but was considerably greater in iron and copper mining where ingenious, large-scale production methods were introduced steadily after 1870. Production increases were smaller for other metals not amenable to large-scale expansion. (It would have been helpful if Herfindahl had reported on Census data for open-pit and underground copper and iron operations in this respect.) His tables show that the ratios of employment to output in mining the same metal differ sharply among states. Such drastic differences are still evident today—although to a lesser extent among states—and reflect substantial differences in the characteristics and riskiness of ore bodies and in the extent to which entrepreneurs are skilled in the introduction of efficient techniques. Greater differences among states in that early period were no doubt due to the inferior communication and dissemination of the more advanced practices compared with those of this century.

The "net rent" question in metal mining is a famous one in economic discussion, but almost no serious quantitative research has been undertaken. The necessary data, of course, are not present in any Census publication but might be derived from different tabulations of mining operations ranked by Census data on cost, such as those which have become available in the *Census of Minerals Industries* since 1954 (and were prepared by the Census for 1939). These retabulations could be made from archival materials and would provide greater insight into the profitability of mining operations in different areas.

The American Petroleum Industry

HAROLD F. WILLIAMSON
NORTHWESTERN UNIVERSITY

RALPH L. ANDREANO
UNIVERSITY OF WISCONSIN

CARMEN MENEZES
NORTHWESTERN UNIVERSITY

Introduction

In this paper the authors have attempted (1) to present a summary of a large volume of previous research on the history of the American petroleum industry and (2) to develop several topics not explored in earlier studies.[1] Within this general framework, the major objective has been to provide a statistical basis for measuring the growth of the industry and its relative importance in the American economy for the period 1859–1914. On the supply side of the industry, we have accordingly indicated only the most significant determinants of outputs and the composition of the flow of related inputs in each of the industry's major sectoral divisions. Discussion of the demand for petroleum products is likewise confined to the principal factors that have affected sales over time. Only the bare essentials of technological developments affecting both supply and demand are presented.

In order to meet the necessary restriction on the length of this paper, many statements of facts and conclusions are presented without the a priori analysis used to arrive at these propositions. In all such cases, however, we have referred the reader to the sources where the analysis is more fully developed. One particular gap, not fully explored here but

[1] The works referred to and on which the bulk of the present paper is based are: Harold F. Williamson and Arnold R. Daum, *The American Petroleum Industry: The Age of Illumination, 1859–1899*, Evanston, 1959; Harold F. Williamson, Ralph L. Andreano, Arnold R. Daum, and Gilbert C. Klose, *The American Petroleum Industry: The Age of Energy, 1900–1959*, Evanston, 1964; and Ralph L. Andreano, "The Emergence of New Competition in the American Petroleum Industry Before 1911," unpublished Ph.D. dissertation, Northwestern University, 1960, available from University Microfilms, Ann Arbor, Michigan.

discussed at length elsewhere, concerns the determinants of vertical integration and their application to changes in the structure of the petroleum industry. Although the Standard Oil complex dominated the industry throughout most of the 1859–1914 period, we have not dealt specifically with the performance of the dominant firm nor have we made any assessment of its impact on the allocation of resources within the industry.

We have organized our presentation as follows: Part I contains aggregate data on the industry related to the general growth of the domestic economy and a sectoral analysis of the industry's output record. Part II presents an outline of the major technical changes which occurred within each sector and the associated changes in input requirements and costs. Part III is a discussion of the principal factors that affected demand for the major refined products. Part IV provides a condensed statement of the dynamics of the industry structure from the late 1870's until the dissolution of the Standard Oil combination in 1911.

I

Physical Outputs, Value Added, and Relative Position of Petroleum Industry in the American Economy

All available time series indicate a remarkable rate of growth for the American petroleum industry from its inception in 1859 to World War I. This section contains a comparison of the rate of growth, in both physical and value terms, of the major sectors of the industry with other relevant sectors in the economy, as well as a discussion of the growing over-all importance of the industry in the domestic economy.

Domestic production of crude oil, as shown in Table 1, was over 100 times as large in 1914 as in 1865. From 1879 (the first year for which comparable data are available) through 1889, output of crude oil grew at approximately the same rate as the domestic production of fuels and minerals. From 1889 through 1914, however, crude oil production increased over 7.5 times compared with a growth of 3.3 times in the output of fuels and 2.9 times in the output of minerals.

Output of refined products expanded even more rapidly than output of crude oil did, reaching a total in 1914 some 138 times larger than production in 1865. Output of refined products also grew at a much faster rate than the physical output of all manufacturing, which increased some 11 times during those years (Table 1).

The growth in physical production was accompanied by an impressive increase in the value of the output in each of the sectors of the industry.

TABLE 1

INDEX OF PHYSICAL OUTPUT OF CRUDE PETROLEUM, FUELS, TOTAL MINERALS, REFINED PRODUCTS, AND TOTAL MANUFACTURING PRODUCTION, 1865-1914
(1889 = 100)

Product	1865	1869	1873-75	1878-80	1883-85	1889	1894	1899	1904	1909	1914
Crude petroleum[a]	7.1	11.9	28.0	58.4	65.9	100.0	140.3	162.3	332.9	520.9	755.8
Fuels[b]				45.5[c]	67.7	100.0	108.1	149.5	207.1	278.8	329.3
Total minerals[b]				51.9[c]	69.2	100.0	105.3	147.4	193.9	261.6	290.9
Refinery products[d]	4.8	11.8	27.7	45.8	67.0	100.0	160.5	174.5	201.6	388.8	662.3
Total manufacturing production[e]	25.7	37.9	43.9	55.4	72.7	100.0	103.0	151.5	183.3	251.5	290.9

[a]*Historical Statistics of the United States, Colonial Times to 1957*, 1960, pp. 360-61.
[b]*Historical Statistics*, 1960 (computed from the Bureau of Mines Indexes of Physical Volume of Mineral Production), p. 353.
[c]Outputs are for the year 1880.
[d]Table A-3.
[e]*Historical Statistics*, 1960 (computed from Frickey's Index of Manufacturing Production), p. 409.

TABLE 2

VALUE OF CRUDE OUTPUT AS PERCENTAGE OF VALUE OF MINERAL FUELS
AND VALUE OF ALL MINERAL PRODUCTS, 1880-1914

	1869	1879	1889	1899	1904	1909	1914
Crude petroleum ($ mill.)	24	17	29	65	101	128	214
Mineral fuels ($ mill.)	--	112	221	392	543	705	936
Total minerals ($ mill.)	--	293	489	900	1,126	1,530	1,813
Crude as per cent of mineral fuels	--	15.3	12.2	16.5	18.6	18.2	23.0
Crude as per cent of total minerals	--	5.8	5.5	7.2	9.0	8.4	11.8
Mineral fuels as per cent of total minerals	--	38.2	45.1	43.5	48.2	46.0	51.6

Source: *Historical Statistics*, 1960, p. 350.

As shown in Table 2, after declining from approximately $24 million in 1869 to $17 million in 1879, the total value (at the wells) of crude production rose to $214 million in 1914. The value of crude as a proportion of the value of total mineral fuels rose from 15.3 per cent in 1879 (the first year comparable data are available) to 23.0 per cent in 1914. Since the relative importance of mineral fuels to total minerals rose from 38.2 per cent to 51.6 per cent during those years, the value of crude as a percentage of the value of total minerals also rose, from 5.8 to 11.8.

The net and gross values added by refining rose, respectively, from $9 million and $27 million in 1879 to $71 million and $147 million in 1914. In general, both the gross and net value added by refining, as indicated in Table 3, rose at a faster rate than the value added by all manufactures during that period.

TABLE 3

GROSS AND NET VALUE ADDED BY REFINING AS PERCENTAGE
OF VALUE ADDED BY ALL MANUFACTURES, 1880-1914

	1879	1889	1899	1904	1909	1914
Net value added by refining ($ mill.)[a]	9	17	21	36	38	71
Per cent of value added by all manufactures	0.45	0.41	0.45	0.59	0.46	0.76
Gross value added by refining ($ mill.)[a]	27	40	43	67	86	147
Per cent of value added by all manufactures	1.36	0.97	0.92	1.11	1.05	1.56
Value added by all manufactures ($ mill.)	1,973	4,102	4,647	6,019	8,160	9,386

Source: *Historical Statistics*, 1960, Table 5, p. 409.

[a] Table 4 figures rounded.

TABLE 4

VALUE OF CRUDE AT WELLS, COST TO REFINERIES OF TRANSPORT AND STORAGE, AND VALUE ADDED BY REFINING, AS PERCENTAGE OF VALUE OF REFINED PRODUCTS AT REFINERY, 1879-1914
(dollars in thousands)

	1879	1889	1899	1904	1909	1914
Value of crude[a] at well prices	16,198	23,579	58,772	57,605	84,543	154,923
Per cent	27.2	27.7	47.4	32.9	35.6	39.08
Net value added by refining	8,706	17,083	21,070	35,618	37,734	71,097
Per cent	14.6	20.09	17.0	20.3	15.9	17.9
Value of materials other than crude	18,658	23,039	22,435	31,900	47,966	75,536
Per cent	31.3	27.1	18.1	18.2	20.2	19.0
Gross value added by refining	27,364	40,122	43,505	67,518	85,690	146,633
Per cent	45.9	47.2	35.1	38.5	36.1	36.9
Transportation and storage costs to refineries	16,000	21,300	21,652	49,882	66,765	94,805
Per cent	26.8	25.1	17.5	28.5	28.5	23.9
Value of refined products at refinery	59,562	85,001	123,929	175,005	236,997	396,361

Source: *Census of Manufactures*, 1914, Vol. II, pp. 581-83; Special Report of the Census Office, *Manufactures*, 1905, Part IV, pp. 569-71; *Historical Statistics*, 1960, pp. 360-61.

[a]Used by refineries.

TABLE 5

VALUE OF OUTPUT AT REFINERIES, AND ESTIMATED VALUE ADDED BY TRANSPORT AND MARKETING, AS PERCENTAGE OF TOTAL VALUE, 1880-1914
(dollars in thousands)

	1879	1889	1899	1904	1909	1914
Value of output at refineries	60,376	86,771	133,425	201,662	275,347	461,136
Per cent of total value	56.9	50.5	52.9	54.1	50.4	51.5
Estimated value added by transport and storage[a], and by wholesaling and retailing	45,752	84,958	118,729	171,144	270,382	433,291
Per cent of total value	43.1	49.5	47.1	45.9	49.6	48.5
Total value	106,128	171,729	252,154	372,806	545,729	894,427

Source: Table A-1.

[a]Other than from well to refinery.

Combined transport and storage costs of shipping crude from wells to refineries, shown in Table 4, rose from an estimated $16 million in 1879 to almost $95 million in 1914. These amounts added between 17.5 per cent and 28.5 per cent to the total value of refined products at refineries, a value which expanded during that period from approximately $59.5 million to over $396 million.

The value of crude (at wells) as a proportion of total value of refined products ranged between 27.2 per cent and 47.4 per cent; in contrast, the net value added by refining varied only between 14.6 per cent and 20.3 per cent. Only the value added by materials (other than crude) showed a sharp drop from about 31 per cent in 1879 to a range varying roughly between 18 per cent and 20 per cent from 1899 through 1914. Despite this trend, expenditures by the industry for purchase of materials outside the industry rose from $18.6 million in 1879 to $75.5 million in 1914.

As shown in Table 5, in 1879 the crude oil, crude transport, and refining sectors contributed almost 57 per cent to the total value of the output of the industry, while an estimated 43 per cent was contributed by distribution and marketing. Over the succeeding years, these proportions remained relatively stable, although there was a slight upward trend in the contribution of marketing and distribution to the total value of the products of the industry from approximately $106 million in 1879 to over $894 million in 1914. (A more detailed breakdown of the data in Table 5 is contained in Table A-1.)

Table 6 provides a rough over-all measure of the growing importance of the petroleum industry in the American economy. At a time when the

TABLE 6

VALUE OF INDUSTRY OUTPUT AS PERCENTAGE OF NET NATIONAL PRODUCT, 1880–1914

Year[a]	Net National Product (mill. dollars)	Year	Value of Industry Output (mill. dollars)	Per Cent
1877–81	8,480	1879	106	1.25
1887–91	11,000	1889	172	1.56
1897–1901	15,000	1899	252	1.68
1902–06	21,200	1904	373	1.76
1907–11	27,200	1909	546	2.00
1912–16	34,600	1914	894	2.58

Source: *Historical Statistics*, 1960, pp. 143 and Table A-1.

[a] Five-year periods are annual averages.

net national product of the United States grew over fourfold, from approximately $8.5 billion in 1879 to $34.6 billion in 1914, petroleum's relative contribution more than doubled, from 1.25 per cent to 2.58 per cent.

II

Sectoral Developments in Crude Oil Production

OUTPUT

Because the basic statistical series on crude production, drilling activity, proved reserves, footage drilled, and the like are available in a number of publications, reproduction of that data here would seem inappropriate.[2] Rather, a few summary comments on the output record will be noted with only brief explanations.

1. Output of crude petroleum grew from approximately 2.5 million barrels in 1865 to 265.8 million barrels in 1914, or at a compound annual rate of about 9.8 per cent.

2. Yearly additions to output showed no clear relationship to the number of new wells drilled. During several periods, in fact, notably between 1900-01 and 1905-06, 1910-11 and 1913-14, the two series moved in opposite directions (see Chart 1). Over the long pull, however, output did grow as more wells were drilled. The short-period relationship is noted here as evidence of the episodic nature of crude oil production in the early years of development.

3. Dry holes as a percentage of total wells drilled fell substantially from the mid-1870's until 1880 and then increased for the remainder of the period. Roughly, about one well in eight was dry in the mid-1870's and about one in five or six after 1900. During that period, as most of the new wells drilled were heuristically, if not actually, of the wildcat variety, the dry-hole rate is probably a rough approximation of the risk burden in drilling decisions (see Chart 2).

A priori, in any given pool the probability of a dry hole falls initially as more wells are drilled; then rises as the deposits either become thoroughly defined or trapping and faulting occur and other technical difficulties develop, making even offset wells a risky venture. It is not suggested that the ratio of dry holes to total wells drilled is an absolute measure of risk but only a relative approximation imputed to each drilling decision. Rank

---

[2] These series may be found in Williamson and Daum, *American Petroleum Industry, 1859–1899;* Williamson, Andreano, Daum, and Klose, *American Petroleum Industry, 1900–1959;* Ralph Arnold and William J. Kemnitzer, *Petroleum in the United States and Its Possessions,* New York, 1931; and various editions of *Petroleum Facts and Figures,* American Petroleum Institute (API), New York.

CHART 1

New Wells Drilled and Total Crude Oil Production, 1865–1914

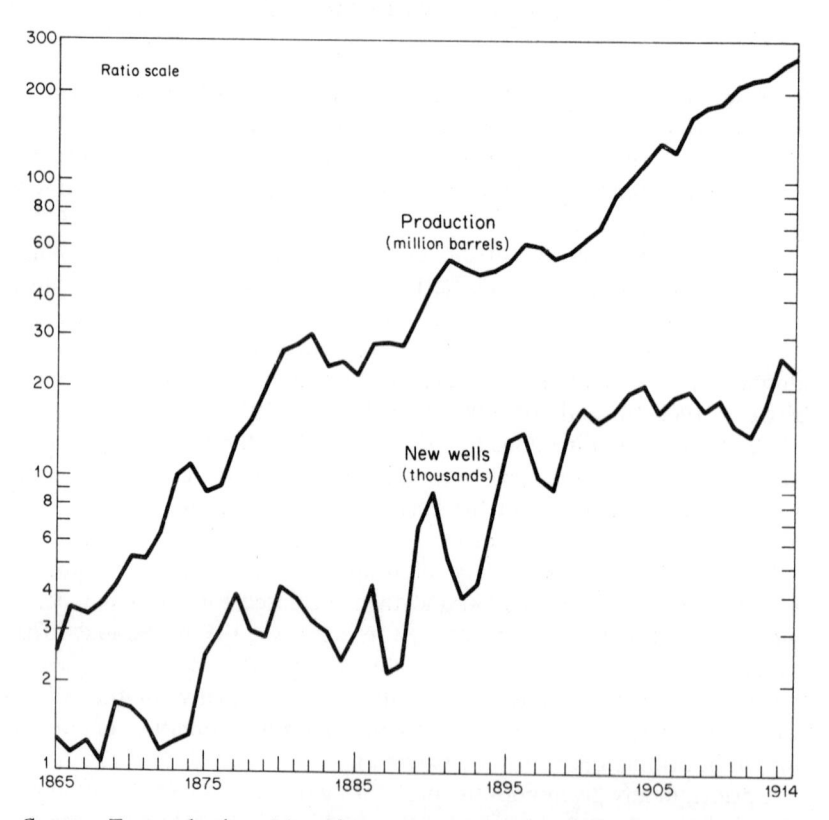

Source: For production, *Mineral Resources of the U.S.*, 1883–1915, and *Derrick's Hand-Book*, I; for new wells, Arnold and Kemnitzer, *Petroleum in the U.S.*, and *Derrick's Hand-Book*, I.

CHART 2

Dry Holes as Percentage of Total Wells Drilled, 1865–1914

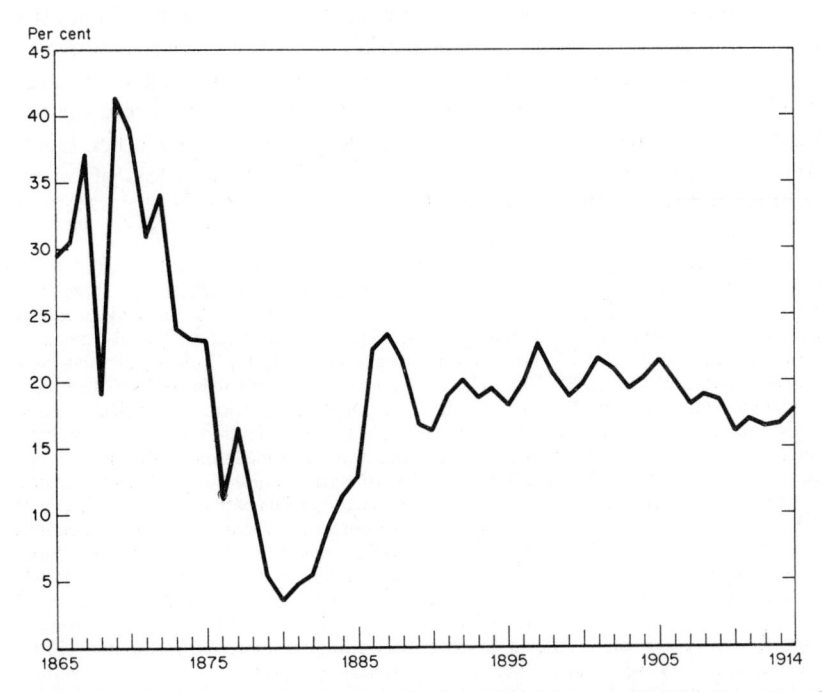

Source: For 1865–82, *Derrick's Hand-Book*, I, Folger Report, Wrigley Report, and Carll Report; for 1883–1914, *Mineral Resources of the U.S.*, 1883–1914.

wildcat wells fall more under the conventional definition of uncertainty rather than risk.[3]

4. The ratio of net well abandonments to new wells completed appeared to widen rather than narrow, over time; between 1900 and 1919, for example, average well life increased (in years) by about $\frac{2}{3}$ over the highest well life estimated for the pre-1900 years. Well life is, in this context, a secular rate between the ratio of wells abandoned to new wells drilled. The reasons for abandonments are, of course, related to new drilling, the most obvious being the prevailing market price for crude. In modern production practices, the cost function of a producing well of average productivity has a large, flat segment—and this was no less true for the flush fields treated in this paper. Abandonments, on the other hand, were also a function of other nonmarket forces, such as loss of oil by faulting and the infiltration of water.[4]

[3] These issues are explored in a number of sources. The technical problem of faulting, etc., and its effect on dry hole rates is discussed in Lester C. Uren, *Petroleum Production Engineering*, New York, 1946, Vol. II, pp. 11–14, 87–88, 101. A useful article with a theoretical decision-making model of short- and long-term production is Paul Davidson, "Public Policy Problems of the Domestic Crude Oil Industry," *American Economic Review*, March 1963, pp. 85–108. The data on dry holes and total wells drilled were compiled from *Mineral Resources of the U.S.*, Dept. of Interior, Geological Survey, beginning in 1883 and published annually until reorganization of the Geological Survey in 1923. Thereafter, the compendium was expanded and published by the Bureau of Mines until 1929, and then issued as *Minerals Yearbook of the United States*. All the basic production data for the crude oil sector in this paper were compiled from the Geological Survey reports. Arnold and Kemnitzer (*Petroleum in the United States*) also used the Geological Survey data in compiling their basic series. A careful check of Arnold and Kemnitzer against the Geological Survey reports revealed that they followed the annual reports on *Mineral Resources*, without exception. The quality of the Geological Survey varied from year to year: the least satisfactory and reliable reports were in the early 1880's; coverage and accuracy improved in the mid-1890's. The reports of 1896-97 and of 1905 were exceptionally useful.

The following sources were also used for all the data relevant to the production section, and were most valuable for the pre-1880 period.

S. F. Peckham, *Production, Technology, and Uses of Petroleum and Its Products*, 47th Cong., 2d Sess., H. R. Misc. Doc. 42, Part 10, 1884 (Peckham Report); *The Derrick's Hand-Book of Petroleum: A Complete Chronological and Statistical Review of Petroleum Developments from 1859 to 1898*, Oil City, Pa., 1898; J. T. Henry, *The Early and Later History of Petroleum, with Authentic Facts in Regard to Its Development in Western Penna.*, Phila., 1873; S. S. Hayes, *Report of the U.S. Revenue Commission on Petroleum as a Source of National Revenue*, Special Report Number 7, 39th Cong., 1st Sess., H. Ex. Doc. Dec. 51, 1866 (Hayes Report); Commonwealth of Pa., *Annual Report of the Secretary of Internal Affairs*, XX, Industrial Statistics, 1892, Part 3, Harrisburg, 1893 (Folger Report); J. F. Carll, *Geological Survey of Pa.*, 7th Report on the Oil and Gas Fields of Western Pa., for 1887 and 1888, Harrisburg, 1890 (Carll Report); H. E. Wrigley, *Second Geological Survey of Pa.*, 1874, *Special Report on the Petroleum of Pa., Its Production, Transportation, Manufacturing and Statistics*, Harrisburg, 1875 (Wrigley Report).

[4] See Davidson, "Public Policy Problems." The data on abandonments were taken from Arnold and Kemnitzer, *Petroleum in the United States*.

5. Variations in new drilling were positively correlated with variations in posted field crude prices. Decisions to drill new wells were probably much more rationally formed than popular notions about the derring-do of "wildcatters" might suggest.

6. Average yields per well per day rose appreciably over time by about 80 per cent between the late 1880's and 1910. This suggests that wells were kept in production longer, that the number of single prolific wells increased, and that technical improvements expanded the recovery of crude from existing wells.

7. The value of crude oil processed as a percentage of the total dollar value of refined products at the retail level (see Table A-1) rarely exceeded 25 per cent and, for the six census dates, averaged closer to 18 per cent.

INSTITUTIONAL FACTORS AFFECTING PRODUCTION[5]

The exploitation of underground deposits of crude petroleum was from the beginning a random process at best. The methods commonly used today for reducing uncertainty—geological and modern technological ways of prospecting and reservoir engineering—were not available to potential oil producers to any great extent before World War I. "Oil science," or petroleum geology, as the field was to be known professionally, was rudimentary and virtually ignored, even when its limited findings might have been of assistance to oil producers in evaluating drilling decisions. Even after an initial oil strike in previously untested areas, predicting the relative probability of success of wells in the same area was virtually impossible. The modern developments noted help to explain why early production practices were, by necessity, random. Among the unknown or imperfectly understood aspects of the new industry, which began as an adventure, a gamble with an uncertain pay-off, were: the nature of underground structures favorable to oil accumulation; the peculiar lithographic properties of subsurface structures favorable to oil accumulations; the high probability that dissipation of reservoir pressure

---

[5] Material for this and the following section is based on a wide variety of sources. An expanded version of this section and a complete list of references may be found in Williamson and Daum, *American Petroleum Industry, 1859–1899*, Appendix E. The best sources on the rule of capture for this period are: R. E. Hardwicke *et al.*, *Legal History of the Conservation of Oil and Gas, A Symposium*, Chicago, 1938; Northcut Ely, "The Conservation of Oil," *Harvard Law Review*, 1938; J. H. Marshall and N. L. Meyers, "Legal Basis of Petroleum Production in the United States," *Yale Law Review*, 1938. The generalizations made on the flush field phenomena are based on Andreano, "The Emergence of New Competition," Appendixes A and B. Technology in crude oil production is summarized in Williamson and Daum, *American Petroleum Industry, 1859–1899*. The most useful and authoritative survey of production techniques is *History of Petroleum Engineering*, API, Division of Production, Dallas, 1961. Isaiah Bowman, "Well Drilling Methods," *Water Supply Paper 257*, Dept. of Interior, Geological Survey, 1911, is also a classic work in oil field technology.

drives—water or gas—would result in trapping underground petroleum; and the many ways of producing underground faulting.

Geological uncertainties were reinforced by the institutional framework in which oil prospecting was carried on. Specifically, the common law "rule of capture," which held that subsurface mineral rights rested in the property owner (or lessee) who reduced such minerals to possession, created peculiar technical difficulties in locating and securing petroleum deposits. Petroleum is a fugaceous mineral; its resting place is not permanent but changes according to underground flow and movement. Two adjacent wells producing petroleum from a common pool may stabilize underground drainage, so that each property owner obtains an output equal to the size of the reserves surrounding his well and the strength of the underground pressure pushing the petroleum to the surface. If only one well is drilled, however, and if the two adjacent locations are separately owned (or leased), the one without a well will suffer a loss of underground oil through drainage to the other well. It is in the interest of the property owner without a well, therefore, to drill one in order to protect his petroleum from drainage and to recover his underground rights vested in his above-ground property right.

The historical effect of the rule of capture was to stimulate the drilling of additional wells to "offset" underground drainage. The rule also tended to favor well drilling in rough proportion to the number of property owners or lessees. A proliferation of wells, each drawing from a common source of underground pressure drives, resulted in underground faulting, quick dissipation of the pressure drive, geologically undesirable drainage and shifting of petroleum deposits, and loss of potentially producible petroleum, which could be recovered only by drilling new wells to relocate the shifts in the underground rocks.

The "rule of capture" and the randomness of locating well deposits were characteristic of the petroleum industry from its inception. They contributed to the phenomenon known as "flush fields" which dominated the pattern of crude petroleum production in the period under review. Strictly defined, a flush field is one where output as a fraction of producible underground reserves attains its peak rate over the life of the field in a relatively short time. Two pieces of evidence may be used to illustrate the occurrence of a flush field: actual production rates of individual pools within a field, and peak production rates and rates of decline over time for individual fields or pools. Data on actual production rates for individual pools are well understood, and virtually any set of pools in any of the major producing oil fields will show periods of from six months to two years of output expansion after the initial opening, a peak in output, and then a decline. Total output of the field will continue to expand and will peak later than an individual pool does as additional deposits within the field are discovered and exploited in the same pattern. Examples of the elapsed time from peak rates for three pools discovered between 1901 and 1910 are shown in Table 7.

The phenomenon of flush fields was not only a function of institutional factors but also had its economic influences. The relative ease of entry, the low marginal costs of production for individual wells, and the nominal total costs of above-ground storage were characteristic of crude petroleum production throughout this period. The combination of economic and

TABLE 7

PEAK PRODUCTION AND DECLINE, THREE OIL POOLS, 1902-11
(average daily production)

Pool	Date of Peak Production	Highest Production (barrels)	Per Cent Decline From Peak After:		
			1 Year	2 Years	3 Years
Spindletop	Oct. 1902	62,548	67	92	94
Humble	June 1905	93,272	84	92	90
Glenn	Mar. 1911	83,370	64	69	66

Source: Andreano, "Emergence of New Competition," p. 309.

institutional factors produced, in the short run, perverse supply elasticity for crude; declines in the price of crude oil would produce proportionately greater increases in the amount of crude produced from a given pool or field. In the short run, the supply curve of crude petroleum could be described as having a large horizontal segment; once a well was brought into production, the marginal cost was zero, or near it.

TECHNOLOGY

Until the development of the rotary drill in the late 1890's, the techniques of well drilling had remained essentially unchanged since the early 1870's. The technique known as cable-tool drilling was adapted to petroleum production from its earlier uses in water- and salt-well drilling. Though the principles of cable-tool drilling remained essentially the same, the quality of equipment improved over time. Application of power, improvements in drilling tools, use of casing, and development of the "torpedo" were the major kinds of technical improvement. Most of the changes resulted from particular problems encountered by oil producers in the fields. Others, such as improvement in metal strengths, use of explosive charges to accelerate the flow of oil, and application of machine operations, were adaptations of techniques developed in other areas of the economy. With one major exception, rotary drilling, the major technical changes in methods and equipment used in drilling and completing wells were developed during the first decade of the industry's history.

Cable-tool drilling is based on the percussion principle of boring. The drilling is generally slow except at shallow well depths and requires a high degree of art. Rotary drilling—a technique first used in drilling water wells in West Virginia, South Dakota, and northern Texas—consists of driving a bit through strata; water, or some other fluid, is passed through the cutter or drill bit under pressure in order to force the upbraided material through the side of a hole outside the bit casing. Rotary drilling was relatively faster than cable-tool drilling, gradually required less of the driller's art, and was specially suited to loose, unconsolidated strata of the type commonly found in the Gulf Coast areas. The later development of a hard rock bit, plus some modifications in derrick design and power systems, made rotary drilling well adapted for harder rock drilling. The method was first used in an oil well at Spindletop in 1901, and with a few minor exceptions its use was confined to wells drilled in Gulf Coast Texas and Louisiana until the 1920's.[6]

INPUTS

Strictly conceived, the final input for crude petroleum output is the drilled well. The geological and engineering peculiarities of crude oil production mentioned previously, however, make such a simplified notion of a production function impracticable and empirically unusable. The collection of inputs must be viewed, therefore, as the flow of land, labor, capital, and entrepreneurial skill in the exploration, drilling, completion, and operation of a well. There are no figures that show the aggregate structure of these separate inputs or of their relative contributions to the final output of crude petroleum. Except for a few Census figures, the aggregate mix of inputs can be discussed only in qualitative terms.

A precise measure of the magnitude of labor and capital employed in the exploration, drilling, completion, and operation of crude petroleum wells for the pre-1900 period is impossible within the present limits of data availability. Lease values were not reported; much of the work force in the early years was part time, floating, and otherwise uncountable;

[6] Williamson and Daum, *American Petroleum Industry, 1859–1899*, pp. 137–158; Williamson, Andreano, Daum, and Klose, *American Petroleum Industry, 1900–1959*, pp. 29–32.

THE AMERICAN PETROLEUM INDUSTRY

TABLE 8

SELECTED CENSUS DATA ON CRUDE OIL PRODUCTION, 1870-1919
(dollars in millions)

Class	1870	1880	1889	1902	1909	1919
Number in work force[a]	4,488	11,477	29,223	50,107	59,085	125,110
Wage bill[b]	4.0	10.9	10.3	29.2	51.6	236.7
Capital (at book value)	--	--	--	330.9	683.3	2,421.5
Value of crude output	19.3	24.6	27.0	71.4	117.7	694.0

Source: Special Reports of the Census Office, *Mines and Quarries*, 1902, p. 721; *Fourteenth Census, 1920*, Vol. XI, *Mines and Quarries*, p. 311; *Thirteenth Census, 1910*, Vol. XI, *Mines and Quarries*, p. 263.

[a] Includes operators, salaried officers, and contract employees, 1902-19; no separation by type earlier.

[b] Includes salaries, wages, and contract payments; contract payments not available for 1870 and 1889.

while equipment in drilling and completing could be reused many times. In Table 8 is a summary of selected input data for the crude petroleum sector for 1870-1919 from the Census reports of those years. Because the pre-1902 data are most unreliable, they are presented only to give some impression of the general direction of expansion and the magnitudes involved.

DRILLING COSTS

The cost of equipment, labor, and materials used in drilling wells can be detailed a bit more than other input components can. But even these data are still only an indication of general trends. The standard list of items used in drilling during the early decades varied for each drilling location depending on the depth of the well, the art of the driller, the terrain, the lithographic properties of the strata being drilled, and other uncontrollable variables. Inability to clarify depreciation charges, to allocate costs of drilling equipment over multiple locations (or for pumping), and wide variability in fuel intake and completeness of materials needed must also be considered as deficiencies of the qualitative estimates of drilling costs presented below.

During the first decade of the industry, drilling costs declined in real terms. It seems most likely that, as a result of equipment improvements, the greater skill of the drillers, and the wider use of power systems in percussion drilling, the total unit cost of putting down a well was reduced appreciably by the early 1870's.

The trend in drilling costs in the Appalachian field after the early 1870's continued downward, although increases in well depths put absolute dollar costs successively higher. Well depths in the early 1870's were on the average from 700 to 900 feet, while for drilling costs, a total of $6,100 probably prevailed as an average.[7]

The pattern of drilling costs and well depths is somewhat clearer to trace after the early 1870's. The estimates of total drilling costs shown in Table 9 include: pipe, casing, tubing, sucker rods, and belting; derrick and rig; engine and boiler; contract drilling (driller plus four hands); torpedoes and storage tanks. Undeflated money values are shown. The modal well depth in the Appalachian field was 1,200 to 1,500 feet

TABLE 9

APPROXIMATE COST OF DRILLING, APPALACHIAN FIELD, 1877-98
(dollars)

Depth (feet)	1877-78	1886-88	1897-98
1,000	3,021	2,119	--
1,400	3,535	2,483	--
1,600	3,822	2,589	2,206
2,000	4,326	2,951	--
2,400	--	--	2,595
2,800	--	--	3,160
3,200	--	--	4,720

Source: Williamson and Daum, *The American Petroleum Industry, 1859-1899*, p. 768; Andreano, "Emergence of New Competition," p. 329. See also footnote 8.

in the late 1870's, 2,000 to 2,200 feet in the late 1880's, and 2,600 to 2,800 in the late 1890's.[8]

[7] This figure is based on a well depth of 900 feet and the use of a derrick, boiler and engine, casing, tubing, sucker rods, a contract driller, a tank, and two torpedoes. On the assumption of a 20-acre drilling location, a sum was added for lease contracts. Prices were obtained for each of the items; for example, a contract driller price of $2.00 per foot was used. Materials used in making the calculations—to be interpreted as only an approximation—are from Cone and Johns, *Petrolia*, p. 463; *Derrick's Hand-Book* I, pp. 153 and 278; and various parts of the Carll, Folger, and Wrigley Reports. The drilling locations developed up to 1875 varied considerably in stratigraphic characteristics, and many wells could be drilled for one-half the average cost presented here. Others, however, presented greater boring difficulties and required more time to drill, with much higher costs.

[8] The calculations presented in Table 9 are a summary of much more detailed data contained in the sources to the table. The calculations are based on the prevailing practice of well drilling in western Pennsylvania for the respective dates. Pipe, casing, tubing, belting, packing, and sucker rods were computed at prices prevailing in each year, and the amounts for wells of various depths used were checked against contemporary practice in the White Sands and Bradford oil areas. Derrick, engine and boiler, and storage tanks were figured at the minimum prices. Only the cost of a 100-barrel storage tank and two torpedoes were included in the calculation; for some wells more

THE AMERICAN PETROLEUM INDUSTRY 365

An impressionistic account of drilling costs, by fields other than the Appalachian area, for the years after 1888 follows.

Ohio-Indiana

The best contemporary estimates suggest that the total cost of bringing wells into production, with the common depth of 2,000 feet, ranged from $2,000 to $2,500.[9]

Gulf Coast

At Spindletop in 1901, the rotary drill was used for the first time successfully in a commercial oil well. Because the rotary was not used to any great extent in other fields, there is no real basis for comparison, and it is not possible to calculate the cost reduction, if any, produced by the innovation. A sample of several of the largest producing companies in the field which used the rotary and drilled thirty-two wells during the year 1906 showed an arithmetic average total cost of $6,331.41. There was a cluster of wells costing $4,000 and $8,000, and the range was from $1,250 to $11,000. Depths were approximately 2,000 feet for half the wells drilled.[10]

Mid-Continent Field

Modal well depths in Oklahoma, Kansas, and north Texas were close to 2,500 feet between 1906 and 1910. The only available estimates placed the total cost of putting down a well (1,900 foot depth) at $5,000. All drilling was done with cable tools.[11]

---

torpedoes and larger storage tanks obviously would be required. Drilling contractors' prices were again checked against actual transactions: $1.00 a foot for 1877–78, $0.60 a foot for 1886–88, and $0.45 a foot for 1897–98 were the figures used. The basic data underlying Table 9 may be found in: *Mineral Resources*, 1888, pp. 316–17; *Engineering and Mining Journal*, Sept. 1, 1888, p. 179; *Derrick's Hand-Book*, II, pp. 177–183; *Peckham Report*, pp. 142–48. Modal well depths were calculated from well records in the Carll Report, pp. 147–323, and *Mineral Resources* for the respective years.

[9] These impressions are based on well records and scattered cost data in Edward Orton, First Annual Report, *Geological Survey of Ohio*, Columbus, 1890.

[10] Gulf Coast data are primarily from *Oil Investor's Journal*, an oil field paper issued from Beaumont, beginning in 1902. Three numbers of the *OIJ* (Nov. 3, 1906, Jan. 18, 1907, and July 6, 1908) contained extensive surveys of well depths and costs, drawn from company records of drilling operations. Also of value was Robert Hill, "The Beaumont Oil Field," *Transactions of the American Institute of Mechanical Engineers*, 1903.

[11] *The Oil Investor's Journal*, July 6, 1908; C. W. Shannon and L. E. Trout, *Petroleum and Natural Gas in Oklahoma*, Oklahoma Geological Survey, Bull. 19, Norman, 1915; *Oil and Gas in Oklahoma*, Oklahoma Geological Survey, Bull. 40, Norman, 1928.

California

The only data available on drilling costs in California are presented in Table 10. The modal well depth was between 2,000 and 2,500 feet. These data are somewhat misleading as they are an unweighted mix of the total costs for all oil pools in California. Consequently, the decline in total cost with increasing depths must not be taken literally.[12]

TABLE 10

TOTAL COST OF DRILLING WELLS, CALIFORNIA FIELD, 1913

Depth (feet)	Cost (dollars)
1,000	11,700
2,000	11,490
3,000	11,080

Source: Andreano, "Emergence of New Competition," p. 334; California State Mining Bureau, Bull. 69, p. 26.

TRANSPORTATION AND STORAGE OF CRUDE AND REFINED PRODUCTS

Early Storage and Transport Facilities

A highly inflammable and volatile liquid, petroleum from the outset posed difficult transport and storage problems. Producers in the oil region initially built wooden tanks or vats to receive the oil from individual wells, although it was not uncommon for the owner of a newly struck flowing well to store his oil in a hastily constructed earthen reservoir.

Even the best wooden storage facilities were more subject to loss through leakage and evaporation and fire than iron tanks introduced during the 1850's by coal oil refiners were. It was not until the wartime shortages of metal were relieved after 1865, though, that iron storage tanks became widely adopted in the region and at major shipping and receiving centers.[13]

Following a long-established practice for transportation of liquids, the industry originally packaged both crude and refined products in barrels for shipment to refining and marketing centers. Because of their cost, weight, and tendency to leak, wooden barrels were not, however, satisfactory or economical for transporting petroleum. One of the major

---

[12] Original data are from *Petroleum Industry of California*, California State Mining Bureau, Bull. 69, Sacramento, 1914. The FTC compiled drilling and completion cost data for 1914–19. See FTC, *Pacific Coast Petroleum Industry*, I, p. 137.

[13] Williamson and Daum, *American Petroleum Industry, 1859–1899*, pp. 183–194.

achievements of the industry during the 1865–1914 period was the gradual substitution of bulk handling methods for special packaging.

Gathering Lines

The first step in the evolution of bulk shipments came early in the 1860's when producers in the oil regions began experimenting with gathering lines. By 1865, the major technical problems had been solved, and the practice of linking producing wells by pipeline to storage facilities and shipping terminals several miles distant was firmly established.

Tank Cars

Until the mid-1860's, producers in the region had no choice but to barrel their product at storage tanks for shipment to market by one of two routes. One was to send the barrels by wagon over rough roads to the nearest railroad station some 25 miles to the north. The other and more popular route was to load the barrels on small craft for the journey down Oil Creek to the Allegheny River, where the cargoes were transferred to flat boats or rafts for transport down river to Pittsburgh.

The relative importance of the river system of transport began to decline when local railroads, closely linked with major truck lines, were extended into the region. Barreling of the crude at shipping terminals continued, however, until the advent of the railroad tank car about 1865. The early models consisted simply of wooden tanks mounted on flat cars and were subject to considerable leakage and the ever-present danger of fire. After 1868, that type of car was gradually replaced by the modern horizontal, boiler type of car, made of steel and equipped with a dome to allow the oil to expand without damaging the tank. Even the earliest tank car brought significant savings to shippers through reduction of losses from leakage and evaporation and, above all, through elimination of barrels costing several dollars each.[14]

Long-Distance Crude Oil Pipelines

By the early 1870's, the major technical bottlenecks in petroleum transportation appeared to have been broken. A system of gathering lines enabled most producers to run their crude from wells to railheads or riversides at one-tenth the cost formerly charged by teamsters. For longer hauls, three major trunk lines had more or less adequately linked the producing fields with refining centers. Even so, transportation costs were still among the industry's largest expense items. It was the possibility

[14] *Ibid.*, pp. 178–183.

of further reducing them that led to the development of the long-distance crude oil pipelines.

Preliminary attempts to construct such a line, beginning in 1874, revealed the difficulties involved. Capital requirements were large, and technical problems requiring solution involved pipelaying and the design of pumps which could provide sufficient pressure to boost the oil over hilly terrain. In addition, the early promoters encountered stubborn opposition from the railroads, which attempted to block pipeline construction in various ways, including purchase or lease of rights-of-way along proposed pipeline routes and the initiation of a variety of legal proceedings in court.

It remained for the Tidewater Pipe Company, backed by independent oil men who had good reason to think Standard Oil was receiving preferential rates from the railroads on its crude shipments, to overcome those obstacles. Five inches in diameter and built at an estimated cost of $750,000, the first crude oil line, stretching over the 100 miles from Corryville to Williamsport, Pennsylvania, began functioning in June, 1879.[15] Once its feasibility had been demonstrated, Standard Oil moved quickly into the construction of long-distance crude oil lines tapping the oil region, as did other independent producing and refining companies.

Pipelines were rapidly introduced into the new fields subsequently brought into production in Ohio, Indiana, the Gulf Coast, California, and the mid-continent region. There was also a trend toward increasing pipeline capacities by enlarging the diameter of pipe from the 5 to 6 inches typical of the 1880's. A growing number of new lines were constructed over the succeeding twenty-five years with 8-inch pipe and in some instances 12-inch pipe.[16] While some crude continued to move by tank car, the great bulk of oil produced from the 1880's on was shipped to refining centers by long-distance lines.

Ocean-Going Tankers

While a few sailing ships fitted with iron tanks began hauling crude oil across the Atlantic during the early 1860's, several factors long prevented bulk transportation from replacing barrels or tins in marine shipments. It was difficult, for example, to prevent movement of oil in tanks from throwing the ship out of balance, while vapor and leakage posed a serious fire hazard. Moreover, the volume of crude oil exports

---

[15] *Ibid.*, pp. 405–412, 440–444.
[16] *Ibid.*, pp. 443, 603; Williamson, Andreano, Daum, and Klose, *American Petroleum Industry, 1900–1959*, pp. 69–71 and 88–92.

was not large enough to attract serious attention of Americans to development of better tank ships for crude oil shipment, while barrels or tins were considered safer and more satisfactory for water-borne transport of refined products.

Actually, it was the attempt of the Russian industry to reduce the cost of transporting refined products to western Europe which led to the introduction of the modern type of ocean-going tankers. Following successful earlier experiments with wooden sailing tankers on the Caspian, the Nobel firm, located in Baku, put into service in 1885 two iron steam tankers, equipped with a multiple system of storage compartments.[17] It was not until after the discovery of the important Gulf Coast fields in 1900 that tankers became important carriers of crude oil to Atlantic coast refining centers. The adaptation of the construction features of the ocean-going tankers to barges also extended the bulk shipment of crude oil on the inland waterways of the United States.

Bulk Handling of Refined Products

All the objections to barreled shipments of crude—the weight and cost of the barrels, the tendency of the contents to leak or evaporate, and the problems associated with their return for reuse—also applied to shipment of refined products. The introduction of bulk shipments by railroad tank cars during the late 1860's was thus a major step in the evolution of a system which ultimately enabled marketers to move refined oil to final consumers without special packaging. Yet, for various reasons, barrels continued to play an important—if diminishing—role in the distribution of refined products throughout the 1865–1914 period.

The expansion of tank-car shipments, for example, necessitated a corresponding construction of storage facilities or bulk stations to receive the oil at major distributing points. Because barrels remained for some time the most convenient means of storing oil by retail outlets and retail customers, the initial effect of tank-car shipments was simply to transfer packaging or barreling operations from refineries to bulk stations. It was not until the 1880's that tank-wagon deliveries from major distributing centers began to extend bulk distribution to retail outlets and—by door-to-door deliveries—to retail customers. Tank-wagon deliveries, however, were economical only in relatively densely populated markets. In other areas, refined products continued to be distributed in barrels or drums.[18]

Overseas distribution of refined products followed much the same pattern. Initially the oil was either barreled or, if destined for shipment

[17] Williamson and Daum, *American Petroleum Industry, 1859–1899*, pp. 637–643.
[18] *Ibid.*, pp. 533–534.

to or through the tropics, put in five-gallon tins at the refineries. The initial effect of tank-car shipments, again, was largely to transfer packaging operations for export sales to the major ports. With the growing use of ocean-going tank steamers after the mid-1880's, bulk handling of refined products was increasingly extended to bulk storage facilities at the principal overseas receiving ports. As in the United States, the further extension of tank-wagon deliveries was primarily a function of the densities of the markets. By 1914, the method was quite widespread in western Europe, but for other parts of the world oil was packaged in barrels or tins at the ports for delivery to retail outlets and retail customers.[19]

In view of the growing significance of long-distance crude oil pipelines during that period, it is at first glance strange that there was no corresponding development of product pipelines before 1914, particularly since kerosene had been successfully transported several hundred miles from its source to Wilkes-Barre, Pennsylvania, through a line completed in 1893.[20] Several factors help explain the delay. One was a general feeling among oil men that piping refined oil over several hundred miles would damage the product. More important, however, were differences in the demography of the markets for crude and refined products. Since a large portion of the industry's refining capacity was located at major marketing centers, operators of crude oil pipelines terminating at such centers were assured of a demand for crude oil sufficiently large to cover operating costs. The fact that refining capacity was so concentrated tended to reduce the volume of refined products demanded at major marketing terminals below the amounts required for the economical operation of long-distance product pipelines.[21] It was not until the 1930's that the domestic demand for refined products was sufficiently large to make long-distance product pipelines economical.

Inputs

Lack of both data and of comparability of statistics make it impossible to provide more than a fragmentary analysis of the growth of inputs of capital and labor in the transport sector of the industry. It was reported, for example, that in 1865 there were "some 2,000 craft" engaged in the transport of crude down Oil Creek and the Allegheny River.[22]

While the size of their holdings is not known, railroads and their affiliated fast freight-forwarding companies apparently owned and operated

[19] *Ibid.*, pp. 653–657.
[20] *Ibid.*, pp. 572–573.
[21] *Ibid.*, pp. 532–533.
[22] *Ibid.*, p. 168.

most of the tank cars available for transport of petroleum for some two decades after the mid-1860's.[23] In 1890, when information on rolling stock was first collected, the number of railroad-owned tank cars—including those used to transport liquids other than petroleum—was 2,056.[24] Although the railroads added substantially to their holdings after 1890, in 1905 Standard Oil's tank-car fleet numbered approximately 10,000 and that of independents 3,000, compared with the 5,163 owned by railroads.[25] A rough estimate suggests that by 1914 there were between 20,000 and 25,000 tank cars available for the transport of petroleum—8,530 owned by railroads and the remainder by refiners and private carriers.[26]

TABLE 11

LENGTH OF GATHERING AND TRUNK PIPELINES, 1879–1910
(miles)

Year	Gathering Lines	Trunk Lines	Total
1879	372	322	694
1880	514	702	1,216
1890	2,457	2,933	8,390
1900	11,152	6,821	17,973
1910	23,804	16,226	40,090

Source: *Petroleum Facts and Figures*, API, 1959, p. 153.

Data on total pipeline mileage, first available in 1879, are shown in Table 11. The total capital investment in pipelines was roughly proportional to the increase in mileage, although construction costs varied considerably with the expense of obtaining rights-of-way, with the terrain, the size of pipe, and with pumping requirements. Estimates of the cost per mile of the early lines into the oil region ranged between $7,500 and $10,000; first lines in the Gulf Coast area cost between $5,400 and $6,500 per mile, compared with costs of approximately $10,000 per mile for lines into Ohio, Indiana, and the mid-continent.[27]

The growth of the world tanker fleet between 1900 and 1914 is shown

---

[23] *Ibid.*, p. 198.
[24] *Report on the Transportation Business in the United States*, Dept. of the Interior, *passim*; Eleventh Census of the U.S., 1890, Part I, *Transportation by Land*, 1895, p. 109.
[25] *Report of the Bureau of Corporations on the Transportation of Petroleum*, Washington, 1906, p. 78; *Statistics of Railways in the United States*, ICC, 1915, p. 25.
[26] *Statistics of Railways in the United States*, ICC, 1915, p. 25.
[27] Williamson and Daum, *American Petroleum Industry, 1859–1899*, pp. 443 and 603; Williamson, Andreano, Daum, and Klose, *American Petroleum Industry, 1900–1959*, pp. 88–92.

in Table 12. Note that in 1900 there were but three tankers under American registry, making up 2 per cent of the world tonnage, while in 1914 the American flag tankers accounted for only about 15 per cent of the total. As a measure of the actual number of tankers owned by American oil companies, these proportions are somewhat misleading. In 1909, for example, two of Standard Oil's European subsidiaries owned and operated 49 tankers.[28] This was a practice followed on a somewhat smaller scale by other American companies.

TABLE 12

PETROLEUM TANKER FLEET, OCEAN-GOING VESSELS OF 2,000 GROSS TONS AND OVER, 1900-14

Year	World Fleet		U. S. Registry		
	Number	Gross Tons (thousands)	Number	Gross Tons (thousands)	Percentage of Total
1900	109	356.5	3	7.6	2.0
1904	153	579.2	19	58.0	10.3
1909	194	778.3	31	108.9	14.5
1914	300	1,317.2	53	197.2	15.5

Source: *Petroleum Facts and Figures*, API, 1959, pp. 174-175.

Storage and Distribution Facilities

As already noted, increased bulk shipments of refined products by tank car or tanker necessitated a corresponding expansion of storage and distribution facilities at principal marketing centers. Standard Oil, for example, increased the number of bulk stations it owned and operated in the domestic market from 130 in 1882 to a reported 3,573 in 1906.[29] Further expansion by the Standard Oil companies and construction by independents must have brought the total to well over 5,000 by 1914. Because of differences in size, the growth of the number of bulk stations provides only a rough indication of capital investments in this sector of the distribution system. Apparently $20,000 was a representative figure during the late 1880's for investment in a metropolitan-area station which, in turn, distributed its products to outlying substations that might cost only a few hundred dollars.[30]

Transport Costs

Information on the relative costs of the various methods of transporting crude is no better than data on inputs. Rates for shipment down Oil

---

[28] Hidy and Hidy, *Pioneering in Big Business*, New York, 1955, p. 456.
[29] Williamson and Daum, *American Petroleum Industry, 1859-1899*, p. 690.
[30] *Ibid.*, p. 690.

Creek during the mid-1860's were reported to range "between $.15–$.75, and occasionally $1.00 a barrel," while the charge by river from Oil Creek to Pittsburgh varied "from $.40 to $1.00 a barrel." Depending on the distance and the condition of the roads, teamster charges for hauling to railroad or river shipping points might run as high as $4.00 per barrel. A fairly reliable estimate for 1862 suggests a total cost for delivering crude oil to New York from the oil region of $8.00 per barrel.[31]

Railroad tank-car shipments could undoubtedly be made at costs much below the combined water and rail costs in the early 1860's. Given the nature of the railroad competition particularly before 1887, however, there was no necessary connection between charges and costs. By the early 1870's the actual cost of rail shipments from the oil region to the Atlantic seaboard was close to 56 cents per barrel, although rates charged varied from 87 cents to $1.50 per barrel. Following the passage of the Interstate Commerce Acts in 1887, rail charges for bulk shipments of crude to the east coast apparently became stabilized at about 55 cents per barrel.[32]

Whatever the rail costs or charges, it is clear that tank-car shipments could not compete for shipments from areas served by pipelines. By the late 1880's, reasonably accurate estimates suggest a pipeline cost of shipping crude from the oil region to tidewater ranging between 21 cents and 25 cents per barrel; a decade later, it had dropped to 11–13 cents per barrel.[33]

How much tankers cut transport costs of refined products relative to barrels during that period is impossible to measure. According to estimates made in 1888, however, the port fees, inspection charges, loading, and other costs for a tanker carrying 20,000 barrels of petroleum products were some $2,300 less than shipping the same amount in barrels. Moreover, compared with a maximum number of three round trips across the Atlantic in one year for sailing ships, tankers by 1892 were averaging seven round trips.[34]

REFINING

Distillation and Treating

The two basic steps involved in converting crude petroleum into finished products during 1859–1914 were distillation and treating. Under the techniques developed by coal-oil refiners and adapted to petroleum, distillation was accomplished essentially by applying heat to crude oil in stills. As temperatures rose, the various fractions were drawn off in

[31] *Ibid.*, pp. 166, 169, 187.
[32] *Ibid.*, pp. 349–350, 436, 586.
[33] *Ibid.*, p. 586.
[34] *Ibid.*, p. 586.

the order of their relative volatility or density, starting with butane and pentane (to use modern terminology), followed in order by the naphtha fractions (gasoline, naphtha, benzine), then kerosene, light fuel oil (gas oil), lubricating stocks, and finally residuum and coke.[35]

The major objectives of refiners over the years after 1859 were to improve the distillation process, by reducing fuel consumption and costs and obtaining closer control over output of the various fractions. Among the highlights of this development were improvements in the design and firing of stills; successful experiments with the introduction of superheated steam into the stills as an alternative to direct firing; vacuum distillation; and, after 1900, introduction of continuous distillation to replace the long-established practice of "batch distillation." [36]

To prepare the distillate fractions for final consumption, it was necessary to subject them to further processing or treating—usually involving use of chemicals—in order to remove impurities and objectionable odors and to improve their appearance or color. By the mid-1860's, for example, the best quality illuminating oil was obtained by first mixing the kerosene fraction with small amounts of sulphuric acid in agitator tanks to rid the product of its bad odor. To eliminate most of the sulphuric acid, the mix was next put through a series of water washes, followed by the addition of caustic soda to further neutralize the acid. The final step was to run the mixture into bleaching tanks where, after several days, the caustic soda tended to settle to the bottom.[37] Naphtha distillates had to be treated by essentially the same processes.

Petroleum lubricants posed somewhat special treating problems. From the start of the industry, some heavy-duty lubricants were produced directly from crude without distillation. This was done by filtering selected types of crude through animal charcoal and bone black, a process long used to purify and improve the color and odor of all kinds of oils. Lighter lubricating oil suitable for use as cylinder oil, however, was produced from distilled lube stocks which had to be treated in much the same way as refined kerosene. Moreover, it was also necessary to remove the wax from the lube stocks by chilling and pressing.[38]

In contrast with the other major refined products, distilled fuel oil required virtually no additional treatment. Care had to be exercised to keep the light fuel oil fraction within acceptable limits of volatility and density, but residuum, also used as a fuel, required no such attention.

[35] *Ibid.*, p. 208.
[36] For a detailed description of these developments, see *ibid.*, Chaps. 9, 10, 18, and 23; and Williamson, Andreano, Daum, and Klose, *American Petroleum Industry, 1900–1959*, Chap. 4.
[37] Williamson and Daum, *American Petroleum Industry, 1859–1899*, pp. 222–223.
[38] *Ibid.*, pp. 242–246.

As with distillation, refiners attempted after 1859 to reduce costs and obtain better results from treating processes. Despite major advances, however, treating tended to remain more of an art than a science during that period—particularly with respect to lubricating oils.

Refining Technology

The state of technology and the improvements in refining techniques, which occurred after the major innovations of the first decade, cannot be discussed in the short compass of this paper. With the exception of the Frasch process and the early experience with continuous distillation, refining remained strictly a matter of separating the components of crude, with the result that product yields were dependent entirely on the characteristics of the available crude. The commercial success enjoyed by continuous distillation after the first decade of the twentieth century indicates that the method, if adopted in the pre-1900 years, would have brought about significant improvements in the size and efficiency of refining plants. The Frasch process, though still committed to a batchstill operation, was developed in direct response to the crude oil discovered at the Lima-Indiana field. This crude, known as "skunk oil" or sour crude, had exceptionally high sulphuric content and was not refinable into kerosene by the known technology used for eastern oils. Frasch, a German-born engineer, developed a refining process for the Standard Oil Company and implemented the management's earlier decision to become heavily engaged in the production and transport of Lima crude.[39]

In any case, refiners entered the new century with a refining system tied to the rigidities of batch operation and mere separation of the components of crude supplies as part of a formula which depended for flexibility primarily upon expansion of facilities and crude supplies. Most of their problems for two decades ahead were concerned with adapting that relatively inflexible refining system, not only to production of four instead of two major products, but also to processing a vast quantity of low-yield crudes from the Gulf Coast, California, and mid-continent crude oil producing areas.

Output Record

While much better than the evidence for other sectors of the industry, the data covering refining are far from complete. The only growth indicators covering the entire 1865–1914 period are estimated rates of refining capacity and outputs of major refined products. Beginning with the Census of 1880, data are also available for the number of plants, average annual employment, book value of capital invested, and gross

[39] *Ibid.*, pp. 475–582, 616–618.

TABLE 13

PETROLEUM REFINING: WORKING CAPACITY AND RATE OF UTILIZATION, SELECTED YEARS, 1864–1911

	1864–65	1872–73	1881	1884	1888	1895–97	1906	1911
Rated capacity, per day (hundred barrels)	11.7	47.6	97.8	124.9	140.3	220.8	331.5	482.3
Rate of capacity utilization (per cent)	—	88	90	78	85	94	104	93

Source: Williamson and Daum, pp. 291, 473, 486; Andreano, pp. 20, 61, 350.

Note: Rated capacity was what the industry claimed as its daily limits of crude intake. The resulting rated capacity was tied to an annual basis (300 days). A deduction of 25 per cent from annual capacity was then made to allow for stills under repair, cleaning operations, and stand-by capacity. That figure was called working capacity. The amount of working capacity above or below annual refining receipts of crude was then taken as a percentage of working capacity.

and net value added by refining. Data on rated refining capacity and rate of utilization for selected years between 1864–65 and 1911 are shown in Table 13. Two summary conclusions may be drawn from this material.

1. Total rated refining capacity doubled about every ten years. Peaks in construction followed, with a short time lag, peaks in crude oil production. This was especially true between 1878 and 1884, with the prolific crude discoveries in Bradford; between 1888 and 1895, as a result of Lima-Indiana; and after 1901 and until 1911, coinciding with the output expansion of petroleum from the Gulf Coast, mid-continent, and California fields.

2. The rate of utilization of capacity, after allowances for stand-by capacity and other technical adjustments, calculated for selected dates from 1873 to 1911, exceeded 94 per cent only in 1906 when the rate was very close to the limits of working capacity.

Total output of major refined products, as already indicated, expanded from 1.33 million barrels in 1865 to 188.88 million barrels in 1914 (see Table A-3). The gross value added by refining, as a percentage of the total value of the products at retail (see Table A-1), ranged roughly between 18 and 25 per cent, while net values varied between approximately 8 and 11 per cent.

Inputs

Data on the capital and labor employed in the refining sector of the industry are shown in Table 14. It should be noted that the data on capital refer to book value as reported in the Census and are not a very satisfactory measure of capital inputs. Similarly, the labor force data refer to the average number of workers employed during the year and therefore correspond only approximately to a total labor input.

Despite these limitations, it is clear from the following tabulation that

TABLE 14

CAPITAL AND LABOR IN REFINING SECTOR
OF PETROLEUM INDUSTRY, 1880-1914

	1880	1889	1899	1904	1909	1914
Book value of capital ($ million)	27.3	77.4	95.3	136.3	181.9	325.6
Wage earners	9,869	11,403	12,199	18,768	16,640	31,097

Source: *Census of Manufactures: 1900*, p. 363; *1919*, p. 757.

refining tended to become much more capital intensive from 1880 through 1914, with a roughly corresponding increase in physical output per worker.

	1880	1889	1899	1904	1909	1914
Capital, per worker (dollars)	2,766	6,788	7,812	7,262	10,931	10,470
Output, per worker (annual, in barrels)	1,282	2,420	3,950	2,964	6,449	5,878

Costs

The cost of physical inputs in the refining sector is a function of prices of inputs, product mix, and size of plant. While precise input costs cannot be found, it is possible to trace unit refining or manufacturing costs for plants of known size. The manufacturing costs were probably equivalent to variable costs, although clearly in some instances the estimates also included contributions to overhead.

From the early 1870's until the late 1890's, manufacturing costs declined by at least 20 to 25 per cent in real terms. After the mid-1880's, data are available for plant size and the mix of crudes as well. While no fundamental change in refining techniques occurred, variations in the quality of the crude changed the mix of inputs for treating and distillation. Manufacturing costs for Standard Oil of 0.534 of a cent per gallon of crude input in 1884 had declined to 0.39 of a cent by the mid-1890's. Greater use of Lima crude subsequently forced Standard's manufacturing costs up to 0.47 of a cent per gallon in 1900. Standard's costs for all plants, sizes, and product mixes declined again to 0.42 of a cent per gallon of crude input by 1911. These cost variations were principally a function of lower material prices, the relative ease in obtaining acceptable quality of refined products from eastern crudes, and the accumulated skill of the refiner's art.[40]

Manufacturing costs for the newer grades of crude other than eastern or Appalachian oil were somewhat higher, on the average, than those noted above. Estimates for 1906 show manufacturing costs of 1.5 cents per gallon of crude input for Gulf Texas crude, 1.8 for Lima crude, 0.8 for mid-continent, and 1.7 for California crude. These higher estimates reflect both increases in raw materials prices and, more importantly, the greater difficulty in obtaining acceptable quality of refined products because of the many technical problems encountered with crude qualities, radically different from the known refining practices used for eastern oils.

[40] These data and succeeding figures on manufacturing costs computed from U.S. *v.* SONJ, *Brief of Facts and Arguments for Petitioners*, I, p. 199; Report of the Commission of Corporations on the Petroleum Industry, Part 2, *Prices and Profits*, p. 54; *Report of the Industrial Commission*, I (testimony of Davis, Westgate, Lee, Emery, and Riu).

The median plant size in the industry increased several times over from the mid-1880's to the midpoint of the first decade after 1900, although wide variation in plant size persisted throughout the period. In the mid-1880's the range of plant capacities was from 40 barrels to 6,500–7,500 barrels per day of crude input. An informed guess of median plant capacities in 1880 put Standard's plants at 4,500 barrels per day and non-Standard at about 1,000 barrels per day.[41]

Economies of scale did not appear to be very extensive despite the highly skewed distribution of plant capacities. The manufacturing cost disadvantage of a non-Standard plant in the pre- and post-1900 periods could not have exceeded the range of 0.12 to 0.25 of a cent per gallon of crude input for crudes of similar grades. The comparison is somewhat misleading, of course, because the persistence of such a wide range of plant sizes can be better explained by the ability of firms to find market niches relatively safe from Standard Oil competition.

Greater refining efficiency was no doubt in part responsible for the drop, shown in the following tabulation, in the value added by materials (chiefly chemicals) other than crude used in the refining process. Most of the decline between 1880 and 1899, however, was apparently a reflection of lower costs of these materials.

Percentage Value Added by Materials (Other than Crude) to Total Retail Value of Refined Products

1880	1889	1899	1904	1909	1914
20.6	13.4	9.2	9.7	9.9	9.5

(Source: Table A-1.)

Construction Costs

In the mid-1880's, the construction costs for a refinery of the most efficient design, using Appalachian crude, were about $100 per barrel of rated daily crude capacity, in current prices. In the late 1880's, construction costs for a plant using Lima crude approximated $141 per barrel of capacity, at current prices. In the 1900–10 period, the only data available refer to the experience of the Standard Oil Company, Gulf Oil Corporation, and the Texas Company. For a refinery using mid-continent crude, the cost per barrel of capacity in 1906 at current prices ranged between $73 and $80; for California crude, $74; for Gulf Coast crude, about $260; for Illinois (about 1910), $176. By multiplying per barrel of capacity cost by median plant sizes of Standard and non-Standard refineries

---

[41] Williamson and Daum, *American Petroleum Industry, 1859–1899*, pp. 282–283, 482–484; Andreano, "Emergence of New Competition," pp. 336–344.

for 1906, it is possible to give rough estimates, shown in Table 15, of the construction costs of refineries of the most efficient design with differing crude mixes.[42]

Construction costs in the present context may also be considered as a first approximation of the total capital requirements a new firm would face at the refining level. Average capital investment at book value per refinery in 1904 was $1.4 million. A sample of 336 firms from a population of 1,344 firms, in *Moody's Investment Manual* for 1906, showed a mean

TABLE 15

ESTIMATES OF CONSTRUCTION COSTS OF REFINERIES, BY CRUDE MIX AND PLANT SIZE, 1906

Plant Size (barrels per day)	Construction Costs, by Type of Crude (thousand dollars)			
	California	Mid-Continent	Gulf Coast	Illinois[a]
1,800	133.2	131.4	468.0	316.0
2,400	177.6	175.2	624.0	422.0
6,500	481.0	474.5	1,690.0	1,114.0
7,500	552.0	547.5	1,950.0	1,320.0

[a] For 1910.

capital investment (total dollar face value of outstanding stocks and bonds) of $1.5 million and a mode of $1.10 million. This comparison suggests that to build *de novo* a plant of median size, total capital requirements were relatively close to the average investment of American manufacturing and industrial firms in general.[43]

III

Demand Structure for Final Products

OVER-ALL DOMESTIC AND FOREIGN DISTRIBUTION

The broad trends in demand for the major refined products of the American petroleum industry between 1865 and 1914 are shown in Chart 3. Illuminating oil was by far the most important refined product marketed by

[42] The capacity construction per-barrel costs were computed by dividing rated capacity into the published construction costs of new refineries. Data were found on contemporary costs of refineries in *Report of the Industrial Commission*, I, pp. 268–69, 278, 355–56, 570–72; *Petroleum Age*, May 1886; *Oil Paint and Drug Reporter*, July 24, 1888; *Moody's Investment Manual*, 1901–1909.

[43] Andreano, "Emergence of New Competition," pp. 346–348.

THE AMERICAN PETROLEUM INDUSTRY

CHART 3
Sales of Major Refined Petroleum Products, 1865–1914

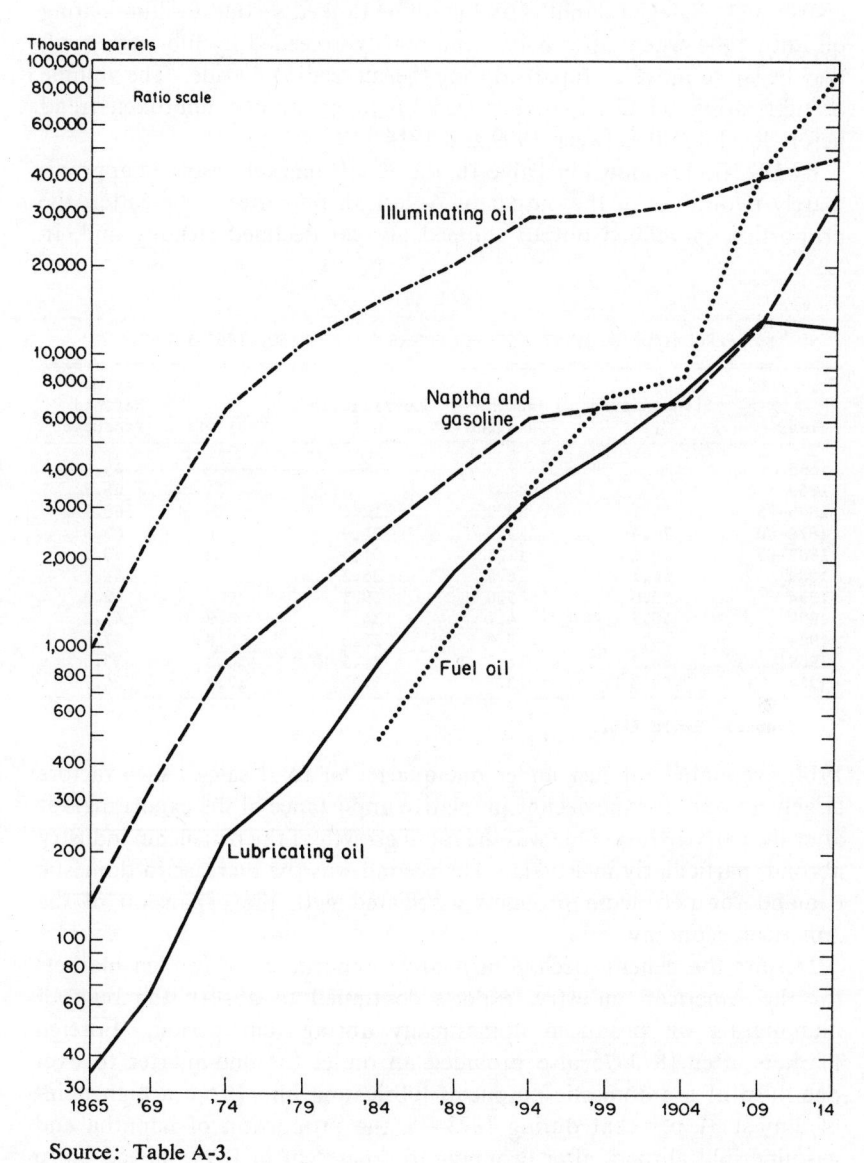

Source: Table A-3.

the industry throughout most of the period. Indeed, it was not until 1909 that the volume of fuel oil—output of which before that date had been relatively small—began to exceed the volume of illuminating oil distributed. Sales of naphtha and gasoline ranked second to illuminating oil until 1904 when, after being temporarily exceeded by lubricating oil, they began to move up rapidly during the succeeding decade. The volume of lubricating oil distributed moved up in an almost unbroken trend before leveling off between 1909 and 1914.

By 1872-73, as shown in Table 16, the export market absorbed approximately two-thirds of the output of American refineries. Thereafter, the proportion of refined output shipped abroad declined steadily and, in

TABLE 16

PERCENTAGE OF MAJOR REFINED PRODUCTS EXPORTED, 1865-1914

Year	Illuminating Oil	Naphtha-Gasoline	Lubricating Oil	Fuel Oil	Total Refined Products
1865					45.6
1869					66.7
1873-75	75.1	30.9	12.1		68.1
1878-80	70.4	25.7	23.9		63.8
1883-85	69.1	12.5	31.5		60.1
1889	65.1	8.4	36.2		51.2
1894	59.0	5.0	29.1		42.3
1899	58.5	4.1	34.7	6.9	41.2
1904	54.3	5.8	27.3	9.8	37.8
1909	64.2	11.6	26.5	10.3	33.0
1914	59.3	13.2	37.5	12.7	24.7

Source: Table A-3.

1914, accounted for just under one-quarter of total sales. Two factors largely account for the decline in relative importance of the export market after the early 1870's. One was the rapid growth of the petroleum industry abroad, particularly in Russia. The second was the increase in domestic demand for petroleum products associated with the expansion of the American economy.

Despite the general decline in relative importance of foreign markets for the American industry, exports continued to absorb the bulk of illuminating oil produced domestically during that period. Foreign markets after 1873-75 also provided an outlet for one-quarter to over one-third of the domestic output of lubricating oil. From a high point of almost 31 per cent during 1873-75, the proportion of naphtha and gasoline sold abroad, after dropping to 4 per cent in 1899, moved up to just over 13 per cent in 1914. Fuel oil exports, practically nonexistent before 1899, also accounted for almost 13 per cent of total fuel oil sales by 1914.

MAJOR FACTORS AFFECTING THE DEMAND FOR REFINED PRODUCTS

Illuminating Oil

The rapid growth and relative importance of illuminating oil production for the American petroleum industry reflected a virtually worldwide demand for a safe, efficient, and comparatively cheap source of artificial illumination. Early developments in the coal oil industry during the 1850's paved the way for kerosene, the petroleum illuminant, to meet that demand. Using processes pioneered in the manufacture of gas from coal, American and British coal oil producers by the end of the decade were turning out a lamp oil less expensive than sperm whale or lard oils, and much safer to use than the cheaper but highly volatile camphene—made from turpentine and alcohol—which had been quite widely adopted in the United States.[44]

Largely because of the necessity of reducing coal under pressure to produce crude coal oil for further processing, refined coal oil (averaging about 75 cents per gallon at wholesale during 1860) was still relatively expensive. Because of the report by Benjamin Silliman that crude petroleum was the approximate equivalent of crude coal oil, George H. Bissell and J. G. Evelyth initiated the series of events which led to the completion of the Drake well in 1859 and marked the birth of the petroleum industry in the United States.[45]

Once the difficult problem of storage and transport from the remote oil region of northwestern Pennsylvania had been solved, a rapid expansion of crude oil production in the area provided an abundant source of raw material for petroleum refineries. Despite various technical obstacles in adapting coal oil refining processes to petroleum, petroleum refining had emerged by 1862 as a firmly established branch of the industry. In that year, the average wholesale price (in gold) of kerosene in New York was approximately 32 cents per gallon.[46]

Because of its excellent illuminating qualities and because it was either cheaper or safer than other liquid illuminants, kerosene found a ready market among the owners of some 4 to 5 million lamps, already in use in the United States in 1862. The demand in the European market, which absorbed some 80–85 per cent of the illuminating oil shipped abroad

[44] Williamson and Daum, *American Petroleum Industry, 1859–1899*, pp. 33–34, 43–60.
[45] *Ibid.*, pp. 59, 63–81.
[46] *Ibid.*, pp. 202, 326.

during the 1860's, also reflected a familiarity with lamps, particularly among the more industrialized and urbanized areas of Great Britain and western Europe.

Several factors account for a subsequent increase in the total distribution of American-made kerosene from just under 1 million barrels in 1865 to 46.5 million barrels in 1914 (see Table A-3). A major contribution to the expansion was an approximate 75 per cent decline between 1865–69 and 1890–94 in the average price of kerosene, as reflected in New York wholesale quotations (Table 17). That decline was in part the result of

TABLE 17

AVERAGE NEW YORK WHOLESALE PRICE OF ILLUMINATING OIL AND NAPTHA-GASOLINE, 1865-1914
(cents per gallon)

Year	Price of Illuminating Oil	Price of Naptha-Gasoline
1865–69	26.6[a]	
1870–74	18.1[a]	
1875–79	13.3	
1880–84	8.1	6.0 (1884)
1885–89	7.3	6.0
1890–94	6.1	6.3
1895–99	6.9	7.5
1900–04	9.4	14.0 (1904)
1905–09	9.1	13.4 (1909)
1910–14	8.1	16.4

Source: Williamson and Daum, *American Petroleum Industry, 1859-1899*, pp. 326, 524, 680; Williamson, Andreano, Daum and Klose, *American Petroleum Industry, 1900-1959*, p. 172.

[a] Price in gold.

lower refining costs and improved handling and storage facilities. More important, however, in affecting delivered prices was the introduction of bulk shipments of crude and refined products.

The adoption of mass-production methods by domestic lamp manufacturers also contributed significantly to the expanding demand for kerosene by making cheap, efficient lamps available to low-income groups both at home and abroad. The evolution of a distribution system and aggressive marketing tactics, which extended kerosene deliveries to thousands of retail outlets throughout the world, provided an additional stimulus to the growing use of kerosene. These forces continued to increase the over-all demand for American-produced kerosene after 1890–94, despite an upward trend in wholesale prices, an expansion of the petroleum industry outside the United States, and growing competition from other sources of artificial illumination, notably manufactured gas and electricity.

Although electricity was eventually to capture the bulk of the illumination markets in the more economically advanced areas of the world, it initially competed with kerosene, as manufactured gas did, chiefly in urban communities in the United States and western Europe. Even in urban centers kerosene remained the "poor man's" illuminant long after 1914, as well as the major source of artificial illumination for a growing proportion of rural populations throughout the world. In other words, kerosene's illuminating qualities plus the requirement of special equipment for its use (lamps) tended to make the demand for the petroleum illuminant relatively inelastic in the short run, but both price and income elastic in the long run.

Naphtha and Gasoline

With the refining techniques available to the industry before 1914, there was relatively little flexibility in the proportion of yields of particular products from inputs of crude oil. As a result, the output of naphtha and gasoline, which grew from 133,000 barrels in 1865 to nearly 35 million barrels in 1914, was primarily a function of the expanding volume of illuminating oil production.

TABLE 18

DOMESTIC DISTRIBUTION OF PETROLEUM FUEL
(million barrels)

	1899	1904	1909	1914
Residuum and gas oil	6.75	7.67	37.21	77.7
Crude[a]	5.16	28.36	50.72	77.0
Total	11.91	36.03	87.93	154.7

Source: Williamson, Andreano, Daum, Klose, *American Petroleum Industry, 1900-1959*, p. 176.

[a] Skimmed or topped.

Initially a "waste" product, either thrown away or used by unscrupulous refineries or distributors to adulterate kerosene, naphtha and gasoline soon found an outlet for use as a solvent and as a cleaning fluid, but before the early 1890's only at prices ranging well below illuminating oil prices, as shown in Table 18. The rise in prices of naphtha-gasoline relative to kerosene between 1895–99 and 1900–04 appears to have reflected the extra refining costs of preparing naphtha-gasoline for special uses.

Until foreign sources were developed, based largely on the Russian industry, exports absorbed nearly one-third of the American output during 1873–75 and about one-quarter during 1878–80 (see Table 16). A further expansion in the domestic demand for solvents and cleaning fluids did much to improve the relative price position of naphtha and gasoline after the early 1890's—a demand further strengthened by the growing popularity of gasoline-burning cook stoves and portable space heaters. It remained for the development of the automobile after 1900, however, to make gasoline rather than kerosene the major product of the American petroleum industry. By 1909, the automotive demand absorbed an estimated 25 per cent of the total domestic distribution of gasoline; by 1914 the proportion was almost 40 per cent.[47]

Lubricating Oil

From the beginning of the industry, the lubricating potentials of petroleum attracted the attention of refiners. By the late 1860's crude petroleum—treated to reduce impurities—was already being used in limited quantities as a lubricant. More important for the future acceptance of petroleum lubricants, however, were Josiah Merrill's refining processes developed during the decade, which rid the petroleum product of its bad odor and improved its lubricating qualities.[48] Initially a high proportion of petroleum lubricants was blended with other oils to meet the requirements of users. This practice continued to be important, although further improvements in refining and treating during the succeeding half-century enabled petroleum lubricants to capture an increasing share of that market.

Petroleum lubricants unquestionably made a significant contribution to general economic growth during this period. There is considerable reason to doubt whether vegetable and animal oils, long used for the purpose, could have met the growing demand for lubricants which accompanied the rapid industrial and commercial expansion, particularly in the United States and western Europe, during those years. The general nature of that demand is reflected in total sales of lubricants, which grew from 35,000 barrels in 1865 to almost 12.5 million barrels in 1914, and in exports (principally to Europe) which, after 1878–80, as already noted, ranged between one-quarter and one-third of the total.

The demand for lubricants was primarily a function of (1) an increase in the use of machines and equipment with moving parts in industry,

[47] Williamson, Andreano, Daum, and Klose, *American Petroleum Industry, 1900–1959*, p. 198.
[48] Williamson and Daum, *American Petroleum Industry, 1859–1899*, pp. 245–246.

agriculture, and transportation and (2) cyclical fluctuation in industrial and commercial activity. Because lubricating costs were relatively small, compared with total operating costs, it is unlikely that price had any significant influence on the quantities of lubricating oil demanded.

Fuel Oil

There were persistent efforts during the decade of the 1860's to promote the use of partially refined crude oil, or the heavier fractions resulting from the distillation processes, as a source of energy to compete with wood and coal. The industry had high hopes that experiments by the United States Navy with petroleum-fueled boilers would establish the greater efficiency of their product. Despite advantages in storage and reductions in labor needed to tend the boilers, the Navy tests revealed that, with the equipment available, operating costs were some eight times greater for petroleum than for coal. Similar results from attempts to utilize petroleum on river steam boats and for industrial plants during that period indicated that cost factors weighed heavily against its use.[49]

An important market did develop for "gas oil," the distillate fraction obtained after the kerosene fraction in the refining process. As the name suggests, gas oil was used an an "enricher" in the production of manufactured gas. While gas oil continued to account for a substantial portion of distilled fuel oil sales, the market for the heavier fuel oils was slow to develop before the late 1880's.

Discovery and development of the Lima field in Ohio and Indiana starting in the mid-1880's marked the beginning of a growing use of petroleum as a fuel in the United States. Heavily laden with sulphur, Lima crude oil proved initially very difficult to refine. Selling (at the well) at prices ranging between 15 and 48 cents a barrel during 1887–94, crude oil "skimmed" or "topped" to evaporate or recover the light ends (gasoline and naphtha) found a ready market among industrial users in the upper Middle West.[50]

The rapid growth of production in California during the late 1890's gave further stimulus to use of petroleum as fuel. California oil, an asphaltic-based crude which was difficult to refine under available techniques and was also skimmed or partially refined, captured an increasing share of the West Coast fuel market, previously largely dependent on expensive imported coal.[51]

[49] *Ibid.*, pp. 240–242.
[50] *Ibid.*, pp. 601–603.
[51] Williamson, Andreano, Daum, and Klose, *American Petroleum Industry, 1900–1959*, p. 175.

By the early 1890's, advances in refining technology—particularly the processes developed by Herman Frasch—made possible more satisfactory yields of illuminating oil from Lima crude. Because of their sulphur content, the heavier fractions could not be used to produce lube stocks, with the result that approximately one-half the output of refineries processing Lima crude was distilled fuel oil.[52] This output contributed substantially to the total domestic distribution of distilled fuel oils, which grew from about 483,000 barrels in 1883–85 to 6.75 million barrels in 1899 (see Table A-3).

Discovery of the prolific field in the Gulf Coast area in 1900 marked a third major impetus to a growing use of petroleum as fuel. Gulf Coast crudes, like those in California, had an asphaltic base initially difficult to refine. Selling at well-head prices ranging between 18 cents and 94 cents a barrel during 1900–14, a high proportion of Gulf crude production, skimmed or partially refined, was also distributed to fuel oil users.[53] As indicated in Table 18, the quantities of unprocessed or skimmed crude oil sold as fuel exceeded residuum or gas-oil sales by substantial margins between 1904 and 1909. By 1914, improvements in refining technology and expansion of refining facilities, particularly in the Gulf Coast area, had combined to reduce the proportion of the domestic fuel oil market supplied by skimmed or partially refined crude.

Aside from its use as a gas enricher, petroleum fuel oil was primarily sold in competition with coal or as an alternative or substitute for boiler fuel to produce heat and energy. Petroleum's ability to compete with coal on a price basis in the domestic market was much greater in certain areas than in others, as already suggested. In 1909, for example, total fuel oil sales were divided by major geographic areas as follows: Pacific Coast, 41 per cent; Southwest, 27 per cent; Atlantic Coast, 16 per cent; Upper Mississippi, 9 per cent; and all other, 7 per cent.[54]

A division of total fuel oil sales by types of major user in 1914 shows the following: industry and commercial, 50 per cent; railroads, 33 per cent; manufactured gas, 10 per cent; Merchant Marine and Navy, 7 per cent.[55] In large part, the high proportion of sales to industrial and commercial users reflected the ready adaptability of their burning equipment to fuel oil in response to a relatively small price differential. While the conversion of coal-burning locomotives to fuel oil was more expensive, the relatively low price of fuel oil, plus greater operational flexibility, prompted the railroads operating west of the Mississippi to make an early switch to

[52] Williamson and Daum, *American Petroleum Industry, 1859–1899*, p. 619.
[53] Williamson, Andreano, Daum, and Klose, *American Petroleum Industry, 1900–1959* pp. 39, 176.
[54] *Ibid.*, p. 177.
[55] *Ibid.*, p. 178.

TABLE 19

ANNUAL ENERGY SUPPLY FROM MINERAL FUELS AND
WATER POWER, SELECTED YEARS, 1899-1914
(trillion British thermal units)

	Total	Water Power	Coal	Petroleum	Natural Gas
1899	7,529	238	6,708	342	240
Per cent	100	3.2	89.1	4.5	3.2
1904	10,680	354	9,291	702	333
Per cent	100	3.3	87.0	6.6	3.1
1909	14,284	513	12,155	1,099	517
Per cent	100	3.6	85.1	7.7	5.6
1914	16,513	636	13,545	1,595	636
Per cent	100	3.9	82.0	9.7	3.9

Source: *Historical Statistics*, 1960, p. 355.

petroleum fuel. Of particular interest was the growing use of petroleum fuel in marine transportation. Experiments beginning in the 1890's had demonstrated the economy of substituting petroleum for coal in modern ocean-going vessels. By 1914, conversion from coal to fuel oil by the American and British navies was already well under way.[56]

Fuel oil sales were the major factor in the growing importance of petroleum as a source of energy for American industry. As shown in Table 19, petroleum's share in the total energy generated from mineral combustion and water power in the United States more than doubled between 1899 and 1914.

CHANGES IN THE RELATIVE VALUE OF MAJOR REFINED PRODUCTS

There was a marked shift in the relative contributions to total value at refineries of each of the major refined products between 1879 and 1914 (Table 20). The relative position of illuminating oil declined from over

TABLE 20

VALUE AT REFINERIES OF MAJOR REFINED PRODUCTS AS A
PERCENTAGE OF TOTAL VALUE, 1879-1914

	1879	1889	1899	1904	1909	1914
Illuminating oil	86.2	74.3	66.9	62.2	44.7	26.5
Gasoline and naphtha	6.8	11.1	14.4	14.5	18.8	33.8
Lubricating oil	7.3	12.7	12.1	17.0	19.2	16.3
Fuel oil	0.7	1.9	6.5	6.3	17.3	23.4

Source: See notes to Table A-1.

[56] *Ibid.*, pp. 181–183.

86 per cent of the total to less than 27 per cent during those years, in contrast to gasoline, which by 1914 contributed nearly 34 per cent, and fuel oil, which contributed over 23 per cent of the total.

IV

Changes in Industry Structure

In 1878, the Standard Oil group achieved the peak of its control over each sector of the industry as it was then structured. At that time, the center of crude oil production was in the western regions of Pennsylvania. To a limited extent, crude was produced elsewhere but some 99 per cent of the crude supplies used to make the major petroleum product—kerosene—came from Pennsylvania. In the mid-1880's, with the discoveries in Ohio and Indiana, the locus of crude oil production moved successively farther west, into mid-continent, California, Gulf Coast, and then Illinois. By 1911, total crude oil production was divided among the major producing areas as follows: Appalachian, 11 per cent; Lima-Indiana, 3 per cent; Gulf Coast, 5 per cent; California, 37 per cent; mid-continent, 30 per cent; and Illinois, 14 per cent.[57] These shifts in the sources of crude production were caused by the dynamic pattern of flush field development, noted earlier, and constituted one of the major factors affecting the structure of the industry.

Quite aside from changes in industry structure attributable to general economic growth, the circumstances leading to entrance of new firms into the industry may be put in three main groups:

1. Growth in the number and location of new flush fields that, because of minimal barriers to entry posed by absolute capital requirements and economies of scale in production and refining, enabled established firms to expand and new firms to obtain crude supplies necessary to begin operations.
2. The quality and quantity of crude oil discoveries, which played an important role in types of product made and sold and, in turn, facilitated entry through the process of market segmentation.
3. Incorrect, or insufficiently rapid, market response on the part of the dominant firm (Standard Oil), which left market opportunities or market space, not only for the exploitation of crude deposits, but also for the distribution of "old" and "new" products.

Changes in Standard Oil's relative position in the industry between 1880 and 1911 are summarized in Table 21. In crude oil, for example,

---

[57] Arnold and Kemnitzer, *Petroleum in the United States*, p. 33.

TABLE 21

SUMMARY OF STANDARD OIL'S POSITION
IN THE AMERICAN PETROLEUM INDUSTRY,
1880-1911

	1880	1899	1906	1911
PERCENTAGE CONTROL OVER CRUDE OIL SUPPLIES				
Appalachian[a]	92	88	72	78
Lima-Indiana[a]	--	85	95	90
Gulf coast[b]	--	--	10	10
Mid-continent[c]	--	--	45	44
Illinois[d]	--	--	100	83
California[e]	--	--	29	29
PERCENTAGE CONTROL OVER REFINERY CAPACITY				
Share of rated daily crude capacity	90-95	82	70	64
PERCENTAGE OF MAJOR PRODUCTS SOLD				
	1880	1899	1906-1911[f]	
Kerosene	90-95	85	75	
Lubes	--	40	55	
Waxes	--	50	67	
Fuel oil[g]	--	85	31	
Gasoline	--	85	66	

Source: Harold F. Williamson and Ralph L. Andreano, "Competitive Structure of the American Petroleum Industry, 1880-1911," *Oil's First Century*, 1960, p. 74.

[a] Share of pipeline run.

[b] Consumption at portside.

[c] Prairie Oil and Gas Company share of crude available (current production plus stock).

[d] Ohio Oil Company pipeline runs, as per cent of total field production.

[e] Standard did not enter the California sector until 1900. Figures are company output as per cent of total field production.

[f] Data available for Standard are 1906 and 1910; all other firms, 1908. Both the lube and wax estimates may thus be overstated.

[g] Include residual fuel oil and unrefined crude sold as fuel.

Standard maintained its predominant control over production in the Appalachian, Lima-Indiana, and Illinois fields between 1880 and 1911. The proportion of crude from California, mid-continent, and Gulf Coast fields going to independents had, by 1911, reduced Standard's share of total domestic production during the period from over 90 per cent to approximately 60–65 per cent. This decline in the company's relative share of domestic crude production was closely matched by a decrease in the proportion of total refining capacity under Standard's control from approximately 90–95 per cent in 1880 to 60–65 per cent in 1911. The impact of these shifts on Standard's general marketing position was varied. By 1911, however, the independents were supplying nearly 70 per cent of the fuel oil, some 45 per cent of the lubricants, one-third of the gasoline and waxes, and about one-quarter of the kerosene distributed by the American petroleum industry.

These changes, coupled with the emergence of a dozen or more integrated companies, plus a large number of smaller ones specializing in one or more phases of the oil business, suggest a competitive structure in 1911 which was a far cry from that of the late 1870's, when Standard Oil and the American petroleum industry were practically synonymous. The circumstances and processes by which new companies entered the industry during the period before 1911 may be illustrated by developments associated with the opening of the major new crude producing areas.

APPALACHIAN

Standard Oil's position in the American industry in the late 1870's was based essentially on ownership (or lease) of some 90 per cent of domestic refining capacity, plus virtually a monopoly control of facilities for gathering and transporting crude oil. But even Standard found it impossible between 1876 and 1882 to expand its gathering lines, storage, and transport facilities rapidly enough to accommodate an approximate trebling of output of crude, largely from Bradford, the first modern flush field, in the Appalachian oil region.

The result was a sufficient supply of crude outside Standard's control to provide the basis for the emergence of two types of firm: (1) fully integrated companies producing a complete line of refinery products, such as the Tidewater Oil Company and a group subsequently merged to form the Pure Oil Company; and (2) companies partially or fully integrated including the Union Petroleum, Crew-Levick, and Pennzoil companies, which specialized in producing lubricating oils and wax.

The immediate impact of Bradford on Standard's position in the industry is reflected in Standard's share of total refining capacity, which fell from about 90 per cent in the late 1870's to approximately 75 per cent by 1884. Some of the loss was subsequently regained and, for the entire 1880-99 period, Standard's share of total refining capacity declined about 10 percentage points; its share of Appalachian crude production over the same period dropped from 92 to 88 per cent. The most important effect of Bradford (supplemented by the growth of production in Lima-Indiana), however, was to reduce Standard's share of the production of lubricants and waxes from 75 per cent or more in 1880 to approximately 40 per cent of lubricants and 50 per cent of waxes by 1899.

LIMA-INDIANA

Production in the Lima-Indiana fields, which began to expand in the mid-1880's and reached a peak in 1896, prompted Standard Oil to move quickly into the area with an extensive system of gathering lines, storage facilities, and crude trunk lines. Even so, Standard's control over output which, measured by pipeline runs, was approximately 93 per cent in 1894, declined to about 85 per cent by 1899. Chiefly because of the quality of the crude (sold largely as fuel oil until the development by Standard of the Frasch process) and the costs of entry, Standard Oil was the only established company to move into the Lima-Indiana fields.

The Lima-Indiana fields did, however, provide an opportunity for new entrants, including the Sun Oil Company, the National Refining Company, and the Paragon Oil Company. These companies were fully integrated, with their own production, pipelines, and refining and marketing facilities. While fuel oil was their major product, adoption of refining techniques similar to the Frasch process enabled them all to produce a full product line by 1899.

While the independents' share of the market was relatively modest—probably no more than 15 per cent of fuel oil and much less of other products—the experience gained in exploiting the industrial demand for petroleum fuel oil was an important factor in subsequent decisions to move into the Gulf area after the turn of the century.

GULF COAST

In contrast to its action in Lima-Indiana, Standard Oil did not formally enter the Gulf Coast region, where crude production reached its peak about 1905. Three reasons have generally been cited in explaining Standard's response: (1) the legal climate in Texas, where an antitrust action had been instituted against the Waters-Pierce Company, Standard's major marketing affiliate in the state; (2) the refining quality of the Texas crude, which yielded relatively small amounts of kerosene, the product Standard was primarily interested in refining; and (3) the feeling on the part of Standard officials that mid-continent and California offered better investment opportunities than Gulf, both in quantity and quality of crudes needed for their operations. Standard's role in the Gulf Coast development was thus limited to the purchase of an estimated 10 per cent of the output from the region.

With some 90 per cent of Gulf Coast crude output available for "outsiders," the impact on the structure of the industry—and on Standard's position—was quite significant. Of the already established firms in the Appalachian and Lima-Indiana regions, Sun Oil Company moved most rapidly to use Gulf crude as a springboard for a highly successful program of expansion, which made the company one of the leading factors in the industry, particularly in the production and distribution of fuel oil and lubricants. Pure Oil was also able to acquire a modest interest in the Gulf fields.

The most significant structural changes stimulated by Gulf Coast developments, however, were attributable to *de novo* entrants into the industry. These were of two types. First, there were fully integrated companies, such as Gulf Oil and the Texas Company, which attempted to produce a full product line but were more successful for several years as distributors of fuel oil. Second, there were the fuel oil companies—about twenty in all—which did not operate refineries (because the crude oil could be sold as fuel after exposure to the sun had evaporated the light ends) but were integrated from production to marketing facilities. The increasing importance of independents in the expanding fuel market was reflected by a decline in Standard's relative market share from about 85 per cent in 1899 to 31 per cent in 1906–11.

MID-CONTINENT

The mid-continent region provided the basis for an even more significant expansion of established independents and the entry of new firms. Not only did the strike at Glenn Pool in 1905 establish the region as a major producing area in the United States, but also the crude was an Appalachian type yielding relatively large proportions of gasoline and kerosene as well as high-quality lube stocks.

Standard Oil understandably was quick to acquire leases and to build gathering lines, storage facilities, and trunk pipelines in mid-continent. Other companies were also quick to move in. Among the leaders were the firms already operating in the Gulf region: notably, Gulf Oil and the Texas Company; National Petroleum and Sun Oil, which had started in Lima-Indiana; Associated Oil Company, a California concern; Union

TABLE 22

SIZE CHARACTERISTICS OF OKLAHOMA-KANSAS INDEPENDENTS, 1906-11

Name of Company	In Production	Daily Crude Oil Capacity of Refineries (42-gallon barrels)	Pipeline Mileage	Number of Railroad Tankers	Number of Distribution Stations
American Refining	Yes	1,500	60	290	33
Chanute Refining	Yes	1,750	128	335	610
Cudahy Refining	Yes	6,000	24	40	1
Great Western Refining	Yes	2,000	43	83	3
Indiahoma Refining	Yes	1,200	---	42	---
Kansas City Refining	No	1,000	60	52	---
Kansas Oil Refining		1,800	6	30	20
Kansas Co-operative Refining	Yes	500	33	41	31
Kanotex Refining	Yes	1,000	65	370	---
Milliken Refining		4,000	18	49	8
Muskogee Refining	Yes	1,440	110	315	50
National Refining	Yes	3,000	7	40	50
Oklahoma Refining		600	170	182	---
Petroleum Producing	Yes	3,500	40	---	---
Paola Refining	Yes	150	35	69	---
Sapulpa Refining	No	3,000		88	7
Uncle Sam Oil	Yes	1,400			

Source: Williamson and Andreano, *Oil's First Century*, p. 78. Basic data compiled from *Oil Investor Journal*; 1906 (Nov. 18, p. 2), 1907 (May 5, p. 4), 1908 (Jan. 14, p. 19, Feb. 5, pp. 3, 4, March 19, p. 12, and April 5, p. 6); and *Oil and Gas Journal*, 1910 (June 16, pp. 23, 24, and July 14, p. 18).

Petroleum; Crew-Levick; and Pure Oil and Tidewater, pioneer processors of Appalachian crudes. In addition, the mid-continent provided the basis for the emergence of almost a score of partially or fully integrated firms (see Table 22).

With Standard's control over mid-continent crude production limited to about 45 per cent during 1905–11, the effect of the growth of independents was to extend competition beyond the fuel oil market into the sale and distribution of kerosene, gasoline, and lubricants. The results of that expansion were most noticeable in the reduction between 1899 and 1906–11 of Standard's share of kerosene distribution from about 85 per cent to 75 per cent and of gasoline from about 85 per cent to approximately 66 per cent.

ILLINOIS

Standard's relative position in the industry would no doubt have deteriorated even more by 1906–11, if the organization had not succeeded in controlling some 85 per cent of the Appalachian-type crude produced in the Illinois fields, which reached their peak during 1907–11. Yet the remaining 15 per cent of the Illinois output was sufficient to make worthwhile Tidewater's acquisition of production in the area and extension of its trunk pipeline from Pennsylvania to southeastern Illinois. That output also enabled the Indian Refining Company (later absorbed by the Texas Company) to emerge by 1911 as a fully integrated concern.

CALIFORNIA

The structure of the petroleum industry in California up to 1911 followed a pattern of development radically different from the evolution of the industry east of the Rockies. It is true that, by extending its marketing organization into California during the 1880's, Standard Oil remained the dominant distributor of kerosene on the west coast until 1911. But the asphaltic-based California crude, ill-suited for production of kerosene or lube stocks, had little attraction for Standard, which acquired no producing properties, pipelines, or refineries in California until 1900. By that date, production in California—which by 1906 was as important quantitatively as output from mid-continent—was already split more or less equally among some seven integrated companies engaged primarily in the production and distribution of fuel oil.

Size Distribution of Firms[58]

Against that broader background of entry of new firms and changes in industry structure, it is possible to give a qualitative assessment of industry

[58] Data in this section from Andreano, "Emergence of New Competition," pp. 280–294.

concentration, by product and function, and to appraise the size of firms in the industry. This impression relates to the situation as it existed just before 1911.

Illuminating Oil. The largest three firms—Standard, Gulf, and the Texas Company—accounted for close to 90 per cent of industry output. The combined output of the largest seven firms—top three, plus Tidewater, National Refining, Crew-Levick Refining, Union Petroleum, and Cudahy Oil and Refining—was approximately 99 per cent. A fringe of twenty-five to thirty-five firms accounted for the remainder.

Lubricants. In all classes and grades of lubricants, the largest three firms—Standard, Sun Oil, and Texas—produced about 75 per cent of total industry output. The seven largest—top three, plus Gulf, Crew-Levick, Union Petroleum, and National—accounted for perhaps 90 per cent, with the remaining 10 per cent divided among thirty or forty firms.

Waxes. Three firms—Standard, Union Petroleum, Crew-Levick—produced 75 per cent of the total output. The national market share of no other firm exceeded 3-4 per cent, while the remainder of the industry output was accounted for by some fifty to sixty firms.

Gasoline. The largest firms—Standard, Gulf, Texas—produced 85–95 per cent of total industry output. The largest seven—top three, plus Cudahy, National, Union Petroleum, and Indian Refining—accounted for 90–98 per cent and thirty to forty firms for the remainder.

Fuel Oil. The top three—Standard, Union Oil of California, and Associated Oil of California—accounted for 50–60 per cent of total industry output. The largest seven firms—top three, plus Sun, Gulf, Texas, Higgins Oil and Fuel of Beaumont—produced 70–90 per cent of the total and a fringe of about fifty firms the remainder.

Crude Oil. The largest three companies—Standard, Gulf, and Union Oil of California—probably produced no more than 20 per cent of total U.S. output of crude in 1911. The largest seven—top three, plus Texas, Southern Pacific Railroad, Sun, and Tidewater—produced 25–30 per cent, and the remaining crude output was accounted for by several thousand separate firms.

Transportation Capacity. Gathering and trunk pipelines, railroad tank cars, river barges, and ocean and coastal tank steamers comprised the bulk of transportation facilities for crude oil and products. Only in trunk pipelines, however, is any impression of firm concentration possible. Nine firms—Standard, Gulf, Texas, Union of California, Associated of California, Tidewater, Cudahy, Sun, and National—owned virtually the total national trunk pipeline mileage.

Refining Capacity. The top three—Standard, Gulf, and Texas—operated or owned 80–85 per cent of total industry-rated daily crude oil capacity.

The largest seven firms—top three, plus Sun, Union Oil of California, National, and Cudahy—accounted for 90–95 per cent of total industry capacity. About forty firms—one of them Indian Refining, only slightly smaller than the smallest of the top seven—owned or operated the remaining capacity.

Over-All Firm Size. Except in a few instances, there are no specific data on the extent of vertical integration by individual firms. It is certain, however, that no refining company could successfully compete in national or regional markets without a minimum degree of vertical depth, backward at least to the extent of owning its own transportation capacity and forward to the ownership of bulk stations. An examination of the total crude inputs supplied from internal sources as a measure of vertical integration would show that the largest firms, ranked by assets, all had high crude-sufficiency ratios. Some, of course, had much higher ratios than others. Of the top four—Standard, Gulf, Texas, and Union Oil of California—for example, the firm with perhaps the least crude oil sufficiency ratio was Standard. The next four—Southern Pacific, Sun, Tidewater, and Union Petroleum—all had relatively high crude oil sufficiency ratios. The next largest five firms were all vertically integrated, although, as shown earlier, the extent of their integration is somewhat uncertain. These were National Refining, Cudahy, Crew-Levick, Higgins Oil and Fuel, and Indian Oil and Refining.

Appendix

TABLE A-1

AMERICAN PETROLEUM INDUSTRY: VALUE ADDED BY SECTORS, 1879-1914
(thousand dollars)

Sector	1879	1889	1899	1904	1909	1914
Value of crude at well, shipped to refineries	16,198	23,579	58,772	57,605	84,543	154,923
Gross value added by refining	27,364	40,122	43,505	67,518	85,690	146,633
Value added by transportation and storage of crude and refined products	37,337	56,751	69,319	115,942	184,310	284,629
Marketing, wholesale	14,378	27,169	36,598	56,695	79,234	118,870
Retail	10,037	22,338	34,464	48,389	73,603	124,597
Total	105,314	169,959	242,658	345,149	507,380	829,652
Value of crude at well,						
Exports	814	1,770	3,666	2,277	2,839	2,405
Fuel	--	--	6,330	24,380	35,510	62,370
Final value of industry output	106,128	171,729	252,154	372,806	545,729	894,427

Source: Data on crude and refining sectors are from *Census of Manufactures*, 1914, Vol. 11, pp. 582-583. Data on marketing and transport are estimated. See Table A-2.

TABLE A-2

AMERICAN PETROLEUM INDUSTRY: AVERAGE COST OF STORAGE AND TRANSPORT OF CRUDE AND REFINED PRODUCTS, 1879-1914
(dollars)

Average Cost of Storage and Transport Per 50-Gallon Barrel	1879	1889	1899	1904	1909	1914
Crude, well to refinery[a]	(1.09)	0.82	0.51	0.88	0.67	0.58
Refined products, refinery to wholesaler[b]	1.25	1.00	0.80	0.80	0.80	0.80
Refined products, wholesaler to retailer[b]	0.60	0.50	0.40	0.40	0.40	0.40

[a] Computed from data in *Census of Manufactures*, 1949. Average costs for 1879 were computed (see explanation in text).

[b] These costs are assumed.

TABLE A-3

PRODUCTION, EXPORTS, AND DOMESTIC DISTRIBUTION OF MAJOR REFINED PRODUCTS, 1865-1914
(thousand barrels)

Product	1865	1869	1873-75	1878-80	1883-85	1889	1894	1899	1904	1909	1914
Illuminating oil											
Production	992	2,450	6,530	10,800	15,171	20,191	29,458	29,500	32,500	40,000	46,500
Domestic			1,626	3,197	4,696	7,054	12,068	12,250	14,842	14,319	18,946
Export			4,903	7,603	10,475	13,137	17,390	17,250	17,658	25,681	27,554
Per cent exported			75.1	70.4	69.1	65.1	59.0	58.5	54.3	64.2	59.3
Naphthas and gasoline											
Production	133	326	895	1,483	2,442	3,916	6,114	6,650	6,900	12,850	34,800
Domestic			618	1,101	2,079	3,583	5,744	6,224	6,498	11,357	30,218
Export			277	382	363	333	370	426	402	1,493	4,582
Per cent exported			30.9	25.7	12.5	8.4	5.0	6.4	5.8	11.6	13.2
Lubricating oil											
Production	35	65	226	376	884	1,835	3,288	4,781	7,470	12,964	12,488
Domestic			199	286	606	1,170	2,332	3,124	5,626	9,531	8,800
Export			27	90	278	665	956	1,657	2,114	3,433	4,688
Per cent exported			12.1	23.7	31.5	36.2	29.1	34.7	27.3	26.5	37.5
Fuel oil											
Production					483	1,195	3,522	7,250	8,500	41,500	89,000
Domestic					483	1,195	3,522	6,750	7,669	37,211	77,687
Export								500	831	4,281	11,313
Per cent exported								6.9	9.4	10.4	12.7
Total production	1,336	3,267	7,650	12,656	18,498	27,600	44,300	48,181	55,640	107,314	182,880
Total domestic	727	2,279	2,442	4,582	7,381	13,465	25,583	28,336	34,635	71,907	137,743
Total exports	609	1,089	5,209	8,074	11,117	14,135	18,717	19,845	21,005	35,407	45,137
Total per cent exported	45.6	66.7	68.1	63.8	60.1	51.2	42.3	41.2	37.2	33.0	24.7

Source: Williamson and Daum, *American Petroleum Industry, 1859-1899*, pp. 489, 633, 678, 737-739; Williamson, Andreano, Daum and Klose, *American Petroleum Industry, 1900-1959*, p. 169.

METHODOLOGICAL NOTE

Although the basic framework underlying the computations is algebraic, a major obstacle to algebraic presentation is the somewhat difficult notation that results from the spatial and multiproduct characteristics of the American petroleum industry.

Value of Output of the Crude Oil Sector

Volume of crude oil shipped to refineries, shipped directly to wholesalers as fuel, and exported were obtained from *Census of Manufactures*, *Historical Statistics*, 1960, and Williamson, Andreano, Daum, and Klose, *American Petroleum Industry, 1900–1959*. These quantities were added and their sum multiplied by the price (average value) of crude at wells to obtain the contribution of the crude oil sector to the industry.

Value Added by Storage and Transport of Crude from Wells to Refineries

The difference between the delivered cost of crude to refineries and the value (at wells) of crude shipped to refineries was used as a substitute for value added by storage and transport. This formula was not used to compute the value added for 1879. In that year, 17.4 million (42-gallon) barrels of crude were shipped to refineries, but value added provided by the formula amounted to only $145,000, or an average cost of storage and transport of less than 1 cent per barrel. To avoid this difficulty, value added for 1879 was assumed to be $16 million. This assumption provided average costs of shipment per barrel (50 gallons) of $1.09. Table A-2 indicates that the average costs for 1879 seem reasonable when compared with similar shipment costs for other years.

Gross Value Added by Refining

This value was obtained by subtracting the cost of crude delivered to refineries from the value of refined products.

Net Value Added by Refining

Costs of materials (other than crude) used in refining were subtracted from gross value added by refining. Data required to obtain both gross and net value added by refining were taken from *Census of Manufactures*.

Value Added by Storage and Transport of Refined Products from Refineries to Wholesalers

Average assumed costs of storage and transport per barrel for shipments from refineries to wholesalers are indicated in Table A-2. The assumption is that these costs, in any particular year, apply to all categories of refined products. The total quantity of refined products multiplied by average shipment costs per barrel was used as a substitute for value added by this sector.

Value Added by Wholesalers

The cost of refined products delivered to wholesalers is assumed to equal the value of refined products at the refinery plus the cost of shipping them to wholesalers. The wholesale value of refined product was then obtained by applying wholesale margins of 18 per cent for all relevant Census years, except 1914 for which a margin of 17 per cent was used. Value added by wholesalers was obtained by subtracting the delivered cost from the wholesale value of refined products.[59]

Value Added by Storage and Transport of Refined Products from Wholesale to Retail Outlets

The quantity of refined exports was subtracted from the total quantity of refined products to obtain an estimate of the volume of refined products shipped to retailers. This amount was multiplied by the average shipment cost per barrel, shown in Table A-2, to determine the value added by storage and transport of refined products from wholesale to retail outlets.

Value Added by Retailers

Value of refined exports was subtracted from the wholesale value of refined products. The value added by storage and transport from wholesalers to retailers was added to this difference to obtain an estimate of the delivered cost of refined products for retail outlets. Retail margins for independent retail stores of 18.5 and 19.0 per cent for 1879 and 1889 and 19.5 for other Census years were applied to obtain the retail value of refined products. The difference between the delivered cost and retail value of refined products[60] was taken to be the value added by retail outlets.

Crude Oil Used Directly as Fuel

A similar procedure was used to derive the wholesale value of crude directly consumed as fuel. Storage and transport charges per barrel were assumed to be: 22 and 36 cents for 1899 and 1904, respectively; 28 and 25 cents for the other Census years.

[59] Harold Barger (*Distribution's Place in the American Economy Since 1869*, Princeton for NBER, 1955, p. 84) reports wholesale margins for gasoline and oil of 18 and 16 per cent for 1909 and 1919, respectively.
[60] *Ibid.*, p. 81.

Crude Exports

The difference between the average value of crude exports and average value at wells was regarded as a measure of the value added per barrel by both transport and marketing sectors. From this amount, 90 cents was assigned to the transport sector as the cost of storage and transport per barrel of crude, while the remainder was assigned to the marketing sector. This procedure was amended for 1909, since in that year the difference between the average value of exports and the average value of crude amounted to less than 90 cents. In that year, the entire difference, or 72 cents, was assigned to the transport sector.

Production, Export, and Domestic Distribution

Production, export, and domestic distribution of illuminating oil, naphthas and gasoline, lubricating oil, fuel oil, and totals of each are shown in barrels and percentages in Table A-3.

COMMENT

Arthur M. Johnson, Harvard University

When a qualitative business historian is asked to comment on a quantitative review of an industry, he is faced with a difficult problem. The approaches that suggest themselves range from challenging the whole concept which underlies the collection of data to questioning the accuracy of specific figures. On the more positive side it might be desirable to suggest that additional categories of data would be helpful, call attention to additional sources of information, and so on. In the present instance, however, none of these approaches would be particularly rewarding in terms of what the authors have set out to do and their competence in executing it.

One function that a qualitative commentator on a quantitative paper can perform is to remind the authors of the authority attributed by qualitatively oriented scholars to studies such as this one. This point is particularly relevant here both because of the authors' stature as historians of the American petroleum industry and because they have distilled in this paper the product of long years of research. In no other place is there such a compact yet comprehensive set of statistics on the American petroleum industry from its birth down to World War I. For this reason it is likely to become a standard reference, and the authors have a special obligation to combine caution in their statements with the greatest possible accuracy in their figures.

While perfection may be unobtainable, the perfectibility of data and their interpretation are constant challenges to both qualitative and quantitative scholars. Certainly, no students of the American petroleum industry are better equipped to meet them than Williamson, Andreano, and Menezes. As prepared for the Chapel Hill meeting in September 1963, however, their preliminary draft of the present paper contained a few sections that needed further refinement, at least as they were interpreted by this qualitative historian. For the most part, therefore, my commentary was addressed to certain sweeping statements and to the need for more clarification and documentation, as well as reference to a wider range of sources.

Without claiming anything for the helpfulness of that commentary, it is clear from a reading of the revised paper that the authors have now made my earlier remarks largely irrelevant. I am delighted that I have been deprived of an opportunity to reiterate them here, and I commend this paper to other students of the American petroleum industry who, like myself, are sure to find it a significant contribution to the literature.

Some Aspects of Development in the Coal Mining Industry, 1839-1918

VERA F. ELIASBERG

RESOURCES FOR THE FUTURE, INCORPORATED

I
Introduction

This paper is concerned with three aspects of the development of the coal industry: (1) growth and regional distribution of production, measured in physical quantities and current values; (2) increase in consumption and use, reflecting the slow shift of the U.S. energy base from wood to coal; and (3) estimates of the labor force engaged in coal mining and of changes and regional differences in the ratio of output to employment. None of these aspects can be treated over a period of some eighty years—from 1839 to 1918—in the detail that would be desirable because reliable information on the development of the industry during the early years is scanty. Even for the period during which coal became closely associated with the process of industrialization and mechanization—roughly from the middle of the nineteenth century—the data are not as precise as one could wish and sometimes are poorly related to the earlier records.

A huge body of literature exists on many economic and technological problems connected with coal mining, but as yet no comprehensive economic history of the coal mining industry, as such, has been compiled. This gap has been filled only partly by comparatively recent studies which deal with the history of coal from special viewpoints—Eavenson, for example, on coal production, and Schurr, Netschert *et al.* on coal as part of the total U.S. energy pattern.[1] The development of the industry during the later part of the period under review—roughly from the mid-1880's when coal overtook wood and became the predominant source of energy—has been documented and analyzed in numerous studies and

---

[1] Howard N. Eavenson, *The First Century and a Quarter of American Coal Industry*, Pittsburgh, 1942; and Sam H. Schurr and Bruce C. Netschert, with Vera F. Eliasberg, Joseph Lerner, and Hans H. Landsberg, *Energy in the American Economy, 1850–1975*, Baltimore, 1960.

surveys. Aspects of changes within the industry and of its changing position within the energy economy and the national economy as a whole have been treated in different detail, from different approaches, and with different emphasis on interrelationships. In fact, as far as statistical records and quantitative analysis are concerned, coal mining from the end of the nineteenth century is possibly one of the best-covered major industries. This does not hold true for the earlier phase. On the contrary, in the period during which coal mining experienced its most rapid expansion, its growth and its contribution to the growth of the national economy were underestimated to an extent that seems surprising in view of the attention devoted to less important industries.

For this reason, in this paper special attention is given to the period 1840 to 1890. The attempt is made to adjust measures other than output in physical quantities in accordance with the improved production series which in recent years has replaced earlier and less accurate data on the growth of coal output. If, as is recognized today, coal production has been underestimated for forty to fifty years, starting with 1840, by anywhere from 10 to 40 per cent, it seems reasonable to assume that value of output in current prices, estimates of value added, the size of the labor force engaged, and income originating in the industry, as originally reported, have also been substantially understated.

At the beginning of the period examined here, during the 1830's, the coal industry was of minor importance, restricted to small mines which produced for local markets. Not yet established on a commercial level, its role as a source of energy was negligible. By the middle of the period, in the 1880's, coal mining had developed into a major basic industry and had become the principal source of energy, supplying approximately one-half the fuel and power used by the United States economy. In the first decade of this century, its share rose to three-quarters of the total, a level which was maintained until after World War I when its relative importance in the total began to decline. Actual production peaked during World War I and again during World War II and its immediate aftermath.

Coal entered the economic scene of this country practically unnoticed and for a long time its potentials were ignored. The coal fields, widely spread throughout the eastern part of the United States, were not strange to many immigrants who came from European countries where coal mining had been well established for several centuries. Thus in some regions coal was mined as soon as settlement began. The earliest records go back to around 1700, from which time through the eighteenth century the mining and use of coal remained largely localized. Production for a "market" was pursued only where coal deposits were close to streams which would provide easy transportation. The year 1758 marks the first recorded commercial shipment: thirty-two tons from the James River district in Virginia shipped mainly to New York. By 1800 coal was mined—or rather gathered from outcroppings—in five states: Pennsylvania, Virginia (including West Virginia), Maryland, Kentucky, and Ohio.

TABLE 1

ESTIMATED ENERGY USE BY PRINCIPAL SOURCES, 1850

		Bituminous Coal Equivalent Required to Supply Same Amount of Heat and Power[a] (million net tons)
Fuel consumed for heat		
Wood (million cords)	96.0	76
Anthracite and bituminous coal (million net tons)	2.6	2.5
Horsepower-hours of mechanical work performed (billions) derived from:		
Windpower	1.4	12
Waterwheels	0.9	8
Steam power from coal	0.7	6
Steam power from wood	0.6	5
Work animals	5.4	48
Total energy use (except human labor)		158

Source: Shurr, et al., *Energy in the American Economy*, Chap. 3 and Appendix, "A Note on the Measurement of Direct Waterpower and Windpower."

[a] The following conversion factors were used: 1 cord of fuel wood = 0.8 net tons of bituminous coal; 1 net ton of anthracite = 0.97 net tons of bituminous coal. It has been estimated that in 1850, on the average, 17.6 pounds of coal were required to obtain 1 horsepower-hour of effectively utilized mechanical work. These estimates indicate an efficiency of converting coal into mechanical energy of 1.1 per cent.

Production estimates for this period vary between 100,000 and 200,000 tons. Apparently coal was used west of the Mississippi as early as 1817. By 1830 the five coal-producing states had increased to ten to include Rhode Island, Alabama, Illinois, Indiana, and Missouri, but the total output was still less than 1 million tons. This wide distribution of a small industry is certainly one reason coal mining attracted so little interest in its initial stages of development, in marked contrast to those metal industries that were more concentrated. Further, there was little need at the time for coal as a source of energy. The quantities of fuel and power used in this country were already substantial in 1850, but fuel was supplied almost entirely by wood and mechanical power, by wind, water, and work animals.

The position of coal in the U.S. economy in 1850 may be put in perspective in the following manner: Estimates of the total amounts of energy utilized by the economy in 1850 (including that provided by work animals) are necessarily crude but if they are converted to comparable tons of coal, about 158 million tons of coal would have been consumed. About 80 million tons of coal equivalent were used for domestic purposes and industrial process heat. Mechanical work performed (measured in horsepower-hours of work) by the utilization of inanimate energy sources would have required some 30 million tons of coal. The work done by animals would have required about 48 million tons of coal (see Table 1). The quantity of coal actually consumed (for heat and mechanical work), however, was 8.6 million tons, or about 5.4 per cent of the total combined sources of energy converted to coal equivalents.

In short, by the middle of the nineteenth century coal was beginning to play a modest role as a source of energy in certain regions of the country, while in others it was unknown, ignored, or regarded with skepticism. Numerous contemporary reports bear evidence of this. When, for instance, the new fuel was offered for sale to steamboats on the Mississippi River, the engineers (or firemen) rejected it because "such fuel would not make steam."[2]

II

Output

MAIN SOURCES OF INFORMATION

Until comparatively recent times the main source of information on the early phases of the coal mining industry on a nationwide scale was the decennial Census reports, supplemented from the mid-1880's by data compiled by the U.S. Geological Survey. Over the years, the Survey attempted to fill in the gaps between Census dates with annual estimates, and to push the data back to 1820.

The *Sixth Census, 1839-40*, in its section on manufactures, includes a set of statistics on coal production, broken down by states. At that time the distinction between the "new" mineral fuel, i.e., coal, and the traditional fuel derived from wood was not always sharply drawn. The term applied could depend on use as much as on source; any fuel

[2] Eavenson, *The First Century*, p. 277.

used in smelting might be called "coal"—even in official statistical data. Anthracite was measured in long tons but bituminous coal in bushels, and in a few instances charcoal was included under bituminous coal.

The *Seventh Census, 1849-50* gives output data in current dollar values but not in physical quantities. However, it is possible to derive estimates of the tonnage of production during that period by means of comparative information included in the 1859-60 Census and also by use of estimates made later by the Geological Survey.

The *Eighth Census, 1859-60* includes somewhat more detailed data on coal output in its volume on manufactures, but ten years elapsed before the first serious effort was made to collect more comprehensive information on the role of mineral industries in the U.S. economy. This was incorporated in a volume entitled *The Statistics of the Wealth and Industry of the United States*.[3] This marked progress. Nevertheless, it was not until the 1879-80 Census that mining was officially recognized and treated as a distinct sector of the U.S. economy, sufficiently important and different from manufacturing to warrant a separate detailed survey.[4]

We are dealing in this section with production, in which, as in many other respects, the early official statistics are frequently so defective that a seriously distorted picture emerges. Fortunately, as far as the growth of coal output is concerned, this state of affairs has been largely corrected by Eavenson's detailed study.[5] Eavenson combed the shipment records of canal companies and railroads, the pioneer railroad and coal trade journals, reports of state legislatures, and numerous other sources, and succeeded in eliminating most of the previous deficiencies inherent in the official statistical data on coal production. In recent years this comprehensive account has been recognized as superior to all formerly available statistics, and his series on coal production has been incorporated into the long-term energy studies of Resources for the Future,[6] the 1960 edition of the Census Bureau's *Historical Statistics*,[7] and the Bureau of Mines' historical data. The same series has been adopted without change as the primary set of statistics for this paper (see Table A-1) and forms the basis for adjustments and corrections of related statistics on value of production, consumption, and employment.

[3] *Ninth Census of the United States, 1870*, Vol. III.
[4] *Tenth Census of the United States, 1880*, Vol. XV, *Mining Industries (Excluding Precious Metals)*.
[5] Eavenson, *The First Century*.
[6] See Schurr *et al.*, *Energy in the American Economy*.
[7] *Historical Statistics of the United States, Colonial Times to 1957*, 1960.

OUTPUT, 1830'S TO WORLD WAR I

A summary of the growth of coal production during the period under consideration is presented in Table 2.[8] In order to gain a clearer view of long-term trends by smoothing out fluctuations caused by business cycles and work stoppages, the output data in Table 2 are shown as annual averages of ten-year periods. In a similar manner, the annual

TABLE 2

AVERAGE ANNUAL COAL PRODUCTION, BY DECADES, 1830-1919
(thousand net tons)

Decade	Bituminous Coal	Anthracite	Total Coal
1830-39	944	707	1,650
1840-49	2,124	2,517	4,641
1850-59	6,927	7,596	14,523
1860-69	12,444	13,669	26,127
1870-79	30,868	23,752	54,620
1880-89	75,923	38,293	114,216
1890-99	138,356	53,425	191,781
1900-09	302,479	71,286	373,765
1910-19	471,644	90,476	562,120

Source: Eavenson, *The First Century*, Part II, pp. 432-434; and U.S. Geological Survey, *Mineral Resources of the United States*, 1921, p. 482.

TABLE 3

AVERAGE ANNUAL INCREASE IN COAL PRODUCTION, BY DECADES, 1830-1919
(per cent)

Period	Bituminous Coal	Anthracite	Total Coal
1830-39 to 1840-49	8.5	13.5	10.9
1840-49 to 1850-59	12.5	11.7	12.1
1850-59 to 1860-69	6.1	6.1	6.1
1860-69 to 1870-79	9.5	5.7	7.7
1870-79 to 1880-89	9.4	4.9	7.7
1880-89 to 1890-99	6.2	3.4	5.3
1890-99 to 1900-09	8.1	2.9	6.9
1900-09 to 1910-19	4.5	2.4	4.2

Source: Table 2.

[8] Throughout this paper the traditionally accepted classification of all coal mined in the United States, that is, the distinction between anthracite and bituminous coal, has been retained.

rates of growth have been computed as averages for decades and are summarized in Table 3.

As may be expected when tracing the development of a major industry almost to its origins, the most striking expansion is concentrated in the earlier period. It took barely fifty years, from the mid-1830's to the 1880's, for coal output to grow one hundredfold, whereas during the following three decades its expansion was considerably less.[9]

Because we are dealing here with only two commodities, anthracite and bituminous coal, for which we have output data in physical units, little additional information would be conveyed by the construction of output indexes based on prices for certain years. It is sufficient to note that

TABLE 4

SHARE OF BITUMINOUS COAL AND ANTHRACITE IN TOTAL COAL PRODUCTION BY DECADES, 1830-1919
(per cent)

Decade	Bituminous Coal	Anthracite
1830–39	57.2	42.8
1840–49	45.8	54.2
1850–59	47.7	52.3
1860–69	47.7	52.3
1870–79	56.5	43.5
1880–89	66.5	33.5
1890–99	72.1	27.9
1900–09	80.9	19.1
1910–19	83.9	16.1

Source: Table 2.

anthracite was the more valuable fuel, except possibly during the very early years, and that its share in total coal output (measured in tons) declined significantly from more than 50 per cent in the midcentury decades to a mere 16 per cent in the decade from 1910 to 1919 (see Table 4).

Although Eavenson's series on coal production has by now become official historical data, related information based on the earlier Censuses has, to the knowledge of this writer, not hitherto been adjusted. The adjustments presented later in this paper may be interpreted more readily if reference is made to a comparison between original Census data on output and the presently recognized production series. This is shown in Table 5.

[9] After a long-lasting steep decline, the 1918 output figure was again reached and surpassed by about 1 per cent in 1944 and 1947, when unusually large exports pushed production up.

It has been known for some time that coal production was underestimated in the early Census reports. Underestimates range from about 20 per cent in the 1839-40 Census to as high as 40 per cent in the official 1859-60 returns. With one exception (1869-70), the discrepancies tend to be more marked for bituminous coal than for anthracite, probably because anthracite mining was always concentrated in a comparatively

TABLE 5

COAL PRODUCTION, ORIGINALLY REPORTED AND ADJUSTED,
SELECTED YEARS, 1839-90
(thousand net tons)

	Reported, Census Year	Adjusted, Calendar Year	Per Cent by Which Original Census Data Have Been Raised
	1839-40	1840	
Anthracite	963	1,129	17.2
Bituminous coal	1,108	1,345	21.4
Total	2,071	2,474	19.5
	1849-50	1850	
Anthracite	4,138	4,327	4.6
Bituminous coal	2,308	4,029	74.6
Total	6,446	8,356	29.6
	1859-60	1860	
Anthracite	8,115	10,984	35.3
Bituminous coal	6,219	9,057	45.6
Total	14,334	20,041	39.8
	1869-70	1870	
Anthracite	17,528	19,958	13.9
Bituminous coal	19,279	20,471	6.2
Total	36,807	40,429	9.8
	1879-80	1880	
Anthracite	28,641	28,650	--
Bituminous coal	42,841	50,757	18.5
Total	71,482	79,407	11.1
	1889	1890	
Anthracite	45,545	46,469	2.0
Bituminous coal	95,685	111,302	16.3
Total	141,230	157,771	11.7

Source: Data reported, Bureau of the Census, *1954 Census of Mineral Industries*; data adjusted, 1840-80, Eavenson, *The First Century*, Part II, production tables; 1890, *Mineral Resources of the United States*, 1921.

small area while bituminous coal fields were worked in numerous widely separated regions of the country. Anthracite, moreover, was produced for markets in the cities of the eastern seaboard at a time when a large portion of bituminous coal was still being mined solely for local use.

This latter point is significant since one of the most valuable sources of information on early coal output is not the mining statistics but the records of shipments carried on canals and somewhat later on railroads, coal "exported" thus being covered much more completely than the quantities consumed locally. But even these records of shipments are far from complete or continuous. It should therefore be kept in mind that, although the adjusted output statistics show substantially larger quantities than the contemporary Census reports do, they are still on the conservative side and are more likely to be understated than exaggerated. To quote the author of the corrected series: "It must be emphasized that while these tables show larger outputs over the years than has heretofore been thought, that [sic] in practically every case where accurate data were found after an estimate had been made, the estimate was found to be too small." [10]

For the sake of balance, one should also take into consideration that, in Table 5, data for Census years are compared with those for calendar years. For the period during which the industry expanded from around 6 to over 15 per cent annually, the transposition by roughly a half-year somewhat exaggerates the discrepancies between the two series.[11]

From the 1880's onward, Census data became increasingly reliable and complete. Disparities between the decennial Census reports and the annual surveys conducted by the Geological Survey are comparatively minor and are caused partly by slight differences in the time period covered (up to six months) and by the fact that the Survey workers make a greater effort than the Census takers to include as many as possible of the multiple small, irregular, or "farm" mines worked by local people only part of the year. For the period from 1885 onward, the data compiled by the Geological Survey have been adopted in this paper without change.

VALUE OF OUTPUT

In view of the size of the underestimates of coal output in the early period, it seems useful to attempt to adjust the original data on value of production which have been carried with only minor changes since they were published in the early Census reports. Information on the value of coal production at the location of the mine, in current prices, was included in all Census reports from 1849-50 on. In Table 6 these value data have been raised by the ratio of adjusted to original output estimates for anthracite and

---

[10] Eavenson, *The First Century*, p. xi.

[11] Yet even here a comparison of Census data with adjusted output data for the average of two calendar years shows underestimates of 16 per cent for 1839-40, 24.4 per cent for 1849-50, and 36.9 per cent for 1859-60—all rather close to the data in Table 5.

TABLE 6

VALUE OF COAL PRODUCTION, ESTIMATED VALUE ADDED, AND COST OF LABOR, ORIGINALLY REPORTED AND ADJUSTED, SELECTED YEARS 1849-90
(current prices)

	Reported, Census Year	Adjusted, Calendar Year	Reported, Census Year	Adjusted, Calendar Year
	1849-50	1850	1859-60	1860
Value of production (mill. $)				
Anthracite	5.3	5.5	11.9	16.0
Bituminous coal	1.9	3.3	8.4	12.2
Total	7.2	8.8	20.2	28.2
Estimated value added (mill. $)				
Anthracite	5.1	5.3	10.2	13.8
Bituminous coal	1.8	3.2	7.3	10.6
Total	6.9	8.5	17.5	24.4
Wages and salaries (mill. $)				
Anthracite	3.0	3.1	5.5	7.4
Bituminous coal	1.1	1.9	4.1	6.0
Total	4.1	5.0	9.7	13.4
Wages and salaries as per cent of value of product				
Anthracite	56.0		46.4	
Bituminous coal	60.0		49.5	
Total	57.0		47.7	
	1869-70	1870	1879-80	1880
Value of production (mill. $)				
Anthracite	38.4	43.7	42.3	42.1
Bituminous coal	35.1	37.3	53.5	63.4
Total	73.5	81.0	95.8	105.5
Estimated value added (mill. $)				
Anthracite	34.8	39.6	35.5	35.4
Bituminous coal	33.0	35.1	48.6	57.6
Total	67.9	74.7	84.2	93.0
Wages and salaries (mill. $)				
Anthracite	23.0	26.1	22.8	22.7
Bituminous coal	21.3	22.7	33.2	39.3
Total	44.3	48.8	55.9	62.0
Wages and salaries as per cent of value of product				
Anthracite	59.8		53.8	
Bituminous coal	60.8		62.0	
Total	60.3		58.4	
	1889	1890		
Value of production (mill. $)				
Anthracite	65.7	66.9		
Bituminous coal	94.5	110.2		
Total	160.2	177.1		
Estimated value added (mill. $)				
Anthracite	52.6	53.5		
Bituminous coal	85.7	99.9		
Total	138.2	153.4		
Wages and salaries (mill. $)				
Anthracite	39.3	40.0		
Bituminous coal	69.9	81.4		
Total	109.1	121.4		
Wages and salaries as per cent of value of product				
Anthracite	59.8			
Bituminous coal	73.9			
Total	68.1			

NOTES TO TABLE 6

Source: Data reported, value of production, wages and salaries, and value added, 1849-50-1889: *Census of Mineral Industries, 1954*, Vol. 1, *Summary and Industry Statistics*. The rough measures of value added were computed by subtracting from the value of production the cost of supplies, the value of coal produced and used at the same establishment and, from 1889 on, the cost of contract work. Data adjusted, reported value of production, value added, and wages and salaries were each increased by the ratio of adjusted to reported coal output (for adjusted coal output see Table A-1).

bituminous coal. Reported estimates of value added and labor cost also have been changed by the ratio of adjusted to original output estimates.[12]

OUTPUT BY REGIONS

The major coal deposits of the country are found in three extensive geological regions: the eastern (Appalachian) region, the Mississippi Valley region, and the Rocky Mountain region. For convenience, the Mississippi area is frequently subdivided into a central interior region east of the river and a western interior region west of it. These large regions are then broken down by states and all regional data are shown on a state-by-state basis. This seems a convenient arrangement for combining the state data to fit any desired regional breakdown.

The regional development of coal mining, summarized in Table 7, shows the adjusted output data for the beginning, the midpoint, and the end of the period under review. The table illustrates the slow extension of coal mining from the Appalachians to the Mississippi Valley and thence to the Rocky Mountain region. More important, it indicates that regional shifts were of comparatively minor importance in the development of the coal mining industry as a whole. Throughout the entire period coal mining remained concentrated in the same five states. In 1840, Pennsylvania, the area of West Virginia, Ohio, Kentucky, and Illinois produced 94 per cent of the total United States output; by 1880 the share of these states was 85 per cent, and in 1918 it still amounted to 79 per cent.[13]

The principal change in regional terms was the slow and persistent decline of the dominant position of Pennsylvania. In 1840 four-fifths of all coal mining was concentrated in this one state; by the end of World War I its share had declined to two-fifths. This was caused partly by the

[12] Our unadjusted (i.e., reported) value-added figures come from the *1954 Census of Mineral Industries*, for which rough estimates of value added were made on a nationwide basis, using data from the early Census reports. The 1954 Census also contains the wage and salary data from the earlier Censuses.

[13] This regional distribution has hardly changed even since World War I; in 1960 the same five states accounted for 84 per cent of the total output. Table A-1 gives production data by states for every tenth year from 1830 to 1910, and for 1918.

TABLE 7

U.S. COAL PRODUCTION, 1840, 1880, AND 1918, BY REGION

	1840		1880		1918	
Region	Net Tons (millions)	Per Cent of Total	Net Tons (millions)	Per Cent of Total	Net Tons (millions)	Per Cent of Total
Total coal	2.47	100	79.41	100	678.21	100
Anthracite						
Pennsylvania	1.13	46	28.65	36	98.82	15
Bituminous coal, total	1.35	54	50.76	64	579.39	85
Eastern (Appalachian)	1.28	52	36.02	45	386.79	57
Pennsylvania	0.70	28	21.28	27	178.55	26
West Virginia	0.31	13	2.18	3	89.94	13
Virginia	0.09	4	0.04	--	10.29	2
Ohio	0.10	4	7.96	10	45.81	7
Kentucky[a]	0.06	2	1.20	1	31.61[b]	5
Maryland	0.01	--	2.23	3	4.50	1
Alabama	c	--	0.25	--	19.19	3
Tennessee	0.01	--	0.72	1	6.83	1
North Carolina	c	--	c	--	c	--
Georgia	--	--	0.16	--	0.07	--
Central interior	0.05	2	8.77	11	121.44	18
Illinois	0.04	2	6.48	8	89.29	13
Indiana	0.01	--	2.16	3	30.68	5
Michigan	--	--	0.13	--	1.47	--
Western interior	0.01	--	4.16	5	29.18	4
Iowa	c	--	1.79	2	8.19	1
Kansas	--	--	0.77	1	7.56	1
Missouri	0.01	--	1.36	2	5.67	1
Nebraska	--	--	0.08	--	--	--
Arkansas	--	--	0.02	--	2.23	--
Oklahoma	--	--	0.12	--	4.81	1
Dakotas	--	--	0.02	--	0.72	--
Rocky Mountain	--	--	1.31	2	35.54	5
Colorado	--	--	0.46	1	12.41	2
Wyoming	--	--	0.59	1	9.41	1
Utah	--	--	0.25	--	5.14	1
Montana	--	--	--	--	4.53	1
New Mexico	--	--	0.01	--	4.02	1
Other regions[d]	0.01	--	0.51	1	6.44	1

Source: Calculated from Table A-1.

[a] Kentucky is the only state which has within its borders parts of two great coal regions. The eastern counties are underlain by Appalachian coal beds, the western district includes part of the central interior coal fields. At the beginning of the period under review coal mining was concentrated in eastern Kentucky, then the center of production shifted to the western district and, during the 1910's, again back to the eastern part of the state.

[b] Of which 20.7 million tons in the eastern and 10.8 million tons in the western district of Kentucky.

[c] Less than 5,000 net tons.

[d] Including Oregon, California, Washington, Texas, Rhode Island, and Massachusetts.

larger increase in the production of bituminous coal than of anthracite. Although Pennsylvania was the leading bituminous coal producer until 1930, long after World War I, its share in bituminous coal production fell from three-fifths around 1840 to one-third in 1918.[14]

III

Consumption

For the 1830–80 period already outlined, information on coal consumption, and especially on the various uses of coal, is scanty; indeed, it is

TABLE 8

APPARENT AVERAGE ANNUAL COAL CONSUMPTION IN THE UNITED STATES, BY DECADES, 1830-1919
(million net tons)

Decade	Bituminous Coal	Anthracite	Total
1830-39	1.1	0.7	1.8
1840-49	2.3	2.5	4.8
1850-59	7.1	7.6	14.7
1860-69	12.6	13.7	26.3
1870-79	31.1	23.4	54.5
1880-89	76.2	37.6	113.7
1890-99	137.3	52.1	189.4
1900-09	296.8	68.9	365.8
1910-19	454.9	86.0	540.9

Source: Apparent consumption equals production (Table 2) plus net imports. 1830-69, net imports assumed to be bituminous coal entirely; 1870-1919, net imports for bituminous coal and anthracite separately are from *Historical Statistics of the United States, Colonial Times to 1957*, Washington, 1960, pp. 356-359.

almost nonexistent for the earlier years. For the period under review, therefore, we shall assume that consumption is equal to adjusted production plus or minus the foreign trade balance.[15] These data are presented in Table 8 as annual averages, by decades.

For nearly one century foreign trade was of comparatively minor importance for the coal industry. During the 1830's, net imports of bituminous coal amounted to approximately 6.5 per cent of total production. The imported coal was burned mainly in the larger cities on the eastern seaboard where coal, frequently brought in as ballast on ships,

---

[14] It was only after 1930 that West Virginia overtook Pennsylvania in bituminous coal output and became the leading state. By 1960 its bituminous coal production had become nearly twice as large as that of Pennsylvania.

[15] Coal imports and exports appear in Table A-2.

was less costly than domestic coal from the Appalachian field. With the rapidly expanding output and improved transportation of the 1860's, net imports of bituminous coal declined to 0.6 per cent of production. In the following decade the United States became a net exporter of coal. Net exports grew from 0.1 per cent of output during the 1870's to 4.3 per cent for the 1910–19 decade.

By the middle of the nineteenth century the total energy consumption of the United States economy was already huge in absolute terms, as can be seen in Table 1. Apparently it was larger in per capita terms than that of any other country, including the industrially more advanced such as Great Britain.[16] Even when the mechanical energy derived from wind, water power, and work animals is disregarded and only fuel material are considered, per capita use around 1850 was the rough equivalent of four tons of bituminous coal per year. Nine-tenths of this (measured in btu content), however, consisted of wood burned at the rate of about four cords per person per year, while only one-third of a ton of coal per year per person was being used.

From around 1850 to 1890, the over-all consumption of fuel materials increased threefold, yet during the same four decades the per capita consumption of fuel materials appears hardly to have increased at all (see Table 9), in spite of the rapid advance in industrialization and construction of a huge railroad network.

Behind this phenomenon lies the fact that clearing the land for cultivation went hand in hand with the lavish use of a seemingly unlimited supply of fuel wood. Most of it went up in smoke through the chimneys of newly established homes, resulting in a large energy consumption in statistical terms in the early years of that period. It was only with the development of larger cities on the eastern seaboard and with advances in industrialization, especially the change from small-scale local iron shops operated by individual blacksmiths to the beginnings of a modern iron and steel industry, that coal slowly began to prove its advantages over wood. The fact that coal is a more compact energy source, both in volume and in weight, became significant at a time when mushrooming centers of population and industry were depleting nearby supplies of wood.

Coal gradually began to replace wood from about the middle of the nineteenth century onward, initially largely as fuel used in industry and transportation. Although substantiated information is lacking, it appears that around 1850 possibly as much as three-quarters of the still rather modest output of coal (some 8.5 million tons) was transformed into steam

[16] Schurr *et al.*, *Energy in the American Economy*, p. 153.

power and mechanical work, while wood continued to fulfill the greater portion of domestic demands.

By the year 1870, coal supplied about one-quarter of total fuel material needs; ten years later it furnished some 40 per cent, and by 1885 it had overtaken wood as a source of heat and mechanical power. Even at that

TABLE 9

PER CAPITA CONSUMPTION OF FUEL MATERIALS, SELECTED YEARS, 1850-1900

Year	Fuel Wood (cords)	Coal (net tons)	Crude Oil (barrels)	Natural Gas (thous. cubic feet)	Total in Bituminous Coal Equivalent (net tons)
1850	4.39	0.36	--	--	3.9
1855	4.16	0.60	--	--	3.9
1860	4.00	0.63	0.02	n.a.	3.8
1865	3.70	0.69	0.05	n.a.	3.6
1870	3.46	1.02	0.05	n.a.	3.8
1875	3.04	1.24	0.04	n.a.	3.7
1880	2.71	1.58	0.34	n.a.	3.8
1885	2.26	1.94	0.13	1.35	3.8
1890	1.90	2.48	0.44	3.79	4.2
1895	1.58	2.74	0.43	1.97	4.2
1900	1.31	3.45	0.52	3.09	4.8

Source: Schurr et al., *Energy in the American Economy*, Statistical Appendix to Part I, pp. 519 and 521.

time, the greater portion of all coal used went into industry and transportation, but nonetheless it was also well established as a domestic fuel, largely because the price of coal was falling relative to the price of wood in the urban centers of consumption. Valuable information on this aspect is included in the 1879-80 Census, which gives the retail prices for coal and fuel wood in various cities located in different parts of the country for the year 1879 (summarized in Table 10). Comparing these prices and keeping in mind that one cord of wood equals about 0.8 tons of coal in heat equivalents (but is nearly twice as heavy and bulky), we see clearly the improved competitive position of coal as the less costly and more convenient fuel material.

The same Census includes a survey of the consumption of wood as fuel for domestic and industrial purposes in 1879. Since this is the only comprehensive study of the use and value of fuel wood ever made, a summary of the information it contains is reproduced in Table A-3. At this point it seems sufficient to note that, while a large portion of all

TABLE 10

RETAIL PRICES OF COAL AND FUEL WOOD, 1879

	Coal ($ per net ton)		Fuel Wood ($ per cord)	
	Bituminous Coal	Anthracite	Hardwood	Softwood
Boston, Mass.	n.a.	4.25-6.50	10.00	8.00
Philadelphia, Pa.	n.a.	4.50-5.25	--	--
New Castle, Pa.	2.80	n.a.	--	--
Reading, Pa.	--	--	4.00	n.a.
New Cumberland, W. Va.	1.75	2.80	2.25	2.00
Springfield, Ohio	3.75	6.50	3.50	n.a.
Cincinnati, Ohio	3.70	6.68	6.00	6.00
Jacksonville, Ill.	2.50	7.00	3.50	n.a.
Indianapolis, Ind.	3.45	n.a.	--	--
Lawrenceburg, Ind.	--	--	4.00	n.a.
Louisville, Ken.	2.86-3.42	n.a.	--	--
Leavenworth, Kan.	3.50	11.00	6.50	n.a.
Cedar Rapids, Iowa	5.00	7.50	5.00	n.a.

Source: *Tenth Census, 1880,* Vol. XX, *Report on the Statistics of Wages in Manufacturing Industries; With Supplementary Reports on the Necessaries of Life and on Trades Societies, and Strikes and Lockouts,* pp. 94-101 of section on Necessaries of Life, Miscellaneous.

wood used in households never entered the market, wood consumed by industry and transportation did, but the quantities sold were rather small—some 5 million cords, or 3.6 per cent of all fuel wood used during that year. During the same year, 1879-80, coal production was about 30 million tons of anthracite and 50 million tons of bituminous coal. Prices per ton at the mine averaged $1.47 for the former and $1.25 for the latter. Even if transportation and delivery costs doubled these prices, coal for use in industry and transportation by that time was less costly than wood in most areas.

The quantity of charcoal used is shown in the 1879-80 Census as 740,000 tons; the average price as $7.13 per ton. The overwhelming part, nearly 700,000 tons, was consumed in the manufacture of iron and steel. The quantity of wood required to produce that amount of charcoal may be estimated at 1.5 million tons, a mere 1 per cent of all wood burned. While wood was still abundant in large areas of the country, pockets of scarcity existed around the manufacturing centers and transportation costs continued to rise. Both factors are reflected in the high price of charcoal. Even before 1850, blast furnaces were frequently abandoned when local wood supplies were exhausted and new ones were built closer to timber sources in order to reduce the cost of fuel transportation.

From the 1830's on, coke from the huge bituminous coal supplies had been used experimentally as blast furnace fuel in the Pittsburgh area, but progress was slow, and during the mid-1850's only some 55,000 tons of coal per year went into coke production in that region. In eastern Pennsylvania, however, where anthracite was readily available, the shift to coal began on a larger scale. By the mid-1860's more than one-half of all pig iron produced was smelted with anthracite, about one-fifth with raw bituminous coal or coke, and still close to 30 per cent with charcoal. By 1880, the share of charcoal as fuel in iron production had declined to 12.5 per cent and coke from bituminous coal had begun to surpass anthracite (see Table A-4).

In the Census year 1879-80, 4.36 million tons of coal went into the production of 2.75 million tons of coke, of which more than 2 million tons were used in blast furnaces. The price of coke f.o.b. cars at ovens averaged $1.95 per ton for the country as a whole, ranging from $1.81 in Pennsylvania, where more than 80 per cent of all coke was produced and consumed, to $5.00 in Colorado. Even with transportation costs added, this is far below the average price of $6.79 paid by the iron industry in the same year for a ton of charcoal. For certain purposes charcoal was still considered the superior metallurgical fuel and, measured by weight, the amount of charcoal required in smelting was about the same as the amount of coke. In terms of volume, however, nearly three times as much charcoal as coke was required for the production of one ton of iron. When, in addition to that disadvantage, the price of charcoal reached a level about three times as high as that of coke, charcoal was priced out of the fast-expanding iron and steel market. Pig iron smelted with coke was first used in large amounts in the manufacture of rails. By 1880, steel production amounted to 1.4 million tons, of which more than half was rolled into rails. At that time, the railroad network had expanded to 90,000 miles and the indirect demand of the railways for coal needed by the iron and steel industry was overshadowed by the larger requirements for coal as locomotive fuel.

During the 1870's bituminous coal production surpassed the output of anthracite; by the turn of the century the latter amounted to only about one-quarter of all coal mined. The role of anthracite as industrial fuel began to decline when bituminous coal in the form of coke became the dominant metallurgical fuel. From the last decades of the nineteenth century anthracite was used mainly for space heating, while the bulk of bituminous coal continued to be consumed in manufacturing and transportation.

Reasonably comprehensive data on bituminous coal consumption by main-use categories became available in the mid-1880's. They are shown

TABLE 11

U.S. CONSUMPTION OF BITUMINOUS COAL, BY CONSUMER CLASS, SELECTED YEARS, 1885-1905

(million net tons)

Year	Railways	Coke	Industrial and Household	Total	RFF Estimate of Total U.S. Consumption
1885	29.4 (43%)	8.1 (12%)	31.1 (45%)	68.6 (100%)	71.9
1890	45.9 (44%)	18.0 (17%)	40.3 (39%)	104.2 (100%)	110.8
1895	50.4 (40%)	20.8 (17%)	54.7 (43%)	125.9 (100%)	134.0
1900	78.4 (40%)	32.1 (16%)	84.3 (43%)	194.8 (99%)	207.3
1905	109.3 (38%)	49.5 (17%)	131.7 (45%)	290.5 (100%)	308.8

Source: H.S. Fleming, *A Report to the Bituminous Coal Trade Association on the Present and Future of the Bituminous Coal Trade*, New York, 1908, p. 10. Resources for the Future estimate, Schurr, et al., *Energy in the American Economy*, p. 508.

in Table 11. In 1885, the year when total coal consumption began to overtake wood, the largest portion, 43 per cent, was burned as locomotive fuel; coke production consumed 12 per cent; and the remainder was distributed among all other industrial and domestic uses. By 1918, at the end of the period under review, the railroad market reached its peak in absolute terms but its share in total bituminous coal consumption had declined to one-quarter. Coke production took some 16 per cent; a fast-expanding new market, electric utility plants, absorbed 6 per cent. Manufacturing industries and retail consumers together accounted for 51 per cent.

Stimulated by war demand, total coal consumption had reached its World War I peak of 651 million tons, about 90 per cent of which was bituminous coal. But by this time it was already faced with a new competitor, fuel oil, which even in 1918 had displaced as much as 8 per cent of total coal consumption.

IV

Employment

ADJUSTMENT OF BASIC DATA

Being of minor importance to the national economy, coal mining did not warrant careful statistical treatment during the early part of the period examined. Although the number of persons employed in the industry was reported in every Census from 1839-40, except for 1849-50, the data are incomplete and vary in definition and detail from one Census to the next.

The 1849-50 Census omits coal miners from a detailed enumeration of more than three hundred occupations and professions of (free) males 15 years of age and over. It does, however, list charcoal burners (159 persons). Coal miners are included under more general classifications, such as miners, laborers, etc. It is only in a footnote to this occupation survey that employment in coal mining is listed for four states: Pennsylvania, Virginia, Kentucky, and Ohio. However, the next Census (1859-60) does include a comparatively detailed section on the coal industry wherein, with other information, total employment is given for the 1849-50 Census year.

The Census reports attempted to show the average number of persons employed during the year, but apparently in many cases either the maximum number or the number employed at the Census date was listed. Further, the treatment is not consistent from one Census to the next. The 1869-70 survey, for example, shows the number of boys under 15 years employed in coal mines—some 17 per cent of the labor force in the anthracite industry but less than 3 per cent in bituminous mining. The 1879-80 Census, however, differentiates between "miners," "laborers," and the "administrative force" for the anthracite industry, but gives no similar detail for bituminous coal mining. Here the labor force was still broken down between men and boys employed above and below ground, although at that time the number of boys under 16 years of age had declined to 1.5 per cent of the total work force.

The 1879-80 Census made a great effort to distinguish between "regular establishments"—i.e., commercial mines that were worked during the entire Census year, or at least throughout most of it—and "irregular establishments," or so-called "farmers' diggings." The latter comprised

TABLE 12

EMPLOYMENT IN COAL MINING ORIGINALLY REPORTED AND ADJUSTED,
SELECTED YEARS, 1839-1918
(thousand persons)

	Reported, Census Years	Adjusted, Calendar Years
	__1839-40__	__1840__
Anthracite	3.0	3.5
Bituminous coal	3.8	5.5
Total	6.8	9.0
	__1849-50__	__1850__
Anthracite	n.a.[a]	10.6
Bituminous coal	n.a.[a]	9.4
Total	15.2	20.4
	__1859-60__	__1860__
Anthracite	25.1	34.0
Bituminous coal	11.4	16.0
Total	36.5	50.0
	__1869-70__	__1870__
Anthracite	53.1	60.4
Bituminous coal	41.7	44.3
Total	94.8	104.7
	__1879-80__	__1880__
Anthracite	70.7	70.7
Bituminous coal	109.0	129.0
Total	179.7	199.7
		__1890__
Anthracite		126.0
Bituminous coal		192.2
Total		318.2
		__1900__
Anthracite		144.2
Bituminous coal		305.0
Total		449.2
		__1910__
Anthracite		169.5
Bituminous coal		555.5
Total		725.0
		__1918__
Anthracite		147.1
Bituminous coal		615.3
Total		762.4

Source: See Tables A-5 and A-6.

[a] See source to Table A-5.

a large number of small producers who had some outcroppings of bituminous coal on their land and did some irregular surface mining when time and weather conditions permitted. In some instances, such producers employed a few helpers, who are enumerated in the Census as some 9,000, compared with 100,000 persons in regular establishments of the bituminous coal industry.

For the latter part of the period examined, from 1890 to 1920, employment data compiled by the Geological Survey have been adopted without change or adjustment. The statistics of the survey refer to the average number of men employed during the year in production and development work, but exclude office personnel, proprietors, and the labor force of coke works connected with bituminous mines.

To arrive at adjusted labor force data for the period from 1839-40 to 1879-80, the Census ratio of employment to output has been applied to the adjusted production series for each of the five most important coal mining states and for all other states as a group. A summary of the original data and the adjusted labor force estimates is shown in Table 12. Details may be found in Tables A-5 and A-6.

EMPLOYMENT BY REGION

There is, of course, a close relationship between the distribution of coal mining employment and the distribution of coal production among the various states. Around 1840, five states—Pennsylvania, Virginia (including West Virginia), Ohio, Kentucky, and Illinois—accounted for 90 per cent of the total adjusted labor force in coal mines. By the end of World War I, the same states still included 77 per cent of the U.S. labor force in the coal mining industry.[17] As was to be expected, Pennsylvania consistently accounted for the bulk of coal mining employment—about 70 per cent of the total in 1840, and still 42 per cent at the end of World War I.

EMPLOYMENT IN COAL MINING IN RELATION TO ALL MINERAL EMPLOYMENT AND TO TOTAL U.S. EMPLOYMENT

Available Census data show that employment in coal mining accounted for more than one-half the labor force engaged in all mineral industries from 1870 to 1939. Between 1840 and 1860, the labor force in metal

---

[17] In contrast with the output data, employment for the earlier years cannot be broken down for the areas which later were to become Virginia and West Virginia. Hence, since Virginia is included in the original and adjusted employment data for the period between 1840 and 1860, the size of the labor force is slightly overstated in relation to output statistics for the five leading coal-producing states.

mining appears to have matched that in the coal industry. By 1870 the share of the coal mining industry in all mineral employment had reached nearly two-thirds, and by 1910 it accounted for three-quarters (on the basis of Census data). The number of workers engaged in coal mining continued to increase until the early 1920's, but their share in the labor force of the total mining sector began to decline about a decade before a decrease in absolute numbers set in.

Measured against the total labor force of the United States, the number of persons engaged in all mining industries[18] was never large—1.2 per cent

TABLE 13

EMPLOYMENT IN COAL MINING AS A PERCENTAGE OF TOTAL EMPLOYMENT IN PRINCIPAL COAL-PRODUCING STATES AND THE UNITED STATES, SELECTED YEARS, 1840-1920

	1840	1850	1860	1870	1880	1890	1900	1910	1920
Pennsylvania				7.7	7.6	9.6	9.7	11.0	9.4
West Virginia and Virginia				1.4	3.1	5.5	8.9	15.3	18.2
Ohio				0.9	2.2	1.6	1.8	2.4	2.1
Kentucky				0.3	0.7	0.9	1.3	2.3	4.6
Illinois				0.9	1.7	2.1	2.2	3.2	3.3
United States	0.2	0.3	0.5	0.8	1.1	1.4	1.5	1.9	1.8

Source: Adjusted coal mining employment data are from Tables A-5 and A-6. Total employment: 1840-60, estimated at 1.2 times the no. of employed (free and slave) males aged 15 to 60 years, listed in *Historical Statistics*, 1960, p. 10. 1870-1920, from Perloff et al., *Regions, Resources and Economic Growth*, Table A-1.
"1920" is 1918 coal employment divided by 1920 U.S. employment.

in 1870 increasing to 2.5 per cent in 1910. Over the same period, the share of workers in coal production relative to total employment expanded at a considerably faster rate—from 0.8 to 1.9 per cent of total employment, as shown in Table 13. Earlier, between 1840 and 1860, coal mining appears to have grown at an even faster pace, from 0.2 to nearly 0.5 per cent of the national labor force. Such a relative increase does not appear unreasonable, but it should be borne in mind that for these early years both sets of data—total U.S. employment as well as the number of persons in coal mining—are crude estimates.[19]

[18] As enumerated by the Census.

[19] For the years 1840, 1850, and 1860, the total U.S. labor force has been estimated to be 1.2 times the number of males (free and slave) aged 15 to 60 years. This is the ratio underlying the total employment data for 1870 in the source used (see H. S. Perloff et al., *Regions, Resources, and Economic Growth*, Baltimore, 1960).

In a few states the work force in coal mining constituted a much larger proportion of total employment. In Pennsylvania employment in the coal industry was between 8 and 11 per cent in the fifty years from 1870 to 1920. In Virginia (including what is now West Virginia) coal mining grew from 1 per cent of total employment in 1870 to 18 per cent in 1920. In Kentucky coal accounted for 5 per cent of the labor force around World War I. In Ohio and Illinois, the remaining leading coal-producing states, coal employment was about 2 to 3 per cent during the coal industry's peak period (see Table 13).

EMPLOYMENT COMPARED WITH OUTPUT

During the entire period under review there were wide variations in the ratio of output to employment, not only between regions and states but also from one mine to another. It was not unusual for production per man to be several times as large in one mine as in another located in the same county. Another obstacle to statistical analysis is variation in the number of days worked per year. For example, anthracite mines in Pennsylvania worked for 200 days in 1890, on the average, while bituminous coal mines in the United States worked 239 days, ranging from 201 days in Ohio (one of the leading coal-producing states) to 289 days in Utah (a state of minor importance for the industry as a whole). Unfortunately, precise information on this aspect is not available in the statistics prior to the 1880's. Still another problem is the varying length of the working day, not only over time but also from one state to another and frequently from one mine to the next. For those years for which reliable statistics on working time have been compiled—namely, from 1890 onward for days worked during the year, and from 1902 onward for the hours comprising an average full working day—the average outputs per man-year, per man-day, and per man-hour in the anthracite and bituminous coal mining industry have been computed and analyzed in several studies and this information need not be repeated here in detail.[20]

For the earlier years, one has to rely on employment data even though changes in the number of persons employed are by no means identical with changes in labor input. Nevertheless, a comparison of the expansion of production with the growth of employment over an eighty-year period can supply a rough approximation of changes in the amount of labor used per unit of output. Such a comparison is shown in Table 14 for the bituminous coal and anthracite industries, based on Census production

[20] See Harold Barger and Sam H. Schurr, *The Mining Industries, 1899–1939: A Study of Output, Employment and Productivity*, New York, NBER, 1944; Vivian E. Spencer, *The Mineral Extractive Industries, 1880–1938*, Philadelphia, 1940; and Geological Survey, *Mineral Resources of the United States*, various issues.

TABLE 14

OUTPUT PER MAN-YEAR IN MAJOR COAL-PRODUCING STATES AND THE UNITED STATES, SELECTED YEARS, 1840-1918
(net tons)

	1840	1850	1860	1870	1880	1890	1900	1910	1918	Annual Increase Between 1880 and 1918 (per cent)
Bituminous coal, U.S.	243	430	567	462	393	579	696	751	942	2.3
Pennsylvania	259	n.a.	579	518	542	690	861	858	1,024	1.7
West Virginia[a]	427	n.a.	397	598	396	604	777	898	1,005	2.5
Ohio	323	n.a.	754	374	363	559	687	733	946	2.6
Kentucky	113	n.a.	378	237	331	514	551	720	804	2.4
Illinois	112	n.a.	509	466	374	535	659	632	1,039	2.7
Anthracite	323	390	323	330	405	369	398	498	672	1.3

Source: Tables A-1, A-5, and A-6. The ratios for individual states are calculated from Census data from 1840 to 1880. The U.S. ratios for bituminous coal for the same years are calculated from Eavenson's output data (*The First Century*) and our adjusted employment data as given in Table A-5.

[a] Including Virginia in 1840, 1850, and 1860.

and employment data. It would seem from this table that during the 1840-1918 period the output per man-year doubled in anthracite mining, but more than quadrupled in the bituminous coal industry.

The increase in productivity was very irregular during the decades from 1840 to 1880. This may be attributable to deficiencies in the data or may be real since atypical conditions may be present in any one Census year. From 1890 on, the decadal increases in output per man-year are more nearly the same in each sector and from decade to decade. These output

TABLE 15

OUTPUT PER MAN-DAY IN MAJOR COAL-PRODUCING STATES, SELECTED YEARS, 1880-1918
(net tons)

	1880	1890	1900	1910	1918	Annual Increase Between 1880 and 1918 (per cent)
Bituminous coal, total	1.90	2.56	2.98	3.46	3.78	1.8
Pennsylvania	2.47	2.97	3.56	3.61	3.81	1.1
West Virginia	1.97	2.66	3.36	3.94	4.22	2.0
Ohio	1.60	2.78	3.20	3.61	4.24	2.6
Kentucky	1.84	2.34	2.43	3.26	3.50	1.7
Illinois	1.65	2.62	2.92	3.95	4.37	2.6
Anthracite	n.a.	1.85	2.40	2.17	2.29	0.8[a]

Source: 1880, *Tenth Census, Mining Industries of the United States*, p. 683. 1890-1918, *Mineral Resources of the United States*, 1921, p. 497.

[a] 1890 to 1918.

per man-year data are, of course, influenced by various short-term factors, such as business cycles, labor disputes, and changes in transportation facilities which are reflected in significant variations in the number of days worked. Thus the number of working days per year ranged in anthracite mining during the 1890-1918 period from a low of 116 to a high of 293; in the bituminous coal industry, for the country as a whole, from 171 to 249.

Available data on output per man-day are given in Table 15. In bituminous coal mining the tonnage rose from 2.56 in 1890 to 3.78 at the end of World War I; the increase in anthracite mining was considerably less—from 1.85 to 2.29 net tons. These figures represent an average annual growth rate in production per man-day of 1.4 and 0.8 per cent, respectively. Since working hours per day declined during that period, the increase in output per unit of labor input was even greater, possibly as much as one-fifth more than is suggested by Table 15.

Among the five leading bituminous coal-producing states listed in Table 14, there has been a substantial but diminishing variability in output per man-year as measured by the coefficient of variation:

1840	.56	1880	.22
1850	—	1890	.12
1860	.30	1900	.17
1870	.32	1910	.14
		1918	.10

For the period after 1880 when data were more reliable than in earlier years, the increases in output per man-year were quite uniform among the leading states other than Pennsylvania, whose increase of 1.7 per cent per year was substantially less than the approximately 2.5 per cent rate of growth enjoyed by the other four states. Output per man-year grew more slowly from 1880 to 1918 for anthracite than for bituminous coal mining, 1.3 per cent against 2.3 per cent.

Growth of bituminous coal output per man-day from 1880 to 1918 showed more variation among the five states than did output per man-year. Also, output per man-day grew at the same or a slower rate than did output per man-year from 1880 to 1918 in each state and also for the United States as a whole. The same was true for anthracite. This result is not attributable solely to factors peculiar to 1880 or 1918, for a similar relation holds more often than not for all the adjacent decadal increases that can be calculated for output per man-day and man-year from 1880 to 1918.

During the first two decades of the twentieth century, trends in output per man-day differed widely between anthracite and bituminous coal mining. In the anthracite industry production per man-day hardly changed. In contrast to the bituminous industry, deteriorating resource conditions of the long-worked deposits were not offset by development of richer or more easily accessible supplies. The average width of the seam declined steadily, while the depth of the anthracite mines increased. The industry was particularly difficult to mechanize, largely because of the steep slope of many of the coal beds. In the bituminous coal industry, on the other hand, one-quarter of the total production was mined by machines by 1900. Ten years later this share had risen to 42 per cent and by the end of World War I to 56 per cent. But even here the most intensive progress in mechanization and the most rapid increase in productivity still lay in the future.

Appendix

TABLE A-1
COAL PRODUCTION, BY MAIN GEOLOGICAL REGIONS AND STATES, SELECTED YEARS, 1830–1918
(thousand net tons)

	1830	1840	1850	1860	1870	1880	1890	1900	1910	1918
Anthracite										
Pennsylvania	235	1,129	4,327	10,984	19,958	28,650	46,469	57,368	84,485	98,826
Bituminous coal, total	646	1,345	4,029	9,057	20,471	50,757	111,302	212,316	417,111	579,386
Appalachian Region										
Pennsylvania	398	700	2,148	4,710	9,224	21,280	42,302	79,842	150,522	178,551
Ohio	25	104	617	1,850	2,959	7,957	11,495	18,988	34,210	45,813
Maryland	9	12	243	438	1,820	2,229	3,358	4,025	5,217	4,497
West Virginia	78	305	348	365	991	2,181	7,395	22,647	61,671	89,936
Virginia	103	88	138	112	90	43	784	2,394	6,508	10,290
Kentucky	19	63	76	131	282	1,203	2,701	5,329	14,623	31,613
Tennessee	---	8	60	165	133	718	2,170	3,510	7,121	6,831
Alabama	a	1	2	15	11	250	4,090	8,394	16,111	19,185
Mississippi Valley Region										
Illinois	4	37	259	858	3,041	6,476	15,292	25,768	45,900	89,291
Indiana	a	9	7	46	559	2,164	3,306	6,484	18,390	30,679
Iowa	---	3	27	48	283	1,792	4,022	5,203	7,928	8,192
Missouri	6	10	100	280	622	1,360	2,735	3,540	2,982	5,668
Kansas	---	---	---	5	33	771	2,260	4,468	4,921	7,562
Rocky Mountain Region										
Colorado	---	---	---	---	14	463	3,077	5,244	11,974	12,408
Utah	---	---	---	---	11	252	318	1,147	2,518	5,137
Wyoming	---	---	---	2	105	590	1,870	4,015	7,533	9,439
Other states (and/or regions)	5	6	4	32	291	1,028	4,127	11,318	18,982	24,294
Total coal	881	2,474	8,356	20,041	40,429	79,407	157,771	269,684	501,596	678,212

Source: 1830–80, based on Eavenson, *The First Century*, Part II, Production Tables, pp. 426–434; 1890–1918, *Mineral Resources of the United States*, 1921, p. 482.

aLess than 1,000 net tons.

TABLE A-2

AVERAGE ANNUAL NET IMPORTS OR EXPORTS OF COAL,
BY DECADES, 1830-1919

Decade	Total Coal Production	Net Imports (thousand net tons)	Net Exports (thousand net tons)	Net Imports (per cent of production)	Net Exports (per cent of production)
1830-39	1,650	107		6.5	
1840-49	4,641	132		2.8	
1850-59	14,523	150		1.0	
1860-69	26,113	145		0.6	
1870-79	54,620		68		0.1
1880-89	114,216		473		0.4
1890-99	191,781		2,404		1.3
1900-09	373,765		7,996		2.1
1910-19	562,120		21,231		3.8

Source: 1830-69, based on Eavenson, *The First Century*, pp. 436-439. 1870-1919, *Historical Statistics*, 1960, pp. 356-359, and Table 2.

TABLE A-4

PERCENTAGE OF ANNUAL PIG IRON PRODUCTION SMELTED WITH DIFFERENT FUELS,
SELECTED YEARS, 1854-1905

Year	Bituminous Coal and Coke	Anthracite	Charcoal
1854	7.4	46.1	46.5
1855	8.0	48.7	43.3
1860	13.3	56.5	30.3
1865	20.4	51.5	28.2
1870	30.6	49.9	19.6
1875	41.8	40.1	18.1
1880	45.4	42.1	12.5
1885	59.1	32.1	8.8
1890	69.4	23.8	6.8
1895	84.2	13.4	2.4
1900	85.0	12.2	2.8
1905	91.2	7.3	1.5

Source: *The Mineral Industry, Its Statistics, Technology and Trade*, ed., R. P. Rothwell, Vol. I, New York, 1892, p. 278; and Fleming, *A Report to the Bituminous Coal Association*, 1908, p. 44.

TABLE A-3

CONSUMPTION AND VALUE OF FUEL WOOD AND CHARCOAL, 1879

Consumer	Cords (millions)	Value Per Cord (dollars)	Bituminous Coal Equivalent (million net tons)[a]	Value of Coal Per Ton at Mine (dollars)
Fuel wood				
Domestic	140.54	2.18	112.43	
Railroads	1.97	2.60	1.58	
Steamboats	0.79	2.30	0.63	
Precious metals	0.36	8.03	0.29	
Other mining	0.27	2.52	0.21	
Brick and tile	1.16	3.44	0.93	
Salt	0.54	0.23	0.43	
Wool	0.16	2.69	0.13	
Total (or average)	145.78	2.21	116.62	1.25

	Net Tons (thousands)	Value Per Net Ton (dollars)	Bituminous Coal Equivalent (thousand net tons)[b]	Value of Coke Per Ton of F.O.B. Oven (dollars)
Charcoal				
Domestic (20 largest cities)	43.19	12.07	66.60	
Iron industry	695.92	6.79	1,073.17	
Precious metals	0.98	30.00	1.51	
Total (or average)	740.09	7.13	1,141.28	1.95

Source: Fuel wood and charcoal, *Tenth Census, 1879-80*, Vol. IX, *Report on the Forests of the United States*, p. 489. Average value of bituminous coal, *Tenth Census*, Vol. XV, *Mining Industries of the United States*, p. xxviii. Average value of coke, *Tenth Census*, Vol. X, *Report on the Manufacture of Coke*, p. 12.

Note: The above cited Census *Report on the Forests of the United States* gives the value of a cord of fuel wood for each state, ranging from $1.21 in North Carolina to $7.16 in the Dakotas. The number of persons using wood as domestic fuel is estimated at 32,375,074, about two-thirds of the total population. This indicates an annual per capita consumption of some 4 1/3 cords for wood users, a figure which does not seem unreasonable compared with additional—though scanty—information for that period. However, the reported value per cord—$2.18 for the country as a whole—implies an expenditure for domestic fuel of more than $9 per person and several times as much for a family. This is out of proportion with the estimates of average per capita income during that period. It must be assumed that the reported value per cord was a retail price which was not paid by all users.

[a] Converted at the rate of 1 cord of fuel wood equals 0.8 net tons of bituminous coal.

[b] Converted at the rate of 1 net ton of charcoal equals 0.976 tons of coke; the production of 1 ton of coke required 1.58 tons of bituminous coal during the Census year 1879-80.

TABLE A-5

REPORTED AND ADJUSTED EMPLOYMENT IN BITUMINOUS COAL MINING
BY MAJOR COAL-PRODUCING STATES,
SELECTED YEARS, 1839-1918
(number of persons)

Census and Calendar Years	Pennsylvania	West Virginia[a]	Ohio	Kentucky	Illinois	Other States	Total
1839-40	1,798	995	434	213	152	242	3,834
1840	2,710	714	322	560	331	890	5,527
1849-50	n.a.	1,044	1,187	453	n.a.	n.a.	n.a.
1850	n.a.	n.a.	n.a.	n.a.	n.a.	n.a.	9,370
1859-60	4,651	1,190	1,678	757	1,430	1,642	11,348
1860	8,141	918	2,452	347	1,685	2,439	15,982
1869-70	16,851	1,140	2,567	714	6,301	9,160	41,733
1870	17,802	1,655	7,901	1,190	6,508	9,275	44,331
1879-80	33,248	4,497	16,331	2,826	16,301	35,811	109,014
1880	39,295	5,504	21,906	3,632	17,337	41,354	129,028
1890	61,333	12,236	20,576	5,259	28,574	64,226	192,204
1900	92,692	29,163	27,628	9,680	39,101	106,711	304,975
1910	175,403	68,663	46,641	20,316	72,645	171,865	555,533
1918	174,306	89,530	48,450	39,342	85,965	177,712	615,305

Source: 1890-1918: Geological Survey, *Mineral Resources of the United States*, various issues; 1840, 1860-80: the Census ratio of employment to output was multiplied by Eavenson's output estimate to get employment for each of the five states listed separately and for other states as a group. In 1840, 28 bushels = 1 long ton, according to Eavenson *(The First Century)*.

State estimates for 1850 would be unreliable. Total employment and the shares of bituminous coal and anthracite are not very solid, either, as is evident from the description of the estimate that follows.

Bituminous coal, 1850 *(E/O)*, for four main states (Pennsylvania, Virginia, Ohio, and Kentucky) was derived by interpolation between 1840 and 1860.

Anthracite, 1860 *(E/O)*, was at the same level as in 1840 (3.10). The Census gives employment for the four states for both bituminous coal and anthracite (see the *1850 Census*, pp. lxxix, 193, 272, 861, and 623), so that a correction factor, λ, which implies that forces peculiar to 1850 acted on *E/O* for both anthracite and bituminous coal:

NOTES TO TABLE A-5 (concluded)

Total coal employment in 1850 (*1860 Census*, p. clxxii)

$$= \frac{\begin{pmatrix}\text{Bit. coal} \\ \text{4 states} \\ (E/O)_{50}\end{pmatrix}\begin{pmatrix}\text{Estimated Census} \\ \text{bit. coal output} \\ \text{of 4 states}\end{pmatrix} + \begin{pmatrix}\text{Bit. coal,} \\ \text{other states} \\ (E/O)_{50}\end{pmatrix}\begin{pmatrix}\text{Estimated bit.} \\ \text{coal output of} \\ \text{other states}\end{pmatrix} + \begin{pmatrix}\text{Anthr.} \\ (E/O)_{50}\end{pmatrix}\begin{pmatrix}\text{Census} \\ \text{anthr.} \\ \text{output}\end{pmatrix}}{2.51(1960) + 4.7(348) + 3.10(4,138)} = \frac{15,118}{} = 0.78.$$

Bituminous coal output for four states in 1850 was estimated by applying their percentage share (0.85, interpolated between 1840 and 1860) to 1850 *Census* bituminous coal production.

Adjusted 4 state bit. coal employment = λ Bit. $(E/O)_{50}\begin{pmatrix}\text{Eavenson bit. coal} \\ \text{output of 5 states}\end{pmatrix}$ = 0.78 (2.51)(3,189) = 6,250.

$\lambda_{os}(E/O)_{50}$ for other states estimated by interpolation (4.7).

Other states bit. coal employment = $\lambda_{os}(E/O)_{50}\begin{pmatrix}\text{Eavenson other} \\ \text{states production}\end{pmatrix}$ = 0.79 (4.7)(840) = 3,120.

Total bituminous coal employment = 6,250 = 3,120 = 9,370.
Anthracite employment = λ anthr. $(E/O)_{50}\begin{pmatrix}\text{Eavenson anthr.} \\ \text{output}\end{pmatrix}$ = 0.79 (3.10)(4,327) = 10,600.

[a]Including Virginia during 1839-60.

TABLE A-6

UNITED STATES ANTHRACITE COAL MINING EMPLOYMENT,
SELECTED YEARS, 1840-1918
(number of persons)

	Census	Adjusted[a]
1840	2,977	3,490
1850	--	10,600
1860	25,138	34,021
1870	53,096	60,402
1880	70,748	70,748
1890		126,000
1900		144,206
1910		169,497
1918		147,121

[a]1840-80: Census employment times the ratio of Eavenson to Census output. But for 1850, see source to Table A-5. 1890-1918: Geological Survey, *Mineral Resources of the United States*, various issues. The data refer to production and development workers.

COMMENT

Paul W. McGann, U.S. Department of Commerce

This paper is a straightforward use of the adjustments of coal production data derived by Eavenson as applied to Census data for the period 1839 through 1890. Subsequent to that period, the data require but minor adjustment. For a half-century period of substantial underreporting of coal mining by the Census, all figures on employment and income from coal mining are adjusted accordingly. The author points out that there exists no comprehensive, detailed economic history of the coal mining industry for that period; and, therefore, no comprehensive set of economic data for the period is available beyond the careful production data work of Eavenson. An even smaller amount of data has been assembled for transportation costs of coal and amounts shipped by different routes and modes, despite the fact that transportation facilities were crucial in the development of energy for industry and that dramatic reductions in transportation costs permitted very large increases in coal production.

There is a trend in the ratio of cost of materials to value of shipments computed from Table 6 where there was a steady rise from 12.5 per cent in 1849-50 to 20 per cent for anthracite in the 1889-90 Census, but a slight decline for bituminous from 12.5 to 9.3 per cent. This interesting statistic apparently reflects the growing difficulty in operating anthracite mines during that period and the corresponding ease in expanding bituminous coal production at new sites in thick, flat, nondeep seams.

Information on average days operated per year in different states can be developed to some extent from Census data on capacity utilization which ostensibly assume 300-day normal years. The coal industry suffered from overexpansion from the earliest date of its development. That is, ease of entry has been so great that there always has been too much coal mining capacity in that large portions of it usually could not be operated profitably. This should be a sobering historical comment to those who have almost as chronically attempted to solve the problems of the coal mining industry.

The author emphasizes the fact that it would take a number of man-years to organize and evaluate the large amount of archives and other research materials on the coal industry which would be necessary to bring price and cost data into a condition of refinement comparable to that for coal mining production by Eavenson. Of course, not all these data would lend themselves to the development of annual data in Census detail as Eavenson developed production data. The author is to be thanked for clarifying the analysis of the impact of coal mining on national income. Coal was the major sector of the mining industry after 1860, but her tables remind us that coal mining was a very modest fraction of all national income generated, while at the same time it was a large industry from the late nineteenth century through World War II.

Several interesting features emerge from the Eliasberg summary of Census data on coal mining. The first is the surprisingly slow growth of coal to a consumption role dominating wood despite very great apparent price advantages. This contrasts sharply with the European experience and is no doubt due to the fact that market success for coal had to await a greater degree of urbanization and transport development (because real prices in rural areas were still much lower for wood, collected and prepared by farmers in the winter when they had no other crop to sell). Another remarkable feature of coal mining is the early and almost unchanging regional dominance pattern with five states as the leading producers, decade after decade. That stability was in contrast with the shifts in metal mining dominance during the period and occurred despite the fact that transportation costs were relatively high for coal compared with many other products. International trade in coal was small during the entire period because the greater relative productivity of U.S. than of European coal mining had not yet been realized. In fact, the growth of output per person employed was much smaller for coal than for almost any other U.S. industrial sector. The Eliasberg data also indicate that the shift after 1850 from anthracite to bituminous coal use depended to a large extent on the use of coal in steel production; that is, the substitution of coke for anthracite.

Harold J. Barnett, Washington University

Vera Eliasberg has made a useful contribution to the statistical economic history of the United States in the nineteenth century. It is attractively presented and interesting to read.

Among her major points are an allegation of extraordinarily high energy consumption per capita in 1850 and a level trend from at least 1850 to 1885. I suspect the possibility of error in both of these. I do not know that she is wrong—indeed I think the probability is high that she will satisfactorily answer my questions and doubts. But my responsibility to her in the minor role of discussant is to state them.

Mrs. Eliasberg writes following Table 8: "By the middle of the nineteenth century the total energy consumption of the United States economy was already huge in absolute terms, as can be seen in Table 1. Apparently it was larger in per capita terms than that of any other country, including the industrially more advanced such as Great Britain. Even when the mechanical energy derived from wind, water power, and work animals is disregarded and only fuel materials are considered, per capita use around 1850 was the rough equivalent of four tons of bituminous coal per year. Nine-tenths of this (measured in btu content), however, consisted of wood burned at the rate of about four cords per person per year, while only one-third of a ton of coal per year per person was being used."

Her description of the availability of virgin forests and their clearing to provide farm lands establishes that wood was plentiful. Wood would be used with a rather free hand, having regard for the cost and effort of cutting, trimming, hauling, sawing, splitting, and handling; the size of homes to be heated; and so on. But why the equivalent of "four tons" of coal per year? I suspect from the paper that the keystone is an 1880 Census volume on forests, which alleged a per capita wood use of about 2.2 tons of coal equivalent. This, when added to the estimated 1.6 tons of coal consumed, yields a figure approximately equal to the 3.9 tons per capita which Mrs. Eliasberg presented for 1850 (Table 9). Do I presume correctly that the 1880 Census per capita figure of wood fuel and total fuel is taken as valid? If so:

1. How does Mrs. Eliasberg view the competence of the group which prepared the particular Census volume in question, *The Forests of the U.S. in Their Economic Aspects*? How was the Census of home use conducted? To what extent was it a survey based on a sensible questionnaire and yielding objective data by state and county, which can be reviewed, and to what extent was it a subjective estimate by professional foresters?

2. The Census figure of wood consumption in homes using wood is the equivalent of about twenty tons of coal per year for a family of six. As Mrs. Eliasberg points out, this yields a very large fuel expense bill (about $50 per year, when valued at the Census unit price) relative to annual family income of the period. And fireplace or stove wood was not free—to the extent that it was not purchased in the market. Twenty-five cords of wood per year (twenty tons of coal equivalent) represent a very considerable labor outlay for the family, in converting living trees into an economic good, and provide an extraordinary amount of heat in found a small home.

Let me now assume that the forest Census figures for 1880 are, however, found to be very satisfactory. Then I have these questions:

1. Mrs. Eliasberg's figures on annual per capita fuel consumption for the preceding thirty years are virtually identical with the 1880 figure.

Did the author project the 1880 figure backward to 1850 at a constant level? If so, what is the basis for the assumption that the increased energy required for increase in GNP per capita from 1850 to 1880 was approximately offset by improved energy efficiency?

2. Can Mrs. Eliasberg reconcile her *level* annual per capita consumption of almost four tons of coal equivalent during the period 1850–85 with the Eavenson study (which she refers to in this paper and quotes in *Energy in the American Economy*[1])? Eavenson, using data reported by Marcus Bull in 1827, found annual fuel consumption in Philadelphia in 1826-27 to total about 150,000 tons of coal equivalent. For the Philadelphia population, then about 75,000, this is an annual consumption for all purposes of about two tons of coal equivalent per capita.

I close with a very brief comment concerning coal output, value of output, and employment. Mrs. Eliasberg has adopted Eavenson's output figures, which are upward revisions of Census data for the period 1840 to 1890. Then, on the assumption that Census coal values and employment were similarly understated, she has raised these figures in proportion to the output increase. From the method of the Eavenson revisions, however, the new output figures may still be seriously underestimated, as Eavenson points out (see quotation from Eavenson in Mrs. Eliasberg's discussion of Table 5). And, therefore, the value and employment figures may be too low. I anticipate that further combing of historical records and review of the implications of improved nineteenth century national statistics of labor force, GNP, and activity in important coal-consuming sectors will ultimately permit further improvement in the coal data.

[1] By Schurr *et al.*, p. 51.

Power and Machines

Growth and Diffusion of Power in Manufacturing, 1838–1919

ALLEN H. FENICHEL
MCGILL UNIVERSITY

Primary-Power Capacity

The data on the growth, diffusion, and changing composition of primary-power capacity in manufacturing presented in this paper are part of a larger and more detailed study in progress. Primary power means the work done by "prime movers," which convert the energy of nature directly into the energy of motion. The study covers prime movers developed or improved in the nineteenth century, the water wheel, the steam engine, the steam turbine, the internal combustion engine, and the electric motor. The list of the prime movers utilizing inanimate sources of energy excludes only the windmill, for which no adequate sources of information have been located. Electric motors are not prime movers from the standpoint of the economy, since they consume electric energy which must first be produced by one of the other prime movers, rather than converting the energy of nature directly into the energy of motion. From the standpoint of the firm, however, when electricity is created by a utility and purchased by the firm to operate an electric motor, the motor is a prime mover.

The basis used for measuring each type of primary power is the capacity of the motor or engine. Capacity refers to the rated ability of a machine to perform tasks requiring motive power; in other words, it is the potential work output of the machine. Changes in the efficiency with which the prime mover is used or in the portion of capacity used represent an important aspect of the growth, diffusion, and changing composition of power. Nevertheless, for some purposes, capacity provides an adequate representation of the role of the various types of primary power in manufacturing. The standard unit for measuring power capacity is horsepower, one unit of which is equivalent to a rate of 550 foot-pounds per second or 33,000 foot-pounds per minute.

NOTE: I am grateful to Richard A. Easterlin of the University of Pennsylvania for his advice during the preparation of this paper.

The period 1838 to 1919 was one in which primary-power capacity in manufacturing expanded substantially in the aggregate. The two basic tables (Tables A-1 and A-2) on power capacity, by geographic division and industry group, show that all primary-power capacity expanded from 2,346,000 horsepower in 1869 to 29,410,000 horsepower in 1919. Breaking down this total by type of power, one finds that steam-power capacity grew from 36,100 hp in 1838 to 1,216,000 hp in 1869 to 13,840,000 hp in 1919; and water-power capacity grew from 1,130,000 to 1,765,000 hp between 1869 and 1919. In addition, purchased electric power, first recorded as a separate source of power in the Census in 1899, contributed 9,348,000 hp to total primary-power capacity in 1919. Finally, there were internal combustion engines, first recorded in 1889, and steam turbines, first recorded separately in 1919, which had a capacity of 1,259,000 and 3,198,000 hp, respectively, in 1919. During the period of that absolute growth in capacity, there were, within each geographic division and industry group, changes in the proportion of total capacity assignable to the different types of power. Furthermore, the shares of various geographic divisions and industry groups in the total of each type of power changed. In the following paragraphs some of these developments are summarized.

The changing relative importance of each type of power in the total is presented in Table B-1, in terms of the varying rates of growth of each power source, and in Table B-2, in terms of the percentage distribution of total primary power by type. Although exact figures are not available on the relative importance of water power in the period before the Civil War, it seems reasonable to infer that water was the major source of power in manufacturing well past the middle of the century. This conclusion is based on the fact that, after its most rapid period of growth, 1838–69, steam was only slightly more important than water as a source of power. From 1869 to 1899, the relative position of steam continued to improve at the expense of water, reaching its peak of importance at the latter date, when it accounted for over four-fifths of total primary-power capacity. Then it too began to lose ground as electric power began to come into use. By 1919, electricity was challenging the dominant position of steam.

Regional Distribution

The share of the northern geographic divisions in total primary power declined from 1869 to 1919 but, as of the latter date, there was still a substantial concentration of total power capacity in these divisions, i.e., 73.4 per cent of the total (Table B-3 and Chart 1). The decline in importance of the North in the share of total primary power is attributable

CHART 1

Percentage Distribution of Total Primary-Power Capacity in Manufacturing, by Geographic Division, 1869–1919

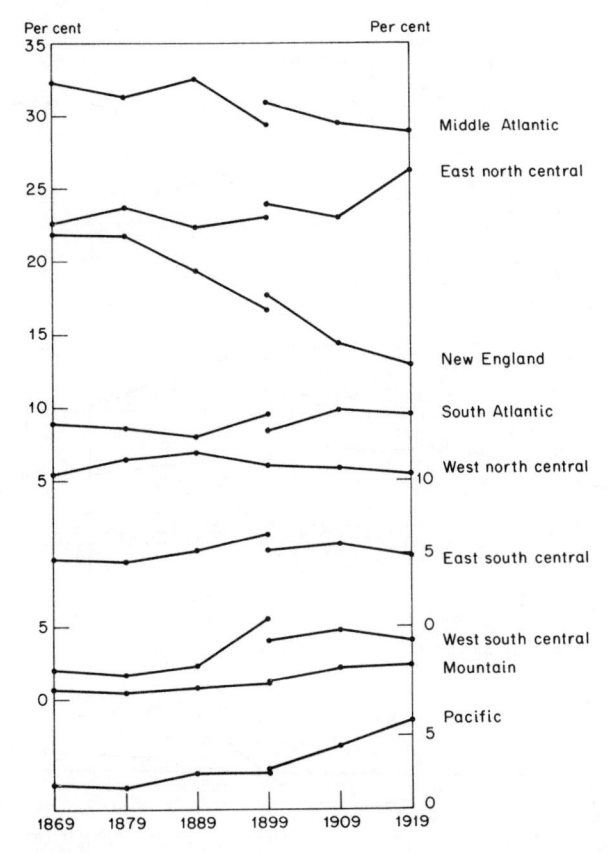

Source: Table B-3.

mainly to the declining importance of the New England division. The Middle Atlantic and east north central divisions together accounted for over 50 per cent of the total capacity throughout the period. Each of the southern and western divisions increased its share between 1869 and 1919, with the Pacific division having by far the largest increase.

A similar pattern appears in the relations between the individual types of power. For water power, the share of the four northern divisions in the total was over 80 per cent in 1869 and had increased slightly by 1919

CHART 2

Percentage Distribution of Total Steam-Power Capacity in Manufacturing, by Geographic Division, 1838–1919

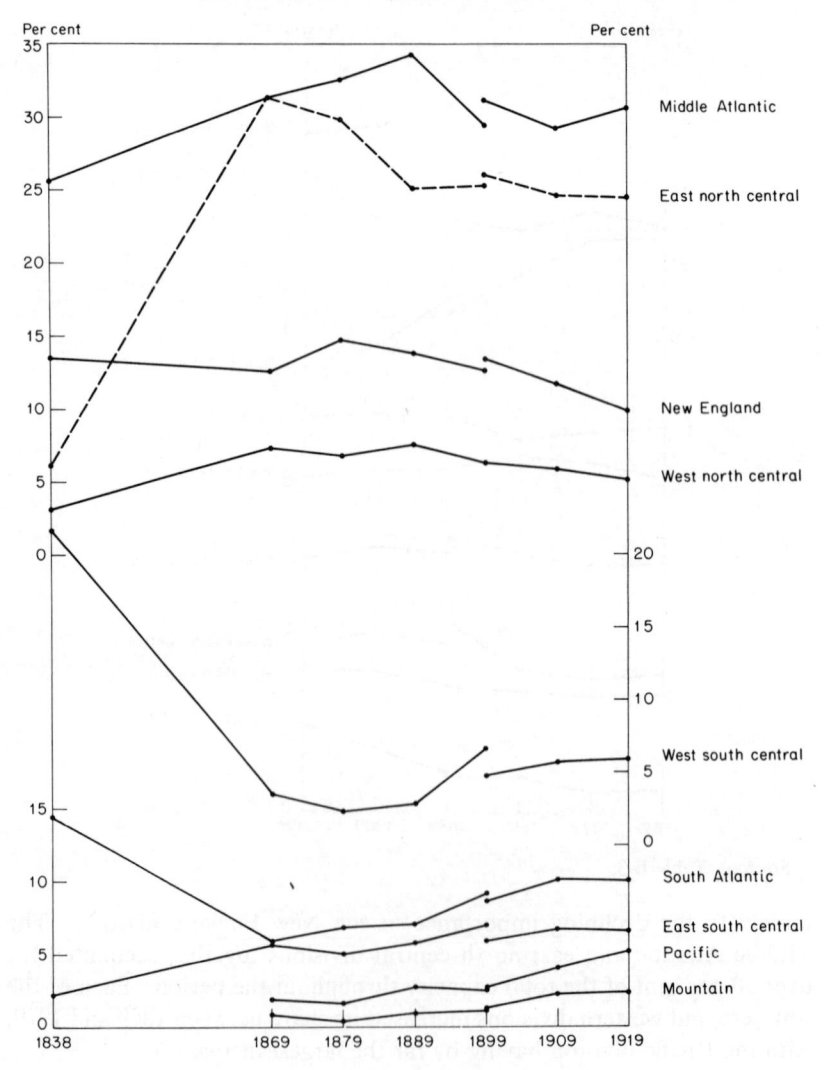

Source: Table B-5.

(Table B-4). The increase was attributable mainly to the New England states which, as will be pointed out again below, were much slower than the other states to reduce their dependence on water power. The Middle Atlantic states suffered a decline in their share, which corresponded to the increase in the share of New England. Given these shifts in their relative importance, the New England and Middle Atlantic divisions combined accounted for over 60 per cent of the total water-power capacity at each Census date. Over this same period, the share of the southern divisions declined slightly, while the western divisions increased their share.

As for steam power, in 1838, the North and the South shared in the total capacity, the northern share being somewhat larger (Table B-5 and Chart 2). Between 1838 and 1869, the major change was a drop in the share of the west south central division and an increase in the share of the east north central division. As a result, by 1869 the North, particularly the Middle Atlantic and east north central divisions, dominated total steam-power capacity, accounting for over 80 per cent of the total. Although subsequently the North's share declined, owing mainly to the drop in capacity in the New England and east north central divisions with a corresponding increase in each of the southern and western divisions, it was still 70 per cent of the total in 1919. By states, the concentration of steam-power capacity is even more evident (Table B-6). The top five states at each of the selected dates had about one-half the total capacity.

The distribution of purchased electric power, which was first recorded separately in the Census of 1899, also shows the relative significance of the North, and particularly the Middle Atlantic and east north central divisions (Table B-7). Although the share of the two divisions combined remained fairly constant, there was a shift in relative importance from the Middle Atlantic to the east north central division over the twenty-year period. The western divisions, especially the Pacific states, increased their share. The West had over 10 per cent of total electric capacity at each Census date, giving further evidence of its increased importance as a user of power.

Industrial Distribution

The distribution of total primary-power capacity, by industry group, shows more significant movements than the distribution by geographic division does. The movements were away from concentration of capacity in a few industry groups and also a greater shift in importance of the various groups, while two geographic divisions were initially, and remained throughout the period, the areas of greatest capacity. The shares of each industry group in total primary-power capacity (Table B-8) reveal that

CHART 3

Percentage Distribution of Total Primary-Power Capacity in Manufacturing, by Selected Industry Group, 1869–1919

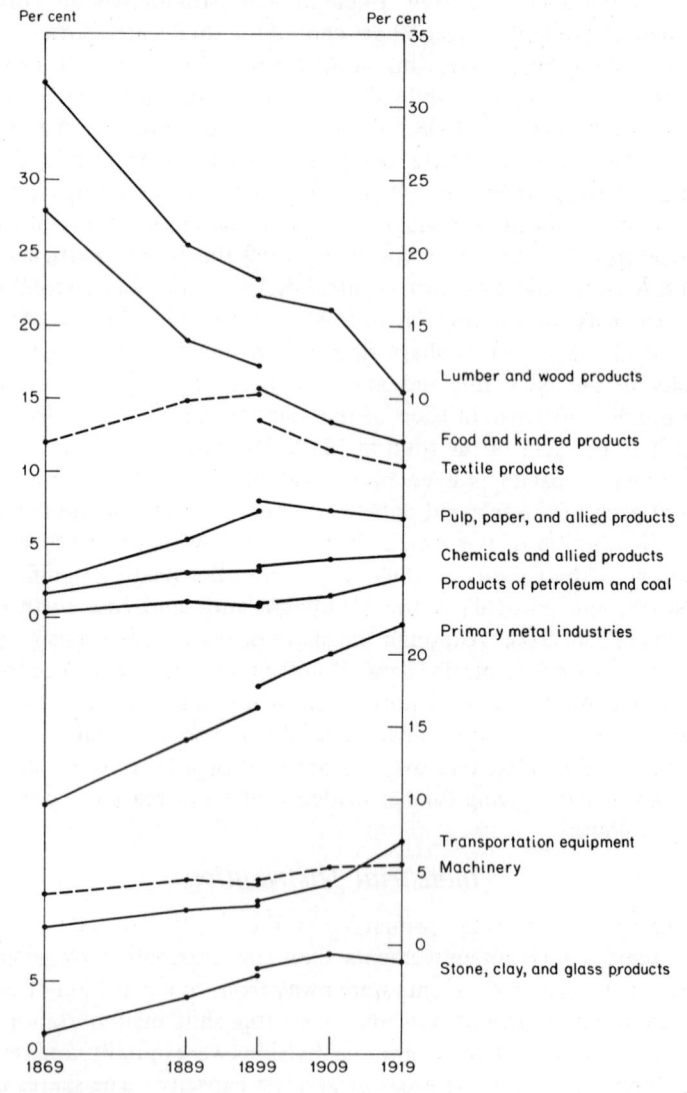

Lumber and wood products
Food and kindred products
Textile products
Pulp, paper, and allied products
Chemicals and allied products
Products of petroleum and coal
Primary metal industries
Transportation equipment
Machinery
Stone, clay, and glass products

Source: Table B-8.

in 1869 the lumber and food groups together had almost 60 per cent of total capacity. By 1919, the share of these two groups was only 22 per cent, while the primary metals group, continuously increasing its share between 1869 and 1919, became the group with the largest capacity. It is also notable that the four leading industry groups in 1919 did not have as large a share of capacity as the lumber and food groups had in 1869. Chart 3 shows the changing shares of total primary-power capacity of certain selected industry groups and points out the substantial shifting which was taking place. The groups presented in the chart are those that showed more than a 2 per cent change between their high and low points during the fifty-year period.

The distribution of total water-power capacity became more concentrated between 1869 and 1919, but there was a radical shift in the industry groups with the largest shares (Table B-9). In 1869 the food and lumber groups were dominant, while in 1919 the pulp and paper group, which had only 4 per cent of total capacity in 1869, had over 50 per cent of the total. The pulp and paper group continued to have a large portion of its total power capacity in the form of water power long after the other groups had come to depend on steam and eventually on electricity. As a result, the pulp and paper group gained an increasing share of total water power.

The shares of the top four industry groups (food, textiles, lumber, and primary metals) in total steam-power capacity remained about the same between 1838 and 1919, although between 1868 and 1919 their shares declined slightly (Table B-10). In addition, during 1838–1919 there were substantial changes in the share of capacity within the top four groups. Chart 4 shows the movements in those four groups, as well as in other industry groups, where the share changed by more than 2 percentage points between the high and low points. The most significant movements were the continuous decline in the share of the food group, the continuous increase in the share of primary metals, and the increase in the lumber group's share between 1838 and 1869, followed by a continuous decline.

The shares of purchased electricity were much more dispersed among the various industry groups than shares of the other types of power were (Table B-11). Two industry groups among the leaders in the share of total electric capacity—machinery and transportation equipment—had only a small share of total steam and water capacity. Two other facts brought out in Table B-11 (also emphasized in the discussion of the shares of each industry group in total primary power, water power, and steam power) are, first, the growing insignificance of the lumber group which, up

CHART 4

Percentage Distribution of Total Steam-Power Capacity in Manufacturing, by Selected Industry Group, 1838–1919

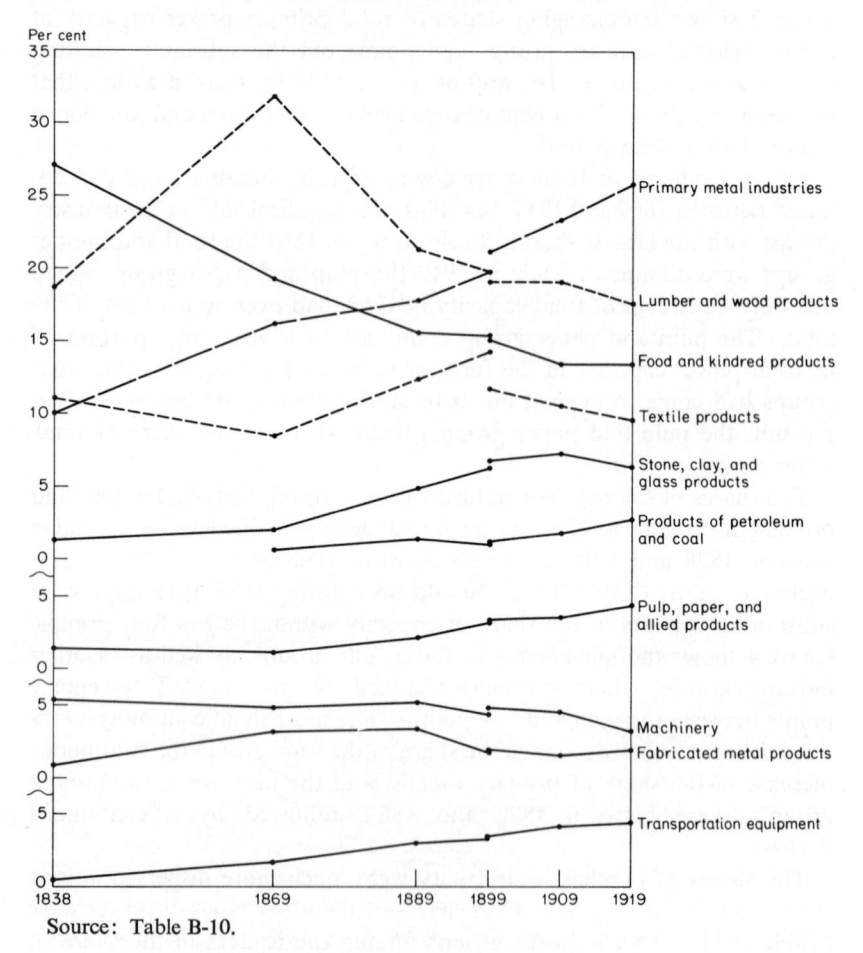

Source: Table B-10.

until 1899, had the largest share of total primary-power capacity[1] and, second, the increasing importance of the primary metals group.

Use of a two-digit industry classification tends to obscure the importance of certain Census industries within the major industry groups, as well as the changes taking place in the relative importance of these industries. There is, for example, the dominance of sawed lumber within the lumber and wood products group and of iron and steel in the primary metals group. There is also the declining relative importance of flour and gristmill products within the food group. To trace particular Census industries through the period is not always possible, because of the changes in definition and classification by the Bureau of the Census. Nevertheless, this aspect of the analysis is significant and will be developed more fully in the larger study.

Distribution by Type of Power

Tables B-12 and B-13 show the percentage distribution of primary-power capacity, by type of power, in each geographic division and industry group. The general pattern described earlier applies for the most part to each of them. Chart 5, for example, shows the percentage of steam-to primary-power capacity in each industry group, during the period 1869 to 1919. The similarities in the general pattern in the various groups is striking. Exceptions to the pattern in steam as well as in water and electricity are discussed below.

The timing of the peak for the relative importance of steam power varies. In certain divisions and industry groups, the peak is in 1889 or 1909 rather than 1899. In addition, the importance of steam, as well as of water and electricity, within each of the divisions and industry groups differs somewhat. The more notable differences by geographic divisions are, first, the continuing importance of water power in the New England states after the other divisions had reduced their share of water to relative insignificance. Next is the very early adoption and exceptionally large relative importance of steam power in the west south central states up to 1919. Finally, there is the importance of electricity in the Pacific states, which was the only group of states to use more purchased electricity than steam as early as 1919.

On an industry basis, in 1869, in only three groups—food, textiles, and pulp and paper—did water account for more than 50 per cent of total power capacity. Of these, only the pulp and paper group continued to depend on water for a substantial portion of its power needs. In fact,

[1] Only part of this decline can be attributed to the change in coverage by the Bureau of the Census after 1899 (see Appendix A).

CHART 5

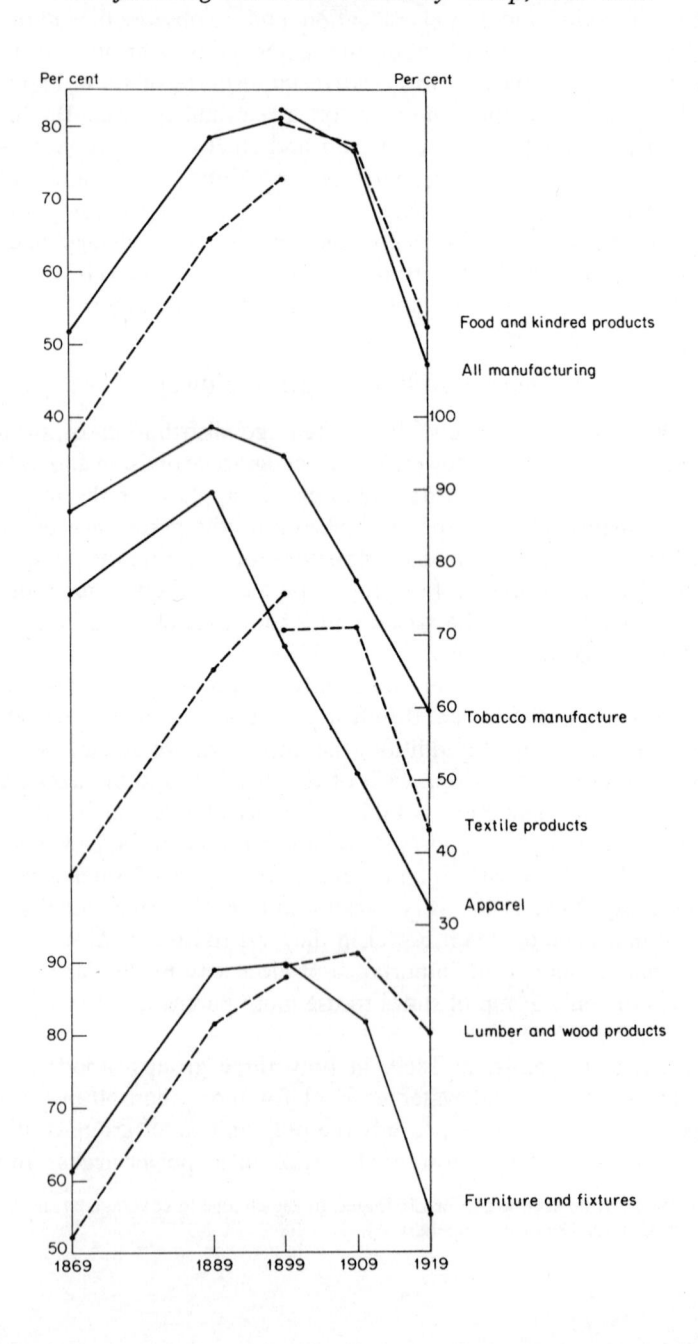

Share of Steam in Primary-Power Capacity in All Manufacturing and in Each Industry Group, 1869–1919

POWER IN MANUFACTURING, 1838–1919   453

CHART 5 (continued)

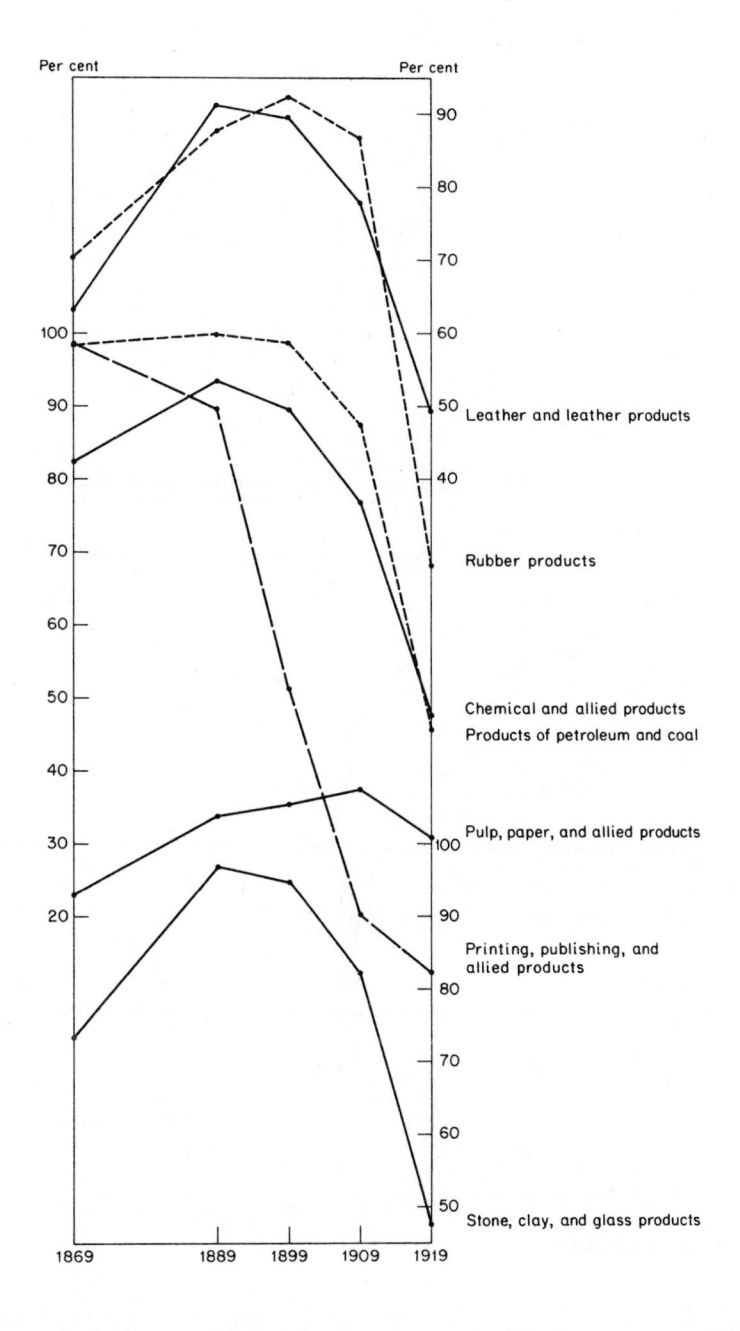

CHART 5 (concluded)

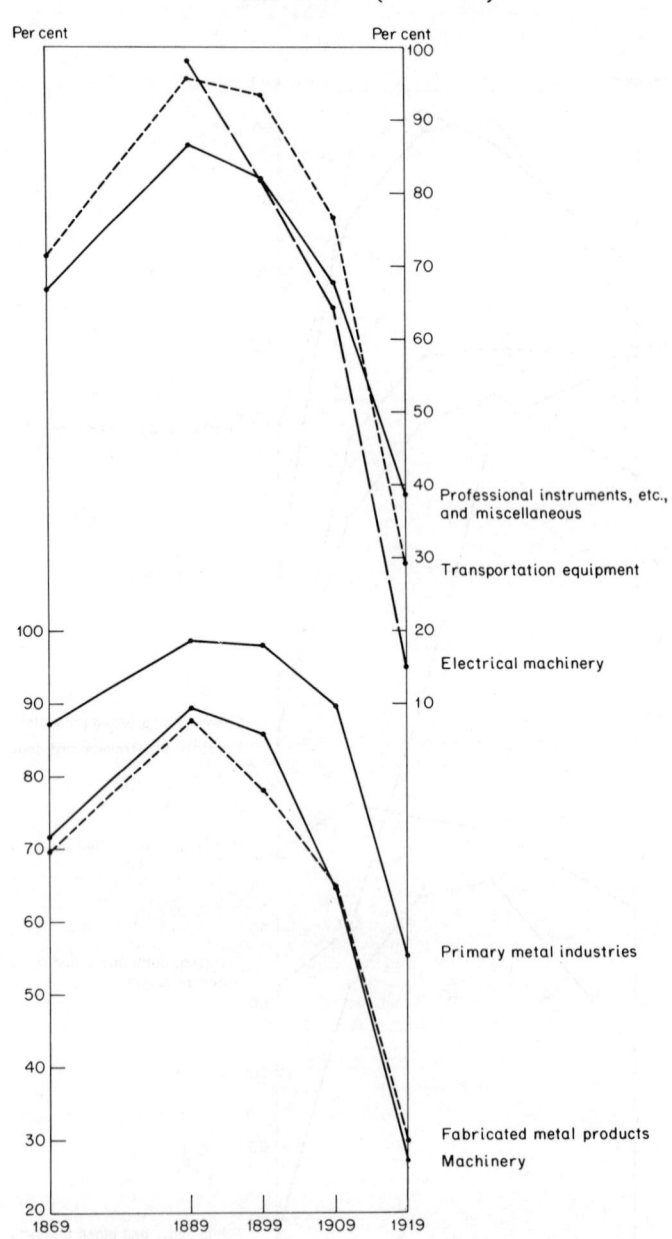

Source: Table B-13.

as late as 1919, pulp and paper still had 46 per cent of its total power capacity in the form of water. At the other extreme were the printing and publishing and primary metals groups, which were almost totally dependent on steam as early as 1869. The primary metals group continued that dependence up to 1909. The printing and publishing group, however, was the first to become dependent on purchased electricity as a source of power. As early as 1899, 30 per cent of its total capacity was purchased electricity, and by 1919 the figure was 82.6 per cent. A number of other industry groups adopted purchased electricity almost immediately, e.g., apparel and electric machinery; and by 1919 they had increased the relative share of purchased electricity within their groups to over 50 per cent. At the other extreme was the lumber group which, in 1919, still used steam for 80 per cent of its primary power, and electricity for only 10 per cent.

Appendix A: Notes on Scope, Sources, and Methods

COVERAGE OF POWER

The figures used for total primary power are based on the sum of steam, water, gas, and purchased electric power. In certain Census years, there are also classifications of "other owned" and "other rented" power. Since the amounts involved are relatively insignificant and not assignable to any particular type of power, they were excluded from the total. To get the actual amount of total primary power capacity at each Census date, the following amounts of horsepower should be added to the figures given in Tables A-1 and A-2.

	Other Horsepower, Rented (thous.)	Other Horsepower, Owned (thous.)
1889	89	5
1899A	137	54
1899B	137	20
1909	124	29
1919	95	–

Horsepower capacity in Alaska was included in the Censuses in 1889 and 1899.

	Horsepower Capacity in Alaska		
	1889	1899A	1899B
Steam	290	1078	954
Water	161	597	117

These amounts were excluded from Tables A-1 and A-2 in order to make them consistent with the other Census dates.

TABLE A-1

PRIMARY-POWER CAPACITY IN MANUFACTURING, BY GEOGRAPHIC DIVISION
(thousand horsepower)

	1838	1869			1879			1889				1899A				
	Steam	Water	Steam	Total	Water	Steam	Total	Water	Steam	Gas	Total	Water	Steam	Gas	Electric (purch.)	Total
United States	36.1	1,130	1,216	2,346	1,225	2,185	3,411	1,255	4,581	9.0	5,845	1,727	8,741	144	184	10,796
New England	4.9	362	153	515	423	320	743	497	634	0.5	1,131	656	1,101	11	32	1,799
Middle Atlantic	9.2	376	380	756	357	710	1,066	332	1,566	3.0	1,901	480	2,572	48	75	3,175
East north central	2.2	150	381	531	158	650	808	155	1,152	3.0	1,310	195	2,205	48	31	2,479
West north central	1.1	37	89	127	71	150	221	60	345	0.8	406	66	550	20	15	651
South Atlantic	5.2	140	70	210	146	149	294	145	319	0.8	466	210	799	7	5	1,022
East south central	0.8	41	68	109	43	110	153	35	268	0.3	303	59	610	3	4	676
West south central	7.8	4	42	46	5	53	58	5	132	0.4	137	7	583	2	4	596
Mountain	0	7	10	17	9	8	17	7	40	0.1	47	20	98	1	4	123
Pacific	0	14	22	36	15	36	51	19	125	0.4	145	34	223	4	15	275
Not assignable	4.9	0	0	0	0	0	0	0	0	0	0	0	0	0	0	0

(continued)

TABLE A-1 (concluded)

	1899B					1909					1919					
	Water	Steam	Gas	Electric (purch.)	Total	Water	Steam	Gas	Electric (purch.)	Total	Water	Steam	Gas	Electric (purch.)	Steam Turbines	Total
United States	1,454	8,140	135	183	9,911	1,823	14,199	751	1,749	18,522	1,765	13,840	1,259	9,348	3,198	29,410
New England	619	1,093	10	32	1,754	757	1,657	42	219	2,675	748	1,356	34	1,130	511	3,780
Middle Atlantic	410	2,529	46	74	3,059	470	4,152	274	569	5,465	396	4,231	410	2,455	1,019	8,513
East north central	171	2,118	45	31	2,365	208	3,491	283	376	4,359	253	3,389	428	2,659	973	7,703
West north central	52	513	18	15	597	86	839	57	115	1,098	93	707	102	596	95	1,593
South Atlantic	125	700	7	5	837	183	1,431	36	171	1,822	173	1,387	137	887	208	2,791
East south central	25	480	2	4	511	29	954	12	39	1,034	25	912	23	311	143	1,414
West south central	3	388	2	4	396	3	806	29	32	870	3	813	81	219	64	1,180
Mountain	18	98	1	4	121	22	307	4	67	399	16	322	10	260	78	685
Pacific	31	221	4	14	270	63	563	12	162	801	59	722	33	830	108	1,752
Not assignable	0	0	0	0	0	0	0	0	0	0	0	0	0	0	0	0

Note: For sources and methods, see Appendix A and source to Table A-2. Calculations were made with unrounded figures. Detail does not necessarily add to totals because of rounding.

TABLE A-2

PRIMARY-POWER CAPACITY IN MANUFACTURING, BY INDUSTRY GROUP
(thousand horsepower)

Major Group Number	Industry Group	1838 Steam	1869 Water	1869 Steam	1869 Total	1889 Water	1889 Steam	1889 Gas	1889 Total
	All Manufacturing	36.1	1,130	1,216	2,346	1,255	4,581	9.0	5,845
20	Food and kindred products	9.8	417	235	652	392	716	1.0	1,109
21	Tobacco manufacture	—	0.4	3	3	0.1	14	0.1	14
22	Textile mill products	4.0	177	104	281	303	568	0.2	871
23	Apparel	0.4	3	11	14	2	26	0.7	29
24	Lumber and wood products	6.7	355	387	742	222	975	0.2	1,198
25	Furniture and fixtures	—	12	18	30	9	76	0.2	85
26	Pulp, paper, and allied products	0.5	44	13	57	205	105	0.2	311
27	Printing, publishing, and allied products	—	0.1	9	9	3	49	3.0	54
28	Chemicals and allied products	0.5	7	31	38	12	172	0.3	184
29	Products of petroleum and coal	0	0.1	7	7	0.1	66	0.1	66
30	Rubber products	—	2	4	6	3	24	—	27
31	Leather and leather products	0.4	16	28	44	7	84	0.6	92
32	Stone, clay, and glass products	0.5	9	25	34	7	223	0.1	230
33	Primary metal industries	3.6	29	197	226	12	814	0.1	826
34	Fabricated metal products	0.3	17	38	55	21	151	0.5	172
35	Machinery	1.9	23	59	82	27	234	0.4	261
36	Electric machinery	0	—	0	—	12	62	—	74
37	Transportation equipment	0.1	8	21	29	6	136	0.2	142
38	Professional instruments, etc., and miscellaneous	0.6	5	10	15	6	44	0.5	50
	Nonmanufacturing	1.7	7	17	23	6	44	0.4	50
	Not assignable	4.9	0	0	0	0	0	0	0
	Adjustments to agree with Table A-1	0	0	0	0	0	0	0	0

(continued)

TABLE A-2 (continued)

Major Group Number	Industry Group	1899A					1899B				
		Water	Steam	Gas	Electric (purch.)	Total	Water	Steam	Gas	Electric (purch.)	Total
	All Manufacturing	1,727	8,741	144	184	10,796	1,454	8,140	135	183	9,911
20	Food and kindred products	462	1,358	31	19	1,871	254	1,232	31	19	1,536
21	Tobacco manufacture	0.4	23	0.4	0.5	24	0.4	23	0.4	0.5	24
22	Textile mill products	388	1,252	3	10	1,653	378	949	3	10	1,339
23	Apparel	4	47	6	13	69	4	47	6	13	69
24	Lumber and wood products	222	1,727	11	4	1,963	168	1,551	11	4	1,734
25	Furniture and fixtures	8	119	3	3	132	8	119	3	3	132
26	Pulp, paper, and allied products	507	279	2	2	790	507	279	2	2	790
27	Printing, publishing, and allied products	3	56	15	35	108	3	56	15	35	108
28	Chemicals and allied products	12	313	3	21	350	12	313	3	21	350
29	Products of petroleum and coal	—	94	1	0.1	95	—	94	1	0.1	95
30	Rubber products	5	65	—	0.2	70	5	65	—	2	70
31	Leather and leather products	6	132	3	7	147	6	132	3	7	147
32	Stone, clay, and glass products	18	554	7	5	584	18	554	7	5	584
33	Primary metal industries	27	1,727	5	4	1,762	27	1,727	5	4	1,762
34	Fabricated metal products	19	155	8	16	199	19	155	8	16	199
35	Machinery	28	387	20	16	451	28	387	20	16	451
36	Electric machinery	1	35	2	5	43	1	35	2	5	43
37	Transportation equipment	7	276	6	6	295	7	276	6	6	295
38	Professional instruments, etc., and miscellaneous	6	72	4	6	88	6	72	4	6	88
	Nonmanufacturing	4	71	13	7	95	4	71	13	7	95
	Not assignable	0	0	0	0	0	0	0	0	0	0
	Adjustments to agree with Table A-1	0	0	0	4	4	0	4	-9	3	-2

(continued)

TABLE A-2 (concluded)

Major Group Number	Industry Group	1909					1919					
		Water	Steam	Gas	Electric (purch.)	Total	Water	Steam	Gas	Electric (purch.)	Steam Turbines	Total
	All Manufacturing	1,823	14,199	751	1,749	18,522	1,765	13,840	1,259	9,348	3,198	29,410
20	Food and kindred products	275	1,921	102	183	2,480	197	1,875	187	1,141	168	3,567
21	Tobacco manufacture	0.2	22	0.8	5	28	0.4	26	0.3	14	2	43
22	Textile mill products	443	1,509	12	160	2,124	448	1,322	12	841	446	3,068
23	Apparel	5	73	10	55	143	4	67	4	129	2	206
24	Lumber and wood products	151	2,709	44	71	2,975	74	2,441	51	314	172	3,052
25	Furniture and fixtures	8	211	9	31	259	5	176	7	115	16	320
26	Pulp, paper, and allied products	794	515	10	57	1,375	916	613	6	305	151	1,990
27	Printing, publishing, and allied products	2	60	33	203	298	1	46	17	311	1	376
28	Chemicals and allied products	18	562	14	137	732	10	599	34	454	168	1,264
29	Products of petroleum and coal	4	249	14	18	285	0.3	368	53	198	188	808
30	Rubber products	5	108	2	9	124	6	120	4	193	106	429
31	Leather and leather products	6	207	12	40	265	5	187	8	143	37	380
32	Stone, clay, and glass products	24	1,042	77	127	1,270	23	884	127	701	123	1,859
33	Primary metal industries	27	3,319	219	135	3,700	22	3,575	554	1,236	1,079	6,465
34	Fabricated metal products	19	217	33	66	335	15	265	38	528	44	890
35	Machinery	28	637	103	217	983	25	438	81	954	120	1,618
36	Electric machinery	1	106	7	51	164	2	72	8	253	144	478
37	Transportation equipment	6	589	34	140	769	3	613	39	1,232	206	2,093
38	Professional instruments, etc., and miscellaneous	8	140	15	43	206	7	141	12	183	22	365
	Nonmanufacturing	0.6	4	0.1	0.7	6	2	13	19	103	2	138
	Not assignable	0	0	0	0	0	0	0	0	0	0	0
	Adjustments to agree with Table A-1	0	0	0	0	0	0	0	0	0	0	0

Note: For sources and methods, see Appendix A. Calculations were made with unrounded figures. Detail does not necessarily add to totals because of rounding. A distinction is made between a rounded-off zero, shown as --, and a real zero, shown as 0.

SOURCE, BY CENSUS YEAR
TABLES A-1 AND A-2

1838: House of Representatives, 25th Cong., 3d sess., *Steam Engines,* H. Ex. Doc. 21, Washington, 1839, pp. 18-367, 379.
1869: Secretary of the Interior, *Ninth Census of the United States, 1870,* Vol. III, *The Statistics of the Wealth and Industry of the United States,* Washington, 1872, pp. 392, 394-398.
1879: Dept. of the Interior, *Tenth Census of the United States, 1880,* Vol. II, *Report on the Manufactures of the United States,* Washington, 1883, p. 501.
1889: Dept. of the Interior, *Eleventh Census of the United States, 1890,* Part I, *Report on Manufacturing Industries in the United States,* Washington, 1895, pp. 758-768.
1899A: Census Office, *Twelfth Census of the United States, 1900,* Vol. VII, Part I, *Manufactures, 1900,* Washington, 1902, pp. 582-595.
1899B: Dept. of Commerce and Labor, Bureau of the Census, *Manufactures, 1905,* Part IV, Washington, 1908, pp. 619-621, 627, 630, 636.
1909: Dept. of Commerce, Bureau of the Census, *Thirteenth Census of the United States, 1910,* Vol. VIII, *Manufactures, 1909,* Washington, 1913, pp. 341, 522-541.
1919: Dept. of Commerce, Bureau of the Census, *Fourteenth Census of the United States, 1920,* Vol. III, *Manufactures, 1919,* Washington, 1923, pp. 123-229.

CLASSIFICATION OF MANUFACTURING INDUSTRIES

One of the problems in analyzing power capacity by industry is to get comparable industry groups for the various Census dates. For 1838 to 1879, the industry classifications used here follow those of Robert Gallman. His two-digit classification scheme was based on the 1945 edition of the *Standard Industrial Classification Manual.* For the Censuses after 1879, the classifications, with adjustments necessary to achieve consistency with Gallman's groupings, were provided by Richard DuBoff, who based his classification scheme on that of Solomon Fabricant.[2]

To establish comparable industry groups, certain Census industries are not included in any of the major groups. Those industries have been included in Table A-2 under the heading of nonmanufacturing and are, for the most part, activities associated with agriculture, services, and mining. Since the objective of this work is to study power capacity in manufacturing, these industries, by definition, should be excluded from the totals, but they were included for two reasons. First, if the industries were excluded from the analysis by industry, then to achieve consistency they would also have to be excluded from the analysis by geographic division. To find the portion of each of the nonmanufacturing items in each state would have been difficult. Second, the amounts involved as a percentage of the total of each type of power at each Census date were

[2] Richard B. DuBoff, "Electric Power in American Manufacturing, 1880-1955," unpublished Ph.D. dissertation, University of Pennsylvania, 1963. Solomon Fabricant, *The Output of Manufacturing Industries, 1899-1937,* New York, NBER, 1940, Appendix C.

insignificant (Tables B-8 to B-11). Therefore, the totals in Tables A-1 and A-2 include data for a few activities not properly assignable to manufacturing.

NATURE OF BASIC DATA AND ADJUSTMENTS MADE FOR EACH CENSUS DATE

1838: The data on steam engines in 1838 is based on a study conducted by the Secretary of the Treasury in response to a June 1837 resolution of the House of Representatives. The House was interested in the use of steam engines in the United States and the accidents and loss of life or property attending their use. The Secretary, in turn, directed the collectors of customs to obtain the necessary information in their respective districts. The accuracy of the data thus depends in large part on the diligence of the individual collectors. There is a strong indication in the original House resolution and in the steps taken by the Secretary of the Treasury that the government made a genuine effort to insure the proper collection of information, although the limited time made available for the study led to incomplete returns.

The results of the study, which included information on steamboats and steam locomotives, as well as steam engines used in manufacturing, was presented in detail in House Executive Document 21.[3] Of the totals of 1,865 stationary steam engines and 36,068 horsepower reported, details were provided for 1,266 engines and 22,593 hp, while 599 engines and 13,475 hp were estimated either by the Secretary of the Treasury or the district collector of customs. Of these estimates, 244 engines and 4,880 hp represented an over-all estimate not assignable to any state or industry group, while 355 engines and 8,595 hp were assigned to particular states but with no additional details provided on the use of the engines. There is no specific reference made in the document to the basis used for the estimates.

Selection of the industry groups to which the engines should be assigned presented two problems. First, a few of the descriptions of the uses of the engines were not clear, and the best possible judgment, given the information presented, had to be made. The second problem was the assignment to industries of the 355 engines and 8,595 hp which were assigned to states, but for which the industry was not reported. The assignment of engines and horsepower made is shown in Table A-3, and the procedure used is described below.

1. For Louisiana and Pennsylvania, the horsepower and engines per industry in 1838, for which the industrial distribution was known, were

---

[3] House of Representatives, 25th Cong., 3d Sess., *Steam Engines*, H. Ex. Doc. 21, Washington, 1839, p. 305.

TABLE A-3

STATIONARY STEAM ENGINES AND TOTAL ESTIMATED HORSEPOWER,
AS ASSIGNED TO INDUSTRIES AND STATES, 1838

Major Group Number	Industry	Louisiana		Pennsylvania		Alabama		Missouri		New York		Other		Total		Percentage of Each Industry's Total Hp, Estimated
		No.	Hp	No.	Hp	No.	Hp	No.	Hp	No.	Hp	No.	Hp	No.	Hp	
20	Food and kindred products	139	4,070	2	38	20	400	28	560	17	384			206	5,452	55.8
21	Tobacco manufacture			--	1									--	1	2.7
22	Textile mill products	8	246	2	82					4	97			14	425	10.7
23	Apparel			--	6					1	17			1	23	5.4
24	Lumber and wood products	28	806	2	40	20	400	28	560	12	228			90	2,034	30.2
26	Pulp and paper products			1	11					--	2			1	13	2.7
28	Chemical and allied products			--	6					3	19			3	25	4.6
31	Leather and leather products			1	8					3	24			4	32	7.6
32	Stone, clay, and glass products			1	12					1	12			2	24	5.1
33	Primary metal industries			3	192									3	192	5.3
34	Fabricated metal products			--	3					2	15			2	18	6.6
35	Machinery			3	45					11	110			14	155	8.1
37	Transportation equipment			1	4					1	10			2	14	9.7
38	Instruments, etc., and miscellaneous			1	5					4	57			5	62	10.1
	Nonmanufacturing			1	52							7	73	8	125	7.5
	Total	175	5,122	18	505	40	800	56	1,120	59	975	7	73	355	8,595	23.8
	Percentage of each state's total hp, estimated	65.7		6.8		100.0		100.0		68.4		--		23.8		

used. The distribution of horsepower in 1869 in these two states indicated that no major Census industries had been left out in 1838. In fact, for Louisiana, the customs collector in 1838 specifically indicates that the estimate represents engines used in sugar mills, saw mills, and cotton gins.[4]

2. For Alabama and Missouri, there was no information on steam capacity by industry in 1838, but the use of steam power in 1870 provided a satisfactory basis for assigning 1838 engines and horsepower. A comparison of the 1840 and 1870 *Census of Manufacturing* showed that two of the major industries in these two states in 1840—flour and gristmill products and sawed lumber—were still the only significant industries, so far as steam-power capacity goes, in 1870. The assignment of 1838 engines and horsepower was based, therefore, on the almost equal distribution of steam power between these industries in 1870.

3. In New York, the basis was the distribution of workers by industry group in 1839 as reported in the 1840 Census.[5] The average steam horsepower available per worker in the United States in each industry group was found by dividing the known steam horsepower in each group in 1838 by the number of workers in each group in 1839. The results were multiplied by the number of workers in each group in 1839 in New York. These figures were then used as the basis for apportioning the 975 horsepower; the percentage of the total in each industry group was computed and applied to the 975 hp. For the industrial distribution of steam in New York in 1838, there was not enough information to supply a basis as in Louisiana and Pennsylvania. Instead, the distribution of engines was based on the 1838 average horsepower per engine in each major group in the United States.

4. In the rest of the states, the number of engines and amounts of horsepower were very small (7 engines, 73 horsepower, in total) and were arbitrarily assigned to the nonmanufacturing group. As mentioned previously, 244 engines and 4,880 horsepower were not assigned to any state or industry in the House document and, without further information, I classed them in Tables A-1 and A-2 as "not assignable."

The data for the decades from 1869 on are based on Census information. Certain problems encountered in use of the Census volumes and the methods used to deal with them are indicated below.

1869: In comparing horsepower capacity in the west central and mountain divisions (Table A-1) and the food and kindred products industry group (Table A-2), in 1869, to horsepower capacity at later

[4] H. Ex. Doc. 21.
[5] *Sixth Census, 1840*, pp. 358–364.

Census dates, it is necessary to take into account certain changes in the treatment of industries by the Bureau of the Census. The changes involve amounts that are not significant in relation to total horsepower capacity in 1869 but are significant within the divisions and industry groups.

In the 1869 Census, 18,296 horsepower of steam capacity in Louisiana was used for making sugar on plantations directly from raw cane. In 1879, steam engines used for that purpose were included by the Census Bureau in agriculture. Quartz milling was treated by the Bureau as part of manufacturing in 1869 and as part of mining after that date. In two

TABLE A-4

HORSEPOWER CAPACITY OF INDUSTRIES IN 1879 CENSUS

Industry Group of Census Industry	Census Industry	Water	Steam	Total
		(thous. hp)		
20	Flour and gristmill products; sugar and molasses, refined	470	322	792
22	Carpets, cotton goods, hosiery, silk, woolen, and worsted goods	218	211	429
24	Lumber, sawed	279	543	822
26	Paper	88	36	124
33	Iron and steel	17	381	397
35	Foundry and machine shop products, and agricultural implements	28	117	145

states, Montana and Nevada, quartz milling accounted for a significant portion of total steam- or water-power capacity in 1869—5,006 steam horsepower in Nevada and 596 steam horsepower in Montana. In addition, 2,168 horsepower of water was available in Nevada in 1869 for quartz milling. The decline in steam-power capacity in the mountain states between 1869 and 1879, shown in Table A-1, can be explained by the treatment of this Census industry.

1879: The 1879 Census gives horsepower capacity in only selected industries rather than for the whole range.[6] This created a comparability problem between the 1879 figures, by industry, and those for the other Census dates. Table A-4 gives the selected industries and the amounts of each type of power for which figures on power were provided. To estimate the percentage of total horsepower capacity in the respective industry group accounted for by these Census industries in 1879, the same industries were selected from the 1869 and 1889 Censuses, and their horsepower capacity for each type of power was expressed as a percentage

[6] *Tenth Census, 1880,* p. 502.

of the total in the industry group. The results are given in Table A-5. While these Census industries accounted for a large percentage of the horsepower capacity in their respective industry groups, perhaps even more important is the fact that, with only three exceptions—steam-power capacity in group 20 and water-power capacity in groups 26 and 33—the percentages were fairly constant over the twenty-year period.

In the computations by industry group in the tables in Appendix B, 1879 data are not included. It is possible, however, to develop an estimate of the horsepower capacity in these six industry groups by using the percentages indicated in Table A-5, which will be done in the larger study.

TABLE A-5

RATIO OF HORSEPOWER CAPACITY IN SELECTED CENSUS INDUSTRIES, REPORTED IN 1879, TO TOTAL HORSEPOWER CAPACITY IN THEIR RESPECTIVE INDUSTRY GROUPS, BY TYPE OF POWER, 1869 AND 1889

Industry Group	1869			1889		
	Water	Steam	All Primary Power	Water	Steam	All Primary Power
20	98.0	82.2	92.5	98.0	55.1	70.2
22	91.9	85.6	89.5	94.6	77.5	83.5
24	92.0	81.4	86.5	89.9	77.6	79.9
26	95.3	88.8	93.8	75.1	85.5	77.4
33	90.3	93.7	93.3	69.8	91.8	91.5
35	92.8	92.2	92.4	96.7	95.7	95.8

1899A-1899B: In 1905, the Bureau of the Census made a major revision in its industry coverage and retabulated 1899 data on a new basis. As a result, there are two sets of figures on power capacity in 1899. The 1899A figures should be used in making comparisons with dates prior to 1899, and the revised figures, 1899B, for comparisons with dates after 1899. The difference between the data in 1899A and 1899B is traceable to the treatment of custom and neighborhood establishments in three Census industries,[7] flour and gristmill (industry group 20), cotton gins (group 22), and lumber and timber (group 24). The effect of the exclusion of these custom and neighborhood establishments on power capacity is to reduce the total figure for steam capacity by about 7 per cent, water capacity by about 16 per cent, and total capacity by about 8 per cent. The exact effect of the exclusion on each geographic division and industry group can be seen by comparing the 1899A and 1899B columns in Tables A-1 and A-2.

[7] *Manufactures, 1905,* Part IV, Bureau of the Census, pp. 619–621.

Corrections were necessary in Table A-2, 1899A and 1899B, to obtain totals corresponding to those in Table A-1. The total electric horsepower figure in 1899A, by industry group, is 179,844 or 3,838 hp less than the total, by geographic division. The total by geographic division is probably correct, since it is higher than the 1899B electric horsepower figure given

TABLE A-6

PERCENTAGE DISTRIBUTION OF TOTAL STEAM-POWER CAPACITY IN MANUFACTURING, ENGINES AND TURBINES, 1919

Geographic Division	Per Cent	Major Group Number	Industry Group	Per Cent
United States	100.0	20	All manufacturing	100.0
North	72.1	20	Food and kindred products	12.0
New England	11.0	21	Tobacco manufacture	0.2
Middle Atlantic	30.8	22	Textile mill products	10.4
East north central	25.6	23	Apparel	0.4
West north central	4.7	24	Lumber and wood products	15.3
South	20.7	25	Furniture and fixtures	1.1
South Atlantic	9.4	26	Pulp, paper, and allied products	4.5
East south central	6.2			
West south central	5.1	27	Printing, publishing, and allied products	0.3
West	7.2			
Mountain	2.3	28	Chemicals and allied products	4.5
Pacific	4.9			
State		29	Products of petroleum and coal	3.3
Pennsylvania	18.6	30	Rubber products	1.3
Ohio	10.5	31	Leather and leather products	1.3
New York	7.6			
Illinois	5.9	32	Stone, clay, and glass products	5.9
Massachusetts	5.7			
Louisiana	1.9	33	Primary metal industries	27.3
Virginia	1.5	34	Fabricated metal products	1.8
Total, top 5	48.3	35	Machinery	3.3
Total, top 7	51.7	36	Electric machinery	1.3
Total, all others	48.3	37	Transportation equipment	4.8
		38	Professional instruments, etc., and miscellaneous	1.0
			Nonmanufacturing	0.1

in the 1905 Census volume—which excludes some custom and neighborhood establishments—rather than lower as the industry total figure is. In 1899B, the adjustment of the total steam-power figure is made necessary by the fact that, in excluding the custom and neighborhood establishments from the three industry groups mentioned earlier, there is a reduction in the total of 605,838 hp, while the actual difference between total steam-power capacity in 1899A and 1899B is only 601,859 hp. The total for gas in 1899B compared with 1899A declined but, since there was no basis given for correcting the individual industry groups, the adjustment item was used. The total for electric horsepower in 1899B was adjusted to agree with the 1899B total purchased electric horsepower figure given in the Census volume.

1909 and 1919: Before 1919, steam turbines were not recorded separately in the Census. Since these turbines were first used after the turn of the century, and the major growth in their use was between 1909 and 1919, the amount of steam power represented by turbines in the 1909 Census figure is relatively insignificant. It would be preferable in studying the changes in the use of steam power to keep steam turbines and steam engines separate. The turbines were used to produce electric energy to run electric motors and thus really represent a movement away from steam, or at least a change in the way steam was used as a source of power. In the Appendix B tables, turbines are not included in the figures on steam in 1919. However, although the relative amount is small, the 1909 figures do contain horsepower representing steam turbines. To indicate the effect on the comparison of 1909 and 1919 data of the inclusion of steam turbines in the 1919 computations, I have recomputed the 1919 columns in Tables B-5, B-6, and B-10. The change in the percentages caused by the inclusion of turbines is relatively small, as is evident in Table A-6.

In Tables B-12 and B-13, it is only necessary to add together the steam and steam turbine columns for 1919 to get the percentage directly comparable with 1909. The inclusion of turbines here does not alter the basic conclusion that steam was declining in importance as a source of power in each geographic division and industry group.

Appendix B

TABLE B-1

AVERAGE PERCENTAGE RATE OF CHANGE PER DECADE OF PRIMARY-POWER CAPACITY IN MANUFACTURING, BY TYPE, 1838-1919

Type of Power	1838-69	1869-79	1879-89	1889-99A	1899B-1909	1909-19
Total primary power	n.a.	45.4	71.4	84.7	86.9	58.8
Steam[a]	223.0	79.8	109.6	90.8	74.4	20.0
Water	n.a.	8.4	2.4	37.6	25.4	-3.2
Electric (purch.)	0	0	0	0	858.0	434.4
Gas	0	0	0	1,511.0	457.5	67.7

Source: Tables A-1 and A-2.

[a] Includes steam turbines.

TABLE B-2

PERCENTAGE DISTRIBUTION OF PRIMARY-POWER CAPACITY IN MANUFACTURING, BY TYPE, 1869-1919

Type of Power	1869	1879	1889	1899A	1899B	1909	1919
Total primary power	100.0	100.0	100.0	100.0	100.0	100.0	100.0
Steam	51.8	64.1	78.4	81.0	82.1	76.7	47.1
Water	48.2	35.9	21.5	16.0	14.7	9.8	6.0
Electric (purch.)	0	0	0	1.7	1.8	9.4	31.8
Gas	0	0	0.2	1.3	1.4	4.1	4.3
Steam turbines	0	0	0	0	0	0	10.9

Source: Table A-1.

TABLE B-3

PERCENTAGE DISTRIBUTION OF TOTAL PRIMARY-POWER CAPACITY IN MANUFACTURING, BY GEOGRAPHIC DIVISION, 1869-1919

Geographic Division	1869	1879	1889	1899A	1899B	1909	1919
United States	100.0	100.0	100.0	100.0	100.0	100.0	100.0
North	82.1	83.3	81.2	75.1	78.5	73.3	73.4
New England	21.9	21.8	19.4	16.7	17.7	14.4	12.9
Middle Atlantic	32.2	31.3	32.5	29.4	30.9	29.5	28.9
East north central	22.6	23.7	22.4	23.0	23.9	23.5	26.2
West north central	5.4	6.5	6.9	6.0	6.0	5.9	5.4
South	15.5	14.8	15.5	21.3	17.6	20.1	18.3
South Atlantic	8.9	8.6	8.0	9.5	8.4	9.8	9.5
East south central	4.6	4.5	5.2	6.3	5.2	5.6	4.8
West south central	2.0	1.7	2.3	5.5	4.0	4.7	4.0
West	2.3	2.0	3.3	3.6	3.9	6.5	8.3
Mountain	0.7	0.5	0.8	1.1	1.2	2.2	2.3
Pacific	1.6	1.5	2.5	2.5	2.7	4.3	6.0

Source: Table A-1.

TABLE B-4

PERCENTAGE DISTRIBUTION OF TOTAL WATER-POWER CAPACITY
IN MANUFACTURING, BY GEOGRAPHIC DIVISION, 1869-1919

Geographic Division	1869	1879	1889	1899A	1899B	1909	1919
United States	100.0	100.0	100.0	100.0	100.0	100.0	100.0
North	81.9	82.3	83.3	80.9	86.2	83.4	84.4
New England	32.0	34.5	39.6	38.0	42.6	41.5	42.4
Middle Atlantic	33.3	29.1	26.5	27.8	28.2	25.8	22.4
East north central	13.3	12.9	12.4	11.3	11.8	11.4	14.3
West north central	3.3	5.8	4.8	3.8	3.6	4.7	5.3
South	16.3	15.8	14.8	16.0	10.5	11.8	11.4
South Atlantic	12.4	11.9	11.6	12.2	8.6	10.0	9.8
East south central	3.6	3.5	2.8	3.4	1.7	1.6	1.4
West south central	0.3	0.4	0.4	0.4	0.2	0.2	0.2
West	1.8	1.9	2.0	3.2	3.3	4.7	4.2
Mountain	0.6	0.7	0.5	1.2	1.2	1.2	0.9
Pacific	1.2	1.2	1.5	2.0	2.1	3.5	3.3

Source: Table A-1.

TABLE B-5

PERCENTAGE DISTRIBUTION OF TOTAL STEAM-POWER CAPACITY
IN MANUFACTURING, BY GEOGRAPHIC DIVISION, 1838-1919

Geographic Division	1838	1869	1879	1889	1899A	1899B	1909	1919
United States	100.0	100.0	100.0	100.0	100.0	100.0	100.0	100.0
North	48.3	82.5	83.8	80.6	73.5	76.8	71.4	70.0
New England	13.5	12.6	14.7	13.8	12.6	13.4	11.7	9.8
Middle Atlantic	25.6	31.3	32.5	34.2	29.4	31.1	29.2	30.6
East north central	6.1	31.3	29.8	25.1	25.2	26.0	24.6	24.5
West north central	3.1	7.3	6.8	7.5	6.3	6.3	5.9	5.1
South	38.2	14.9	14.2	15.7	22.8	19.3	22.5	22.5
South Atlantic	14.4	5.8	6.8	7.0	9.1	8.6	10.1	10.0
East south central	2.2	5.6	5.0	5.8	7.0	5.9	6.7	6.6
West south central	21.6	3.5	2.4	2.9	6.7	4.8	5.7	5.9
West	0	2.6	2.0	3.6	3.6	3.9	6.2	7.5
Mountain	0	0.8	0.4	0.9	1.1	1.2	2.2	2.3
Pacific	0	1.8	1.6	2.7	2.5	2.7	4.0	5.2
Not assignable	13.5	0	0	0	0	0	0	0

Source: Table A-1.

TABLE B-6

PERCENTAGE DISTRIBUTION OF TOTAL STEAM-POWER CAPACITY
IN MANUFACTURING, BY STATE, 1838-1919
(top 5 states at each date)

State	1838	1869	1899B	1919
United States	100.0	100.0	100.0	100.0
Pennsylvania	20.6	18.3	19.5	18.8
Ohio	5.0	10.7	9.0	10.2
New York	4.0	10.4	8.1	7.6
Illinois	0	6.0	6.2	5.6
Massachusetts	6.2	6.5	7.1	5.0
Louisiana	21.6	2.1	2.3	2.3
Virginia	4.3	.7	1.3	1.5
Total, top 5	57.7	51.9	49.9	47.2
Total, all 7	61.7	54.7	53.5	51.0
Total, all others	38.3[a]	45.3	46.5	49.0

Source: Data for individual states will appear in subsequent study.

[a] Includes item noted in Tables A-1 and A-2 as not assignable.

TABLE B-7

PERCENTAGE DISTRIBUTION OF TOTAL PURCHASED
ELECTRIC-POWER CAPACITY IN MANUFACTURING,
BY GEOGRAPHIC DIVISION, 1899-1919

Geographic Division	1899B	1909	1919
United States	100.0	100.0	100.0
North	82.8	73.1	73.2
New England	17.3	12.5	12.1
Middle Atlantic	40.7	32.5	26.3
East north central	16.8	21.5	28.4
West north central	8.0	6.6	6.4
South	7.1	13.8	15.1
South Atlantic	2.8	9.8	9.5
East south central	2.1	2.2	3.3
West south central	2.2	1.8	2.3
West	10.1	13.1	11.7
Mountain	2.2	3.8	2.8
Pacific	7.9	9.3	8.9

Source: Table A-1.

TABLE B-8

PERCENTAGE DISTRIBUTION OF TOTAL PRIMARY-POWER CAPACITY
IN MANUFACTURING, BY INDUSTRY GROUP, 1869-1919

Major Group Number	Industry Group	1869	1889	1899A	1899B	1909	1919
	All Manufacturing	100.0	100.0	100.0	100.0	100.0	100.0
20	Food and kindred products	27.8	19.0	17.3	15.7	13.4	12.1
21	Tobacco manufacture	0.1	0.2	0.2	0.2	0.2	0.1
22	Textile products	12.0	14.9	15.3	13.5	11.5	10.4
23	Apparel	0.6	0.5	0.6	0.7	0.8	0.7
24	Lumber and wood products	31.6	20.5	18.2	17.1	16.1	10.4
25	Furniture and fixtures	1.3	1.5	1.2	1.3	1.4	1.1
26	Pulp, paper, and allied products	2.4	5.3	7.3	8.0	7.4	6.8
27	Printing, publishing, and allied industries	0.4	0.9	1.0	1.1	1.6	1.3
28	Chemicals and allied products	1.6	3.1	3.2	3.5	4.0	4.3
29	Products of petroleum and coal	0.3	1.1	0.9	1.0	1.5	2.7
30	Rubber products	0.3	0.5	0.7	0.7	0.7	1.5
31	Leather and leather products	1.9	1.6	1.4	1.5	1.4	1.3
32	Stone, clay, and glass products	1.4	3.9	5.4	5.9	6.9	6.3
33	Primary metal industries	9.6	14.1	16.3	17.8	20.0	22.0
34	Fabricated metal products	2.3	2.9	1.8	2.0	1.8	3.0
35	Machinery	3.5	4.5	4.2	4.5	5.3	5.5
36	Electric machinery	--	0.1	0.4	0.4	0.9	1.6
37	Transportation equipment	1.2	2.4	2.7	3.0	4.1	7.1
38	Professional instruments, etc., and miscellaneous	0.6	0.9	0.8	0.9	1.1	1.2
	Nonmanufacturing	1.0	2.0	.9	1.0	--	0.5
	Adjustment	0	0	--	--	0	0

Source: Table A-2.

TABLE B-9

PERCENTAGE DISTRIBUTION OF TOTAL WATER-POWER CAPACITY
IN MANUFACTURING, BY INDUSTRY GROUP, 1869-1919

Major Group Number	Industry Group	1869	1889	1899A	1899B	1909	1919
	All Manufacturing	100.0	100.0	100.0	100.0	100.0	100.0
20	Food and kindred products	36.9	31.2	26.8	17.5	15.1	11.2
21	Tobacco manufacture	--	--	--	--	--	--
22	Textile products	15.6	24.1	22.5	26.0	24.3	25.4
23	Apparel	0.3	0.2	0.2	0.3	0.3	0.2
24	Lumber and wood products	31.4	17.7	12.9	11.5	8.3	4.2
25	Furniture and fixtures	1.0	0.7	0.4	0.5	0.4	0.3
26	Pulp, paper, and allied products	3.9	16.4	29.3	34.8	43.5	51.9
27	Printing, publishing, and allied industries	--	0.2	0.2	0.2	0.1	0.1
28	Chemicals and allied products	0.6	0.9	0.7	0.8	1.0	0.6
29	Products of petroleum and coal	--	--	--	--	0.2	--
30	Rubber products	0.2	0.3	0.3	0.4	0.3	0.3
31	Leather and leather products	1.4	0.6	0.3	0.4	0.3	0.3
32	Stone, clay, and glass products	0.8	0.6	1.1	1.3	1.3	1.3
33	Primary metal industries	2.5	0.9	1.5	1.8	1.5	1.2
34	Fabricated metal products	1.5	1.7	1.1	1.3	1.0	0.8
35	Machinery	2.0	2.1	1.6	1.9	1.5	1.4
36	Electric machinery	--	--	0.1	0.1	0.1	0.1
37	Transportation equipment	0.7	0.5	0.4	0.5	0.3	0.2
38	Professional instruments, etc., and miscellaneous	0.4	0.5	0.3	0.4	0.4	0.4
	Nonmanufacturing	0.6	1.4	0.3	0.3	--	0.1

Source: Table A-2.

TABLE B-10

PERCENTAGE DISTRIBUTION OF TOTAL STEAM-POWER CAPACITY
IN MANUFACTURING, BY INDUSTRY GROUP, 1838-1919

Major Group Number	Industry Group	1838	1869	1889	1899A	1899B	1909	1919
	All Manufacturing	100.0	100.0	100.0	100.0	100.0	100.0	100.0
20	Food and kindred products	27.1	19.3	15.6	15.5	15.1	13.5	13.5
21	Tobacco manufacture	0.1	0.2	0.3	0.3	0.3	0.2	0.2
22	Textile products	11.0	8.5	12.4	14.3	11.7	10.6	9.6
23	Apparel	1.2	0.9	0.6	0.5	0.6	0.5	0.5
24	Lumber and wood products	18.7	31.8	21.3	19.8	19.1	19.1	17.6
25	Furniture and fixtures	0.1	1.5	1.7	1.4	1.5	1.5	1.3
26	Pulp, paper, and allied products	1.4	1.1	2.2	3.2	3.4	3.6	4.4
27	Printing, publishing, and allied industries	0.1	0.7	1.1	0.6	0.7	0.4	0.3
28	Chemicals and allied products	1.5	2.6	3.7	3.6	3.9	4.0	4.3
29	Products of petroleum and coal	0	0.6	1.4	1.1	1.2	1.8	2.7
30	Rubber products	0.1	0.4	0.5	0.7	0.8	0.8	0.9
31	Leather and leather products	1.2	2.3	1.8	1.5	1.6	1.5	1.4
32	Stone, clay, and glass products	1.3	2.0	4.9	6.3	6.8	7.3	6.4
33	Primary metal industries	10.0	16.2	17.8	19.8	21.2	23.4	25.8
34	Fabricated metal products	0.8	3.1	3.3	1.8	1.9	1.5	1.9
35	Machinery	5.3	4.8	5.1	4.4	4.8	4.5	3.2
36	Electric machinery	0	0	0.1	0.4	0.4	0.7	0.5
37	Transportation equipment	0.4	1.7	3.0	3.2	3.4	4.1	4.4
38	Professional instruments, etc., and miscellaneous	1.7	0.8	1.0	0.8	0.9	1.0	1.0
	Nonmanufacturing	4.6	1.4	2.2	0.8	0.9	---	0.1
	Not assignable	13.5	0	0	0	0	0	0
	Adjustment	0	0	0	---	---	0	0

Source: Table A-2.

TABLE B-11

PERCENTAGE DISTRIBUTION OF TOTAL PURCHASED ELECTRIC-POWER
CAPACITY IN MANUFACTURING, BY INDUSTRY GROUP, 1899-1919

Major Group Number	Industry Group	1899B	1909	1919
	All Manufacturing	100.0	100.0	100.0
20	Food and kindred products	10.6	10.5	12.2
21	Tobacco manufacture	0.3	0.3	0.2
22	Textile products	5.3	9.2	9.0
23	Apparel	7.0	3.1	1.4
24	Lumber and wood products	2.4	4.1	3.4
25	Furniture and fixtures	1.8	1.8	1.2
26	Pulp, paper, and allied products	1.1	3.2	3.3
27	Printing, publishing, and allied industries	19.0	11.6	3.3
28	Chemicals and allied products	11.7	7.9	4.9
29	Products of petroleum and coal	0.1	1.0	2.1
30	Rubber products	0.1	0.5	2.1
31	Leather and leather products	3.6	2.3	1.5
32	Stone, clay, and glass products	2.6	7.3	7.5
33	Primary metal industries	2.3	7.7	13.2
34	Fabricated metal products	9.0	3.8	5.7
35	Machinery	8.5	12.4	10.2
36	Electric machinery	2.7	2.9	2.7
37	Transportation equipment	3.4	8.0	13.2
38	Professional instruments, etc., and miscellaneous	3.3	2.5	2.0
	Nonmanufacturing	3.8	0	1.1
	Adjustment	1.1	0	0

Source: Table A-2.

TABLE B-12

PERCENTAGE DISTRIBUTION OF PRIMARY-POWER CAPACITY IN MANUFACTURING,
BY TYPE AND GEOGRAPHIC DIVISION, 1869-1919

| Geographic | 1869 | | 1879 | | 1889 | | | 1899A | | | |
Division	Water	Steam	Water	Steam	Water	Steam	Gas	Water	Steam	Gas	Electric (purch.)
United States	48.2	51.8	35.9	64.1	21.5	78.4	0.2	16.0	81.0	1.3	1.7
New England	70.3	29.7	56.9	43.1	43.9	56.0	0.1	36.5	61.2	0.6	1.8
Middle Atlantic	49.7	50.3	33.4	66.6	17.5	82.4	0.2	15.1	81.0	1.5	2.3
East north central	28.2	71.8	19.5	80.5	11.9	88.0	0.2	7.8	89.0	1.9	1.2
West north central	29.4	70.6	32.2	67.8	14.8	85.0	0.2	10.1	84.5	3.1	2.3
South Atlantic	66.6	33.4	49.5	50.5	31.2	68.6	0.2	20.6	78.2	0.7	0.5
East south central	37.3	62.7	28.0	72.0	11.5	88.4	0.1	8.7	90.3	0.4	0.6
West south central	7.7	92.3	8.0	92.0	3.3	96.4	0.3	1.2	97.7	0.4	0.7
Mountain	43.3	56.7	51.2	48.8	14.4	85.3	0.2	16.4	79.3	1.0	3.3
Pacific	38.6	61.4	30.0	70.0	13.3	86.4	0.3	12.3	81.1	1.3	5.3

| Geographic | 1899B | | | | 1909 | | | | 1919 | | | | |
Division	Water	Steam	Gas	Electric (purch.)	Water	Steam	Gas	Electric (purch.)	Water	Steam	Gas	Electric (purch.)	Steam Turbine
United States	14.7	82.1	1.4	1.8	9.8	76.7	4.1	9.4	6.0	47.1	4.3	31.8	10.9
New England	35.3	62.3	.6	1.8	28.3	61.9	1.6	8.2	19.8	35.9	.9	29.9	13.5
Middle Atlantic	13.4	82.7	1.5	2.4	8.6	76.0	5.0	10.4	4.7	49.7	4.8	28.8	12.0
East north central	7.2	89.6	1.9	1.3	4.8	80.1	6.5	8.6	3.3	44.0	5.6	34.5	12.6
West north central	8.7	85.9	3.0	2.4	7.9	76.4	5.2	10.5	5.8	44.4	6.4	37.4	5.9
South Atlantic	14.9	83.6	0.8	0.6	10.1	78.6	2.0	9.4	6.2	49.7	4.9	31.8	7.4
East south central	4.9	93.9	0.4	0.7	2.8	92.2	1.2	3.7	1.8	64.5	1.6	22.0	10.1
West south central	.7	97.8	0.5	1.0	.4	92.6	3.4	3.7	.3	68.9	6.9	18.5	5.4
Mountain	15.0	80.7	1.0	3.3	5.4	76.8	1.0	16.8	2.3	47.0	1.5	37.9	11.3
Pacific	11.4	81.9	1.3	5.3	7.9	70.3	1.5	20.3	3.3	41.2	1.9	47.4	6.2

Source: Table A-1.

TABLE B-13

PERCENTAGE DISTRIBUTION OF PRIMARY-POWER CAPACITY IN MANUFACTURING, BY TYPE, ALL MANUFACTURING AND EACH INDUSTRY GROUP, 1869-1919

Major Group Number	Industry Group	1869			1889				1899A			
		Water	Steam		Water	Steam	Gas		Water	Steam	Gas	Electric (purch.)
	All Manufacturing	48.2	51.8		21.5	78.4	0.2		16.0	81.0	1.3	1.7
20	Food and kindred products	63.9	36.1		35.3	64.6	0.1		24.7	72.6	1.7	1.0
21	Tobacco manufacture	13.0	87.0		0.4	98.6	1.0		1.8	94.7	1.5	1.9
22	Textile products	63.0	37.0		34.7	65.2	0.1		23.5	75.8	0.2	0.6
23	Apparel	24.4	75.6		7.8	89.8	2.4		5.4	68.4	8.0	18.2
24	Lumber and wood products	47.9	52.1		18.6	81.4	0		11.3	87.9	0.5	0.2
25	Furniture and fixtures	38.9	61.1		10.9	88.9	0.2		5.8	89.6	2.1	2.5
26	Pulp, paper, and allied products	77.0	23.0		66.0	33.9	0.1		64.1	35.3	0.3	0.3
27	Printing, publishing, and allied industries	1.5	98.5		4.8	89.8	5.3		2.6	51.2	14.3	31.9
28	Chemicals and allied products	17.8	82.2		6.4	93.4	0.2		3.5	89.6	0.8	6.1
29	Products of petroleum and coal	1.6	98.4		0.1	99.8	0.1		—	98.7	1.1	0.2
30	Rubber products	29.7	70.3		12.1	87.9	—		7.5	92.2	—	0.3
31	Leather and leather products	36.8	63.2		8.1	91.2	0.6		3.9	89.6	2.1	4.4
32	Stone, clay, and glass products	26.7	73.3		3.1	96.9	—		3.2	94.8	1.2	0.8
33	Primary metal industries	12.8	87.2		1.4	98.6	—		1.5	98.0	0.3	0.2
34	Fabricated metal products	30.4	69.6		12.0	87.7	0.3		9.7	78.0	4.0	8.3
35	Machinery	28.3	71.7		10.3	89.5	0.2		6.2	85.8	4.5	3.5
36	Electric machinery	100.0	0		1.9	98.1	—		2.1	81.9	4.4	11.7
37	Transportation equipment	28.6	71.4		4.1	95.7	0.2		2.3	93.5	2.2	2.1
38	Professional instruments, etc., and miscellaneous	33.3	66.7		12.3	86.6	1.0		6.6	82.0	4.5	6.9

(continued)

TABLE B-13 (concluded)

Major Group Number	Industry Group	1899B				1909				1919				
		Water	Steam	Gas	Electric (purch.)	Water	Steam	Gas	Electric (purch.)	Water	Steam	Gas	Electric (purch.)	Steam Turbine
	All Manufacturing	14.7	82.1	1.4	1.8	9.8	76.7	4.1	9.4	6.0	47.1	4.3	31.8	10.9
20	Food and kindred products	16.5	80.2	2.0	1.2	11.1	77.4	4.1	7.4	5.5	52.6	5.2	32.0	4.7
21	Tobacco manufacture	1.8	94.7	1.5	1.9	0.9	77.4	2.8	18.9	0.9	59.6	0.8	33.3	5.4
22	Textile products	28.2	70.8	0.2	0.7	20.8	71.0	0.6	7.5	14.6	43.1	0.4	27.4	14.5
23	Apparel	5.4	68.4	8.0	18.2	3.6	51.0	7.3	38.1	1.8	32.3	2.0	62.8	1.0
24	Lumber and wood products	9.7	89.5	0.6	0.3	5.1	91.0	1.5	2.4	2.4	80.0	1.7	10.3	5.6
25	Furniture and fixtures	5.8	89.6	2.1	2.5	2.9	81.6	3.4	12.1	1.7	55.0	2.3	36.0	5.0
26	Pulp, paper, and allied products	64.1	35.3	0.3	0.3	57.7	37.4	0.7	4.1	46.0	30.8	0.3	15.3	7.6
27	Printing, publishing and allied products	2.6	51.2	14.3	31.9	0.8	20.1	11.0	68.2	0.3	12.2	4.6	82.6	0.3
28	Chemicals and allied products	3.5	89.6	0.8	6.1	2.5	76.8	1.9	18.8	0.8	47.4	2.7	35.9	13.3
29	Products of petroleum and coal	—	98.7	1.1	0.2	1.2	87.5	5.1	6.2	—	45.5	6.6	24.6	23.3
30	Rubber products	7.5	92.2		0.3	4.1	86.9	1.9	7.0	1.3	28.1	0.9	45.0	24.7
31	Leather and leather products	3.9	89.6	2.1	4.4	2.2	77.9	4.7	15.2	1.2	49.3	2.2	37.6	9.8
32	Stone, clay, and glass products	3.2	94.8	1.2	0.8	1.9	82.1	6.0	10.0	1.2	47.6	6.8	37.7	6.6
33	Primary metal industries	1.5	98.0	0.3	0.2	0.7	89.7	5.9	3.7	0.3	55.3	8.6	19.1	16.7
34	Fabricated metal products	9.7	78.0	4.0	8.3	5.6	64.8	9.8	19.7	1.7	29.8	4.3	59.4	4.9
35	Machinery	6.2	85.8	4.5	3.5	2.8	64.7	10.4	22.0	1.5	27.1	5.0	59.0	7.4
36	Electric machinery	2.1	81.9	4.4	11.7	0.7	64.4	4.1	30.8	0.4	15.0	1.6	53.0	30.0
37	Transportation equipment	2.3	93.5	2.2	2.1	0.8	76.6	4.4	18.2	0.1	29.3	1.8	58.9	9.8
38	Professional instruments, etc., and miscellaneous	6.6	82.0	4.5	6.9	3.8	67.9	7.2	21.1	1.9	38.6	3.3	50.1	6.1

Source: Table A-2.

Changing Production of Metalworking Machinery, 1860-1920

ROSS M. ROBERTSON
INDIANA UNIVERSITY

Because of insufficient data, neither McDougall nor I could stick by our original resolve to treat the growth in output of nonelectrical machinery in general. But we need not apologize for restricting our investigations to the category of metalworking machinery, or, to use McDougall's somewhat more common and less inclusive term, machine tools.[1] Except for a few expensive toys of do-it-yourself addicts, metalworking machinery is a pure capital good, sold always in the producer-goods market, and the fluctuations in sales (output) of these products are a marvel to behold. It is common knowledge that the first industrial revolution, to say nothing of the second, would have been impossible without metalworking machinery.

Because of their obvious importance, machine tools have long since attracted historians, who have done yeoman service in tracing their evolution.[2] A marvelously inventive group of innovators started devising

NOTE: The author gratefully acknowledges the financial assistance of both the Interuniversity Committee on American Economic History and Indiana University in support of this research. I am indebted to my research assistants, Gerald W. Kuhn, Martha Ann Eppley, Kent Tool, Jules Levine, and George Wing and to a former student, D. J. Clinch of the University of Sussex, for their help. My colleague, Irvin M. Grossack, kindly advised me on the many statistical questions that came up from time to time.

[1] As we proceed, the reason for making the inconsiderable distinction between metalworking machines and machine tools will become apparent. At the moment it is enough to observe that machine tools are usually defined as power-driven machines that *cut* metal, excluding the shaping or forming machines that press, forge, hammer, etc. Metalworking machinery includes both types. (See Duncan McDougall's paper which follows.)

[2] This literature is voluminous. At one time or another every economic historian has dipped into Joseph Wickham Roe, *English and American Tool Builders* (New Haven, 1916). For excellent historical summaries containing substantial bibliographical notation, see Bertold Buxbaum, "Der amerikanische Werkzeugmaschinen und Werkzeugbau im 18. und 19. Jahrhundert," *Beiträge zur Geschichte der Technik und Industrie*, Vol. X, 1920, pp. 121-154, and W. Paul Strassmann, *Risk and Technological Innovation*, Ithaca, 1959, pp. 116-157. Detailed studies of the development of the several types of machine tools are to be found in Robert S. Woodbury's little books, all published by the Technology Press at Massachusetts Institute of Technology. See, for example, his *History of the Gear Cutting Machine*, 1958; *History of the Grinding Machine*, 1959; *History of the Milling Machine*, 1960.

metal-cutting and metal-shaping machines before anyone had invented a generic term for them. Roe remarks that they were a well-established class of machines by 1830, but the earliest allusion that I have found to tools that are at the same time machines appears in a report of a select committee of the British House of Commons in 1841.[3] The first U.S. Census designation of machine tools is found in the Eighth Census (1860) as "machinists' tools." The Ninth Census actually counted the output of two kinds of machine tools but did not refer to the category as a total. The Tenth Census (1880) contained an elegantly descriptive article on "machine tools" and their uses but included no quantitative information about them.[4] Not until the Twelfth Census (1900) did "machine tools" receive the careful attention of enumerators, though "metalworking machinery" appeared as a category in export-import figures as early as 1898.

Another reason why it is so hard to obtain data on the machine tool industry is that, more than most, it was a "peel-off" industry. Beginning in the latter half of the eighteenth century, manufacturers of products requiring machine tools often had no realistic alternative to making their own. Even after English and American firms began to emerge as specialists in machine tool manufacture, thus dispensing the largesse of Marshallian external economies to their customer firms, many businesses continued to make metalworking machines for their own use. In the late nineteenth century many a Cincinnati firm requiring a lathe or a planer would order the tool from a reputable manufacturer and, needing others, would proceed to make copies. Even with the best definitions and the most conscientious census-taking techniques, this kind of production would surely have gone uncounted.

Whatever the difficulties of estimation, the problem of calculating output and price changes over a meaningful span of years remains. When the data are so much better at the end of the series than they are at the

---

[3] *First Report from Select Committee Appointed to Inquire into the Operation of the Existing Laws Affecting the Exportation of Machinery; with the Minutes of Evidence and Appendix*, ordered by the House of Commons, to be printed April 1, 1841, p. 96.

Question 1314: "Much ignorance appears to prevail as to the nature and extent of this department of industry; can you give the Committee some outline of the extent of the manufacture of tools, in what degree they are now used in machine-making, and what are the descriptions of tools ranging under this general name?"

Witness W. Jenkinson: "What used to be called tools were simple instruments, as I should call them, such as hammers and chissels [sic] and files; but those now called tools are in fact machines, and very important machines; they are not only important but they are now made at very great cost, from £100 up to £2,000 each...."

[4] See *Tenth Census of the United States, 1880*, F. R. Hutton, "Report on Machine Tools and Woodworking Machinery," Vol. 22, pp. 5–294.

beginning, it seems best to start at the end and work backward. We first take a look at the output of metalworking machinery from 1900 to 1920 before examining the evidence for the two decades 1880–1900.

Output in Current Dollars and Real Output, 1900–20

For the period 1900–20 there are Census figures of metalworking machinery output for four years—1900, 1905, 1914, and 1919. We also have estimates of the National Machine Tool Builders' Association for the years 1901–20.[5] The estimates of the trade association are for metal-cutting tools as a component of the larger category of metalworking machinery.

Getting the Census definitions straight is a job in itself. The Census of 1900 gives the total value of all products manufactured by firms making metalworking machinery as $44.385 million. From this figure has been subtracted the value of products not considered metalworking machinery, $16.376 million, plus the amount received for custom work and repairs, $3.271 million, the result being $24.738 million. According to an explicit statement of the Office of the Census, the 1905 figure, $32.409 million, is comparable to the 1900 figure. The 1914 Census figure for machine tool output is $31.447 million, and for metalworking machinery other than machine tools $17.420 million, which comes to a total of $48.867 million. The 1919 machine tool figure was $212.400 million and the figure for metalworking machinery other than machine tools $57.541 million. Subtracting from this sum the value of all other products, $32.653 million, yields an output of metalworking machinery for 1919 of $237.200 million.[6]

The problem then becomes one of interpolation of time series.[7] The approach taken is the common-sense one of relating the benchmark figures for metalworking machinery to the National Machine Tool Builders' Association estimates of the output of metal-cutting tools. The

---

[5] This series has been continued to the present. See Release F-A40a, dated February 21, 1962, of the National Machine Tool Builders' Association. This is the same series cited by McDougall.

[6] The foregoing data are taken from the following sources: *Twelfth Census of the United States, 1900*, Vol. X, Manufactures, Part IV, Selected Industries, p. 381; *Special Reports of the Census Office*, 1905, Manufactures, Part IV, Selected Industries, p. 227; *Census of Manufactures, 1914*, Vol. II, p. 272; *Fourteenth Census of the United States, 1920*, Vol. X, Manufactures, 1919, pp. 373 and 385; *Biennial Census of Manufactures, 1921*, p. 419.

[7] At this point we embark on a venture beset by obvious pitfalls. For the kind of problems involved, see Milton Friedman, *The Interpolation of Time Series by Related Series*, NBER Technical Paper 16, New York, 1962.

relevant data are set forth in Table 1. It was assumed that the ratios of output of metalworking machinery to output of metal-cutting machine tools changed gradually between benchmark years. The ratios of the Census figures to the NMTBA figures were computed for 1905, 1914, and 1919, along with the ratio of the 1900 Census figure to an extrapolated NMTBA figure for 1900. On the assumption of linearity of change between benchmark years, expansion ratios were then computed for intervening years. Multiplying the dollar value of output in the NMTBA series by the computed ratios, we obtained the estimates of total output of metalworking machinery in current dollars, as shown in column 4 of Table 1.

It was felt that this series for 1900–20 was reliable enough to permit the computation of an index of real output of metalworking machinery. Since no index of prices of metalworking machinery was available for the entire period, a deflator was constructed by splicing the American Appraisal Company's index of machine tool prices for the period 1914–20 to the wholesale price index of the Bureau of Labor Statistics for the period 1900–13. (For alternative procedures, see Table A-2.) The resulting index of real output (1914 = 100) is shown in column 7 of Table 1.

Comparisons of output of metalworking machinery by regions are not altogether satisfactory because of odd and unexplained gaps in the data, especially for the Census of 1919. It is not far off, however, to assign the output for the year 1900 approximately as follows: 30 per cent to the New England states, 26 per cent to the Middle Atlantic states, and 34 per cent to the Midwestern states, with Ohio alone accounting for 29 per cent. The remaining 10 per cent was scattered among all other producing states, including Vermont, Michigan, and Wisconsin, for which data were not available in the 1900 Census. Over the ensuing two decades there was an appreciable shift in the relative importance of the Middle Atlantic and Midwestern states. In 1919, New England states (including Vermont) accounted for 35 per cent of machine tool output, the Middle Atlantic states for 14 per cent, and the Midwestern states of Ohio, Illinois, Michigan, and Wisconsin for 47 per cent.[8] In 1919, metalworking machinery other than machine tools was produced for the most part in six states, Connecticut accounting for 12.2 per cent of this category of output, New Jersey for 5.1 per cent, New York for 18.7 per cent, Pennsylvania for 16.4 per cent, Ohio for 17.7 per cent, and Illinois for 16.4 per cent. On the whole, the Midwestern states of Ohio and Illinois were dominant in the total production of metalworking machinery, but the New England

[8] Unfortunately, these data include the category "all other products" in addition to machine tools.

TABLE 1

PRODUCTION OF METALWORKING MACHINERY IN CURRENT AND IN CONSTANT DOLLARS, 1900-20
(million dollars)

Year	Shipments of Metalcutting Tools[a] (current dollars) (1)	Output of Metalworking Machinery[b] (current dollars) (2)	Expansion Ratios (3)	Output of Metalworking Machinery (current dollars) (4)	Index of Metalworking Machinery Prices (1926=100)[c] (5)	Estimated Output of Metalworking Machinery (1926 dollars) (6)	Index of Real Output of Metalworking Machinery (1914=100) (7)
1900	(17.3)	24.7	142.8	24.7	41.4	59.7	55.7
1901	17.9		136.8	24.5	40.8	60.0	56.0
1902	22.8		130.8	29.8	42.9	69.5	64.8
1903	23.7		124.8	29.7	43.9	67.7	63.2
1904	18.3		118.8	21.7	44.0	49.3	46.0
1905	28.7	32.4	112.9	32.4	44.3	73.1	68.2
1906	36.4		115.7	42.1	45.6	92.3	86.1
1907	41.3		118.5	48.9	48.1	101.7	94.9
1908	16.8		121.3	20.4	46.4	45.3	42.3
1909	33.5		124.1	41.6	49.8	83.5	77.9
1910	44.3		126.9	56.2	51.9	108.3	101.0
1911	32.8		129.7	42.5	47.8	88.9	82.9
1912	44.4		132.5	58.8	50.9	115.5	107.7
1913	44.6		135.3	60.3	51.5	117.1	109.2
1914	35.3	48.9	138.4	48.9	45.6	107.2	100.0
1915	103.4		140.2	145.0	49.4	293.5	273.6
1916	141.4		142.0	198.2	62.0	319.6	298.0
1917	168.5		143.8	242.3	82.3	294.4	274.5
1918	220.6		145.6	321.2	101.2	317.4	295.9
1919	161.0	237.2	147.3	237.2	106.3	223.1	208.0
1920	151.5		149.1	225.9	116.5	193.9	180.7

[a] Estimates of National Machine Tool Builders' Association.
[b] Census figures include both cutting and forming tools.
[c] For alternative computations of price indexes, see Table A-2.

states were still major producers of metal-cutting tools, and the Middle Atlantic states even surpassed Ohio and Illinois in the production of metalforming tools.

In general, attempts to estimate foreign trade in metalworking machinery for the entire period 1840–1920 met with frustration. One set of statistics emerged that ought to be included. These are the exports of metalworking machinery, with major countries of destination, for 1898–1920. A comparison of domestic output with exports for 1908 and 1914 suggests how important foreign sales were to the metalworking machinery industry during lean years at home (see Table A-1).

Evidence of the Level of Output, 1860–1900

A quantitative study of metalworking machinery output in this or any other country for the latter half of the nineteenth century faces formidable data problems. McDougall takes the view that it was not until the end of the period "that a distinct machine tool industry emerged in the sense of a group of firms whose principal product was machine tools." I would place this emergence some twenty years earlier, or about 1880. If it is true that a church denomination has come of age when it creates a seminary, it is equally true that a group of firms begins to constitute an "industry" in some formal sense when a trade magazine is established to minister to it. *American Machinist* began publication in 1878 and in its issue of June 28, 1879, published a list of machine tool builders that constituted, in the editor's words, a "distinct branch of industry." Again in its issue of November 11, 1882, a list of machine tool builders, purporting to be exhaustive, is offered.[9] One reason, then, for selecting 1880 as the earliest date for estimating a nineteenth century time series for metalworking machinery production is that by this time an industry had clearly formed. But a more compelling reason is that before 1880 it is possible to estimate output only for 1860 and 1870, and there is presently no really adequate basis for interpolating figures for the interim years in the two decades of 1860–80.

Standard sources for data before 1880 were combed with unusual care. In addition to such ordinarily productive periodicals as *Hunt's Merchants' Magazine*, *DeBow's Review*, *Niles' National Register*, and *Hazard's United States Commercial and Statistical Register*, much of the nineteenth century literature on manufacturing and technological change was examined. Relevant public documents were combed, including some

[9] For these lists, see *American Machinist*, June 28, 1879, pp. 8–9, and November 11, 1882, p. 7.

exciting material in the Pennsylvania State Library.[10] British Parliamentary reports from 1830 on were examined, as were British Consular reports, largely in the hope of finding knowledgeable English estimates of the proportion of machine tool output accounted for by various major cities.[11] But except for a few gems to be noted later, these researches produced little of value, though several bits and pieces will ultimately be useful in reconstructing a production series for nonelectrical machinery as a whole.

As nearly as I can determine, estimates of output of metalworking machinery in the nineteenth century must begin with the Eighth (1860) Census of the United States. Here for the first time there is specific reference in an official American document to machine tools. "Machinists' tools employed 17 manufactories, a capital of $536,150, and 455 hands, and the value of the manufacture was $540,292, of which $205,000 was the product of one establishment in Philadelphia, having a capital of $280,000 and employing 190 hands and turning out machinists' tools of acknowledged excellence. Nine establishments in Massachusetts reported a value of $165,600 made, and two in New Haven, Connecticut, a product of $71,600. Three in New York, made tools of the value of $47,950; one in New Jersey, $2,800; and one in Delaware, $22,142." [12]

The separate amounts cited do not add up to the total Census figure, but there is another difficulty. From two other sources, one unofficial and the other official, there are estimates of the output of metalworking machinery in Philadelphia, both of them suggesting that the Census summary is in error. Edwin T. Freedley, a contemporary observer, estimated that for 1857 the two principal machine tool manufacturing

---

[10] Among the U.S. documents, the biggest disappointment was the Special Census of Manufactures ordered by the House of Representatives in 1832. Had the census of the several states been properly taken, the data compilation would have been invaluable, but as far as I can tell only the Massachusetts enumerators did a creditable job. In other states data are sketchy and incomplete; not all firms were polled, and some that were refused to respond. See *Executive Documents*, 1st Sess., 22d Congress, Serial Numbers 222 and 223, *Documents Relating to the Manufactures in the United States*, collected and transmitted to the House of Representatives in compliance with a resolution of January 19, 1832, by the Secretary of the Treasury, 2 vols., 1833. The Pennsylvania documents referred to are *Census of the United States, Original Returns of the Assistant Marshals, Products of Industry*, housed in the Law Library, Pennsylvania State Library, Harrisburg, Pennsylvania.

[11] In the context of the present study, the most valuable information obtained from the British Parliamentary Papers was a list of prices of American machine tools and allied products quoted English buyers by Mr. Ames of Chicopee, Massachusetts, and Messrs. Robbins and Lawrence of Windsor, Vermont. See *Report of the Committee on the Machinery of the United States*, 1855, pp. 75–79.

[12] *Eighth Census of the United States, 1860*, Statistics of Manufactures, Vol. 3, p. clxxxviii.

firms of William Sellers & Co. and Bement & Dougherty turned out a product of $350,000, employing together about 300 hands.[13] More important, the original Census returns for Philadelphia for the years 1860 and 1870 give the following information:

William Sellers & Co., Inc.
1600 Hamilton Street
Philadelphia, Pa.
15th Ward, 43rd District

1860 Census:
  Product: Machinists & Foundry
  Hands employed: 190
  Value of product: Machine Tools, $80,000
                            Total (all goods), $205,000
1870 Census:
  Hands employed: 423
  Value of product: Machine Tools, $273,962
                            Total (all products), $707,542

Bement & Dougherty
15th Ward, 43rd District
Philadelphia, Pa.

1860 Census:
  Product: Machinery & Tools
  Hands employed: 180
  Value of product: Machinery & Castings, $198,000

1870 Census:
  Hands employed: 375
  Value of product: Machine Tools, $401,000
                            Total (all products), $511,918

Comparison of the original reports with the summary from the 1860 Census quoted above reveals the inaccuracies of the Census writer. He took the *total* output of Sellers for 1860 as the output of machinists' tools only, and he failed to include the machine tool output of Bement & Dougherty and other Philadelphia firms. If Bement & Dougherty devoted 80 per cent of its production to machine tools in 1860, as it did in 1870, then the two firms, Sellers and Bement & Dougherty, produced $240,000 worth of machine tools rather than the $205,000 assigned to Philadelphia

[13] Edwin T. Freedley, *Philadelphia and Its Manufactures*, Philadelphia, 1858, pp. 314-316. This same figure was picked up by *Hunt's Merchants' Magazine* and quoted in the issue of July-December, 1958, p. 629.

by the 1860 Census. The adjusted figures fall considerably short of the Freedley estimate for 1857, the discrepancy doubtless being the result of the ambiguity of the term "machine tools." At least three other Philadelphia firms were then in business, but original Census returns could not be found for them. They were certainly much smaller than the two major firms, for neither Freedley nor Buxbaum mention them by name; together they require an increase in the adjusted Census total of at least 10 per cent to give an estimate for Philadelphia production in 1860 of $264,000. The actual value of output probably lay between this figure and the Freedley estimate, and I have taken a Philadelphia production figure of $300,000 as plausible for 1860.

For 1870 the Census figure of machine tool output of the two reporting firms totaled $675,000. We have no information on their share of Philadelphia production, but Buxbaum indicates no reason for thinking that their share changed during the Civil War decade. At their estimated 1860 shares, a production for Philadelphia of approximately $750,000 is implied for 1870.

Because the Censuses of 1860 and 1870 are not reliable as a guide to estimation of production by geographical regions, these estimates must be inferred from three major sources that span the decades of the 1860's and 1870's.[14] A count of the known firms in the industry as of 1860,

[14] The count of firms in the several regions is based on companies mentioned in the works by Roe and Buxbaum (especially pp. 132–140) mentioned in footnote 2, with some adjustment in the light of firms contained in lists of the *American Machinist* for 1879 and 1882 (see *American Machinist*, June 28, 1879, pp. 8–9, and November 11, 1882, p. 7).

In 1879 the *American Machinist* listed nine tool-producing firms in the Midwestern group of a total of seventy-three that had been in the business within the past ten years; but only forty-six of this group were manufacturing tools at the time of the count, of which six were in the western group. The 1882 list contained 132 firms, twenty-one of them in the Midwest. Many of the firms in the later list were new and still small, and at this date the western tool builders could not have accounted for more than 10 or 12 per cent of the country's production of metalworking machinery.

As nearly as I can determine, Philadelphia in 1882 still held its dominant position as a producer of large tools, although the city's proportion of total output may have begun to fall. In his report on the industries and products of the Consular District of Philadelphia for 1883 and a part of 1884, British Consul Clipperton made the following comment:

"Lovin Blodget, Esq., the eminent statistician, has for many months been engaged, assisted by the mayor and the entire police force of the city, in collecting industrial returns for 1882 and 1883, a brief summary of which is herein given by permission."

The Blodget estimates for the city of Philadelphia in 1882 report a category of "iron working machine tools." Ten firms, employing 1289 hands, turned out a value of product of $2,255,750 (see *Accounts and Papers, Commercial Reports*, Vol. LXXXI, pp. 1933–1938). This figure is close to 30 per cent of my estimated 1880 output of metalworking machinery for the country.

TABLE 2

ESTIMATE OF CINCINNATI MACHINE TOOL OUTPUT, 1880-1900

Year	Number of Firms Established in Year Cited (1)	Total Number of Cincinnati Firms (2)	G.A. Gray Co. Actual Output (dollars) (3)	Estimate of Cincinnati Machine Tool Output (dollars) (4)
1880	2	6	--	250,000
1881	2	8	--	360,000
1882	0	8	--	410,000
1883	0	8	--	510,000
1884	2	10	--	600,000
1885	0	10	--	650,000
1886	2	12	--	870,000
1887	4	16	80,500	1,270,000
1888	1	17	91,800	1,330,000
1889	2	19	113,100	1,600,000
1890	0	19	176,700	1,830,000
1891	0	19	169,600	1,790,000
1892	2	21	161,400	1,770,000
1893	1	22	86,300	1,300,000
1894	0	22	58,600	820,000
1895	0	22	123,400	1,380,000
1896	1	23	111,300	1,340,000
1897	2	25	137,200	1,800,000
1898	3	28	185,900	2,100,000[a]
1899	1	29	288,600	3,340,000[a]
1900	0	29	362,500	3,375,000[b]

[a] Obtained from "Leading Industries of Cincinnati," published by the *Cincinnati Enquirer*, 1900, p. 26.

[b] *Twelfth Census of the United States, 1900.*

adjusted for type of output,[15] indicates that New England tool builders were turning out at least 40 per cent of total output of metalworking machinery, and I estimate a figure of 25 per cent for the Middle Atlantic states, including Pennsylvania outside Philadelphia, Delaware, and Maryland. Not more than 5 per cent of total output was accounted for by tool builders west of Pennsylvania.[16] Thus, 30 per cent of output is attributable to the part of the industry located in Philadelphia, and there

[15] There seems little question that by 1860 Philadelphia firms were specializing in heavier, higher-priced machines, most of them in the metal-forming category. New England manufacturers, on the other hand, were doubtless making a much smaller proportion of metalworking machines and were even then specializing in the light production machine tools that McDougall believes were not quantitatively important in the period of rapid American industrialization.

[16] As early as 1860, Cincinnati machinery manufacturing firms were beginning the production of commonly used tools, such as lathes, but output was almost negligible (see Fredrick V. Geier, *The Tool Builders of Cincinnati*, New York, 1949). In 1868 the Niles Tool Company, then of Cincinnati, was specializing in the production of machine tools, and the Cincinnati firm of John Steptoe was certainly building a few tools by 1870.

is no evidence to indicate much change in these geographical ratios for 1870. Assuming that Philadelphia output was 30 per cent of the total in both 1860 and 1870, we may take U.S. totals of approximately $1 million and $2.5 million as plausible for these two dates.

For the period 1880–1900, it is possible to add to the Brown and Sharpe series given by McDougall a series for Cincinnati production. Estimates of production of metalworking machinery in Cincinnati before 1880 would have dubious validity, for up to that date the producing firms, with the possible exception of the H. Bickford Company, also manufactured a variety of other machinery that could not be classed as metalworking. But on the basis of the known output of the largest Cincinnati firm existing in 1880—Lodge, Barker & Company—Cincinnati production for 1880 was reckoned at $250,000 for the six firms then in existence. George Wing, Xavier University, Cincinnati, has made year-by-year estimates of output up to 1898 (see Table 2); local Census information is available

· TABLE 3

ESTIMATED PRODUCTION OF METALWORKING MACHINERY IN CINCINNATI, 1880-1900

Year	Cincinnati Metalworking Machinery Output Trend (dollars) (1)	Number of Machine Tool Shipments[a] (2)	Number of Machine Tool Shipments Trend (3)	Shipments Divided by Trend (4)	Adjusted Cincinnati Output[b] (5)
1880	250,000	264	264	100.0	250,000
1881	353,333	319	270	118.1	417,286
1882	456,666	303	277	109.4	499,593
1883	560,000	302	285	106.0	593,600
1884	663,333	241	293	82.3	545,923
1885	766,666	197	302	65.2	479,166
1886	870,000	508	510	99.6	866,520
1887	967,500	587	567	103.5	1,001,363
1888	1,070,000	702	625	112.3	1,201,610
1889	1,172,500	868	683	127.1	1,490,248
1890	1,275,000	1,011	760	133.0	1,695,750
1891	1,377,500	905	832	108.8	1,498,720
1892	1,480,000	955	925	103.2	1,527,360
1893	1,582,500	795	1,030	77.2	1,221,690
1894	1,685,000	523	1,140	45.9	773,415
1895	1,787,500	1,081	1,250	86.5	1,546,188
1896	1,890,000	1,345	1,380	97.5	1,842,750
1897	1,992,500	1,571	1,520	103.4	2,060,205
1898	2,100,000	1,963	--	--	2,100,000
1899	3,340,000	2,591	--	--	3,340,000
1900	3,375,000	2,295	--	--	3,375,000

[a] Number of machines shipped by Brown and Sharpe and Bullard, 1880-85, and by Brown and Sharpe, Bullard, and Gray, 1886-1900 (1886 estimated for Gray).

[b] Column 1 times column 4 for 1880-97. Amounts for 1898 and 1899 are from a Cincinnati census, and the 1900 figure is from the U.S. Census for that year. The values in this column reflect cyclical variations.

for 1898 and 1899, and the U.S. Census count serves for 1900. The annual estimates are shown in Table 2.[17] Because Wing made no allowance for cyclical variation in 1880–86 and because it seemed best to add more firms before computing irregular and cyclical variation from 1887 to 1898, a linear trend of Cincinnati output was computed (column 1 of Table 3). The Cincinnati time series was then constructed by adjusting the trend values of output between the benchmark years of 1880 and 1898 for irregular and cyclical disturbances. To compute the index of irregular and cyclical variation, exponential trend lines were fitted to the time series of shipments of machine tools by (1) Brown and Sharpe and (2) Bullard for 1880–85, and by (1) Brown and Sharpe, (2) Bullard, and (3) Gray for 1886–1900.[18] The indexes of irregular and cyclical variation were derived by computing the ratios of actual shipments to shipment trend values. An 1886 shipment value for Gray was estimated in order to obtain a better fit.

Reflections on the Foregoing Estimates

Except for inconsequential details, the Cincinnati data are consistent with McDougall's findings. Output of this essential category of capital goods was extremely volatile over the period 1880–1920, as indeed it has been since 1920.[19] The rate of increase of output was likewise remarkable, in current dollars approximately doubling in the 1860's, trebling in the 1870's, and probably quadrupling in the 1880's.[20] Although this study did not develop a measure of pre-1900 price changes of metalworking machinery, large gains in real output, especially during the 1870's and 1880's, may be inferred from the current dollar figures. During the nineteenth century, manufacturers of metal-cutting and metal-forming tools set carefully reckoned prices on their wares and were not inclined to change them over a long period of years. Prices of machine tools in Cincinnati, for example, did not rise more than 10 or 15 per cent during the 1880's and were very

[17] Details of this process of estimation may be obtained from Wing's Ph.D. dissertation at Indiana University.

[18] The data for Brown and Sharpe and Bullard were obtained from McDougall's paper. McDougall has pointed out that Brown and Sharpe specialized in lighter machines and Bullard in heavier. Gray's output was similar to that of Brown and Sharpe. The year-to-year movements in shipments of the three companies showed remarkable similarity in amplitude and direction.

[19] For recent data in convenient form, see Milton H. Spencer, "Demand for Machine Tools," *California Management Review*, Vol. 5, No. 4, 1963, pp. 75–84.

[20] My estimates of national output for 1880 and 1890 are, respectively, $7.5 and $32 million, the estimates being based largely on changes in Cincinnati and New England production over these decades. With the depressed years of the 1890's, the industry fell on bad days, estimated output for 1894 dropping to a mere $11 million for the nation.

nearly constant during the 1890's. The rapid growth of the industry after 1860 was probably halted by the onset of depression in the 1890's; the industry was struck a tremendous blow by the slump in general activity. The automobile industry doubtless accounted in large part for the doubling of real output of metalworking machinery between 1900 and 1913. Stimulated by war demand both overseas and at home, real output approximately trebled between 1914 and the years of peak output, 1916 and 1918.

The small magnitudes of the output figures, even in the World War I period of soaring production, furnish evidence to support the proposition with which Nathan Rosenberg introduces his discussion of technological change in the machine tool industry.[21] Even when the industry was booming, annual production of metalworking machinery amounted to but a small fraction of 1 per cent of the country's annual product. Clearly the importance of metalworking machinery as a variable in U.S. industrialization was out of all proportion to the dollar value of production of this crucial form of capital. The foregoing data furnish a telling example of the way in which qualitative change in the stock of capital dwarfs quantitative change as an influence on economic growth.

The output figures finally derived may indeed be biased on the low side. Any process of counting is almost certain to miss some items, and this likelihood becomes greater the farther back we go in time. Moreover, as both McDougall and I have observed, metalworking tools were for a long time made on an *ad hoc* basis by clever craftsmen who needed them, and the most diligent census-taker would never have included this part of output in his totals.

Any substantial improvement in nineteenth century time series must come from detailed investigation of narrowly defined product categories. Future research will require better cooperation from the Bureau of the Census. Inquiries addressed to Census officers invariably receive a polite response, but, with the single exception noted above, I have yet to obtain historical information from official Census sources that is not already available in published form. Indeed, on at least three occasions I was informed that original nineteenth century Census returns relevant to this study had been destroyed, only to discover a batch of them, filed in an apparently routine way, in a state library.

But the last great untapped resource of the historian dealing in economic quantities lies in the files of business firms. The problems of developing a national estimate before 1900 would be substantially reduced by the

[21] Nathan Rosenberg, "Technological Change in the Machine Tool Industry," *The Journal of Economic History*, December 1963, pp. 414–443.

papers of only two firms in Worcester and two in Philadelphia.[22] Even such modest results as were achieved would have been impossible without access to business records. It is only from such sources that time series like the one presented in this paper can be extended and refined. To be really useful, the points should be quarterly instead of annual, and they should reach back to the 1850's. The tenuous basis of inferring national output, particularly from 1860 to 1880, should be strengthened. If, as Friedman has suggested, time series are highly manufactured products, there is no reason why more of them, including this one, should not be given a better finish.

[22] After more than two years of correspondence a few papers have been discovered in the files of the Farrel Corporation, successor firm to William Sellers and Company, and the Sellers Family Association of Ardmore, Pennsylvania, may provide some help.

TABLE A-1

EXPORTS OF METALWORKING MACHINERY WITH MAJOR COUNTRIES
OF DESTINATION, 1898-1920
(thousand dollars)

Year Ending June 30		Exports	Total	Year Ending June 30		Exports	Total
1898	Germany	1,670		1910	Germany	1,805	
	U.K.	1,461			U.K.	1,363	
	France	577			France	691	
	Belgium	252	4,619		Canada	336	5,976
1899	Germany	2,638		1911	Germany	2,524	
	U.K.	1,681			U.K.	2,319	
	France	741			France	963	
	Belgium	338	6,492		Canada	766	9,627
1900	Germany	2,480		1912	Germany	2,953	
	U.K.	1,883			U.K.	2,687	
	France	1,090			Canada	1,362	
	Belgium	656	7,193		France	1,268	12,152
1901	U.K.	1,482		1913	U.K.	3,418	
	Germany	1,035			Germany	3,175	
	France	441			Canada	2,326	
	Belgium	246	4,054		France	1,937	16,097
1902	U.K.	1,706		1914	U.K.	3,179	
	France	307			Germany	2,167	
	Germany	259			France	1,771	
	Belgium	117	2,977		Russia	1,335	14,011
1903	U.K.	1,309		1915	U.K.	12,295	
	France	375			France	8,696	
	Germany	318			Russia	2,489	
	Belgium	185	2,826		Canada	1,813	28,163
1904	U.K.	1,122		1916	U.K.	20,438	
	Germany	887			France	13,317	
	France	369			Russia	12,333	
	Belgium	282	3,717		Canada	6,464	61,315
1905	U.K.	1,038		1917	France	29,254	
	Germany	913			U.K.	16,300	
	Belgium	592			Russia	15,329	
	France	392	4,333		Italy	8,771	84,935
1906	Germany	1,814		1918	France	20,271	
	U.K.	1,361			U.K.	18,396	
	Italy	737			Italy	5,077	
	France	654	6,446		Canada	3,751	58,408
1907	Germany	2,245		1918[a]	U.K.	9,833	
	U.K.	1,937			France	6,331	
	France	1,304			Canada	3,072	
	Italy	1,146	9,369		Japan	2,254	25,183
1908	Germany	1,935		1918[b]	U.K.	19,296	
	U.K.	1,642			France	15,351	
	France	1,063			Canada	4,814	
	Belgium	702	8,696		Japan	4,047	51,620
1909	U.K.	952		1919[b]	France	15,785	
	Germany	943			U.K.	15,210	
	France	307			Japan	5,383	
	Austria-Hungary	255	3,640		Canada	4,035	58,508
				1920[b]	U.K.	10,999	
					France	7,596	
					Canada	5,815	
					Japan	4,251	44,312

Source: *Commerce and Navigation of the United States.*

[a] July 1 to Dec. 31.   [b] Calendar year.

TABLE A-2

ALTERNATIVE COMPUTATIONS OF REAL OUTPUT OF METALWORKING MACHINERY, 1900–20

Year	Output of Metalworking Machinery in Current Prices[a] (mill. doll.) (1)	Cincinnati Price Index (Alternative 2)[b] (1900=100) (2)	Output of Metalworking Machinery in 1900 Prices (mill. doll.) (3)	Index of Real Output (Alternative 2) (1914=100) (4)	U.S. Price Index (Alternative 3) (1926=100)[c] (5)	Output of Metalworking Machinery in 1926 Prices (mill. doll.) (6)	Index of Real Output (Alternative 3) (1914=100) (7)
1900	24.7	100	24.7	66.8	37.6	65.7	61.2
1901	24.5	100	24.5	66.2	37.0	66.2	61.8
1902	29.8	100	29.8	80.5	39.0	76.4	71.3
1903	29.7	103	28.8	77.3	39.9	74.4	69.4
1904	21.7	103	21.1	57.0	40.0	54.3	50.7
1905	32.4	104	31.2	84.3	40.3	80.4	75.0
1906	42.1	118	35.7	96.5	41.4	101.7	94.9
1907	48.9	118	41.4	111.9	43.7	111.9	104.3
1908	20.4	122	16.7	41.5	42.1	48.5	45.2
1909	41.6	125	33.3	90.0	45.3	91.8	85.6
1910	56.2	125	45.0	121.6	47.2	119.1	111.1
1911	42.5	131	32.4	87.6	43.5	97.7	91.1
1912	58.8	131	45.0	121.6	46.3	127.0	118.5
1913	60.3	140	43.1	116.5	46.8	128.8	120.1
1914	48.9	132	37.0	100.0	45.6	107.2	100.0
1915	145.0	138	105.1	284.1	49.4	293.5	273.6
1916	198.2	186	106.6	288.1	62.0	319.6	298.0
1917	242.3	254	95.4	257.8	82.3	294.4	274.5
1918	321.2	311	103.3	279.2	101.2	317.4	295.9
1919	237.2	333	71.2	192.4	106.3	223.1	208.0
1920	225.9	380	59.4	160.5	116.5	193.9	180.7

NOTES TO TABLE A-2

^aTaken from Table 1.

^bPrice index of metalworking machinery constructed from list prices of firms in the Cincinnati area. See George Wing's "History of the Cincinnati Machine-Tool Industry," unpublished Ph.D. dissertation, Indiana University. This method of computation was considered Alternative 2.

^cFor 1900-13, the Bureau of Labor Statistics Wholesale Price Index--all commodities; for 1914-20, the American Appraisal Company's machine tool price index. The base year 1926 was used for both indexes. The BLS Wholesale Price Index was spliced to the American Appraisal Company's index by dividing each BLS index number from 1900 to 1913 by the ratio of the 1914 BLS price index number to the 1914 American Appraisal Company's index number, the ratio being 149.3 This method of computation was considered Alternative 3. However, the constant used for splicing was the average ratio of BLS to American Appraisal Company price indexes for 1915-20. These ratios ranged between 130 and 140, with a mean of 135.7. It was thought that the mean better represented the relationship between the two price indexes, and this second splicing alternative was used to adjust the BLS index numbers before computing the index of "real" output that was carried to Table 1.

Machine Tool Output, 1861–1910

DUNCAN M. McDOUGALL
CARLETON UNIVERSITY

Introduction

A quantitative study of the American machine tool industry in the latter half of the nineteenth century faces a data problem of formidable proportions. This results not only from the lack of statistical information, common to many areas of research of that period, but primarily from the fact that it was not until the end of the period that a distinct machine tool industry emerged in the sense of a group of firms whose principal product was machine tools. Until about 1900, the machine tool industry could not be called a large, independent sector of the economy like many of the other groups of companies that led in American industrialization.

A reading of the classic work by Roe on the early American machine tool manufacturers gives the opposite impression.[1] A whole chapter is devoted to Joseph Brown and the Brown and Sharpe Manufacturing Company of Providence, Rhode Island, as one of the leading New England machine tool companies. Founded on the invention of the universal milling machine and the precision devices invented and produced by Brown, Brown and Sharpe was undoubtedly a leading company. On the other hand, examination of the financial records of the company reveals that the value of its output of sewing machines, manufactured under license from Wilcox and Gibbs, exceeded the value of machine tool shipments until 1885, twenty-five years after the invention of the universal milling machine. The year 1898 was the first in which the value of machine tool shipments accounted for more than 50 per cent of the total sales of Brown and Sharpe and, although that percentage was approached in

---

NOTE: The author wishes to acknowledge financial assistance from the Ford Foundation, the Interuniversity Committee on American Economic History, and Purdue University; the cooperation of many persons in the machine tool industry; and the assistance of George A. Wing and Arvid M. Zarley; all of whom contributed greatly to this paper, but none of whom are responsible for the conclusions, opinions, or statements contained herein. That responsibility remains the author's alone.

[1] J. W. Roe, *English and American Tool Builders*, New Haven, 1916.

subsequent years, it was not reached again before 1904 when the data end (Table A-1).[2]

A further indication of the relatively small and unspecialized nature of the industry is the fact that the National Machine Tool Builders Association was founded by seventeen firms as late as 1902, and the Association has published no industry data for the period before 1900. Finally, it was not until the 1914 Census of Manufactures that the federal government published separate data for industrial machinery output, including machine tools, but the report is far from complete. The earlier decennial Census reports on manufacturing have a section listed as machinery, but machine tools are not treated separately. The sector as reported includes such things as agricultural implements, sewing machines (including presumably the total output of Brown and Sharpe until at least the Census of 1880), pumps, engines, professional and scientific instruments, and an omnibus category of foundry and machine shop products, n.e.s.—by far the largest category of all.

This paper deals only with metal-cutting machines and includes no information on metal-forming types. A lathe or milling machine would be an example of the first type, a press or hammer would be an example of the second. Furthermore, the paper deals with general-purpose standard machines of the light variety. Neither machines made on special order for a particular job nor industry-specialized machines such as textile or mining machinery are included. The machines considered here are the versatile, primary machines used in the production of other machinery as well as in the production of such goods as sewing machines and typewriters, not generally considered to be machines. It has proved impossible to quantify the size of the special-order output of machine tools in the nineteenth century. Manufacture of such tools was apparently concentrated in Philadelphia, particularly in the William Sellers Company and in the group of firms eventually combined as the Niles-Bement-Pond Company.[3] All attempts to uncover any records of the early Philadelphia companies met with failure.

The period covered in this paper is 1861–1910. The beginning date was chosen because no reliable quantitative data were found for any earlier years. Hubbard, in a series of articles in the *American Machinist* about the early beginnings of machine tool manufacture in New England, lists an impressive number of persons and companies involved in machine

---

[2] There is, of course, a difference between sales and shipments, but the relative importance of machine tools in total sales is nonetheless represented accurately.

[3] Roe, *English and American*, pp. 249–260.

tool design and production before 1860.[4] Yet the impression gained is of a series of small undertakings to meet specific needs, such as government musket contracts, or construction of special machinery to meet individual industry demands. For example, a mechanic in a textile mill or a small shop might build a lathe for his own use, but there is no evidence in Hubbard's work that there were machine tool companies as such. Roe, although more concerned with technical developments in machine tool design than with quantitative measures of the industry, uses 1850 as the beginning date of the American machine tool industry. As subsidiary evidence, it might be noted that none of the charter companies of the National Machine Tool Builders Association existed in 1860, as far as can be determined from the various regional genealogies of the industry.

The primary source of the data used in the paper is company records. The best records, and the only ones for 1861–81, were obtained from the Brown and Sharpe Company for 1861–1905. The records list a total of 23,658 machine tools shipped by the company from September 1861 to June 1905.[5] Each shipment is recorded by type and size of machine, date of shipment, name and location of consignee, and price. Year-end financial statements of the same company were also obtained for 1869–1905. A second shipment series was obtained from the Bullard Company of Bridgeport, Connecticut, covering a total of 6,535 machines shipped between April 1881 and December 1912. Those records did not include price, but a separate price record found for 1895–1912 made possible calculation of dollar sales for the shorter period. Sales figures have also been obtained for the G. A. Gray Company, Cincinnati, for 1886–1910, and for the Cincinnati Shaper Company for 1899–1910. A 20 per cent sample of the shipment records of the two Cincinnati companies showing region of destination, by various time periods, is also available.[6] Finally, a series of the dollar value of new orders was obtained from the Warner and Swasey Company of Cleveland for 1880–1910, and a series of sales figures from the same company beginning in 1903. A total of twenty-one existing machine tool companies with roots extending back into the nineteenth century were communicated with in the study. All requests were met with offers of assistance, but all except the above-mentioned

[4] G. Hubbard, "Development of Machine Tools in New England," *American Machinist*, Vols. 59 and 60, 1923 and 1924.

[5] Shipments differ from production by the number of machines produced but retained for the company's use and by the net change in machine inventory. Data for shipments only were available.

[6] The information on the Cincinnati companies was made available to the author through the kindness of George A. Wing.

few reported that early records had been destroyed, frequently as a result of consolidation or reorganization of firms.[7]

The subsequent analysis in this paper is based, therefore, on the records of a relatively small number of firms. There is no way of telling how representative of the industry the sample is. We do know that Brown and Sharpe was an early technical leader in the machine tool field and one of the leading producers of light machine tools in New England. The Brown and Sharpe Company began in 1853 when Lucien Sharpe, who proved to be the businessman of the combination, was brought into the company formed in 1833 by David and Joseph R. Brown, father and son, to make and repair clocks. In 1850 Joseph Brown developed the first automatic linear dividing engine for graduating rules, a machine Roe mentions as still being in use in the shop in 1916. The next efforts were production of protractors and calipers with the vernier scale attached. Standards and accuracy of measurement continued to be an important concern of the firm—as they still are today—but, from the point of view of this paper, the developments must be considered peripheral to the main subject. The first machine tool produced for sale by the company was a turret screw machine of a general design well known at the time, sold in 1861 to the Providence Tool Company which had a contract to manufacture Springfield muskets for the government. Joseph Brown's significant contribution to machine tool design came in 1861-62, when he invented the universal milling machine. The first of these new machines was shipped in 1862, again to the Providence Tool Company.

Throughout the period covered by the available records, Brown and Sharpe concentrated on the production of light, standard machine tools, screw machines, grinding machines, and the universal milling machine. In this respect, it was apparently similar to Pratt and Whitney, whose product line was much the same in nature although more extensive in both kind and number of machines produced. Brown and Sharpe is therefore probably a good sample of the light machine tool operations in New England as a whole.

The Bridgeport Machine Tool Company, which was to become the Bullard Company, was established in 1880 as a producer of machine tools in Bridgeport, Connecticut, by E. P. Bullard. Bullard was a mechanic who had become a machine tool agent in New York City and, from his experience in selling tools, he recognized the need for a more accurate

---

[7] The greatest disappointment in this project was the discovery that the first volume of the shipment records of the Pratt and Whitney Company of West Hartford, Connecticut, covering the period up to about 1904 had been destroyed eighteen months before I wrote to the firm.

engine lathe than those currently available. He engaged a mechanic to produce the lathes in Bridgeport but, within a year, had taken full control of the operation. The first shipment of the new company was a total of twenty-five 16-foot by 5-inch engine lathes between April and June 1881, consigned to the Westinghouse Airbrake Company of Pittsburgh. In 1883, Bullard made a significant contribution to the technical progress of the machine tool art with the invention of a small boring mill capable of accurate production work. The new machine was publicized for the first time in 1883, but no shipments were made until 1885. With the movement into the boring mill and the larger engine lathe, the Bullard Company developed a product line of machines that were larger, as measured by average price per machine, than those developed by Brown and Sharpe.

There is no way of assessing the representativeness of the sample collected. The records of Brown and Sharpe are extremely valuable because it is fairly certain it was quantitatively a large firm and was certainly a leader in the technology of the light machine tool field. Bullard is also a valuable sample because, while not quantitatively so large a firm as Brown and Sharpe, its output encompassed the larger type of production machine tool. The evidence available for the Ohio companies is only meager and, while implications can be drawn from the information available, it will not support much analysis.

Little more can be said beyond affirming that the information gathered for this paper exceeds anything previously available, and that an effort has been made to track down as much relevant material as possible. What follows, therefore, while strictly speaking a quantitative study of selected firms, represents the growth pattern of the American type of production cutting machine tool through the first decade of this century.

Growth of Output

The literature on the history of the American machine tool industry never fails to point to the supreme importance of the machine tool in the development of what came to be called the American system of manufacture. There can be no doubt that mass production, the standardization of parts, and precision manufacturing all stemmed from advancing technology embodied in machine tools. What seems clear from the record, however, is that these accouterments of an industrial society came at a fairly late stage in the process of development and were associated with a particular change in output mix.

Perhaps the most impressive aspect of the quantitative records available is that the volume of shipments was so small (Table A-2). The first machine tool was shipped by Brown and Sharpe in 1861. It was not, however, until 1875 that the cumulative total of machines shipped passed the 1,000 mark, and not until 1883 that it reached 2,000. Between September 1861 and June 1905, the records show that Brown and Sharpe shipped a total of 23,658 machine tools, but 12,447 of the total number, or 52.6 per cent, were shipped between January 1899 and June 1905. The Bullard Company began shipping machines in April 1881, but it was not until 1890 that the cumulative number of shipments passed the 1,000 mark. Up to the end of 1910, Bullard shipped a total of 6,162 machines, but 3,229 of the total number, or 52.4 per cent, were shipped between January 1901 and December 1910.

There is ample evidence that the acceleration in the rate of growth of commodity output which marked the initial period of industrial development in the United States began before the Civil War. Whether or not one wishes to call it the period of "take-off," following Rostow, it is clear from Gallman's figures that the high decade rates of growth shown for the period just before 1860 must have marked a sharp change from the rates existing in the early decades of the nineteenth century.[8] Gallman's figures also show evidence of a decline in the growth rates of commodity output, and particularly in the growth rate of value added by manufacturing, in the latter decades of the nineteenth century. As Gallman pointed out, his results are roughly consistent with the trend-cycle dating determined by Burns from his study of production trends, which shows a period of rapid increase of nonagricultural industrial output between 1875 and 1885 and again between 1895 and 1905.[9]

Table 1 presents rates of change calculated from the available long-term records of machine tool output along with selected rates of change derived from Gallman. The Gallman rates of change are based on single-year figures, while those for the machine tool shipments are three-year averages centered on the years available to Gallman. Presumably, therefore, the Gallman figures would show the influence of business cycles more sharply than the shipment figures do, but the differences are so clear that the conclusions drawn from the table are unlikely to be affected by cyclical variations in the underlying figures.

The conclusion derived from the table is that the light production machine tool was not quantitatively important in the period of American industrialization but became important at a later period. This conclusion

[8] R. E. Gallman, "Commodity Output, 1839–1899," *Trends in the American Economy in the Nineteenth Century*, Studies in Income and Wealth 23, Princeton for NBER, 1960, pp. 15–17.

[9] A. F. Burns, *Production Trends in the United States since 1870*, New York, NBER, 1934, pp. 215–220.

could result from the fact that only one company is considered. The explanation of the difference in the aggregate and single-company growth figures could merely reflect a sudden increase in the prominence of Brown and Sharpe, or a technical development that permitted a sharp increase in the output of the firm. Both explanations appear unlikely. Brown and Sharpe had a wide reputation as a machine tool producer, if its exhibition of tools at international expositions as early as 1869 is any indication.[10] Also, the second hypothesis seems unlikely in view of the fact that the company's product line, which represented inputs to its own productive process, shows no apparent marked technical change over time.

TABLE 1

RATES OF CHANGE IN MACHINE TOOL PRICES AND OUTPUT, SELECTED VARIABLES, DECENNIAL OR QUINQUENNIAL, 1864-69
(per cent)

Year	Value Added in 1879 Prices of Manufacturing (1)	Change in Value in 1879 Prices of Manufactured Producer Durables Output (2)	Change in Number of Brown and Sharpe Shipments of Machines		
			Total (3)	Domestic (4)	Foreign (5)
1864					
1869	26	72			
1874			25	4	
1879	82	67	25	35	-12
1884	90		50	121	67
1889	112	117	376	300	654
1894	71		229	201	400
1899	51	48	221	170	320

Source: Col. 1 from Gallman, "Commodity Output," Table 3, p. 24; col. 2 calculated from *ibid.*, Table A-12, p. 65; cols. 3-5 from company records.

The more likely explanation is found in the relation between machine tool output and the industrial destination of the machines, which can be inferred from the major product of the buyer. The name of the buyer does not always appear in the available records showing the consignee of shipments. Where the consignee was an agent and only the agent's name was listed, it was impossible to determine the eventual destination of the machine. In other records, the name of the consignee gave no clue to the industry to which it belonged, although in some such cases the firm could be allocated by reference to other sources. Of a total of 22,478 machines shipped by Brown and Sharpe to December 1904, 8,499 or

[10] Brown and Sharpe published a catalogue in French in August 1867, and one in German in April 1868.

37.8 per cent were consigned to domestic and foreign agents or allocated to unknown buyers. Of this number, 4,469 or more than half were consigned to foreign agents. Of the total of 6,162 machines shipped by Bullard to December 1910, 2,498 or 40.5 per cent were consigned to domestic and foreign agents or allocated to unknown buyers. Of this number, 1,284 or more than half were consigned to foreign agents. The problem of the agent comes up again with respect to machine tool companies as consignees. We know, for example, that Pratt and Whitney established agencies in many parts of the United States fairly early in its history, and acted as agents for Brown and Sharpe as well. The very minor number of machines consigned to Pratt and Whitney indicates that, in spite of the affiliation, final consignee was specified on sales through the agent. There is no way of telling, however, how many local machine shops, especially those begun by Brown and Sharpe apprentices, might have acted as agents.

Table 2 presents quinquennial totals of part of the industrial distribution of shipments by Brown and Sharpe and by Bullard. There are two significant points shown by the table. The first is the clear association between the expanding output of machine tools and the growth of industries producing fairly complex and technically sophisticated final products. The bicycle, the cash register, and the electrical equipment industries, and government arsenals all placed substantial orders for machines after 1885. Second, the difference in the relative size of the machines in the Brown and Sharpe and Bullard product lines shows up in the table. Bullard shipped no machines to companies producing cash registers between 1881 and 1910, only eighteen machines to sewing machine companies, and only seven machines to bicycle companies. On the other hand, the percentage of total Bullard shipments consigned to iron foundries and iron and steel mills was six times the corresponding percentage for Brown and Sharpe. Sales by the two companies to automobile producers show an interesting difference. Bullard shipped its first machine to an automobile company in 1901 by which time Brown and Sharpe had shipped 101 machines but, as the automobile increased in size and complexity, heavy tools became more important as inputs, and shipments by Bullard increased markedly.

A machine tool of the type considered here is a fairly versatile input. It is true that the machines produced before 1910 were specific in the types of operations they could perform; an engine lathe, for example, was quite limited in the kinds of operations it could be set up to accomplish at one time, but these machines could be used to produce many different products. The reasonable conclusion seems to be that, until after 1884

TABLE 2
INDUSTRIAL DISTRIBUTION OF MACHINE TOOL OUTPUT, QUINQUENNIAL, 1861-1909
(number of machines)

Purchasing Industry	1861-64[a]	1865-69	1870-74	1875-79	1880-84[b]	1885-89	1890-94	1895-99	1900-04	1905-09
A. SOLD BY BROWN AND SHARPE										
Machine tools	16	22	16	8	63	90	194	439	649	
Rifles and ammunition	100	11	41	10	5	9	62	85	259	
Sewing and shoe machines	22	62	135	35	120	88	93	140	253	
Calculators and cash registers	0	0	0	0	9	13	101	183	716	
Professional and scientific instruments[c]	0	10	6	9	35	30	65	144	282	
Government arsenals	3	14	14	1	6	379	113	409	601	
Electric equipment	0	0	0	3	60	49	157	538	797	
Bicycles	0	0	0	0	0	27	32	267	78	
Automobiles	0	0	0	0	0	0	0	55	268	
Railroads and R.R. equipment	4	9	19	11	69	71	95	117	182	
Sum of cols. as per cent of total machines[d]	95.4	80.0	68.5	61.1	71.3	67.8	58.2	63.3	65.4	
B. SOLD BY BULLARD										
Machine tools					36	87	99	66	101	91
Railroads and R.R. equipment					112	65	104	77	207	212
Iron and steel					3	31	85	81	65	91
Electric equipment					4	53	54	52	192	167
Government arsenals					1	10	49	31	24	17
Bicycles					0	0	0	0	5	2
Automobiles					0	0	0	0	19	88
Sum of cols. as per cent of total machines[d]					76.1	70.3	68.2	74.2	71.9	65.6

Source: Company records.

[a] September 1861 to December 1864, Brown and Sharpe.

[b] April 1881 to December 1884, Bullard.

[c] Includes companies making watches, clocks, cameras, optical equipment, and dental equipment.

[d] These are column totals as a percentage of the total number of machines that could be allocated by user.

or more noticeably after 1895, the demand for light machine tools was relatively limited. The large demand for machine tools beginning at the turn of the century can be associated with the development of a new technology, and with the beginnings of what Rostow has called the period of "high mass consumption."

Market for Machine Tools

While the volume of shipments from machine tool companies was apparently quite modest until nearly the end of the nineteenth century, the market was anything but local. During the Civil War the market was dominated by domestic demands from armament makers. Between September 1861 and the end of 1864, Brown and Sharpe shipped a total of 201 machines, 100 of them to armament makers in New England. As soon as the war demand ended, the market area expanded considerably. The first shipments by Brown and Sharpe to foreign customers were made in 1865, when two machines were shipped to Canada and two to France. From 1865 on, the foreign market accounted for a significant proportion of Brown and Sharpe shipments, never accounting for less than 10 per cent of the total in any one year, and in some years rising to over 50 per cent. For the Bullard Company, foreign shipments were not as significant a proportion of the total until after 1896 but, of the ninety machines shipped in the first year, 1881, two were consigned to foreign customers. In the fifteen years between 1881 and 1895, there were only five years when no foreign shipments were made, and a total of thirty-seven machines were shipped abroad. In 1896, 25.9 per cent of the machines shipped went to foreign buyers, and in the four years 1897–1900 the proportion was between 50 and 60 per cent. In the decade 1901–10 an average of 23 per cent of Bullard shipments was consigned to foreign customers (Table A-3).

A sample of the shipments of the two Cincinnati companies shows that the New England toolmakers did not monopolize the foreign trade. A 20 per cent sample of the shipments by the G. A. Gray Company between 1884 and 1907 (sample size, 799 machines) shows 165 machines, or 20.8 per cent of the total, consigned to foreign customers. The same type of sample taken from the records of the Cincinnati Shaper Company between 1899 and 1907 (sample size, 554 machines) shows that 110 machines, or 19.9 per cent of the total, were shipped abroad. Finally the sales records of the Warner and Swasey Company of Cleveland show an average share of foreign in total sales of 18.8 per cent between 1903 and 1910.

As one would expect, northwest Europe dominated the foreign market. The number of machines shipped by Brown and Sharpe and by Bullard to foreign regions, by quinquennia, is shown in Table 3. On the reasonable

TABLE 3

FOREIGN SHIPMENTS OF MACHINE TOOLS, BY REGION OF DESTINATION, QUINQUENNIAL, 1861-1909
(number of machines)

Period	Northwest Europe	South Europe[a]	Russia and East Europe[b]	Canada	Central and South America	Australia	Asia	Africa
	A.	BROWN	AND	SHARPE				
1861-64[c]	0	0	0	0	0	0	0	0
1865-69	27	3	1	16	0	0	0	0
1870-74	60	10	5	34	0	0	0	0
1875-79	55	0	0	5	3	0	0	0
1880-84	84	0	2	29	7	1	2	0
1885-89	633	0	2	13	5	2	0	0
1890-94	431	6	38	25	8	7	0	7
1895-99	2,560	43	257	39	16	7	16	2
1900-04	2,349	49	127	140	18	2	331	5
Percentage distribution of totals	82.9	1.5	5.8	4.0	0.8	0.2	4.7	0.2
	B.	BULLARD						
1881-84[d]	7	0	0	1	1	0	0	0
1885-89	13	0	0	0	0	0	0	0
1890-94	7	0	0	4	0	0	0	0
1895-99	446	0	8	2	1	0	0	0
1900-04	394	3	1	30	1	0	2	0
1905-09	341	20	9	29	4	1	9	0
Percentage distribution of totals	90.6	1.7	1.3	4.9	0.5	0.0	1.2	0.0

Source: Company records.
[a] Italy, Spain, Portugal, Greece, Turkey.
[b] Poland, Rumania, Bulgaria.
[c] Sept. 1861 to Dec. 1864.
[d] Apr. 1881 to Dec. 1884.

assumption that foreign companies purchasing American machine tools were not tied to particular American producers, this table can be taken as representative of the regional distribution of foreign shipments by the American machine tool industry as a whole. There are differences between the two panels of the table. The Brown and Sharpe figures show a smaller proportion of total foreign shipments to northwest Europe and a larger proportion to Russia (including eastern Europe) and Asia than the

Bullard figures do. The differences are accounted for by relatively large shipments by Brown and Sharpe to government arsenals in Russia between 1895 and 1901 (over 300 machines) and to Japanese government arsenals and shipyards in 1904 (260 machines). Since the Bullard line of products was not suitable for armaments manufacture at that time, the company did not share the market.

That northwest Europe was the major foreign market for shipments is, of course, not surprising. Nor is the distribution by country of destination within Europe. Table 4 presents the distribution of shipments to northwest

TABLE 4

SHIPMENTS BY BROWN AND SHARPE TO NORTHWEST EUROPE, QUINQUENNIAL, 1865-1904
(number of machines)

Period	United Kingdom	France	Germany	Sweden, Denmark	Belgium, Netherlands	Switzerland, Austria
1865-69	4	12	4	0	0	0
1870-74	28	1	17	13	0	1
1875-79	23	20	11	0	1	0
1880-84	56	7	20	1	0	0
1885-89	181	352	67	23	3	7
1890-94	182	60	92	47	14	36
1895-99	975	511	621	243	80	130
1900-04	1,106	587	323	121	77	135

Source: Company records.

Europe, by country of destination, for Brown and Sharpe only.[11] The United Kingdom was clearly the largest single purchaser of machines from Brown and Sharpe, not only in total but also in all subperiods except 1865-69 and 1885-89. The large volume of shipments to France in the latter quinquennium is made up primarily of shipments to French government arsenals, which received nearly 200 machines between 1886 and 1888.

In the total number of machines shipped, France was the second largest customer and Germany the third largest. Most, although not all, of the difference is accounted for by the large shipment to French arsenals noted above. The time distribution of machines to other regions generally coincides with what is known about their periods of industrialization.

[11] In Tables 3 and 4, the distribution was derived from records that generally list only one consignee. Thus, a London or Antwerp agent might subsequently ship a machine consigned to him to another region or country. In view of the number of agents scattered throughout Europe, this possibility is unlikely to affect the distribution appreciably.

Expectably, the destination of the machine shipments moved across Europe in general conformity with the eastward progress of the industrial revolution.

The same sort of geographic distribution of machine shipments based on the level of industrial activity in a region is evident in Table 5, where

TABLE 5

SHIPMENTS TO UNITED STATES REGIONS, QUINQUENNIAL, 1861-1909
(number of machines)

Period	New England	Middle Atlantic	East North Central	West North Central	South Atlantic	East South Central	West South Central	Mountain	Pacific
			A.	BROWN AND SHARPE					
1861-64[a]	112	81	3	0	5	0	0	0	0
1865-69	95	122	15	0	5	0	0	0	2
1870-74	161	208	20	2	0	2	1	0	4
1875-79	69	90	35	4	2	0	0	0	8
1880-84	272	419	138	17	24	4	2	0	9
1885-89	396	491	221	26	52	8	6	3	7
1890-94	699	784	373	26	90	18	3	4	10
1895-99	1,252	1,270	853	35	93	23	7	10	23
1900-04	1,840	2,219	1,695	151	250	16	12	19	83
Percentage distribution of totals	32.6	37.9	22.4	1.7	3.5	0.5	0.2	0.2	1.0
			B.	BULLARD					
1881-84[b]	90	214	20	0	3	0	3	0	0
1885-89	206	276	85	0	14	1	1	0	1
1890-94	173	351	37	0	109[c]	0	0	1	3
1895-99	126	296	14	4	49	0	0	0	47
1900-04	188	521	241	20	69	8	11	4	25
1905-09	206	500	421	36	36	19	14	6	31
Percentage distribution of totals	22.1	48.3	18.3	1.3	6.3	0.6	0.4	0.2	2.4

Source: Company records.

[a] September 1861 to December 1864.

[b] April 1881 to December 1884.

[c] Between 1890 and 1892, 46 machines were shipped to the Navy Department in the District of Columbia.

the distribution of shipments to United States regions (Census definition) is given for Brown and Sharpe and for Bullard. The dominant position of the middle Atlantic region is clear from the table. In all but the first subperiod 1861-64, that region received more machines than any other from both companies. The New England and east north central regions received some machines in all subperiods, and while neither was as quantitatively important as the middle Atlantic region, the inference certainly is that the domestic market for the tools of these New England companies was widespread. The transportation costs—probably fairly substantial, at least in the early period—did not limit Brown and Sharpe to a local market.

That transportation cost was not a strong deterrent is shown most strikingly by the increase in the volume of shipments to the east north central region after 1880. As Roe has pointed out, "prior to 1880 practically all of the tool building in the United States was done east of the Alleghenies," but that "good tool building appeared in Ohio in the early eighties, and within the ten years its competition was felt by eastern tool builders." [12] It is clear, however, that growth of manufacturing and of a machine tool industry in the Ohio Valley had only good effects upon the New England companies. Increased industrial activity meant increased machine tool inputs, and New England production machine tools were among the best available.

The table does indicate that, whereas from the subperiod 1885–89 on Brown and Sharpe shipped machines to all regions, Bullard did not consistently ship to all regions until the period 1900–04. That the difference resulted from the higher transportation costs incurred by the larger and heavier Bullard machines is unlikely. The more likely explanation is that the heavier production machine is not required as an input until a region reaches a certain threshold of industrial sophistication.

The information available from the Cincinnati firms shows that they also enjoyed a wide geographic market for their output. The 20 per cent sample of the shipments of G. A. Gray Company for the period 1884–1907 contains 631 domestic shipments. The percentage distribution of the destination of the shipments shows that 8.2 per cent went to New England, 26.9 per cent to the East,[13] 53.2 per cent to the north central region,[14] 4.6 per cent to the South, and 7.0 per cent to the West. The percentage distribution of the 444 domestic machine shipments in the 20 per cent sample of the Cincinnati Shaper Company between 1899 and 1907 shows that 5.2 per cent went to New England, 45.7 per cent to the East, 26.8 per cent to the north central region, 10.4 per cent to the South, and 11.9 per cent to the West. It appears from these figures that Roe's statement that the Ohio companies competed with the New England firms was correct. It is also true that such competition, carried on in a rapidly expanding market, had beneficial effects on the firms in both regions.

[12] Roe, *English and American*, p. 261.
[13] Defined as New York, New Jersey, Pennsylvania, Maryland, Delaware.
[14] Defined as Ohio, Indiana, Illinois, Wisconsin, Michigan, Minnesota.

The relative size of the firms in both regions should be noted. Roe mentions that the Gray Company started in 1883 to build lathes, but soon specialized on planers and "is now [1916] one of the foremost firms in the country specializing in this type of tool." [15] Yet the 20 per cent sample taken from Gray's shipment records shows a sample size of only 987 machines shipped between 1884 and 1915, 767 of which were domestic shipments.

Feast or Famine Industry

Discussions of the machine tool industry mention in some form or other the fact that fluctuations in output of the industry tend to be much wider than fluctuations in general industrial output or in the index of economic activity. It might be argued, however, that this characteristic of the machine tool industry is part of a well-known phenomenon of an industrial economy which has acquired and is using a large stock of producer durable equipment. During a period of industrialization, sufficient momentum might be generated by new, rapidly growing industries so that the output of machine tools is little affected by fluctuations in general commodity output.

Whatever the theoretical merits of such a hypothesis, it is clear from the available records of machine tool output and sales (Table A-2) that the firms were subject to substantial fluctuations in demand. Unfortunately, the National Bureau reference cycle chronology does not give measures of the severity of cycles for the period under review here, and it is impossible therefore to say whether machine tool demand fluctuated more widely than aggregate demand.

Comparison of the National Bureau reference cycle chronology with the measures of machine tool output shows three periods of nonconformity.[16] The reference cycle peak of 1869 and trough of 1870, the peak of 1887 and trough of 1888, and the peak of 1895 and trough of 1896 are not reflected clearly in the output measures. For the first period, 1869-70, only the output series for Brown and Sharpe is available, and so perhaps no great importance should be attached to the nonconformity. For the second period, 1887-88, four series are available and only the new-orders series of Warner and Swasey shows a contraction. For the third period, 1895-96, four series are available and only the Gray Company sales series shows a contraction. Both 1886–90 and 1895–99 were periods

---

[15] Roe, *English and American*, p. 273.
[16] A. F. Burns and W. C. Mitchell, *Measuring Business Cycles*, New York, NBER, 1946, p. 78.

of substantial increase in the level of machine tool output. The nonconformity with reference cycle dating might then serve as evidence that, during periods of vigorous demand associated perhaps with rapid technical change, the output of machine tools is unaffected by cyclical contractions in aggregate demand.

The evidence almost disappears, however, when the output series are separated into domestic and foreign shipments, which is possible with the Brown and Sharpe and the Bullard data (Table A-3). There is no contraction in the Brown and Sharpe domestic shipment series in 1869-70, or in the Bullard domestic shipment series in 1887-88. But there is a clear contraction in the Brown and Sharpe domestic shipment series in 1887-88, and the domestic shipment series of both companies show a contraction after 1895, Bullard in 1897, and Brown and Sharpe in 1896-97.

The differences in the cyclical behavior of the total and domestic shipment series is perhaps to be expected from the differences in the reference cycle dates in the United States, France, Great Britain, and Germany. There is very little evidence of an inverse cycle in domestic and foreign shipments, but the foreign shipments of Brown and Sharpe reached a peak in 1888, and there was a strong foreign demand for both Brown and Sharpe and Bullard output between 1895 and 1901 which submerged the domestic contraction of 1895-96.

Summary

The data collected for this paper are undoubtedly far from satisfactory as a basis for an analysis of the nineteenth century machine tool industry, but company records are the only source from which a quantitative record of the industry can be established.

The data assembled suggest that the demand for light, metal-cutting machine tools was relatively small during the initial period of industrialization, when technical development was embodied in fairly large and crude systems. Sewing machine manufacturers were a steady component of the demand for Brown and Sharpe machines back to 1861. It was not until the 1890's, however, that manufacturers of electrical equipment, calculators, cash registers, and bicycles provided a rapidly expanding market for light tools.

The data also show that the American machine tool industry enjoyed a truly worldwide market during the nineteenth century. Much has been written about the technical superiority of the American machines during the period, and certainly the records indicate that quality was recognized by expanding export markets in six continents.

For this particular industry, 1910—the end date of the analysis here—is significant because it can serve as a dividing line in the history of machine tools. After 1910, the assembly line called for more specialized and special-order machines than before, and faster cutting speeds and heavier machines were in demand. The automobile industry alone created a revolution in machine tool building.

Appendix

In addition to background data for the text tables, the appendix presents some additional information collected from machine tool companies during the study but not used directly in the earlier pages.

Table A-1 presents information relating to Brown and Sharpe only. The figures on total sales were taken from annual financial statements, while the amount of sewing machine work and "other work done" came from other records. The same basic data must have been used in both records, because the breakdown of the sales figures adds to total sales. Other work done is not further specified, but it must have included machine tool sales and probably also repair work and miscellaneous products. The shipment records mention production of core ovens, foundry rattlers, and soda kettles, which would serve to utilize the foundry facilities of the company. Also noted are grindstones and grindstone troughs which would be the forerunners of the grinding machines, an important part of machinery shipments after the late 1870's. Between 1873 and 1881, the company produced 199 cylindrical sewing machines or seamers with an aggregate value of $26,891 that were shipped to domestic and British print works and bleacheries. No description of this machine was found, but the name suggests that Brown and Sharpe's experience in sewing machine production enabled it to produce the machine as a stopgap measure during the depression following 1873.

The profit series in Table A-1 is taken from a set of financial statements found in the company's files. The statements apparently were put together as a hybrid balance sheet and income statement for the information of the owners of the business to show the position of the firm on January 1 of each year. They were probably used by the owners to determine the total dividends to be paid each year.

To arrive at the profit figure, a figure of capital value on January 1 was first derived. Capital value was computed as the sum of cash, notes, accounts, value of land, plant, and equipment, and what was called "stock," which may or may not include the inventory of finished and unfinished goods. The profit during the year (column 6) was derived by taking the difference between two successive capital values, adding the dividends taken out during the year, and subtracting the amount received from Darling, Brown, and Sharpe, the precision instrument subsidiary until 1893 when Darling's share was purchased by the parent company. Causes of the fluctuations in what is called profits could not be determined by examination of the records. The substantial figures for the period 1869–73 appear to be largely the result of an upward valuation of the land, buildings, and tools owned by the company, and the losses of 1874 and 1875 of a downward revaluation of buildings and fixtures.

The financial statements are clearly unacceptable in terms of good accounting practice. Business decisions probably were made in part on the basis of them, however, and on that ground the profit series is relevant information. That other factors also influenced decisions is clear from the lack of correspondence between the dividend series and the profit series. Dividends do not move in the same direction as profits in as many as half of the years shown.

TABLE A-1

SUMMARY OF FINANCIAL STATEMENTS, BROWN AND SHARPE, 1863-1905
(dollars, current prices)

Year	Total Sales (1)	Sewing Machine Work (2)	Other Work Done (3)	Machine Tool Shipments (4)	Dividends Paid (5)	Profit or Loss (6)
1861[a]				3,590		
1862				17,771		
1863	89,827	59,728	30,099	15,294		
1864	168,437	108,130	60,307	37,436		
1865	159,059	109,649	49,410	29,248		
1866	254,874	201,607	53,267	33,459		
1867	231,547	194,387	37,159	28,018		
1868	210,720	155,940	54,780	40,930	24,000	
1869	292,571	215,832	76,739	52,410	27,200	165,225
1870	331,366	261,273	70,092	54,350	32,000	128,348
1871	393,781	297,863	95,918	74,841	32,000	158,477
1872	288,138	193,057	95,080	75,094	32,000	70,984
1873	225,331	131,111	94,219	77,493	16,000	85,647
1874	156,309	111,823	44,486	23,463	--	-49,472
1875	188,793	137,910	50,884	20,005	--	-781
1876	179,330	134,534	44,977	33,613	--	10,928
1877	158,676	104,109	54,567	36,450	67,875	14,157
1878	173,450	124,153	49,297	20,346	27,150	35,167
1879	243,524	164,867	78,656	40,890	76,925	70,513
1880	334,866	183,806	151,061	74,504	54,300	78,063
1881	436,036	190,774	245,262	122,872	--	132,124
1882	495,993	187,173	308,811	124,010	36,200	150,526
1883	433,903	162,075	271,827	120,230	76,925	136,470
1884	440,698	194,847	245,850	95,565	76,925	163,992
1885	401,001	197,751	203,250	66,949	31,675	78,999
1886	557,195	160,689	396,506	162,400	72,400	124,176
1887	592,246	149,261	442,985	189,305	22,625	139,915
1888	772,439	237,386	535,052	243,843	27,150	199,421
1889	881,455	183,530	697,925	298,580	45,250	156,254
1890	960,840	117,176	843,665	312,322	40,725	222,265
1891	881,055	105,182	775,874	247,236	49,775	174,870
1892	892,481	140,598	751,883	259,698	63,350	208,525
1893	836,695	147,696	689,000	239,660	36,200	33,179
1894	701,395			185,455	27,150	103,824
1895	1,029,160			378,942	54,300	221,587
1896	1,098,710			467,997	45,250	174,657
1897	1,270,082			612,216	36,200	501,272
1898	1,541,354			799,885	54,300	421,254
1899	2,070,859			1,015,093	99,550	522,434
1900	1,963,382			966,341	72,400	492,953
1901	1,961,215			845,374	72,400	526,888
1902	2,426,404			1,197,145	90,500	502,461
1903	2,540,331			1,166,546	181,000	359,256
1904	2,339,047			1,035,161	181,000	830,022
1905	3,604,377				208,150	

Source: Company records. See text for derivation of col. 6.
Note: Details may not add to total because of rounding.

[a] September to December 1861.

TABLE A-2

MEASURES OF OUTPUT OF SELECTED MACHINE TOOL COMPANIES, 1861-1910

Year	Brown and Sharpe Shipments (number) (1)	Bullard Shipments Number (2)	Bullard Shipments Dollars (3)	G.A. Gray Sales (dollars) (4)	Cincinnati Shaper, Sales (dollars) (5)	Warner and Swasey New Orders (dollars) (6)	Total Output, Metal-Cutting Tools[a] (dollars) (7)
1861	17[b]						
1862	64						
1863	53						
1864	67						
1865	45						
1866	49						
1867	42						
1868	63						
1869	85						
1870	95						
1871	124						
1872	111						
1873	133						
1874	44						
1875	31						
1876	57						
1877	63						
1878	37						
1879	83						
1880	184					2,210[c]	
1881	229	90[d]				28,403	
1882	207	96				36,042	
1883	213	89				38,521	
1884	177	64				44,712	
1885	135	62				40,846	
1886	331	107		41,416		53,294	
1887	374	123		80,552		118,656	
1888	457	135		91,811		59,610	
1889	568	170		113,177		111,705	
1890	574	192		176,708		118,438	
1891	500	155		169,603		163,778	
1892	541	169		161,453		85,764	
1893	522	113		86,290		122,584	
1894	392	56		58,590		73,329	
1895	815	96	116,476	123,382		155,120	
1896	1,015	135	121,150	111,349		280,488	
1897	1,206	155	177,326	134,159		300,544	
1898	1,508	273	499,620	185,906		313,813	
1899	1,962	334	396,618	288,561	42,000	290,893	
1900	1,666	319	422,631	362,510	84,000	334,663	
1901	1,476	314	440,760	242,159	84,000	338,469	17,900,000
1902	2,126	362	544,826	330,659	137,000	328,413	22,800,000
1903	2,152	358	537,359	320,774	152,000	327,044	23,700,000
1904	1,885	167	219,140	169,952	97,000	200,804	18,300,000
1905	1,180	383	509,408	294,800	158,000	496,110	28,700,000
1906		469	733,908	266,098	139,000	756,521	36,400,000
1907		444	719,082	339,776	182,000	546,879	41,300,000
1908		108	185,868	89,044	114,000	309,408	16,800,000
1909		278	510,549	168,601	187,000	990,502	33,500,000
1910		346	694,337	313,352	240,000	739,429	44,300,000

Source: Cols. 1-6 from company records; col. 7 from National Machine Tool Builders Association, Washington, release F-A40a, Feb. 21, 1962.

[a] These figures are said by the NMTBA to include more than 90 per cent of total industry shipments, to exclude repair work, and to include parts shipped with machines.

[b] September to December 1861.

[c] August to December 1880.

[d] April to December 1881.

TABLE A-3

DOMESTIC AND FOREIGN DESTINATION OF OUTPUT, BROWN AND SHARPE AND BULLARD, 1861-1910

Year	Brown and Sharpe Shipments			Bullard Shipments		
	Domestic (number) (1)	Foreign (2)	Foreign as Per Cent of Total (3)	Domestic (number) (4)	Foreign (5)	Foreign as Per Cent of Total (6)
1861[a]	17	0	0.0			
1862	64	0	0.0			
1863	53	0	0.0			
1864	67	0	0.0			
1865	41	4	9.9			
1866	43	6	12.2			
1867	31	11	26.2			
1868	55	8	12.7			
1869	67	18	21.2			
1870	72	23	24.2			
1871	100	24	19.4			
1872	87	24	21.6			
1873	107	26	19.5			
1874	32	12	27.3			
1875	28	3	9.7			
1876	36	21	36.8			
1877	51	12	19.0			
1878	22	15	40.5			
1879	71	12	14.5			
1880	169	15	8.2			
1881	208	21	9.2	88[b]	2[b]	2.2
1882	174	33	15.9	90	6	6.2
1883	182	31	14.6	89	0	0
1884	152	25	14.1	63	1	1.6
1885	115	20	14.8	62	0	0
1886	200	131	39.6	106	1	0.9
1887	284	90	24.1	113	10	8.1
1888	241	216	47.2	135	0	0
1889	370	198	34.9	168	2	1.2
1890	445	129	22.5	188	4	2.1
1891	396	104	20.7	149	6	3.8
1892	450	91	16.8	168	1	0.6
1893	434	88	16.9	113	0	0
1894	282	110	28.1	56	0	0
1895	638	177	21.7	92	4	4.2
1896	560	455	44.8	100	35	25.9
1897	444	762	63.2	64	91	58.7
1898	719	789	52.3	129	144	52.7
1899	1,205	757	38.6	151	183	54.8
1900	928	738	44.3	143	176	55.2
1901	1,123	353	23.9	216	98	31.2
1902	1,669	457	21.5	330	32	8.8
1903	1,593	559	26.0	283	76	21.2
1904	972	913	48.4	115	52	31.1
1905				307	66	17.7
1906				332	137	29.2
1907				293	152	34.2
1908				82	26	24.1
1909				245	33	11.9
1910				273	73	21.1

Source: Cols. 1, 2, 4, and 5--from company records; col. 3--col. 2 as per cent of col. 1, Table A-2; col. 6--col. 5 as per cent of col. 2, Table A-2.

[a] September to December 1861. [b] April to December 1881.

COMMENT ON ROBERTSON AND McDOUGALL

W. Paul Strassman, Michigan State University

It is basic to machine tool technology that almost any object can be polished off or hammered out in remarkably dissimilar ways. So it is with these papers. Duncan McDougall insists that a quantitative record can only be exacted from company accounts and with unflinching consistency limits his observations to the data. Ross Robertson supplements company accounts with information from trade publications and government sources and aims at constructing estimates of the real national output.

Gratitude is due to Robertson for his useful appraisal of the sources of information he encountered and the difficulties of interpretation. I suspect that his estimates of metalworking machinery output for 1900–20 will have a greater durability than those of the machinery made during that period. I have only one question about his 1900–20 calculations. Why is the wholesale price index for all commodities better for adjusting 1900–13 prices than the Cincinnati metalworking machinery price index of George Wing? I would have expected Robertson to splice the American Appraisal Company's index to Wing's. Cincinnati prices went up three times as fast as the adjusted wholesale price index during 1900–14, 32 per cent compared with 10 per cent. But they rose only 21 per cent faster than the American Appraisal Company's index during 1914–20, or 188 per cent compared with 156 per cent.

The patterns McDougall found in the accounts of Brown and Sharpe and Bullard suggest once more that what seems plausible with casual hindsight is not necessarily what happened in history. Of course, once it is established that light machine tools were not quantitatively important until after the main push of industrialization, explanations for the lag come cheaper than September tomatoes, and each of us can provide his own. Since it is difficult to quarrel with what he documents so well, I have decided to pick mostly on things I imagine McDougall might have added.

But first I wonder if the late, accelerated importance of machine tools might not be understated by limiting the analysis to "general-purpose standard machines," as McDougall has. If special-purpose machines were developed after standard machines and then spread partly at the expense of standard machines, their omission means underestimating the trend. The more specialized users there are, the greater the chance of developing a special tool for them. Indeed, I was struck by McDougall's

Table 2 which shows Brown and Sharpe supplying nine industries after 1900 compared with four during the Civil War; and yet the four had accounted for 95.4 per cent of output compared with 65.4 per cent for the nine. A quarter of this 65.4 per cent, moreover, consisted of those diversified clusters, electrical manufacturers and professional and scientific instruments, which includes watches, cameras, and dental equipment, hardly a homogeneous outlet. According to the Census of 1905, the following types of machine tools accounted for only 66.3 per cent of production: lathes of all types; boring and drilling machines; milling machines; planers; stamping, flanging, and forging machines; and punching and shearing machines (*Bulletin 67*, pp. 9–13). Presumably some of these and much of the other third were special-purpose machines. Can one ignore the trend toward specialized machines for making gears, files, chains, and dozens of other products and components?

On the other hand, it may be that the advantages of copying declined during these decades, so that omitting copying means comparatively greater understatement of the earlier years, thus suggesting a higher rate of growth than the actual one. As the years passed, the machine tool producers accumulated enough tricks and specialized tool-building tools to discourage small-scale copying. During the Second World War, in Latin America and India a few metalworking shops likewise made lathes and planers for themselves but went back to importing afterward.

Speaking of Latin America, I was struck by Table 3 which shows that almost 12 per cent of Brown and Sharpe's exports after 1900 went to Latin America, Asia, and Africa. Were there buyers other than arsenals, shipyards, and sewing machine factories? From an industry as strategic as machine tools much can be learned about industrial development in other sectors. By the same token, much can be learned about machine tool production from the records of companies making sewing machines, electric motors, and the like, for example, from their changing inventories of machinery. I believe that this is one source not yet tapped. It is obvious how valuable it would be to know the changing durability of machine tools, and an approximation of the stock available in given years compared with the annual additions.

Finally, I should like to ask McDougall if anything can be done with the price per general-purpose lathe or milling machine? Do the accounts permit an estimate of how prices changed compared with cutting speed and capacity? If prices are available, perhaps an engineer could estimate "best-practice production functions" from surviving catalogues and other specifications.

Sources of Productivity Change

Productivity Growth in Grain Production in the United States, 1840–60 and 1900–10

WILLIAM N. PARKER

AND

JUDITH L. V. KLEIN

YALE UNIVERSITY

In nineteenth century America, productivity growth in grain production derived largely from two characteristic features of the century's history: westward expansion and technological change. We know that productivity increased and we know that both westward expansion and technological change occurred. By averaging the dispersed and fragmentary data we may even guess at the extent of the productivity increase—at least in the use of labor. Can we also assess the relative importance of the factors which produced this effect? This paper contains the results of an effort to do this, for the two elements just mentioned and for the three major grain crops: wheat, oats, and corn. These crops accounted for about 55 per cent of the land harvested for crops in the United States in 1910.[1]

To estimate the effect of one factor in history is an exercise in imagination controlled and guided by the available data. It requires a mental

NOTE: This study is one of a series carried out under the auspices of the Inter-University Project for American Economic History. Financial help was provided by The Ford Foundation through the Inter-University Project and through other avenues. Resources for the Future, Inc., the University of North Carolina, and the Social Science Research Council also gave financial support. Library space and facilities were generously furnished by the National Agricultural Library (U.S. Department of Agriculture) and The Brookings Institution. Our efforts were assisted by a staff at Washington and at regional libraries which included from time to time Martin, Max, and Marlene Primack, Chaplain W. Morrison, Fred Bateman, Jr., Don C. Schilling, Mary Lee Spence, Miles F. Frost, Diane Light, and Mary G. Bullock. Studies on cotton, hay, corn fodder, and dairy and meat products are now in preparation.

[1] *Historical Statistics of the United States, Colonial Times to 1957*, U.S. Bureau of the Census, Washington, 1960, Series K-98, K-265, K-269, K-274.

experiment in which one tries to conceive how the history would have developed if one element in a situation had varied while the others remained unchanged. In this study the effort is made to conceive how productivity would have moved from the levels of 1840, if westward movement had taken place without technological change, and if technological change had occurred without westward movement. The principal difficulty encountered arises from the fact—common enough in the nineteenth century economy—that while production was shifting westward and techniques were changing, total output was undergoing an enormous expansion. To do our work completely, then, we should like to know how productivity would have moved in the East under the same demand pressures for output expansion which in fact found their outlet in the western development. To estimate this, however, we would need data on the shape of the supply curve of the eastern regions at those levels of output which were in fact never attained. A close enough investigation into the production conditions in the East might permit such curves to be estimated, but such an effort lies far beyond the data and research resources at our disposal. As will appear in the course of the report, the interpretation of the indexes developed here is restricted by this deficiency.[2]

Even within these limits, a question arises about the independence of the two variables whose influence is under examination. Would movement into the Middle West and the shift in grain production which accompanied it have occurred without the improvements in farm machinery? And would those improvements, so well adapted to western conditions, have been available in the East, if the stimulus to invention offered by labor "scarcity" and abundant land in the West had not appeared? Such questions go behind the immediate influences on productivity to the causes of the changes from which productivity growth was produced. The evidence developed in this paper is designed to throw light on first-order causes of the productivity growth; the causes of those causes are left to further investigation.

Against these limitations may be set one opportunity that our data and technique appear to offer. That great, half-inbred family of ideas known as technological change has affected production conditions in agriculture through two main lines of descent. The succession of mechanical inventions—from simple tools to complex power-driven machinery in the field and barn—has affected most directly the changes in labor inputs in the

[2] A second deficiency in this study is the neglect of capital inputs and, with it, the lack of a production function and of a measure of total factor productivity. The measurement of agricultural capital is so uncertain that an attempt to include it would, in our opinion, add more to the study's range of error than to its completeness or precision.

operations on the soil or in the handling and processing of the crop. At a given level of mechanical techniques, however, the operations on the soil—plowing, sowing, and harrowing—use labor in a relatively fixed proportion to the area under cultivation. The productivity of this labor, measured in man-hours per bushel of harvested grain, depends then upon the yields of land, and these in turn are influenced less by mechanical techniques than by another branch of technology, agricultural chemistry and plant biology. Invention in biology and chemistry was, by modern standards, far less advanced in the nineteenth century than invention of mechanical equipment; and the levels of practice both in the development of seeds and in the management of the land depended less upon organized experiment than upon the process of natural selection, both in nature among plants and in the economy among innovating farmers. The movements of land yields reflect both the margin of cultivation and the results of the social processes of invention in this area, while the movements of labor inputs per acre are most strongly affected by the use of mechanical equipment.[3] These two results of the two compartments of technology—land yields and labor per acre—are available separately in the data here, and our analysis can consider the effect of holding one constant while varying the other by the same statistical operation as that by which the independent effect of the interregional shifts is identified.

Quantitative methods of a limited and partial character, such as those used in this study, veer rather rapidly into nonsensical results. Yet, used with sufficient caution they may yield meaningful statements about the causes or mechanisms of historical change and indicate where important lines of further investigation may lie. In the present study, thanks to much generous assistance, they have the advantage of being based upon a rather wide survey of the remaining quantitative materials from nineteenth century sources. They represent an effort to make precise, through numerical estimates, those general impressions about productivity and its sources which a reading of the history inevitably gives.

Definitions and Assumptions

The variables whose separate influences on labor productivity are assessed in this study are:

1. The weights of regions in total acreage and output, weights whose shifts form part of what the historian knows as westward movement.

[3] The correspondence of these two variables to the two types of technological change is not complete, of course. Use of mechanical equipment affects yields per acre, and a chemical invention such as fertilizer requires a large per acre labor input. These cross-effects do not appear to be strongly present in our period, and we have omitted labor in manuring from the calculations.

2. The mode of carrying out operations on the soil and on the standing and harvested crop, whose improvement reflects mechanical invention and changes in the amount and form of capital.

3. The choice of soils and plant strains, the timing of operations, and other bits of biological and chemical knowledge, mostly not incorporated into mechanical equipment and affecting the yield of the land.

The technique used here—essentially the "partitioning" technique familiar from the studies of Kuznets, Abramovitz, and Denison—has been selected with an eye both to the objective of exposing the influence of these variables and to the severe limitations of the available numerical data. To derive conclusions from the data by this technique requires certain definitions and assumptions.

First, we define the two periods of study: period 1, 1839, and period 2, 1907–11. These are terminal periods, so the study is confined to a comparison of two benchmarks. The periods are set by the availability of Census production data from the first agricultural Census, 1840, and the possibility of a five-year average of the Department of Agriculture's revised estimates of acreage and output around 1909. Data on labor inputs must be gathered over a somewhat wider period. For period 1 inputs, sources from the 1830's to the 1860's were used, and for period 2, reports from the late 1890's to mid-1910's. Production conditions are not uniform within either of these wider terminal periods. In the first period, rural settlement between the Ohio and the Mississippi rivers, begun forty years earlier, is intensified and the Mississippi is crossed. After 1845 the mechanical reaper begins to be sold, and used in the 1850's in significant numbers, and the horse- or mule-drawn cultivator in corn assumes increasing importance. Still, in comparison with the shifts to the Great Plains between 1860 and 1900 the movement—both geographical and technological—is relatively small. Data were selected from the earlier period to represent production conditions around 1840, and the estimate of yields and acreage distribution is taken as of 1839. In the 1907–11 period, both the westward movement and the development of the horse-drawn mechanical technology are almost complete. Between these terminal dates it has not seemed possible to assemble sufficient evidence on the diffusion of the mechanical techniques to permit labor input estimates to be drawn up.

The use of 1839 as the initial date is of course partly determined by the availability of the Census production data. In addition, it has the advantage that the West, by that time, was a distinct region and had achieved substantial development. In wheat and oats, on the basis of homogeneity in land yields, the United States appears to fall into two great regions: North and South. But in the North, labor inputs per acre in period 1 may be divided between the Northwest, with inputs resembling those of the South, and the Northeast, with higher levels of labor cost. Combining land yields and labor inputs per acre, then, gives three regions as follows:

	Wheat and Oats Output Per Acre	Labor Input Per Acre
Northeast	high	high
West	high	low
South	low	low

In corn, the border states separate from both North and South on the basis of land yields, and, with the division between Northeast and West on the basis of labor inputs, four regions may be delineated. The statistical basis for defining these regions is described more fully in Appendix A. To permit comparison, the same regions must be retained in period 2. However, for grains the effective boundary of the West in period 2 is extended from the Mississippi to the Pacific Ocean. The impact of the large intraregional shift within this extended West is analyzed separately (in the next to last section of this study).

In addition to period and region, it has been necessary to group the labor-using agricultural operations into the three customary groups: those on the soil before harvest, those on the standing crop (harvesting, including shocking of wheat), and those on the cut crop (threshing and shelling). The shelling of corn, as well as the transport of all the crops to the barn and to market, has been omitted from the calculation. These three groups of operations correspond to three engineering problems inventors faced in the effort to mechanize. The preharvest operations, involving contact of a tool with the soil (except for broadcast seeders), required knowledge of the soil's physical characteristics and the relation of plant growth to them. It required also the invention of implements—plows, cultivators, and harrows—to produce the required tilth, and their adaptation to soils of varying composition. This problem was especially acute as the prairie soils came under the plow. The harvest operation in wheat, handling not the soil but the unthreshed crop, required the appropriate adaptations of mowing and raking tools. In the humid areas it involved also the problem, never satisfactorily solved by mechanical means in this period, of protecting, ripening, and drying the crop between reaping and threshing. The postharvest (threshing) operation in wheat involved, when mechanized, the substitution of the beating motion of the flail or the hoof by a stripping motion not unlike that of the cotton gin. In corn, the problem of mechanical harvesting was never satisfactorily solved during this period. In the first two groups of operations movement over the field was required, and so they were suitable to improvement by animal power or self-propelled engine. In the second and third groups, mechanization required power transmission and the activation of movable parts.

These engineering requirements gave the production function in each group of operations its economic characteristics. In the first group for all the crops, under any technique, labor time varied with the character of the soil and the area planted. The yield of the land may depend partly on the care, timing, and skill in the operations—particularly in corn where cultivation is required. Rich land and hot weather grow weeds as well as crops and the tasks of hoeing and cultivating are heavier. But the connection between cultivation and yield in corn was not known in any scientific sense, and opinions on this relation were almost as numerous as the farmers holding them. The functional relation between yield and tillage as actually practiced cannot be defined, while the variation with acreage at any level of cultivation is quite apparent. In harvesting, time spent by laborers in travel over the fields depends partly on the area harvested and partly on the thickness of the stand. But in addition to travel time, the time in place during the gathering of the crop varies with the kind of crop and the yield. In wheat and oats it is secondary to the travel time; a thick crop can be reaped almost as fast as a thin one. In corn, a heavier crop involves larger ears as well as closer planting. Hence, in the grains, harvest labor is assumed to be proportional to acreage. The operations then are grouped as follows:

Preharvest (assumed fixed per acre)

Wheat and Oats	Corn
Plowing	Field preparation[a]
Sowing	Planting
Harrowing	Cultivating
	Hoeing

Harvest (assumed fixed per acre)

Wheat and Oats	Corn
Reaping	Picking or husking
Raking	
Binding	
Shocking	

Postharvest (assumed fixed per bushel)

Wheat and Oats	Corn
Threshing	Shelling[b]
Winnowing	

[a] Depending on regional variations in practices.
[b] Not included in this study.

The Statistical Technique

For each region (as defined above), and at each of the two periods, an estimate of average labor input is made for each of the three groups of operations. For groups 1 and 2, these estimates are made in man-hours per acre, and divided by an estimate of average regional yield to give man-hours per bushel. This estimate is then added to the estimate of man-hours per bushel in group 3 to give the total labor input per bushel. These estimates for each region are weighted by regional output to give national averages at each period. The productivity change is then the change in this national estimate (or rather in its reciprocal) between the two dates.

Symbols and the formulas are as follows:

L_1 = Preharvest labor (man-hours)
L_2 = Harvest labor (man-hours)
L_3 = Postharvest labor (man-hours)
R_1 = Northeast
R_2 = South (wheat and oats)
R_{2a} = Middle east (corn)
R_{2b} = South (corn)
R_3 = West
A = Area (acres planted)
O = Output (bushels of threshed grain)
$a = L_1/A$ $y = O/A$
$b = L_2/A$ $v = O/\Sigma O$
$c = L_3/O$ $w = A/\Sigma A$
1 = period 1 (1839)
2 = period 2 (1907–11)

$$\sum_{R_1}^{R_3}\left(\frac{a+b}{y}+c\right)v = \text{national average labor input per bushel}$$

To analyze the effect of change in regional weights, labor inputs, and land yields, it is possible to calculate an index with changes in the subscripts of the variables v, abc, and y. Indexes which use the formula with period 1 labor inputs as a base, and with numerators in which the subscripts of the variables are changed one at a time from 1 to 2, show the effect of the change of each variable in isolation. Change in the subscripts in groups of two at a time shows the effect of combined changes in the variables. For the three variables there is thus generated a family of eight indexes on the period 1 base as follows: $v_2abc_2y_2$; $v_2abc_2y_1$; $v_2abc_1y_2$; $v_2abc_1y_1$; $v_1abc_2y_2$; $v_1abc_2y_1$; $v_1abc_1y_2$; $v_1abc_1y_1$. If we then consider each variable and each grouping of variables as a factor with measurable independent effect, a plausible solution to the index-number problem may be found by representing the relative effects of the factors by the relative algebraic sums of the rows of indexes with signs indicated in the following matrix:

Factors	$v_2abc_2y_2$	$v_1abc_1y_1$	$v_2abc_1y_1$	$v_1abc_2y_1$	$v_1abc_1y_2$	$v_2abc_2y_1$	$v_2abc_1y_2$	$v_1abc_2y_2$
v	+	−	+	−	−	+	+	−
abc	+	−	−	+	−	+	−	+
y	+	−	−	−	+	−	+	+
$v(abc)$	+	+	−	−	+	+	−	−
vy	+	+	−	+	−	−	+	−
$(abc)y$	+	+	+	−	−	−	−	+
$v(abc)y$	+	−	+	+	+	−	−	−

The source of this method and the formulas to prove the signs of the indexes are shown in Appendix C.

Our principal conclusions are obtained by applying this technique to the three variables just cited. It may also be applied within the variables v (regional weights) and abc (labor inputs) to indicate the relative importance of factors working in and through them. The change in regional weights has an effect on labor inputs, both directly in the differing labor inputs per acre of the period 1 regions, and indirectly through the different land yields of the period 1 regions; the relative importance of the effect through each of these channels can be calculated. Within labor inputs, too, indexes may be computed for wheat and oats, which vary each of the three groups of labor separately and in combination, while holding yields and regional weights at their period 1 levels. Both these subsidiary families of indexes were computed and the importance of the factors assessed, as shown above. Finally, indexes were computed using the period 2 subregional values of that portion of the western region included in the period 1 West. This was done to show the effect on the indexes of the intraregional shift from the old Northwest to the Plains and Pacific Coast states. A separate index showing the effect of the intraregional shift within the South was computed for corn.

The Data

Some understanding of the data for this study is necessary for a sympathetic—or an unsympathetic—appraisal of the conclusions. These data are presented in summary fashion in Appendixes B and D, and what follows here is only a general discussion of their sources and reliability.

The most dubious use of the data lies in the combination of the estimates of land yields by state, based on the USDA revised estimates, with an average of labor inputs per acre derived from a sample of contemporary evidence which, particularly for the early period, is drawn largely from the most productive farms. The labor data for period 1 are drawn largely from premium reports to state agricultural societies in the Northeast and, in the South, from the records of plantations. There is reason, however, to argue that labor inputs on such farms are less biased than a similar sample of yields would be. For one thing, if bias exists, it is not certain which way it would run. Are the premium farms and record-keeping plantations those which use more labor per acre to get higher yields, or less labor because of greater efficiency? Furthermore, in wheat and oats, the case is particularly strong since high yields are obtained by superior seed, soil, and skill rather than by the labor-intensive operations of weeding and cultivating. In plowing and planting, there is no clear reason that labor input should vary with yield, while in the harvest and postharvest operations it has been generally possible to take account of the techniques employed. In the few cases where a statement about average conditions or from a professedly average farm is obtained, the figure falls well within the range of the distribution.

The data for labor inputs by region in period 2 are much more plentiful and specific than those for the mid-nineteenth century. They were drawn from cost studies of the USDA and the Agricultural Experiment Stations and from a retrospective study in the 1930's by the WPA, National Research Project (NRP).

With respect to the two kinds of data, the usual—but perhaps not more than the usual—uncertainties appear. For land yields, the USDA estimates for 1866–75 were projected back by state to 1839, without allowance for trend. This procedure was thought to be justified in view of the small trend in the state series throughout the whole half-century after 1866. However, for grains in the Ohio Valley states, these levels differ from the higher contemporary estimates for the 1840's and 1850's, and an alternate estimate of all the indexes, based on a higher set of yields for these states in this period, is presented in Appendix B. In the data on labor inputs by operation and region, the variety of sources and the numerous qualifying considerations to individual figures make summary statement difficult. The scatter is greatest in the period 1 estimates for plowing, as might be expected, and the cases are least numerous for the West in period 1. In each case, the regional data have been examined to see whether a significant difference in the means is present and, where this is not the case, data for regions have been combined. Similarly, in a few cases it has been possible to combine data for the same operation in wheat and oats. In general, data have been collected to the limit of the time and resources of money and patience available to the project and—unlike the agricultural operations which are the subject of this paper—it was apparent at several points that diminishing returns had set in.

The averages and dispersion of the data, and count of cases, are shown in Appendix D.

The Principal Findings

LABOR INPUTS, LAND YIELDS, AND REGIONAL WEIGHTS

Table 1 shows averages of land yields and labor inputs for each of the regions at the two periods of this study, with national averages determined by the output weights of the regions. From these data the principal conclusions about the extent and sources of the productivity increase are

TABLE 1

LABOR INPUTS, LAND YIELDS, AND WEIGHTS, PERIODS 1 AND 2

Region	a		b		y		c		v		w	
	1	2	1	2	1	2	1	2	1	2	1	2
WHEAT												
R_1	19.1	11.6	15.0	3.0	14.5	17.5	.73	.19	.334	.046	.259	.037
R_2	11.3	10.7	12.5	3.0	8.4	12.3	.73	.29	.342	.075	.459	.085
R_3*	12.4	4.7	15.0	2.3	13.0	14.0	.73	.19	.324	.879	.282	.878
U.S.	13.6	5.5	13.9	2.4	11.3	14.0	.73	.20				
OATS												
R_1	14.3	9.3	12.8	3.4	28.5	29.7	.40	.23	.422	.087	.316	.077
R_2	8.8	9.5	11.0	4.5	13.9	17.0	.40	.24	.332	.044	.506	.068
R_3*	8.8	3.9	12.8	2.6	29.3	26.5	.40	.10	.246	.869	.178	.855
U.S.	10.5	4.7	11.9	2.8	21.3	26.1	.40	.12				
CORN												
R_1	98.3	46.4	13.0	13.0	33.5	36.8			.097	.029	.057	.020
R_{2a}*	52.0	26.7	10.1	10.1	21.8	24.4			.344	.099	.310	.106
$_{2b}$*	67.3	21.3	4.3	4.3	11.8	16.1			.279	.175	.465	.285
R_{3a}*	46.2	14.2	13.0	7.6	32.7	31.0			.280	.697	.168	.589
U.S.	60.8	18.2	8.1	7.0	19.6	26.2						

Note: See text for explanation of the symbols.
Source: v, w, y: Table 10 below, and sources listed there.
 a, b, c: Appendix D.

*Appendix B contains alternate yields (y) for period 1 for all crops in the western region and for corn in the South and Middle East.

derived by the indexes of Table 2. The first, and most apparent, conclusion is the index of labor productivity for each crop: wheat, 417; oats, 363; corn, 365. In Table 2, this total index is shown as one of the family of indexes generated by combining the three main variables, using all possible combinations of period 1 and period 2 values. From this it is apparent that mechanization, in combination with the regional shift (i_6), is responsible for nearly the whole effect in wheat and corn, and more than the whole effect in oats, and that even leaving the distribution of production in its period 1 proportions, mechanization within each region taken alone produces an effect which is large (i_3). Applying the technique described above, values indicating relative importance of the factors are derived as follows:

	Wheat	Oats	Corn
v	.170	.287	.207
abc	.598	.506	.562
y	.082	.005	.084
$v(abc)$	.158	.234	.120
vy	−.049	−.026	−.007
$(abc)y$	.046	.004	.032
$v(abc)y$	−.005	−.010	.002

The economic meaning to be given to these results is discussed in the concluding section of this paper.

TABLE 2

LABOR REQUIREMENTS (U.S. AVERAGE AND INDEXES) AS AFFECTED BY INTER-REGIONAL SHIFTS, REGIONAL YIELDS, AND REGIONAL LABOR INPUTS PER ACRE

Index	Period for Values of			Labor Requirement (L/O)*			Productivity (O/L) Index (i_1/i_n x 100)		
	v	y	abc	Wheat (1)	Oats (2)	Corn (3)	Wheat (4)	Oats (5)	Corn (6)
i_1	1	1	1	3.17	1.45	3.50	100	100	100
i_2	1	2	1	2.68	1.37	2.94	118.3	105.8	119.0
i_3	1	1	2	1.29	0.78	1.54	245.7	185.9	227.3
i_4	2	1	1	2.90	1.18	2.70	109.3	122.9	129.6
i_5	1	2	2	1.05	0.72	1.32	302.1	201.2	265.2
i_6	2	1	2	0.84	0.39	1.06	377.3	371.7	330.2
i_7	2	2	1	2.69	1.23	2.45	117.8	117.9	142.9
i_8	2	2	2	0.76	0.40	0.96	416.7	362.6	364.6

*$\Sigma \left(\frac{a+b}{y} + c\right) v.$

TYPES OF MECHANIZATION

The effect of mechanization alone (Table 2, i_3) can be broken down into its effects on each of the three groups of operations in each region. The averages derived from our data for this calculation are given in Table 1. Here the relatively greater importance of the mechanization in harvesting and threshing, as compared with that of the improvements in power and implements in plowing and planting, is clear. By a similar calculation to that for Table 2, we produce for wheat and oats the indexes of Table 3 in which, at the land yields and distributions of period 1, each element is varied independently and in conjunction with each of the others. In corn, the lack of significant mechanization in harvest and postharvest operations makes such a calculation of no importance. The values indicating the relative importance of the factors are as follows:

	Wheat	Oats
a	.165	.126
b	.384	.456
c	.214	.250
ab	.074	.045
ac	.043	.026
bc	.093	.086
abc	.026	.011

TABLE 3

LABOR REQUIREMENTS (U.S. AVERAGE AND INDEXES) AS AFFECTED BY CHANGES IN REGIONAL LABOR INPUTS

Index	Period for Values of			Labor Requirement (L/O)*		Productivity (O/L) Index $(i_1/i_n \times 100)$	
	a	b	c	Wheat (1)	Oats (2)	Wheat (3)	Oats (4)
i_1	1	1	1	3.17	1.45	100	100
i_2	1	2	1	2.19	1.08	144.7	134.2
i_3	1	1	2	2.66	1.26	119.2	115.1
i_4	2	1	1	2.79	1.36	113.6	106.7
i_5	1	2	2	1.68	0.88	188.7	164.7
i_6	2	1	2	2.28	1.16	139.1	125.0
i_7	2	2	1	1.80	0.98	176.1	147.9
i_8	2	2	2	1.29	0.78	245.7	185.9

*$\Sigma \left(\dfrac{a+b}{y_1} + c\right) v_1.$

THE TWO EFFECTS OF REGIONAL SHIFTS

The regions of our study are defined with respect both to labor per acre and to the yields of land. This definition of regions is discussed in Appendix A. It is apparent that interregional shifts will affect average labor per bushel through an effect on the national average labor input per acre and on the average yield in bushels per acre. It is possible then to break down the effect shown in Table 2 into an effect by way of each of these components. Moreover, the effect through labor per acre is further divisible into an effect by way of each type of labor input which shows interregional differences in period 1. In Table 4, these subsidiary indexes are given, holding the values of labor inputs and yields at the period 1 level, but weighting them by acreage rather than output weights to permit variation of the weights of yield and labor input variables independently.

TABLE 4

LABOR REQUIREMENTS (U.S. AVERAGE AND INDEXES) AS AFFECTED BY INTERREGIONAL SHIFTS

Index	Period for Values of			Labor Requirement $(L/O)^*$			Productivity (O/L) Index $(i_1/i_n \times 100)$		
	a_1w	b_1w	y_1w	Wheat (1)	Oats (2)	Corn (3)	Wheat (4)	Oats (5)	Corn (6)
i_1	1	1	1	3.17	1.45	3.50	100	100	100
i_2	2	1	1	3.07	1.39	3.15	103.3	104.3	111.1
i_3	1	2	1	3.25	1.49	3.61	97.6	97.3	97.0
i_4	1	1	2	2.90	1.20	2.69	109.3	120.8	130.1
i_5	2	2	1	3.15	1.43	3.26	100.6	101.4	107.4
i_6	2	1	2	2.82	1.15	2.42	112.4	126.1	144.6
i_7	1	2	2	2.97	1.22	2.77	106.7	118.9	126.4
i_8	2	2	2	2.89	1.18	2.50	109.6	122.9	140.0

* $(\frac{\Sigma a_1 w + \Sigma b_1 w}{\Sigma y_1 w}) + c_1.$ For corn, $c = 0$. For i_1, this reduces to the formula of Table 2.

Intraregional Shifts

Nearly all of the acreage increase between 1840 and 1910 occurred in the region we have labeled West. In wheat, this expansion occurred in the 1860's, 1870's, and 1890's; in oats, in the 1880's; in corn, in the 1850's and 1870's. Acreage of wheat and oats remained fairly steady in the other two major regions of period 1, and their acreage and output weights (Table 1) had all fallen to less than 10 per cent in period 2. The behavior of the yields and labor-per-acre coefficients for the western region is thus of decisive importance in controlling the average productivity change in the nation as a whole. As might be expected from the changes in the regional weights, the western region exhibits a sharper fall in labor per acre than the other regions do (Table 1), and the enormous expansion does not seriously affect the level of land yields.

In the foregoing sections, the three regions have been treated as if each were a homogeneous unit whose yields and labor input coefficients changed only as a result of technical change. The stability of land yields in the presence of absolute changes of acreage was taken as evidence for this assumption. Within each region, however, it is possible that the stability of yields and the fall in labor costs per acre were due in part to intraregional shifts to acres of similar yield, but of a physical aspect more favorable to farming. A study of labor costs in land clearing shows, for example, that the shift from forested area to prairie was a major factor in saving labor in those operations.[4] In the Northeast and South, the period 2 weights for wheat and oats are too small to make further intraregional analysis useful. But for the West, and, in the case of corn in the South, a closer look at the yields and labor coefficients of the major geographic subregions must be obtained.

For this purpose, the West may be divided in period 2 into the major subregions used in the basic NRP reports of the WPA. These subregions with their output and acreage weights are shown in Table 5. The period 2 subregional weights show where the great acreage expansions in the West occurred. For wheat, the growth of the "small grain" tier of states (Montana, the Dakotas, Nebraska, and Kansas) was the most important development, but the secondary growth of acreage on the West Coast (Northwest and California), the Great Lakes states (WD) and in the corn belt (C) accounted together for an almost equally large new acreage. Of a 38-million-acre expansion in the West, about 20 million is accounted for by the new acreage in the small grain subregion. In oats, the expansion was more heavily concentrated in the corn belt and the western dairy states. For corn, as might be expected, almost 60 per cent of the acreage increase in the West was concentrated in the corn belt, and only 33 per cent in the small grain subregion. The differentials among these subregions in period 2 labor inputs are thus of considerable interest. If these differentials

[4] M. Primack, "Land Clearing under Nineteenth Century Techniques: Some Preliminary Calculations," *Journal of Economic History*, XXI, 1962, pp. 484–497.

TABLE 5

OUTPUT AND ACREAGE WEIGHTS OF WESTERN SUBREGIONS, PERIOD 2

Subregions	Output ($O/\Sigma O$)			Acreage ($A/\Sigma A$)		
	Wheat	Oats	Corn	Wheat	Oats	Corn
W	.879	.869	.697	.878	.855	.589
W:C	.235	.413	.490	.207	.379	.359
W:WD	.108	.190	.051	.098	.185	.042
W:SG + W:WC	.408	.212	.151[a]	.477	.249	.181[a]
W:R	.020	.015		.015	.012	
W:NW + Cal	.108	.079	.005	.081	.030	.007

Source: USDA, revised estimates, 1907-11.
Key: W = West; W:C = corn belt; W:WD = western dairy; W:SG = small grain; W:WC = western cotton; W:R = range; W:NW = Northwest; W:Cal = California. See Appendix D for states included in each subregion.
[a] W:SG only.

are large, much of the productivity improvement attributed above to mechanization might instead be attributed simply to the intraregional shift within the West.

In Table 6, the period 2 yields and labor inputs for the subregions of the West are compared. The region labeled W:C includes the states within whose limits most of the West's acreage of period 1 appears. The other subregions were almost wholly untouched in the 1840's. With

TABLE 6

YIELDS AND LABOR INPUTS IN THE WESTERN SUBREGIONS, PERIOD 2

	Yields (y)			LABOR INPUTS								
				Preharvest (a)			Harvest (b)			Postharvest (c)		
Subregions	Wheat	Oats	Corn	Wheat	Oats	Corn	Wheat	Oats	Corn	Wheat	Oats	Corn
W	14.0	26.5	31.0	4.7	3.9	14.2	2.3	2.6	7.6	.193	.098	0
W:C	15.8	28.4	35.8	5.5	3.3	15.2	3.0	2.6	7.4	.224	.092	0
W:WD	15.3	26.8	31.7	6.1	6.1	17.3	3.0	3.5	13.0	.165	.108	0
W:SG + W:WC	12.0	22.2	21.9[a]	4.2	3.0	11.4[a]	1.8	1.7	6.8[a]	.190	.094	0
W:R	18.6	32.7		6.0	7.6		7.5	8.3		.226	.129	
W:NW + Cal	19.4	33.8		3.2	3.2		2.0	2.0		.165	.095	

Source: Yields, from USDA, revised estimates. For labor inputs, see Appendix Tables D-10, D-12, D-13, D-15, D-17, D-18. For explanation of subregion symbols, see Table 5.
[a] W:SG only.

respect to land yields, it is apparent that the new subregions W:WD and W:SG were lower, and the northwestern coastal regions higher, than the old Northwest. That the lower level of W:WD was not a persistent trend is shown in Table 11 below; in previous decades and—in wheat—under somewhat larger acreages, the yields in the western dairy states (W:WD) were as high as, or higher than, those in the older states to the south of them (W:C). In the small grain states (W:SG) the yields in 1909 were much lower relative to the other subregions than in earlier years, and this is partly accounted for by the rapid rise in the subregion's acreage in the decade 1899–1909. The movement to W:SG did tend to depress yields in the West; for wheat this is partly balanced in the regional average by some rises in intrasubregional yields in the 1900's, and for all crops by the growth in the other, somewhat higher yielding regions. However, in the small grain subregion of the West, somewhat lower yields were balanced by easier conditions of cultivation and harvesting. This differential, shown in Table 6, is clearly significant for harvesting wheat and oats, where the dry weather at harvest time permitted the labor of binding and shocking to be eliminated on a large portion of the acreage. But it is also large for corn, where the simpler harvesting method of husking from the standing stalk was universally used (Appendix Table D-10).

To indicate the net effect of this intraregional shift within the West on the indexes calculated in Table 2, one may assume that the westward movement took place under the period 2 yields and labor inputs, shown for the period 1 western region (W:C in period 2). The effect of replacing the period 2 W values by W:C values may be seen by comparing the first two lines of Table 6 and by the calculations of Table 7. In all our indexes where y_2 values appear, the replacement would raise the productivity index by less than 10 per cent, and the portion of the rise attributed to intraregional changes in land yields would be raised. The difference between this index and the one calculated in Table 2 is a measure of the depressing effect of intraregional shifts on productivity in the West by way of their effect on land yields. In wheat, the replacement of W by W:C values for labor inputs would reduce the productivity index by about $12\frac{1}{2}$ per cent, and reduce the portion to be attributed to mechanical improvements within the regions. In oats the reverse would be true and in corn the change would be negligible. The net effect on labor per bushel in the West is shown in Table 7; the indexes of Table 2, with the W values altered to W:C values in period 2, are shown in Table 8. From the comparison it appears that in wheat the total productivity index (i_8) is depressed by about 8 per cent because of the higher labor inputs in W:C (i_3) and especially through the interaction of this with the regional

TABLE 7

PERIOD 2 VALUES OF CORN BELT AND WEST COMPARED

Variables	Wheat		Oats		Corn	
	W:C	W	W:C	W	W:C	W
a	5.5	4.7	3.3	3.9	15.2	14.2
b	3.0	2.3	2.6	2.6	7.4	7.6
$a + b$	8.5	7.0	5.9	6.5	22.6	21.8
y	15.8	14.0	28.4	26.5	35.8	31.0
$\frac{a+b}{y}$	0.538	0.500	0.208	0.245	0.631	0.703
c	0.224	0.193	0.092	0.098	0	0
$\frac{a+b}{y} + c$	0.762	0.693	0.300	0.343	0.631	0.703

Source: Table 6.

shift (i_6). In oats the index is raised about 10 per cent largely through the interaction between the small reduction in labor inputs per acre and the massive interregional shift (i_6). In corn, the rise of about 5 per cent in i_8 caused by the substitution of the W:C values for the W values comes from the effect on land yields in conjunction with the regional shift (i_7). It is evident then that the indexes would not have been strongly affected if the intraregional shifts within the West—which permitted the growth

TABLE 8

EFFECT ON TABLE 2 INDEXES OF SUBSTITUTION OF CORN BELT FOR WEST

Index	Period for Values of			Wheat		Oats		Corn	
	v	y	abc	Table 2	W=W:C	Table 2	W=W:C	Table 2	W=W:C
i_1	1	1	1	100	100	100	100	100	100
i_2	1	2	1	118	122	106	107	119	122
i_3	1	1	2	246	236	186	189	227	226
i_4	2	1	1	109	109	123	124	130	130
i_5	1	2	2	302	297	201	207	265	270
i_6	2	1	2	377	326	372	400	330	324
i_7	2	2	1	118	127	118	123	143	154
i_8	2	2	2	417	385	363	403	365	385

TABLE 9

OUTPUT WEIGHTS, ACREAGE WEIGHTS, YIELDS AND LABOR INPUTS FOR CORN IN THE SOUTHERN SUBREGIONS, PERIOD 2

Variables	South (S)	Eastern Cotton (S:EC)	Delta Cotton (S:DC)	Western Cotton (S:WC)
$A/\Sigma A$	0.285	0.112	0.062	0.111
$O/\Sigma O$	0.175	0.058	0.041	0.076
Yields (y)	16.1	13.7	17.1	18.1
Labor				
Preharvest (a)	21.3	24.75	28.0	10.2
Harvest (b)	4.3	4.3	4.3	4.3
Postharvest (c)	0	0	0	0
$\dfrac{a+b}{y} + c$	1.590	2.120	1.888	0.801

Source: Yields from USDA, revised estimates. For labor inputs, see Appendix Tables D-10, D-12, D-14.

of output to occur—had taken place under conditions similar to those on the acreage under wheat in period 2 in the corn belt.

In the case of corn, significant intraregional shifts occurred among the southern as well as the western subregions. The relevant data for the southern subregions are shown in Table 9. Recalculation of the indexes, as shown in Table 2, substituting the S:EC values for the South affects the indexes as follows:

Table 2

	O/L Index	S = S:EC
i_1	100	100
i_2	119	111
i_3	227	215
i_4	130	130
i_5	265	239
i_6	330	315
i_7	143	135
i_8	365	331

It will be noted that this substitution reduces index i_8 by about 10 per cent, whereas the calculation in Table 8 substituting the W:C data for the West raises the final index by about 5 per cent. The combined effect of both substitutions would thus be less significant than that of either one taken alone.

Conclusions and Speculations

Over the seventy years between 1839 and 1907–11, output of wheat, oats and corn in the United States each increased about $7\frac{1}{2}$ times, and acreage between 5 and 6 times. The growth of the West relative to the low-yield border states of the upper South raised average land yields about 25 per cent. Within the West, the relative growth of the plains states between 1890 and 1910 exercised a downward pressure on regional yields, balanced by yield rises in the other major western subregions (Table 11). Otherwise, taking the regions individually, the yields of land given in Table 10 showed no marked or continuous trend. Acreages planted in the Northeast and—for wheat and oats—in the South also changed very little. Nearly the whole rise in acreage and output in wheat and oats was accounted for by the expansion North and West of the Ohio River, where the share of national output rose from 32 per cent in wheat and 25 per cent in oats in 1839 to about 87 per cent in both crops in 1907–11. In corn, about two-thirds of the acreage rise occurred in the West and about one-quarter in the South. Rising yields in the South were a result of the rapid growth of the western cotton subregion (Table 12).

Even before the Civil War, the West had some advantage over the South in land yields and over the Northeast in labor costs per acre. Table 13 shows the extent of this advantage with the techniques prevailing in that period. At these differentials, the rise in the West's share to its period 2 level, without changes in yields or labor inputs per acre within the regions, would have raised productivity only 9 per cent in wheat, 23 per cent in oats, and 30 per cent in corn (Table 2, index i_4). Table 4 indexes i_4 and i_5 show that most of this improvement would have been attributable to the shift from the relatively lower yielding lands of the South. Had the land yield changes within each region occurred simultaneously with the regional shifts, the improvement would have been a bit more in wheat and corn (Table 2, index i_7) but less in oats.[5] On the other hand, if the land yield changes had not occurred, but if the regional shifts had been accompanied by improvements in labor input per acre (Table 2, index i_6), resulting largely from mechanization, virtually the whole productivity change in both crops would have been achieved without change in average yields of land. Alternatively, if the mechanical equipment of 1900–10 could have been used in the output of 1840, then labor in wheat and corn

[5] The lower level of the oats index, with yield changes (Table 2, indexes i_7 and i_8 compared with i_4 and i_6), is a result of the unusually low western yield in 1907–11. At the levels around 1900, this result would not be obtained.

TABLE 10

ACREAGES AND YIELDS, BY REGION, 1839-1909
(acres in millions and yields in bushels per acre)

	Northeast		Middle East		South		West	
	Acres (A)	Yield (O/A)	Acres (A)	Yield (O/A)	Acres (A)	Yield (O/A)	Acres (A)	Yield (O/A)
WHEAT								
1839	2.0	14.5			3.5	8.4	2.1	13.0
1849	2.2	14.4			2.8	8.8	3.5	12.8
1859	1.7	14.3			5.3	8.4	8.2	12.7
1869	2.2	15.0			3.7	8.6	14.1	13.4
1879	2.4	15.8			5.9	8.2	26.0	13.7
1889	2.1	16.1			5.2	9.9	29.8	14.2
1899	2.1	16.4			6.5	11.5	40.6	14.0
1909	1.7	17.5			3.9	12.3	40.2	14.0
OATS								
1839	1.8	28.5			2.9	13.9	1.0	29.3
1849	2.1	28.5			3.2	13.7	1.4	30.0
1859	2.7	28.5			2.1	14.0	2.2	30.6
1869	2.6	28.6			2.1	14.5	4.9	32.1
1879	3.0	29.1			3.6	13.0	9.4	31.7
1889	3.1	26.2			4.0	12.9	20.6	29.9
1899	2.8	27.3			2.6	15.4	24.4	30.0
1909	2.7	29.7			2.4	17.0	30.4	26.5
CORN								
1839	1.1	33.5	5.9	21.8	8.9	11.8	3.2	32.7
1849	1.7	33.3	7.3	22.1	12.6	12.2	6.8	32.5
1859	2.0	33.4	7.8	22.1	15.0	12.8	12.6	32.3
1869	2.5	33.2	7.1	21.9	10.0	13.4	17.1	33.8
1879	2.7	31.4	9.0	20.7	14.5	13.2	35.1	31.1
1889	2.8	34.6	9.8	22.3	17.0	14.7	46.9	31.2
1899	2.9	35.2	10.2	22.1	22.8	14.7	56.4	29.6
1909[a]	1.9	36.8	10.1	24.4	27.1	16.1	56.0	31.0

Source: 1839, 1849, and 1859, based on 1866-75 average yield by state and census production data. 1869-1909, USDA, revised estimates.

[a] Total acreage reduced by 3.9 per cent to allow for corn grown for silage. Estimates for regions and subregions on the basis of 1919 ratios are as follows: Northeast, 29 per cent; Corn belt (W:C), 4 per cent; Western dairy (W:WD), 25 per cent; Small grain (W:SG), 2 per cent. (See USDA-AMS Crop Reporting Board, *Corn Acreage, Yield and Production*, June 1954, and USDA Monthly Crop Report, July 15, 1915.)

GRAIN PRODUCTION, 1840–60 AND 1900–10

TABLE 11

ACREAGES AND YIELDS IN MAJOR WESTERN SUBREGIONS, 1839–1909
(acres in millions and yields in bushels per acre)

	W:C		W:WD		W:SG	
	Acres (A)	Yields (O/A)	Acres (A)	Yields (O/A)	Acres (A)	Yields (O/A)
WHEAT						
1839	2.0	12.8	0.2	14.4		
1849	2.8	12.4	0.7	13.9		
1859	5.7	12.1	1.9	13.8		
1869	8.8	12.8	3.8	14.3	0.3	12.9
1879	13.1	14.0	6.5	13.8	3.4	11.0
1889	10.8	14.4	6.2	14.5	8.5	13.3
1899	10.5	14.1	8.1	14.2	14.8	12.7
1909	9.5	15.8	4.5	15.3	20.1	12.0
OATS						
1839	1.0	29.1	0.1	32.3		
1849	1.2	29.3	0.2	32.5		
1859	1.5	30.0	0.5	32.9		
1869	3.6	31.4	1.1	34.4	0.2	31.8
1879	6.1	31.6	2.2	34.7	0.6	27.7
1889	11.4	30.2	4.2	31.8	4.0	27.6
1899	12.4	32.1	5.5	32.3	4.6	25.9
1909	13.5	28.4	6.6	26.8	7.6	22.1
CORN						
1839	3.2	32.7	0.1	33.2		
1849	6.6	32.6	0.2	33.1		
1859	11.6	32.4	0.7	32.8	0.3	29.7
1869	15.2	33.8	1.3	34.4	0.6	33.4
1879	27.5	31.2	2.4	33.5	5.0	29.8
1889	31.4	32.6	3.2	30.2	12.1	27.9
1899	35.4	32.9	4.5	32.5	16.3	21.8
1909[a]	34.2	35.8	4.0	31.7	17.2	21.9

Source: See source to Table 10.

[a] See note a to Table 10.

would have been more than twice—and labor in oats just under twice—as productive as it was (Table 2, index i_3).

Mechanization, then, was the strongest direct cause of the productivity growth in the production of these grains. It accounted directly for over half the improvement, according to the values derived from Table 2 (wheat, .598; oats, .506; corn, .562). The effects of mechanical harvesting and threshing were felt with equal strength in all the regions, as the averages of Table 1, columns *b* and *c*, indicate. In the operations of plowing, harrowing, and planting, improvements were relatively less effective in the East and South, but were strongly felt in the West. Hence, a significant interaction appears between the interregional shifts in acreage weights and the changes in labor inputs per acre. The values following Table 3, showing the relative importance of the different types of mechanization, indicate that the traditional emphasis on the reaper and the thresher is not misplaced. Alone or in interaction, these accounted for over 80 per cent of the improvement due to mechanization in both wheat and oats. In corn, on the other hand, nearly all of the fall in labor inputs per acre, shown in Table 1, occurred in preharvest operations, principally from the abandonment of hoe cultivation. In view of these different sources of

TABLE 12

CORN ACREAGES AND YIELDS IN THE SOUTHERN SUBREGIONS, 1839–1909

	Eastern Cotton (S:EC)		Delta Cotton (S:DC)		Western Cotton (S:WC)	
	Acres (A)	Yield (O/A)	Acres (A)	Yield (O/A)	Acres (A)	Yield (O/A)
1839	7.3	11.1	1.6	14.9		
1849	9.5	11.1	2.8	15.0	0.3	20.6
1859	10.1	11.1	4.2	15.3	0.8	20.6
1869	6.6	11.2	2.4	16.3	0.9	21.5
1879	8.3	10.7	3.6	15.6	2.5	18.3
1889	9.4	11.6	4.2	16.6	3.4	21.0
1899	10.9	11.2	5.6	15.5	6.3	20.0
1909	10.7	13.7	5.9	17.1	10.5	18.1

Source: See source to Table 10.

the productivity growth in the crops, it is interesting to note how close the productivity indexes in corn come to those in wheat and oats.

Mechanization had a direct effect, appearing at once in the statistical evidence. The influence of improvements in nonmechanical technology are more deeply hidden—although not for that reason more fundamental. In these crops, the most important contribution of nonmechanical technology lay in the adaptation of practices and seed to the new conditions of soil and climate in the West. Little is known about the causes of the stability of land yields in the East. Some intraregional shifting of the crops was involved, and some benefits were obtained from the proliferation of varieties of improved seed. It seems likely, however, that a satisfactory set of practices for maintaining yields in these crops had been evolved in the East by the mid-nineteenth century, and little may have been added to this knowledge until the rises in yields in the late 1930's. But how might techniques have altered to maintain yields and reduce labor costs in the East under the pressure of a growing demand, if the western lands had not been available? To answer this question requires, first, a guess as to the shape of cost curves in the East under rapidly growing output, and then, an inquiry into the feed-back from rising costs to technological change.

TABLE 13

LABOR INPUTS AND LAND YIELDS, PERIOD 1

Region	Preharvest and Harvest Labor			Postharvest Labor (L/O) (4)	Total (L/O) (5)
	L/A (1)	O/A (2)	L/O (3)		
WHEAT					
Northeast	34.1	14.5	2.35	0.73	3.08
South	23.8	8.4	2.83	0.73	3.56
West	27.4	13.0	2.11	0.73	2.84
U.S.	27.5	11.3	2.44	0.73	3.17
OATS					
Northeast	27.1	28.5	0.95	0.40	1.35
South	19.8	13.9	1.42	0.40	1.82
West	21.6	29.3	0.74	0.40	1.14
U.S.	22.4	21.3	1.05	0.40	1.45
CORN					
Northeast	111.3	33.5	3.32	0	3.32
Middle east	62.1	21.8	2.85	0	2.85
South	71.6	11.8	6.07	0	6.07
West	59.2	32.7	1.81	0	1.81
U.S.	68.8	19.6	3.51	0	3.51

Source
Col. 1: Table 1.
Col. 2: Output from U.S. Census of 1840. Acreage obtained by dividing 1839 state output by estimated state yield. Yields, by state, estimated at 1866-75 average in USDA, revised estimates. For alternative yield estimates, see Appendix B.
Col. 3: Column 1 divided by column 2.
Col. 4: See discussion in Appendix D, "Period 1, Postharvest Labor."
Col. 5: Sum of columns 3 and 4.

One may indeed speculate more broadly on alternative growth paths for the American economy in the nineteenth century. Our statistical analysis takes us a short direction along three such paths: (1) westward movement without technological change; (2) technological change without westward movement; (3) westward movement and technological change, as they actually occurred.

Without technological change, westward expansion would have been accompanied by very little rise in productivity in agriculture. Shipment of crops and movements of population away from farms would perforce have been less rapid and less complete. The effects on economic welfare would have depended strongly upon the effects of a growing density of rural settlement upon the rate of rural population growth. The second alternative, technological change without westward movement, might have occurred under different political arrangements or land policies beyond the Appalachians. Confined to an eastern region with sharply different factor proportions, technology might have appeared quite a different animal from the labor-saving, land-using creature that emerges in our statistics. Could the land-saving developments in agricultural chemistry and biology that have raised yields since 1940 have emerged a century earlier in place of the labor-saving cultivator, reaper, and thresher? Even a passing look at the state of scientific knowledge in the fields of mechanical and chemical invention casts great doubt upon the plausibility of such an alternative. Despite the external economy of transport derivable from a less extensive agriculture, neither alternative appears likely to have produced by itself any large portion of the productivity gain actually achieved. The great opportunity for American agriculture, and the economy growing from it in the nineteenth century, derived from the simultaneous presence of many factors—technological change, empty lands, a growing population, a means of increase in the capital stock, improvements in transport, and the expansion of markets. Nor is the task of economic history finished with a simple suggestion of the relative weights of these factors, even if the suggestion be accompanied by a zealous use of quantitative data and techniques of measurement. The crucial question is: why did all these opportunities appear so close to one another in time? One is led thus, by all routes, out of factorial analysis back to examination of the complicated process of change in social behavior, of which these specific developments—great as they are—are manifestations. Scientific historiography has substituted the mind of man for the mind of God in which the ultimate explanations of an earlier age could come to rest. The substitution has not made the work simpler or the final end less elusive.

Appendix A: Definition of the Regions

The definition of a region required for the analysis of the effect of interregional shifts on productivity is implicit in the choice of variables measured. For these purposes, a region is an area over which man-hours per output unit, under the techniques of period 1, remain grouped in a stable distribution around a constant mean, distinct from that of other similarly defined regions during rather wide variations in acreage. On this definition, a region forms, on the curve relating productivity and acreage for the country as a whole, a plateau over which output may expand or contract at more or less constant per unit labor costs. Such statistical plateaus are not, of course, necessarily geographically contiguous pieces of land. Ideally, we should define labor costs under the given technique for every acre on which a grain might be produced, and group together those, wherever located, with a common cost coefficient.

In practice, the only bases for such regionalization in the mid-nineteenth century are the series of land yields by state, beginning in 1866 in the USDA revised estimates, and the labor input data presented in this study. The USDA estimates are given for each state and year from 1866 on, and their averages by state for the earliest ten years (1866–75) are arrayed as shown in Table A-1. We can break these series between North and South, and, moving Delaware and Maryland below the line in the wheat series, we produce the same two major regions for all three crops on the basis of yield. In corn, another break occurs between the so-called "border" states and the lower South.

To regionalize on the basis of the labor/output relation, it is necessary next to examine the variation within the yield regions with respect to labor per acre in various operations. The labor data are too scarce and too variable to define regions even as roughly as the land yield data can do. However, we observe that in period 2 (1895–1915) those northern states producing corn, wheat, and oats in period 1 are divided into two regions: a northeast region, including New England, New York, and Pennsylvania; and a western region, including Missouri and the states north and west of the Ohio River.[6]

Our period 1 labor data do appear good enough to test whether these divisions, as well as the North-South division made for land yields, are significant at mid-century. In Appendix D the labor data are arrayed for major preharvest operations, singly and together, and for harvesting in three regions for wheat and oats, and in four for corn. Table A-2 gives means of these distributions for preharvest and harvest labor totals. Perceptible differences occur between the means of the series for preharvest labor between the Northeast and the other northern region, and the differences in a given operation among the three crops in each region probably reflect differences in practice. In the harvest operations, our

[6] These period 2 divisions are adopted from the WPA/NRP report (see Appendix E, U.S. Document 5).

TABLE A-1
AVERAGE YIELDS BY STATE, 1866-75

State	Wheat Yield (O/A)	First Difference	State	Oats Yield (O/A)	First Difference	State	Corn Yield (O/A)	First Difference
Connecticut	17.2		Iowa	34.4		New Hampshire	37.6	
Massachusetts	17.0	0.2	Vermont	33.4	1.0	Vermont	37.4	0.2
Vermont	16.7	0.3	Wisconsin	33.0	0.4	Iowa	37.3	0.1
New York	16.1	0.6	Michigan	32.0	1.0	Ohio	35.7	1.6
New Hampshire	16.1	0	New Hampshire	31.1	0.9	Massachusetts	34.8	0.9
Michigan	14.6	1.5	Illinois	30.5	0.6	Pennsylvania	34.7	0.1
New Jersey	14.1	0.5	New York	29.9	0.6	Connecticut	33.7	1.0
Maine	13.6	0.5	Ohio	29.8	0.1	Indiana	33.5	0.2
Ohio	13.4	0.2	Connecticut	28.8	1.0	Michigan	33.3	0.2
Pennsylvania	13.2	0.2	Rhode Island	28.2	0.6	Wisconsin	32.7	0.6
Wisconsin	13.1	0.1	Massachusetts	28.2	0	New York	32.1	0.6
Missouri	12.4	0.7	Pennsylvania	27.4	0.8	New Jersey	31.7	0.4
Indiana	12.2	0.2	Indiana	27.3	0.1	Rhode Island	31.5	0.2
Maryland	12.1	0.1	Maine	26.9	0.4	Nebraska	31.1	0.4
Iowa	12.1	0	Missouri	26.8	0.1	Minnesota	31.0	0.1
Delaware	12.0	0.1	New Jersey	23.9	2.9	Illinois	30.6	0.4
Illinois	11.3	0.7	Delaware	19.1	4.8	Maine	30.2	0.4
Virginia–West Virginia	9.5	1.8	Maryland	18.6	0.5	Kansas	29.4	0.8
Kentucky	9.2	0.3	Kentucky	17.6	0.9	Missouri	29.2	0.2
Arkansas	9.2	0	Arkansas	17.6	0.1	Kentucky	25.1	4.1
Tennessee	7.4	1.8	Tennessee	15.6	2.0	Maryland	22.8	2.3
Mississippi	7.3	0.1	Louisiana	14.2	1.4	Tennessee	21.6	1.2
Alabama	6.7	0.6	Virginia–West Virginia	12.3	1.9	Virginia–West Virginia	19.0	1.6
North Carolina	6.3	0.4	Mississippi	12.1	0.2	Arkansas	18.9	0.1
Georgia	6.0	0.3	Georgia	11.9	0.2	Delaware	18.6	0.2
South Carolina	5.7	0.3	North Carolina	11.8	0.1	Louisiana	14.8	3.8
			Alabama	11.6	0.2	Mississippi	13.9	0.9
			South Carolina	10.9	0.7	North Carolina	12.4	1.5
			Florida	10.3	0.6	Alabama	12.3	0.1
						Georgia	9.9	2.4
						Florida	9.9	0
						South Carolina	9.8	0.1

Source: USDA, revised estimates (see Appendix E below, U.S. Documents 7, 8, 9).

TABLE A-2

LABOR INPUT COEFFICIENTS: PREHARVEST AND HARVEST LABOR FOR PERIOD 1

	Northeast		Middle East		South		West	
	No.of Cases	Man-Hours Per Acre	No.of Cases	Man-Hours Per Acre	No.of Cases	Man-Hours Per Acre	No.of Cases	Man-Hours Per Acre
Wheat								
Preharvest	32	19.1			43	11.3	21	12.4
Harvest	40[a]	15.0			b	12.5	40[a]	15.0
Oats								
Preharvest	25	14.3			51[c]	8.8	51[c]	8.8
Harvest	17[a]	12.8			b	11.0	17[a]	12.8
Corn[d]								
Preharvest		98.3		52.0		67.3		46.2
Harvest		13.0		10.1		4.3		13.0

Source: Appendix D. Preharvest labor, Tables D-1, D-1a; harvest labor, Tables D-8, D-9, D-10 and notes.

[a] Northeast and West treated as a single region.
[b] See Appendix D, period 1, harvest labor, for derivation of this figure.
[c] South and West treated as a single region.
[d] On number of cases, see Appendix D.

data for the southern region are deficient but the differences between the two northern regions do not appear significant.

On the basis of these considerations, we may separate Northeast from the rest of the North as defined by land yields, and produce the regions shown in the text.

Appendix B: Alternative Land Yield Estimates and Indexes

The average land yields shown in Table A-1 are taken from the 1866–75 USDA revised series, by state. Period 1, defined by our output and labor input data, however, relates to the period around 1839. Can the USDA 1866–75 state average yields be extrapolated back to the antebellum decades without allowance for trend? Shifts in the distribution of acreage between states of differing average yields within each region imply that extrapolation of regional averages might involve avoidable error. But the relative absence of strong trends in the state data from 1866–96 suggests that extrapolation by state may be justified. Fortunately, the *Annual Reports* of the Commissioner of Patents for the years 1843–55 contain estimates of county yields at several points within many of the states, and in a few states, state censuses provide similar information. Table B-1 compares these estimates, in summary form, with the average 1866–75 USDA revised estimate. Except in the frontier states, where acreages

TABLE B-1

ESTIMATES OF COUNTY LAND YIELDS, 1843-61, AND USDA AVERAGE, 1866-75
(bushels per acre)

State or Region	County Estimates, 1843-55			USDA 1866-75 State Average (4)
	Number (1)	Range (2)	Median (3)	
WHEAT				
New England	20	10-25	15	16
New York	23	10-25	18	18
Pennsylvania	22	8-30	15	13
Delaware	6	14-20	15	12
New Jersey	5	12-30	20	14
Maryland	4	6-20	15	12
Virginia	13	8-15	9	9
South[a]	17	5-20	10	7
Ohio	18	10-35	15	13
Michigan	8	12-30	21	15
Wisconsin	12	12-40	20	13
Illinois	14	10-20	16	11
Indiana	16	9-25	16	12
Iowa-Missouri	9	10-20	15	12
OATS				
New England	31	17-50	30	30
New York	19	28-50	35	30
Pennsylvania-New Jersey	13	30-50	40	26
Delaware-Maryland	8	9-50	20	19
Virginia	6	10-28	16	12
South[a]	12	12-60	16	14
Ohio, Indiana, Illinois, Iowa	14	20-55	40	30
Michigan-Wisconsin	13	30-60	40	33
CORN				
New England	36	20-50	35	
New York	28	25-50	30	32
New Jersey	3	35-50	45	32
Ohio	22	25-60	39	36
Pennsylvania	12	30-50	38	35
Indiana	17	30-60	40	34
Michigan	9	15-40	30	34
Wisconsin	6	25-50	45	33
Iowa	7	22-40	35	37
Illinois	8	40-50	43	31
Missouri	7	30-60	40	29
Kentucky	5	40-60	50	25
Maryland	1		30	23
Tennessee	3	30-35	35	22
Texas	8	30-70	50	21
Virginia	21	10-40	20	19
Delaware	6	22-50	35	19
Arkansas	1		20	19
Louisiana	4	10-50	30	15
Mississippi	7	15-50	25	14
Alabama	6	20-37	30	12
North Carolina	4	10-20	15	12
Georgia	5	5-40	18	10
South Carolina	3	15-50	25	10

Source: Cols. 1-3, *Annual Report: Agriculture*, U.S. Commissioner of Patents, 1843-56; col. 4, USDA, revised estimates.

[a] States south of Ohio and Maryland, except Virginia.

are small in any case and yields high for a few years, the median of the 1843–55 county reports, by state, taken without regard to year, falls surprisingly close to the 1866–75 USDA state average.[7] The state estimates from contemporary sources shown in Table B-2 also fall close to the 1866–75 USDA state averages. It is probably justifiable then to extrapolate the state averages back to the 1840–60 period. Taking these yields, we calculate acreages by state from the Census production data for 1840, 1850, and 1860, and produce regional average yields (shown in Table 1).

Such extrapolation does indeed result in yields in the western region (R_3) decidedly below those shown in Tables B-1 and B-2. We have therefore prepared a set of alternative indexes based on an assumption of western yields (O/A) in 1839 (with a corresponding reduction in the regional acreage estimate) as follows:

	Wheat	Oats	Corn
Ohio	15	40	40
Indiana	16	40	40
Illinois	16	40	40
Iowa	15	40	38
Missouri	15	40	40
Michigan	21	40	33
Wisconsin	20	40	40
Minnesota	20	40	40

For corn, the presumption of higher yields arising from the newness of the region exists in the South and border states as well. In the Southeast, there is also the possibility that yields were abnormally depressed during the "reconstruction" period. On the basis of Table B-2, it seems desirable to allow for yields in the South and middle-east regions 50 per cent higher than the 1866–75 state averages.

These assumptions produce a set of regional yields in period 1 which compare with those of text Table 1 as follows:

	Wheat		Oats		Corn	
	Table 1	Alternative	Table 1	Alternative	Table 1	Alternative
R_1	14.5	14.5	28.5	28.5	33.5	33.5
R_{2a}	8.4	8.4	13.9	13.9	21.8	32.6
R_{2b}					11.8	17.7
R_3	13.0	15.3	29.3	40.0	32.7	39.8

[7] Surprise is reduced and confidence in the result perhaps raised, if we consider that the USDA series was itself derived from county crop reporters, though the number of such reporters was greater and the data supplied more explicit after 1870 than under the Patent Office.

TABLE B-2

ESTIMATES OF STATE LAND YIELDS, 1844-62, COUNTY LAND
YIELDS, 1843-56, AND USDA AVERAGE, 1866-75
(bushels per acre)

State	1866-75 USDA Revised Series (1)	1843-56 Median County Estimates (2)	1844-62 State Land-Yield Estimates (3)	
WHEAT				
United States			1842	15-20
			1845	18
Massachusetts	17	15	1853	17
			1855	15
New York	16	18	1844	14
			1854	11
Virginia	9	9	1847	8-10
Tennessee	7	11	1849	15
Ohio	13	15	1845	13
			1850-57	8-17
Michigan	15	21	1845-49	15-20
Indiana	12	16	1844-48	20
OATS				
United States			1845	35
Northeast			1847	30-40
Northwest			1847	30
South			1847	10
Massachusetts	28	30	1847	35
			1853	30
			1855	21
			1861	26
New York	30	35	1844	26
			1854	20
Delaware	19	20	1848	26
Georgia	12		1848	12-15
Tennessee	16	16	1848	25-30
Wisconsin	33	40	1849	35
Michigan	32	40	1848	40
			1849	35
CORN				
New England	35	35	1847	25-30
New Hampshire	38		1862	38
Vermont	37		1862	35
Massachusetts	35		1854	25
			1855	29
			1862	37
Connecticut	34		1862	32
Rhode Island	32		1862	37
New York	32	30	1844	25
			1845	24
			1855	21
			1862	35
New Jersey	32	45	1862	37
Pennsylvania	35	38	1847	25-30
			1862	36

(continued)

TABLE B-2 (concluded)

State	1866-75 USDA Revised Series (1)	1843-56 Median County Estimates (2)	1844-62 State Land-Yield Estimates (3)	
Ohio	36	39	1847	43
			1845-48 (average)	38
			1850	36
			1852	34
			1853	40
			1854	26
			1855	40
			1856	28
			1857	37
			1858	28
			1862	33
Indiana	34	40	1862	42
Illinois	31	43	1848 (prairie)	40-50
			1862	40
Michigan	33	30	1847	40
			1854	23
			1856	18
Wisconsin	33	45	1860	40
Iowa	38	35	1862	38
Minnesota	31		1862	45
Missouri	29	40	1862	38
Kentucky	25	50	1847	30-40
Delaware	19	35	1862	20
Louisiana	15	30	1854	15-25
Alabama	12	30		
North Carolina	12	15	1847	20-35
Georgia	10	18		

Source

Cols. 1 and 2: Table 3.
Col. 3: *Annual Report: Agriculture*, U.S. Commissioner of Patents, 1844-62; and, for New York, State Censuses of 1845 and 1855; Ohio, 1850-58 from State Censuses of Agriculture; Massachusetts, 1855 State Census.

Since the period 1 acreage weights in Table 1 are derived by dividing estimated yields into Census production data for 1839, change in the yield estimate alters acreage weights as follows:

	Wheat		Oats		Corn	
	Table 1	Alternative	Table 1	Alternative	Table 1	Alternative
R_1	.259	.273	.316	.331	.057	.080
R_{2a}	.459	.482	.509	.521	.310	.291
R_{2b}					.465	.435
R_3	.282	.246	.178	.138	.168	.194

TABLE B-3

PRODUCTIVITY INDEXES BASED ON ALTERNATIVE ESTIMATES
OF WESTERN YIELDS IN PERIOD 1

	Text Table 2,[a] Estimate	Variant	Text Table 3,[b] Estimate	Variant	Text Table 4,[c] Estimate	Variant
			WHEAT			
i_1	100.0	100.0	100.0	100.0	100.0	100.0
i_2	118.3	113.8	144.7	143.3	103.3	103.1
i_3	245.7	242.1	119.2	119.6	97.6	97.1
i_4	109.3	117.8	113.6	112.5	109.3	118.6
i_5	302.1	290.7	188.7	188.3	100.6	100.3
i_6	377.3	401.6	139.1	138.7	112.4	122.5
i_7	117.8	113.4	176.1	172.4	106.7	115.1
i_8	416.7	401.6	245.7	242.1	109.6	119.2
			OATS			
i_1	100.0	100.0	100.0	100.0	100.0	100.0
i_2	105.8	102.9	134.2	134.2	104.3	105.3
i_3	185.9	185.5	115.1	116.1	97.3	97.9
i_4	122.9	139.7	106.7	106.8	120.8	141.0
i_5	201.2	195.7	164.7	165.8	101.4	102.1
i_6	371.7	414.9	125.0	125.9	126.1	146.8
i_7	117.9	114.7	147.9	146.8	118.9	138.3
i_8	362.6	352.1	185.9	185.5	122.9	142.5
			CORN			
i_1	100.0	100.0			100.0	100.0
i_2	119.0	85.7			111.1	112.2
i_3	227.3	226.4			97.0	96.1
i_4	129.6	124.3			130.1	118.0
i_5	265.2	191.6			107.4	107.3
i_6	330.2	316.4			144.6	132.4
i_7	142.9	103.0			126.4	113.7
i_8	364.6	261.6			140.0	127.0

Source: See text accompanying Tables 2, 3, and 4.

[a] See cols. 4, 5, and 6 of Table 2, for wheat, oats, and corn, respectively.

[b] See cols. 2, and 4 of Table 3 for wheat and oats, respectively.

[c] See cols. 4, 5, and 6 of Table 4 for wheat, oats, and corn, respectively.

GRAIN PRODUCTION, 1840–60 AND 1900–10   555

The productivity (O/L) indexes derived by inserting those values in the formulas of Tables 2, 3, and 4 are shown, with the text indexes for comparison, in Table B-3. It is apparent that the levels and relative values of the indexes are not strongly affected by this alternative assumption.

Appendix C: Measurement of Relative Importance of the Factors

The assessment of the relative importance of a factor is derived from Yates (see Appendix E, book 9). Gratitude is owed to Leo Katz of Michigan State University for calling attention to the method, though he bears no responsibility for the transfer of it to the present context. To derive the signs shown in the matrix in the text, let the variables of Table 2 (v, abc, y) be represented by a, b, c (period 1 values) and A, B, C (period 2 values). Then the independent effect of A is taken as the mean difference between the indexes where it appears in a period 2 value (A) and those where it appears in a period 1 value (a). The independent effect of AB is the sum of two differences $(ABC - abC) + (ABc - abc)$, with each reduced by the indexes of the independent effects of the other period 2 values occurring in them. The whole equation is

$$AB = (ABC - abC) - [(AbC - abC) + (aBC - abC)]$$
$$+ (ABc - abc) - [(Abc - abc) + (aBc - abc)].$$

The independent effect of ABC takes the whole difference $ABC - abc$ less a similar allowance for the effects of the other indexes.

$$(ABC - abc) - [(ABc - abc) - [(Abc - abc) + (aBc - abc)]$$
$$- [(aBC - abc) - [(aBc - abc) + (abC - abc)]$$
$$- [(AbC - abc) - [(Abc - abc) + (abC - abc)]$$
$$- (Abc - abc) - (aBc - abc) - (abC - abc)$$

These formulas reduce to the terms with their signs shown in the text matrix.

Appendix D: Summary Tables for Labor Requirements by Type of Operation

The tables are in four groups: period 1, preharvest; period 1, harvest and postharvest; period 2, preharvest and harvest; period 2, postharvest.

SYMBOLS

n = number of cases; $\bar{X}$ = arithmetic mean; M = median; σ = standard deviation; s = standard error of the mean.

REGIONAL ABBREVIATIONS AND DEFINITIONS

NE, Northeast (Pa., N.J., N.Y., Vt., N.H., Mass., Conn., Me., R.I.)
S, South[8]
  S:ME, South: Middle east (Md., Del., Va., Ky., W.Va., Tenn., Ark.)
  S:EC, South: Eastern cotton (N.C., S.C., Ga., Fla., Ala.)
  S:DC, South: Delta cotton (Miss., La.)
W, West
  W:C, West: Corn (Ohio, Ind., Ill., Mo., Iowa)
  W:WD, West: Western dairy (Mich., Wis., Minn.)
  W:SG, West: Small grain (Nebr., Kans., S.Dak., N.Dak., Mont.)
  W:WC, West: Western cotton (Tex., Okla.)
  W:R, West: Range (N.Mex., Ariz., Colo., Utah, Nev., Wyo.)
  W:NW, West: Northwest (Idaho, Ore., Wash.)
  W:Cal., West: Calif.

PERIOD 1: PREHARVEST LABOR, CORN

For period 1, total preharvest labor for each region is composed of the sum of the average man-hours per acre for the standard operations as follows:

R_1 Plowing, harrowing, planting, cultivating, and hoeing

R_{2a} Plowing, harrowing, running off rows, planting, cultivating, and hoeing

R_{2b} Clearing and cutting stalks, plowing, running off rows, planting, cultivating, and hoeing

R_3 Plowing, harrowing, planting, cultivating and hoeing.

For all regions except R_{2a} (middle east), there were also a number of cases where complete preharvest totals were available (see Table D-1). In the case of the Northeast and the West, these were remarkably close to the totals obtained by the above method which is based on the prevailing techniques within each region (Table D-1a). Where our data indicated that time differences for the same operation were insignificant, regional averages were combined. In the case of "running off rows," regions R_{2a} and R_{2b} (middle east and South) were combined, and for cultivating and hoeing, middle eastern and western regional averages were combined. In addition, an allowance for cutting and clearing stalks in Tennessee was made by weighting the R_{2b} labor coefficient for this operation by Tennessee's acreage weight in the R_{2a} region. This was due to the fact that the corn was harvested in Tennessee by the southern method of pulling off the

---

[8] For corn, the middle east (S:ME) becomes a separate region (ME) and the South (S) includes the eastern cotton (S:EC), delta cotton (S:DC), and western cotton (S:WC) regions, with Arkansas included in the delta cotton states.

TABLE D-1

PERIOD 1: PREHARVEST LABOR, TOTAL

	NE	S:ME	S:EC	S:DC	S	W:C	W:WD	W
				WHEAT				
n	32.0	36.0	7.0		43.0			21.0
$\bar{X}$	19.1	11.3	10.9		11.3			12.4
M	20.0	10.2	11.6		10.2			10.9
σ	7.3	4.1	2.3		3.8			7.2
s	1.3	0.7	0.94		0.6			1.6
				OATS				
n	25.0	23.0	11.0	7.0	41.0			10.0
$\bar{X}$	14.3	9.4	7.2	10.5	9.0[a]			8.7[a]
M	13.7	9.0	7.1	10.5	8.6			7.9
σ	3.7	2.4	2.4	1.9	2.6			1.9
s	0.75	0.5	0.76	0.84	0.4			0.63
				CORN				
n	61.0		28.0	6.0	34.0	19.0	2.0	21.0
$\bar{X}$	97.8		55.0	68.2	57.3	48.4	47.5	48.3
M	88.0		54.3	68.9	56.7	42.0		42.0
σ	42.5		23.9	8.0	22.5	18.0		18.0
s	5.5		4.6	3.5	3.9	4.3		4.0

[a] South and West treated as a single region for Table 1.

TABLE D-1a

PERIOD 1: PREHARVEST LABOR FOR CORN, BY OPERATION

	NE	ME	S	W
Clearing and cutting stalks		2.5[a]	7.5	
Plowing	11.9	7.4	8.9	7.5
Harrowing	5.4	1.2		2.9
Running off rows		3.5	3.5	
Planting	14.5	7.5	8.8	5.9
Cultivating	14.7	12.0	17.8	12.0
Hoeing	51.8	17.9	20.8	17.9
Totals	98.3	52.0	67.3	46.2

Source: Tables D-2 to D-7, and Appendix D text.

[a] Allowance for cutting and clearing stalks in Tennessee.

TABLE D-2

PERIOD 1: PREHARVEST LABOR, PLOWING

	NE	S:ME[a]	S:EC	S:DC	S	W:C	W:WD	W
			WHEAT					
n	32.0	24.0	3.0		27.0			22.0
$\bar{X}$	12.4	6.3	7.3		6.4			6.5
M	10.0	5.8	5.6		6.0			5.7
σ	6.7	2.1			2.1			2.8
s	1.2	0.45			0.4			0.62
			OATS					
n	25.0	12.0	1.0	7.0	20.0			10.0
$\bar{X}$	7.1	5.7	7.27	7.0	6.2			5.4
M	7.1	5.6		6.7	6.1			5.0
σ	2.2	2.1		1.5	1.8			1.2
s	0.45	0.64		0.6	0.41			0.41
			CORN					
n	72.0	12.0	29.0	12.0	41.0	44.0	4.0	48.0
$\bar{X}$	11.9	7.4	7.9	11.4	8.9	7.5	6.9	7.5
M	10.0	6.3	7.5	9.5	8.0	7.1	6.5	7.1
σ	5.2	4.8	4.14	5.1	4.7	2.9	2.2	2.81
s	0.62	1.5	0.8	1.6	0.75	0.4	1.27	0.40

[a] For corn, the region is ME.

TABLE D-3

PERIOD 1: PREHARVEST LABOR, HARROWING

	NE	S:ME[a]	S:EC	S:DC	S	W:C	W:WD	W
				WHEAT				
n	32.0	26.0	2.0		28.0			23.0
$\bar{X}$	6.2	3.0	2.8		3.0			4.9
M	5.0	2.5	2.8		2.5			3.3
σ	3.4	1.5			1.5			3.4
s	0.61	0.31			0.28			0.71
				OATS				
n	22.0	13.0	1.0	1.0	15.0			11.0
$\bar{X}$	4.7	2.0	11.1	5.0	2.8			2.3
M	5.0	1.8			1.8			2.5
σ	2.0	0.9			2.5			1.0
s	0.44	0.27			0.67			0.32
				CORN				
n	44.0	5.0				27.0	2.0	29.0
$\bar{X}$	5.4	1.2				2.8	3.75	2.9
M	5.0	1.0				2.5		2.5
σ	3.7	0.5				1.8		1.79
s	0.6	0.25				0.3		0.34

[a] For corn, the region is ME.

TABLE D-4

PERIOD 1: PREHARVEST LABOR, SOWING AND PLANTING

	NE	S:ME[a]	S:EC	S:DC	S	W:C	W:WD	W
WHEAT								
n	9.0	11.0	1.0					12.0
$\bar{X}$	1.68	1.5	1.1					1.2
M	1.65	1.5						1.3
σ	0.6	0.34						0.55
s	0.21	0.11						0.17
OATS								
n	9.0	4.0						1.0
$\bar{X}$	1.2	1.5						1.42
M	1.0	1.5						
σ	0.6	0.26						
s	0.21	0.15						
CORN								
n	68.0	23.0	40.0	23.0	63.0	40.0	6.0	46.0
$\bar{X}$	14.5	7.5	8.8	8.6	8.8	5.6	7.5	5.9
M	12.5	7.7	8.66	7.0	7.9	5.0	5.6	5.0
σ	9.7	2.9	4.7	5.0	4.9	3.8	5.8	4.1
s	1.2	0.62	0.76	1.1	0.62	0.6	2.59	0.62

[a] For corn, the region is ME.

TABLE D-5

PERIOD 1: PREHARVEST LABOR, CORN CULTIVATING

	NE	ME	S:EC	S:DC	S	W:C	W:WD	W
n	44.0	16.0	41.0	11.0	52.0	41.0	2.0	43.0
$\bar{X}$	14.7	12.1	18.5	15.3	17.8	11.7	16.0	11.9
M	12.0	11.9	16.7	14.8	16.4	10.0		10.0
σ	9.9	3.8	9.6	6.3	9.1	6.9		6.9
s	1.5	0.975	1.5	2.0	1.3	1.1		1.1

TABLE D-6

PERIOD 1: PREHARVEST LABOR, CORN HOEING

	NE	ME	S:EC	S:DC	S	W:C	W:WD	W
n	48.0	8.0	37.0	10.0	47.0	19.0	2.0	21.0
$\bar{X}$	51.8	11.5	20.0	23.7	20.8	23.1	17.5	20.4
M	48.0	10.8	14.25	21.3	16.5	20.0		20.0
σ	30.7	5.64	20.68	10.6	19.1	14.7		13.9
s	4.5	2.1	3.4	3.5	2.8	3.5		3.1

TABLE D-7

PERIOD 1: PREHARVEST LABOR FOR CORN, MISCELLANEOUS OPERATIONS

	ME	S:EC	S:DC	S	ME+S
	CLEARING AND CUTTING STALKS				
n					11.0
$\bar{X}$					7.5
M					2.3
σ					8.4
s					2.65
	RIDGING, BEDDING, OR FURROWING				
n		15.0	1.0	16.0	
$\bar{X}$		8.0	3.9	7.8	
M		7.06		6.9	
σ		4.3		4.3	
s		1.1		1.1	
	RUNNING OFF ROWS				
n	3.0	17.0	3.0	20.0	23.0
$\bar{X}$	4.4	3.0	5.5	3.37	3.5
M	6.0	2.25	5.4	3.65	3.8
σ		1.73	1.0	1.9	1.97
s		0.43	0.7	0.4	0.42

ears, necessitating the removal of stalks the following spring before land could be plowed.

We have assumed that plowing was standard practice in all regions, although in the South, the cornfield was frequently ridged or bedded instead. Sixteen cases for this operation gave us an average of 7.8 man-hours per acre, which is somewhat less than the plowing average of 8.9 man-hours per acre obtained from forty-one cases, and almost the same as the plowing time for the eastern cotton area (twenty-nine cases, 7.9 man-hours per acre) from which fifteen of our sixteen cases originated. It therefore did not appear necessary to make an allowance for the alternative operation.

In the Northeast, instead of being harrowed, the field was occasionally dragged or rolled. In only two cases out of ten was rolling carried out in addition to harrowing, but in both cases the combined time was less than the regional average for harrowing. Four cases gave a combined figure for rolling, dragging, and harrowing, of which two were below the regional average. It would appear that either the cornfield was harrowed or dragged and rolled, but that harrowing was the usual operation. In the South the harrow was rarely used before planting, although occasionally a harrow was used to "cover" after planting or for the first cultivation. In place of harrowing, we find that "running off rows" precedes planting. This appears to be a somewhat more elaborate operation than "marking" in the Northeast and West before the days of the corn planter. But whereas "marking" time was included with planting in those two regions, running off rows was taken into account separately in the South, and, combined with the three middle-eastern cases, applied to both regions.

The planting operation in all regions includes an allowance for replanting, where the records show that this was done. No allowance has been made for manuring.

PERIOD 1: HARVEST LABOR, WHEAT AND OATS

To estimate harvest labor requirements in period 1, we have three bodies of evidence: (1) a number of cases for total harvest labor in the North and for reaping alone in the South; (2) a few cases for individual operations in the North and operations other than reaping in the South; (3) Rogin's estimates (Appendix E, book 7), based on a few cases of labor in each operation by various techniques.

From (1), shown in Table D-8, we derive an average figure for Northeast

GRAIN PRODUCTION, 1840-60 AND 1900-10 563

and West, in the same way as for preharvest operations. Since the averages for each region are close to one another, the two are combined into a single sample (col. 3). This is the more plausible since harvest labor is affected less by natural conditions, other than yield and regional differences in farm practice, than the preharvest operations are. The lower figure for oats, despite very high yields in many cases, is probably due to less careful handling of the crop after cutting.

The problem then is to derive a figure for the South based on the reaping data, summarized in Table D-8, and the few figures on other operations shown in Table D-9. These scattered figures may be summarized as follows:

	Operation							Wheat		Oats	
	Reaping	Raking	Binding	Taking Up	Setting Up	Shocking	Stacking	Total	$\bar{X}$	n	$\bar{X}$	n
1.	x	x	x						7.2	5	10.0	3
2.		x	x						4.3	1	5.0	1
3.					x				3.5	3		
4.						x			5.8	2	9.0	3
5.		x	x		x				5.3	1		
6.			x			x			11.3	3		
7.				x							3.3	1
8.			x			x					5.3	5
9.	x		x								6.0	1
10.							x		12.2	8	11.0	3

The estimates for wheat (12.5) and oats (11.0) in Table 1 in the text are derived from these data in several ways. For wheat, the estimates are as follows:

1. 12.2, the total figure (line 10). This, however, is a simple average of four figures from region S:EC (10.0) and region S:ME (14.4) and so may be too low.

2. 12.7, the average of line 1 (7.2) plus the average of lines 4 and 5 (5.5).

3. 12.3, the sum of the estimate for reaping, Table D-8 (3.6) plus line 2 (4.3) plus average of lines 3 and 4 (4.4).

4. Rogin's estimate of 11.0-13.5. This estimate appears a little too low for the North, and Rogin states that binding labor was probably

higher in proportion to cradling in Virginia than in the West. The lower yields of the South, however, make a lower total figure plausible.

For oats, the figure in Table 1 is derived from these data by several routes:

1. 11.1, reaping (4.0), Table D-8, plus line 2 (5.0) plus 2.0 for shocking, estimated from the Northeast figures of Table D-9.
2. 12.0, line 1 plus 2.0 for shocking.
3. 11.0, line 10.
4. 11.3, line 8 plus line 9, as a maximum, with "taking up" double-counted.

TABLE D-8

PERIOD 1: HARVEST LABOR FOR WHEAT AND OATS

	Harvesting, Total			Reaping
	NE (1)	W (2)	NE and W (3)	S (4)
	WHEAT			
n	21.0	19.0	40.0	48.0
$\bar{X}$	15.3	14.8	15.0	3.6
M	14.6	13.3	13.7	3.25
σ	7.2	5.6	6.4	1.3
s	1.6	1.3	1.03	0.19
	OATS			
n	12.0	5.0	17.0	36.0
$\bar{X}$	13.1	12.1	12.8	4.1
M	12.5	10.0	12.5	4.14
σ	3.3	3.1	3.3	1.4
s	0.99	2.3	0.81	0.23

Other combinations of the data, involving the high stacking figure for oats (line 4), would yield higher estimates. But these three cases are all from one Virginia plantation in one year and it is not clear what the term stacking here means. The small differences between oats and wheat are explainable in the same terms in the South as in the North, and between southern and northern oats in the same terms as for wheat. That yields affect both the wheat and oats figures is indicated by the relatively low reaping figure for the South compared with Rogin's estimate of 5.0 for cradling.

TABLE D-9
PERIOD 1: MISCELLANEOUS HARVESTING FIGURES FOR WHEAT AND OATS
(man-hours per acre)

PART A. WHEAT

Region	Reaping	Reaping, Raking, and Binding	Raking and Binding	Setting Up	Shocking	Stacking	Raking, Binding, and Shocking	Binding and Stacking	Total Harvest
NE	5.0								*15.3*
S:ME	*3.6*		4.3		2.8		5.3		10.0
		8.4			4.5				17.5
		6.3			3.1				14.5
		7.7							15.6
		5.7							
		8.1							
S:EC						4.8		8.6	8.3
						6.8		16.2	9.0
								9.1	9.9
									13.1
W		15.0					4.7	5.2	*14.8*
US (Rogin)	5.0		5.0–7.5		1.0				11.0–13.5

PART B. OATS

Region	Reaping	Reaping, Raking, and Binding	Raking and Binding	Setting Up	Shocking	Stacking	Taking Up and Stacking	Reaping and Taking Up	Reaping, Binding, Taking Up, and Shocking	Total Harvest
NE	5.0		5.0		2.5					*13.1*
					2.2					
					2.1					
S:ME	4.1[a]	8.8	5.0	3.3		8.8	4.5[b]	6.0[c]	10.0[c]	11.1[d]
		8.6				7.9	8.9[b]			11.9[d]
						10.3	4.3[b]			
							6.4[b]			
							2.5			
S										*12.1*
US (Rogin)	5.0	10.0	5.0–7.5		1.0					11.0–13.5

Source: Italicized figures are averages, derived from Table D-8. The Rogin figure is the summary estimate from Rogin (Appendix E, B-7), pp. 125-37. The source of the other figures is given in Appendix Tables S-25 and S-26 (not shown).

[a] includes S:EC region.
[b] Figures from North Carolina.
[c] Figures from Georgia.
[d] Figures from Mississippi.

PERIODS 1 AND 2: HARVEST LABOR, CORN

Harvesting of the corn grain consists of picking or snapping the ear from the stalk. Throughout the North, husking is generally performed as the ears are picked. The data, therefore, give single figures for the picking and husking operations combined. Operations on the corn plant—topping, pulling leaves, or cutting and shocking—are not part of grain harvesting, and the considerable amounts of time devoted to them, especially in the Northeast, are chargeable to fodder or stover production. Since the acreage thick-sown and cut for silage in period 2 is not included in the acreage figures of Table 10, it is not necessary to estimate labor costs in these operations.

Three methods of harvesting the corn grain were in use in the United States in the period 1840–1910: (1) The plant was cut and shocked, and the ears picked and husked from the shocks in the field or barn before storage. (2) The ears in the husks were picked from the standing stalks, stored in the husks, and husked as used during the year. (3) The ears were picked and husked from the standing stalks in one operation. These methods are known respectively as: (1) husking from the shock; (2) snapping from the standing stalk; (3) husking from the standing stalk. In the two periods, these methods appear to have prevailed in the various regions as follows:[9]

	Period 1 Method	Period 2 Method
R_1 plus Ohio, W:WD and North Dakota	1	1
R_{2a} except Tennessee	1	1
R_{2b} plus Tennessee	2	2
R_3 less Ohio, W:WD and North Dakota	1	3

In the mid-nineteenth century, all our contemporary figures are for cutting and shocking in the North and border states and for snapping in the South below Kentucky and Virginia. In the West, some corn land was harvested, grain and all, by livestock, but the method of going through the field to husk from the standing stalks did not appear until the 1870's. In 1840, the weight of the eastern portion of the corn belt

---

[9] In period 2, the division between method 1 and methods 2 and 3 is shown in USDA, *Yearbook, 1917* (Appendix F, U.S.-1d), pp. 566–567. Here, in fact, method 1 extends across most of North Dakota and Montana, and occupies a large island in central Missouri. Methods 2 or 3 include the western quarter of Kentucky and the southern third of Minnesota. These adjustments partly balance out along the boundaries as we have drawn them. The distinction between the regions using methods 2 and 3 is shown roughly in the NRP Report A-5, Appendix Tables D-5 and D-10 (Appendix E, U.S.-6).

where cutting and shocking were universal, and the evident predominance of the practice even in Illinois, justifies the use of the method 1 figure for the region as a whole.

For methods 1 and 3, a basic source is USDA, *Bulletin 3*, 1913 (see Appendix F, U.S.-1b), giving results for an extensive survey and indicating standard coefficients for the United States in these operations as follows:

Method	Yield in Bushels of Ears	Bushels per $9\frac{1}{2}$-hour day
1	1–40	35
	41–60	42
	61 and over	50
3	1–40	60
	41–60	70
	61 and over	75

Taking two bushels of husked ears as the equivalent of one bushel of shelled corn, it appears that the yields in the North—25–35 bushels per acre—would require 10–14 hours per acre by method 1, and 7–9 hours per acre by method 3. These results, however, are based on returns heavily weighted from the North Central states and "adjusted" arbitrarily by the compilers to compensate for assumed biases in the farmers' reports.

For method 1, several pieces of evidence indicate that the estimate should be placed at the upper limit of *Bulletin 3*'s range of 10–14 man-hours per acre. For western New York, a survey reported in USDA, *Bulletin 412*, 1913 (see Appendix F, U.S.-1b) indicates a range of 12–14 man-hours per acre for a 30-bushel (shelled) yield. A similar survey in 1915, for Chester County, Pennsylvania,[10] also yields an estimate of 13.5 man-hours per acre for a yield of 50 bushels.

In West Virginia, in 1913–15, a survey of fifteen to thirty farms in each of twelve counties[11] gave an average of 15.6 man-hours per acre for an average yield of 40 bushels, with a county range of 13.2–17.7. Finally, in two counties in Minnesota, estimates of 12.8 and 14.0 were obtained as an average during the 1910's.[12] The figure used in Table 1, therefore, is taken to indicate conditions outside the corn belt, for the range of yields in the states where method 1 was prevalent.

The *Bulletin 3* range for method 3, on the other hand, appears a little high on the basis of other evidence. Here the sources of evidence are:

1. NRP, Report A-5, giving county estimates for sixteen counties scattered through the central West.

---

[10] USDA, *Bulletin 528*, 1915, p. 13 (Appendix F, U.S.-1b).
[11] West Virginia AES *Bulletin 163*, 1917, p. 6 (Appendix F, S-23).
[12] Minnesota AES *Bulletin 179*, p. 31, and *Bulletin 157*, p. 29 (Appendix F, S.-12).

2. Illinois AES *Bulletin 50*, 1896 (Appendix F, S.-6), p. 50, giving figures for 16,600 acres, including a sampling of every county in the state.

3. A survey of several thousand acres in Iowa, Illinois, and Indiana (USDA, *Bulletin 1,000*, 1919, pp. 5–11; see Appendix F, U.S.-1b).

4. Scattered reports (five) given by farmers.

By averaging these data taken by counties or region and omitting the scattered reports, the following values are obtained (see Table D-15):

n	$\bar{X}$	σ	s
28	6.67	1.6	0.31

The labor time required in method 2 is less than that in method 1 largely because time is not used in pulling the shock apart to get to the ears of corn. In method 2, husking time is not included in our data, since the corn is husked as used over the year. Storing corn in the husks in the South is said to help protect it against insect damage.[13] In any case, the husking time must be accounted negligible. The labor at odd moments over the year has little value, the husks are thoroughly dried, and the rather small variation of labor time in methods 1 and 3 with yield per acre indicates that even when the corn is only partially dried, husking itself takes only a small part of the time spent in moving over the field, finding and pulling the ears from the plant. The omission of the husking operation and the much lower yields of shelled corn in the South, reduce the harvesting time in method 2 below the western (method 3) standard. The average of thirteen NRP sample counties in the cotton region[14] shows (see Table D-15):

n	$\bar{X}$	σ	s
13	4.3	0.6	0.2

In Table D-10, the period 2 coefficients are used without change for the regions to which the methods apply in period 1. Hence, the only improvement shown is the substitution of method 3 for method 1 in the West between the two periods. Even this improvement is partly offset in our figures by inclusion of a cost item for clearing the field of stalks in method 3, while this operation in method 1 is charged to the value of

---

[13] USDA Office of Experiment Stations, *Bulletin 173*, p. 33 (Appendix F, U.S.-1k).

[14] USDA, *Bulletin 1181*, 1924, p. 15 (Appendix F, U.S.-1b), gives a higher figure (6.1 as an average of seven counties) for a standardized yield of 25 bushels per acre in Arkansas in 1924, but the higher yield and restricted area of this study indicate that it should not be included in our data.

TABLE D-10

PERIOD 1 AND 2: HARVEST LABOR FOR CORN[a]

	Period 1		Period 2	
	Method[b]	Man-Hours Per Acre	Method[b]	Man-Hours Per Acre
R_1	1	13.0	1	13.0
R_{2a} (excl. Tennessee)	1	13.0	1	13.0
Tennessee	2	4.3	2	4.3
R_{2b}	2	4.3	2	4.3
R_3 (excl. Ohio, W:WD, and North Dakota)	1	13.0	3	6.7
Ohio, W:WD, and North Dakota	1	13.0	1	13.0

[a]The harvest labor coefficients b in Table 1 for R_{2a} in periods 1 and 2 and for R_3 in period 2 are the weighted regional averages.

[b]Method 1, husked from shock; method 2, snapped; method 3, husked from standing stalk.

fodder production. The direct data from period 1 for the Northeast give much higher figures per acre, but include hauling and housing, and in some cases probably cutting and shocking as well. They are based largely on premium reports with yields two to four times the regional average. In Maryland and Tennessee, seven cases from plantation manuscripts show costs of 0.8 to 1.8 man-hours per bushel for unknown yields, presumably for harvesting from the shock, and probably including hauling.

In the eastern cotton region, thirty-two manuscript cases for the operation of "gathering" or picking give an average of 6.6 man-hours per acre (Table D-11), including in at least one case, and probably in others, hauling and housing.

TABLE D-11

PERIOD 1: HARVEST LABOR, CORN

	ME	S:EC	S:DC	W:C
n	2.0	32.0	2.0	14.0
$\bar{X}$	20.1	6.6	31.5	11.3
M		6.1		11.6
σ		3.02		3.7
s		0.54		1.03

In the corn region of the West, fourteen cases from contemporary sources designated harvesting, gathering, or husking (including hauling and cribbing in some cases) average 11.3 man-hours per acre. The method of harvesting is not known, and seven extreme cases of 20–40 man-hours per acre are omitted. Yields in all these cases run two to four times the regional average.

PERIOD 1: POSTHARVEST LABOR, WHEAT AND OATS

In postharvest labor (threshing) only the size of the crop and method of threshing affect labor cost. For threshing, the flail, the treading floor, and small hand- or horse-powered machines were all in common use. For 1838–61, our contemporary sources for wheat, based largely on Virginia and Illinois manuscripts, yield a bimodal series of fifteen items bunched between 0.47–0.74 and 1.00–1.50 man-hours per bushel. If these two groups correspond to the ranges of treading floor and machine techniques, or the hand method, then the median (0.73) is the upper limit of methods[15] other than flail and winnowing sheet.

For oats, the mix of techniques may have been somewhat less labor-saving since a portion was grown for feed on farms which grew no wheat. But the lighter weight and larger grain of oats make it easier to thresh by any technique. One source for the flail and sheet method gives the same labor for an acre of oats at 40 bushels per acre as for an acre of wheat at 20 bushels.[16] On this basis, an estimate of 0.40 man-hours per bushel for oats appears to be comparable to one of 0.73 for wheat.

[15] Though stated somewhat unclearly, Rogin's data appear to yield the following rough estimates for the various techniques:

	Technique		Man-Hours Per Bushel of Wheat		
	Threshing	Winnowing	Threshing	Winnowing	Total
1.	flail	sheet	1.00	0.30	1.30
2.	flail	hand-mill	1.00	0.06	1.06
3.	treading	sheet	0.37	0.30	0.67
4.	treading	hand-mill	0.37	0.06	0.43
5.	machine-horsepowered				0.30
6.	machine-horsepowered				0.11

Source: Rogin (See Appendix E, B-7), pp. 176–191.
The figure 0.73 is just the midpoint of the range of techniques 1 to 5, as estimated by Rogin.

[16] Source: See Appendix E, U.S.-4, Vol. II, pp. 446, 470. The estimate used here for winnowing with a sheet is considered high by Rogin.

TABLE D-12

PERIOD 2: PREHARVEST LABOR FOR WHEAT, OATS, AND CORN

	NE	ME	S:EC	S:DC	S[a]	W:C	W:WD	W:SG + WC[b]	W:R	W:NW + Cal.	W:NW + Cal.
					WHEAT						
n	4.0				9.0	14.0	10.0	78.0	7.0	23.0	29.0[c]
$\bar{X}$	11.6				10.7	5.5	6.1	4.2	6.0	3.1	3.2[c]
M	12.5				10.0	5.4	6.2	4.1	5.7		3.3[c]
σ	2.2				2.8	1.7	1.5	1.6	1.7		1.1[c]
s	1.2				0.97	0.47	0.49	0.19	0.7		0.2[c]
					OATS						
n	10.0				10.0	15.0	15.0	12.0	8.0	6.0	
$\bar{X}$	9.3				9.5	3.3	6.1	3.0	7.6	4.3	
M	9.8				9.3	2.7	6.0	2.95	7.8		
σ	1.8				2.4	1.9	1.8	0.9	2.3		
s	0.6				0.8	0.5	0.47	0.27	0.86		
					CORN						
n	34.0	12.0	10.0	11.0	8.0	56.0	18.0	21.0			
$\bar{X}$	46.4	26.7	26.9	29.1	11.4	15.2	17.3	11.4			
M	34.4	25.6	24.75	28.0	10.2	13.75	14.75	9.5			
σ	30.9	5.6	5.0	4.8	4.6	4.9	5.8	5.3			
s	5.4	1.7	1.7	1.5	1.7	0.6	1.4	1.2			

[a] For corn, the region is S:WC.
[b] For corn, the region is W:SG.
[c] For wheat and oats combined.

TABLE D-13

PERIOD 2: PREHARVEST AND HARVEST LABOR, DERIVATION OF W FROM WESTERN SUBREGIONS

	Wheat			Oats			Corn	
	Preharvest		Harvest	Preharvest		Harvest	Preharvest	
Subregion	w (Ar/Aw)	a (L/A)	b (L/A)	w (Ar/Aw)	a (L/A)	b (L/A)	w (Ar/Aw)	a (L/A)
W:C	0.236	5.5	3.0	0.443	3.3	2.6	0.610	15.2
W:WD	0.112	6.1	3.0	0.216	6.1	3.5	0.072	17.3
W:SG + W:WC[a]	0.543	4.2	1.8	0.291	3.0	1.7	0.307	11.4
W:R	0.017	6.0	7.5	0.014	7.6	8.3	0.011	11.4[b]
W:NW + Cal.	0.092	3.2	2.0	0.035	3.2	2.0		
West (Σaw)		4.66			3.88			14.2
(Σbw)			2.33			2.5		

Source: Weights, from USDA, revised estimates; labor requirements from Tables D-12, D-15.

[a] For corn, W:SG only.
[b] No data available. W:SG figure.

TABLE D-14

PERIOD 2: PREHARVEST LABOR FOR CORN, DERIVATION OF S FROM SOUTHERN SUBREGION

Subregion	w (Ar/Aw)	a (L/A)
S:EC	0.393	26.9
S:DC	0.218	29.1
S:WC	0.389	11.4
South (Σaw)		21.3

Source: Weights, from USDA, revised estimates; labor requirements, from Table D-12.

TABLE D-15

PERIOD 2: HARVEST LABOR

	NE	S	W:C	W:WD	W:SG + W:WC[a]	W:R	W:NW + Cal.
WHEAT
n	7.0	8.0	21.0	10.0	44.0	7.0	10.0
$\bar{X}$	3.0	3.0	3.0	3.0	1.8	7.5	2.0
M	2.9	3.2	2.7	2.1	1.7	8.2	1.9
σ	0.8	1.5	1.0	1.5	0.8	2.1	0.8
s	0.26	0.59	0.23	0.5	0.12	0.85	0.26
OATS
n	10.0	10.0	14.0	15.0	11.0	9.0	10.0
$\bar{X}$	3.4	4.5	2.6	3.5	1.7	8.3	2.0
M	2.9	4.4	2.2	2.6	1.4	8.3	1.9
σ	1.4	1.9	0.9	2.0	0.5	1.4	0.8
s	0.35	0.64	0.25	0.53	0.17	0.5	0.26
CORN
n		13.0[b]			28.0[c]		
$\bar{X}$		4.3			6.7		
M		4.3			6.2		
σ		0.6			1.6		
s		0.2			0.3		

[a] For corn, the region is W:C + W:SG.
[b] Method 2.
[c] Method 3. See notes on period 1 harvest labor.

PERIOD 2: POSTHARVEST LABOR, WHEAT AND OATS

The estimates of labor per acre, shown in Table D-16, present some superficial anomalies. In wheat, the lower average for W:SG + W:WC than for the eastern regions is due to the lower yield and to a number of cases where threshing was done from the header stack rather than the shock. The NRP sample cases (Appendix E, U.S.-5) show nearly uniform use of the header in western Kansas and a noticeable proportion (10–30 per cent) throughout W:SG and west of it. Technical manuals disagree on whether grain harvested by a header threshes more easily than reaped grain, but the AES studies seem unmistakable. Though methods are not stated, the use of the header in Minnesota probably accounts for the low figure in W:WD. Use of the combine in wheat is confined largely to a portion of W:NW, and is neglected in these estimates. In W:R, the high figure is not easy to explain, but the weight is negligible.

In oats, the higher figure for NE is due largely to use of NRP figures based on barn threshing. If, as seems likely, that was a common practice in this region, it would seem desirable to charge the labor time in storing the grain to this operation. In the South, barn threshing was the major technique in the two NRP survey counties, but the lower yields may partly account for the lower per acre figure.

The use of regional average yields with per acre figures from sample studies is a weak feature of these estimates. To check them, we have compared them with three other estimates:

1. Average labor per bushel, by region, in those of our cases where both yield and labor per acre are given.
2. NRP reports of data by county, divided by counties' "normal yield."
3. NRP estimates by region, based on its cases, divided by regional yield.

The results were as follows:

	Wheat					Oats		
	NE	S	W:C	W:SG + W:WC	W:NW + Cal.	NE	W:C	W:SG + W:WC
Table D-17, line 3	0.195	0.287	0.224	0.189	0.165	0.234	0.092	0.094
1.	0.075	0.230	0.215	0.125				
2.	0.334	0.256	0.221	0.195	0.110	0.258	0.116	0.085
3.	0.334	0.256	0.219	0.213	0.092	0.250	0.103	0.078

TABLE D-16

PERIOD 2: POSTHARVEST LABOR PER ACRE

	NE	S	W:C	W:WD	W:SG + W:WC	W:R	W:NW + Cal.
				WHEAT			
n	5.0	8.0	20.0	8.0	31.0	1.0	7.0
$\bar{X}$	3.42	3.59	3.54	2.53	2.28	4.2	3.21
M	3.3	3.5	3.4	2.4	2.1		3.9
σ	0.9	1.1	1.1	0.5	0.9		1.1
s	0.45	0.41	0.25	0.12	0.16		0.46
				OATS			
n	6.0	8.0	9.0	11.0	8.0	1.0	a
$\bar{X}$	6.96	4.07	2.62	2.91	2.35	4.23	
M	7.5	3.95	2.5	3.2	2.4		
σ	1.9	1.5	1.0	0.8	0.8		
s	0.87	0.55	0.35	0.26	0.34		

[a] No figures available. Wheat figure used in our calculations.

TABLE D-17

PERIOD 2: POSTHARVEST LABOR, DERIVATION OF L/O

	NE	S	W:C	W:WD	W:SG + W:WC	W:R	W:NW + Cal.
				WHEAT			
L/A	3.42	3.59	3.54	2.53	2.28	4.20	3.21
O/A	17.5	12.3	15.8	15.3	12.0	18.6	19.4
L/O	0.195	0.292	0.224	0.165	0.190	0.226	0.165
				OATS			
L/A	6.96	4.07	2.62	2.91	2.35	4.23	3.21
O/A	29.7	17.0	28.4	26.8	22.2	32.7	33.8
L/O	0.234	0.239	0.092	0.108	0.094	0.129	0.095

Source: L/A, Table D-16. O/A, USDA, revised estimates, 1907-11. L/O, $L/A \div O/A$.

TABLE D-18

PERIOD 2: POSTHARVEST LABOR, DERIVATION OF WESTERN REGION FROM SUBREGIONS

	Wheat		Oats	
Region	v (O_R/O_W)	c (L/O)	v (O_R/O_W)	c (L/O)
W:C	0.267	0.224	0.475	0.092
W:WD	0.123	0.165	0.219	0.108
W:SG + W:WC	0.465	0.190	0.244	0.094
W:R	0.023	0.226	0.017	0.129
W:NW + Cal.	0.122	0.165	0.045	0.095
$W = (\Sigma cv)$		0.193		0.098

Source: v, USDA, revised estimates, 1907-11.
c, Table D-17.

Except for the first column, our estimates are not implausible, considering the small number of cases and the possibilities of variation. In the first column, the NRP figure (lines 2 and 3) is based on cases from a single county (Lancaster County, Pennsylvania) and may be disregarded.

Appendix E: Source Bibliography for Period 1

U.S. DOCUMENTS

1. Commissioner of Patents, *Annual Report*, "Agriculture," Washington, 1841-61.
2. Commissioner of Agriculture, *Annual Report*, Washington, 1862-81.
3. Dept. of Agriculture, *Annual Report*, Washington, 1882.
4. Commissioner of Labor, *13th Annual Report*, Washington, 1898.
5. Works Progress Administration, National Research Project (NRP), Changes in Technology and Labor Requirements in Crop Production, Report A10, *Wheat and Oats*, Phila., Apr. 1939.
6. *Idem.*, Report A5, *Corn*, Phila., June, 1938.
7. Dept. of Agriculture, Bureau of Agricultural Economics, *Revised Estimates of Wheat Acreage, Yield and Production, 1866-1929*, Washington, July 1934.
8. *Idem.*, *Revised Estimates of Oats Acreage, Yield and Production, 1866-1929*, Washington, July 1934.
9. *Idem.*, *Revised Estimates of Corn Acreage, Yield and Production, 1866-1929*, Washington, May 1934.
10. Censuses of Agriculture, 1840-1910, Washington.

STATE DOCUMENTS

1. Illinois: *Transactions*, Ill. State Agricultural Society, Springfield, 1853–60.
2. Indiana: *Annual Report*, Ind. State Board of Agriculture, Indianapolis, 1848–60.
3. Maine: 1853–55, *Transactions*, Agricultural Societies in the State of Maine, Augusta.
   1856–75, *Annual Report*, Secretary of the Maine Board of Agriculture, "Abstract of Returns of Agricultural Societies."
4. Massachusetts: 1837, *Report on the Agriculture of Massachusetts*, Boston; 1853–75, *Annual Report*, Secretary of the Board of Agriculture, with Reports of Committees Appointed to Visit the County Societies.
5. Michigan: *Transactions*, Mich. State Agricultural Society, Lansing, 1849–61.
6. New York: *Transactions*, N.Y. State Agricultural Society, Albany, 1842–75.
7. Ohio: 1850–56, *Annual Report*, Board of Agriculture; 1856–75, *Annual Report*, Board of Agriculture, with an abstract of Proceedings, County Agricultural Societies, Columbus.
8. Tennessee: *Biennial Report*, State Agricultural Bureau, Nashville, 1855–58; *Transactions*, 1854–59.

PERIODICALS

(1) *Agriculturist*, Nashville; (2) *American Farmer*, Baltimore; (3) *Carolina Planter*, Columbia (S.C.); (4) *Country Gentleman*, Albany; (5) *Cultivator*, Albany; (6) *Farmer and Gardener*, Baltimore; (7) *Farmer and Planter*, Pendleton; (8) *Farmers' Cabinet*, Philadelphia; (9) *Illinois Farmer*, Chicago; (10) *Journal of Agriculture*, St. Louis; (11) *Maine Farmer*, Augusta; (12) *New Genesee Farmer*, Rochester; (13) *Pennsylvania Farm Journal*, Lancaster; (14) *Prairie Farmer*, Chicago; (15) *Soil of the South*, Columbus (Ga.); (16) *Southern Agriculturist*, Laurensville (S.C.); (17) *Southern Cultivator*, Columbia (Tenn.); (18) *Southern Planter*, Richmond; (19) *Tennessee Farmer*, Jonesborough (Tenn.); (20) *Union Agriculturist*, Chicago; (21) *Valley Farmer*, St. Louis; (22) *Western Farmer*, Cincinnati.

MANUSCRIPTS

Library	Title	County and State
1. Virginia Historical Society, Richmond	Diary of Robert Henderson Allen	Lunenburg, Va.
2. Southern Historical Collection, Univ. of N.C., Chapel Hill	Plantation Journal of John D. Ashmore	Anderson and Sumter, S.C.
3. Department of Archives, Louisiana State Univ., Baton Rouge	Eli J. Capell Plantation Diaries and Record Books	Amite, Miss.
4. Maryland Historical Society, Baltimore	H. D. G. Carroll, "The Perry Hall Farm Journal"	Baltimore, Md.
5. Virginia Historical Society, Richmond	Papers of Phillip St. George Cocke, "Belmead Plantation"	Powhatan, Va.

Library	Title	County and State
6. Univ. of Georgia, Athens	Wm. J. Dickey Diaries, "Birdsong Plantation"	Hancock, Ga.
7. Univ. of Illinois, Urbana	M. L. Dunlap's Ledger	Cook, Ill.
8. Louisiana State Univ., Baton Rouge	Ferchaud Papers	St. James Parish, La.
9. South Carolina Library, University of S.C., Columbia	Samuel P. Gaillard Plantation Journal	Sumter, S.C.
10. College of William and Mary, Williamsburg, Va.	James Galt: Diary and Plantation Memoranda	Fluvanna, Va.
11. Maryland Historical Society, Baltimore	Gittings Account Book, "Roslin Farm"	Baltimore, Md.
12. Georgia Dept. of Archives and History, Atlanta	Seaborn Hawks Farm Journal	Jasper, Ga.
13. Univ. of Georgia, Athens	John B. Lamar Plantation Book	Sumter, Ga.
14. Louisiana State Univ., Baton Rouge	Liddell Plantation Book, "Llanada Plantation"	Concordia Parish, La.
15. Louisiana State Univ., Baton Rouge	The Marston (Henry W. and family) Papers, Plantation Diary	Feliciana Parish, La.
16. Southern Historical Collection, Univ. of N.C., Chapel Hill	Farm Journal of Nicholas Massenburg	Franklin, N.C.
17. Virginia State Library, Richmond	Farm Journal of William Massie	Nelson, Va.
18. The Filson Club, Louisville	The Howard Miller Diary	Jefferson, Ky.
19. Southern Historical Collection, Univ. of N.C., Chapel Hill	The Norfleet Diaries	Bertie, N.C.
20. Library of Congress, Washington, D.C.	The Physick Family Papers	Cecil, Md.
21. Univ. of Maryland, College Park	The John Piper Family Record and Time Book	Allegany, Md.
22. Virginia State Library, Richmond	The Ruffin Papers	Hanover, Va.
23. The Filson Club, Louisville, Ky.	Robert W. Scott Diary	Franklin, Ky.
24. Hall of Records, Annapolis, Md.	John H. Sellman Diary	Anne Arundel, Md.
25. The Filson Club, Louisville	Deacon's Journal of the Shaker Community	Mercer, Ky.
26. National Agricultural Library (USDA), Washington	E. J. Tayloe Agricultural Journal	King George, Va.
27. Virginia State Library, Richmond	L. W. Tazewell Plantation Book	James City, Norfolk, Northampton, Va.
28. Georgia Dept. of Archives and History, Atlanta	Diary, Ledger, and Account Book of James Washington Watts	Bartow (formerly Cass), Ga.

BOOKS

1. John Spencer Bassett (ed.), "The Westover Journal of John A. Seldon, Esq., 1858–1862," *Smith College Studies in History*, Vol. VI, No. 4, Northampton, Mass., July 1921.
2. P. D. Coburn, *Swine Husbandry*, New York, 1877.
3. Edwin A. Davis, "Plantation Life in the Florida Parishes of Louisiana, 1836–1846 as Reflected in the *Diary of Bennet H. Barrow*," New York: Columbia University Press, 1943.
4. F. Gerhard, *Illinois as It Is*, Chicago, Philadelphia, 1857.
5. George N. Lamphere, "History of Wheat Raising in the Red River Valley," *Minnesota Historical Society Collections*, Vol. 10, pt. 1, St. Paul, 1905.
6. Frank L. Riley (ed.), "Diary of a Mississippi Planter," *Publications of the Mississippi Historical Society*, X, Oxford, 1909.
7. Leo Rogin, *The Introduction of Farm Machinery in its Relation to the Productivity of Labor in the Agriculture of the United States during the Nineteenth Century*, Univ. of California Publications in Economics, Vol. 9, Berkeley, Univ. of California Press, 1931.
8. Edmund Ruffin, *Essays and Notes on Agriculture*, Richmond, 1855.
9. F. Yates, *The Design and Analysis of Factorial Experiments*, Imperial Institute of Soil Science, Technical Communication 35, Harpenden, Eng., 1937.

Appendix F: Source Bibliography for Period 2

U.S. DOCUMENTS

1. Department of Agriculture
    a. Department Circular 183, Washington, 1922.
    b. Department Bulletins 3, 214, 218, 219, 412, 482, 528, 595, 757, 814, 917, 943, 961, 1000, 1181, 1198, 1296, 1421, 1446, Washington.
    c. Bureau of Agricultural Economics, Divisions of Cost of Production and of Farm Management, Preliminary Report, *Cost of Wheat Production and Incomes for Farming*, Washington, 1923.
    d. *Yearbook*, 1917, 1923, Washington.
    e. Division of Statistics, Bulletin 20, Washington, 1901.
    f. Office of Farm Management, *Farm Management Monthly*, Vol. II, Washington, 1914.
    g. Office of Farm Management, *Farm Management Circular 3*, Washington, 1919.
    h. Special Report 40, Washington, 1882.
    i. Agricultural Marketing Service, Crop Reporting Board, *Corn Acreage, Yield and Production*, Washington, 1954.
    j. Monthly Crop Report, Washington, 1915.
    k. Office of Experiment Stations, Bulletin 173, Washington, 1907.
2. WPA-NRP: See Appendix E, items U.S.-5, and U.S.-6.
3. Commissioner of Labor, *13th Annual Report*, Washington, 1898.
4. *Hearings Before the Senate Committee on Agriculture and Forestry*, 65th Cong., 2d Sess., Washington, 1918.
5. *Hearings Before the Joint Commission of Agricultural Inquiry*, 67th Cong., 1st Sess., Washington, 1922.

STATE DOCUMENTS

1. Alabama: Alabama Polytechnic Institute Extension Service, Circular 33, Auburn.
2. California: Univ. of California, College of Agriculture Agronomy Project 337, Berkeley.
3. Colorado: State Agricultural College of Colorado, AES Bulletins, Fort Collins.
4. Georgia: Georgia State College of Agriculture, Extension Division, Bulletins 270, 273, and 428, Athens.
5. Idaho: Univ. of Idaho, AES Bulletins 123 and 195, Moscow.
6. Illinois: Univ. of Illinois, AES Bulletins 50 and 277, Urbana.
7. Iowa: (1) Iowa Department of Agriculture, *Sixth Annual Iowa Year Book of Agriculture*, Des Moines, 1905. (2) Iowa AES Bulletin 16, Ames. (3) Iowa Bureau of Labor Statistics, Fourth Biennial Report of the Commissioner of Labor, Des Moines, 1891.
8. Kansas: (1) *Report*, State Board of Agriculture, Topeka, 1920. (2) *Ninth Annual Report*, Bureau of Labor Statistics, Topeka, 1893.
9. Maine: (1) Maine Board of Agriculture, *Twenty-Second Annual Report of the Secretary*, Augusta. (2) Univ. of Maine, College of Agriculture, M.D. Jones, *The Cost of Producing Sweet Corn*, Orono, 1920.
10. Massachusetts: Massachusetts State Board of Agriculture, Abstract of Returns of the Agricultural Societies of Massachusetts with the *Twenty-Eighth Annual Report* of the Secretary, Boston, 1880.
11. Michigan: *19th* and *27th Annual Report*, State Board of Agriculture, Lansing.
12. Minnesota: Univ. of Minnesota, AES Bulletins 157 and 179, St. Paul.
13. Missouri: Univ. of Missouri, AES Bulletins 125 and 165; Research Bulletin 6; Circular 100, Columbia.
14. Montana: (1) Univ. of Montana, AES Bulletins 116 and 122, Bozeman. (2) Montana Extension Service in Agriculture and Home Economics, Bulletin 71, Bozeman.
15. Nebraska: Univ. of Nebraska, AES Bulletin 29, Lincoln.
16. New Jersey: (1) *Thirty-Third Annual Report*, New Jersey State AES, Union Hill, 1912. (2) New Jersey AES Bulletin 312, New Brunswick.
17. New York: (1) State College of Agriculture, Experiment Station Bulletins, Cornell Station, 377, 414, and 475, Ithaca. (2) New York State College of Agriculture, Dept. of Agricultural Economics and Farm Management, *Farm Cost Accounting Project*, Ithaca, 1931. (3) New York Dept. of Agriculture, Bulletin 86, Albany.
18. North Dakota: (1) North Dakota Agricultural College, AES Bulletins 142 and 144, Agricultural College. (2) *Third Biennial Report*, Commissioner of Agriculture and Labor, Bismarck.
19. Ohio: (1) Ohio State Univ. Agriculture Extension Service, Vol. 18, No. 5, Colombus. (2) Ohio AES Bulletin 266, Wooster. (3) AES Bimonthly Bulletin, Vol. 12. (4) Monthly Bulletin Vol. 3, Wooster.

20. South Carolina: South Carolina Agricultural Experiment Station of Clemson College, Bulletin 221, Clemson.
21. Utah: Utah Agricultural College Experiment Station, Bulletin 165, Logan.
22. Washington: State College of Washington, AES Bulletins 175 and 244, Pullman.
23. West Virginia: West Virginia AES Bulletins 163 and 187, Morgantown.
24. Wisconsin: Bureau of Labor and Industrial Statistics, *9th Biennial Report*, 1897–98, Madison.

PERIODICALS

(1) *Atlantic Monthly*, Boston; (2) *Breeders Gazette*, Spencer, Indiana; (3) *Indiana Farmer*, Huntington, Indiana; (4) *National Stockman and Farmer*, Pittsburgh, Buffalo; (5) *Orange Judd Farmer*, Chicago; (6) *Pennsylvania Farmer*, Mercer, Philadelphia, Pittsburgh; (7) *Practical Farmer*, Camden, New Jersey; (8) *Rural New Yorker*, New York; (9) *Southern Planter*, Richmond; (10) *Tribune Farmer*, New York; (11) *Utah Farmer*, Salt Lake City; (12) *Wallace's Farmer*, Des Moines, Ames, Iowa.

BOOKS

1. *North Dakota State Historical Society Collections*, Vol. III, Bismarck, 1910.
2. Howard M. Eliot, *The Elementary Principles of Farm Management*, Ann Arbor, Mich., 1923.
3. Melville L. Bowman and Bruce W. Crossley, *Corn*, Ames, Iowa, 1908.
4. George D. Leavens, *Corn*, New York, 1915.
5. Herbert Myrick, *The Book of Corn*, New York, Chicago, 1903.

COMMENT

Glen T. Barton, U.S. Department of Agriculture

Output per man-hour in the production of wheat and oats in the early 1900's was about four times as great as in the 1840's. In their paper the authors set as their objective the allocation of this increase in labor productivity among three broad sources: (1) westward movement of the crop; (2) changes in yields per acre; and (3) improvements in mechanization and other practices which reduce labor inputs per acre. Three broad regions, the Northeast, South, and West, were delineated for the analysis. The authors concluded that the bulk of the rise in labor productivity was due to mechanization of harvesting operations in all the producing regions and that very little improvement resulted from westward movement of the crop or from change in yields.

I have no major quarrel with the authors' conclusions. They are to be complimented on their careful, painstaking analysis, and especially for

NOTE: This comment was prepared for an earlier version of the Parker paper in which the data on corn were omitted.

the development of methodology which should prove useful to other research workers.

Study of the paper raised several points in my mind, some by way of constructive criticism of the paper per se, others regarding degree of emphasis or points made by the authors and, perhaps most important, facets omitted or not developed fully by the authors.

1. Despite the careful and thorough assembly of data and the rigor of the authors' analysis, quantitative conclusions reached in the paper should be regarded as broad indications, rather than as precise measurements. As the authors recognize, the basic data used, especially those on labor inputs per acre for wheat and oats, leave much to be desired. There is a strong suspicion that the labor coefficients assembled generally reflect operations on farms with above-average management practices. Also, it is generally recognized that estimates of acreage and yield of the two grain crops in the early part of the period studied are not as accurate as most of our agricultural data today.

2. From a statistical point of view, the authors are quite correct in concluding that little change occurred in yields of the two grains during the period of analysis. However, I want to give even more emphasis than they did to the importance of technological improvement in crops during the period. Even today, the small grain crops pointedly illustrate the crucial importance of variety improvements in preventing sharp reductions in yields because of persistent, potential ravage by diseases and insects.

3. In view of my own research background in the field of "productivity" measurement, I believe the authors were somewhat hasty in their ready acceptance of average labor productivity as a measuring device. Admittedly, lack of data would have prevented any major attempt to examine changes in output-input ratios of other factors of production. The authors' conclusion that mechanization of harvest operations was by far the dominant influence in raising production of wheat and oats per man-hour directly implies that substitution of capital for labor may have been an important influence during the period analyzed.

4. The authors rigidly adhered to their stated purpose of measuring *sources* of improvement in labor productivity in small grain production. However, I wish they had engaged in some speculation as to *causes* of these changes in terms of probable economic, social, and institutional factors. This observation leads me to my final point.

5. Granting the authors' thorough and workmanlike job of data collection and analysis and the general accuracy of their conclusions, just what does the paper contribute to our body of useful economic intelligence?

More attention to causes of changes in labor productivity might have provided some basis for projections of what lay ahead in the twentieth century. Data available in the Economic Research Service, for example, indicate that production per man-hour of wheat in 1961-62 was 9 times that prevailing in 1910–14. Present-day output of oats per man-hour is 5.5 times that of the early 1900's. These improvements in production per man-hour are substantially greater than those that occurred from the 1840's to the early 1900's.

If as a result of our analysis of given historical periods we can develop better frameworks for looking ahead, both economists and historians can better serve their function as social scientists. Such analysis also should contribute to a more complete understanding of the forces behind economic growth. Obviously, contributions of this sort are badly needed in providing better guidelines for policy decisions in the many underdeveloped and developing economies throughout the world.

Productivity and Technological Change in the Railroad Sector, 1840–1910

ALBERT FISHLOW
UNIVERSITY OF CALIFORNIA, BERKELEY

Schumpeter once expressed the opinion that the economic history of the United States in the second half of the nineteenth century could be written solely in terms of the railroad sector. Clearly an exaggeration, the observation still has the merit of focusing upon one of the dynamic forces in American economic development. Yet our measure of that force is severely restricted. Our only continuous railway series is mileage operated, and such relevant magnitudes as ton and passenger mileage, employment, and capital stock are not available, even in imperfect form, until the 1880's, well after the phase of most rapid growth.

The first and principal part of this paper seeks to repair these omissions by setting out a continuous quantitative record of development for the railroad sector from its beginnings in the 1830's to its peak just before the First World War. The detailed statistics compiled by the Interstate Commerce Commission carry the record forward more than adequately from that time on.

Four series are constructed: output, employment, capital stock, and fuel consumption. Despite their obvious diversity, as well as the differences in the quality of the underlying information, there is a unity in the approach that should be noted. In the great majority of instances, the estimates have been reached by two alternative routes. The first is by direct manipulation of physical units—ton- and passenger-miles, tons of coal, track mileage, and so forth—whereas the second starts from the financial accounts reflecting these magnitudes—traffic receipts, fuel expenditures, book value of road and equipment—but has the same goal. The obvious

NOTE: I gratefully acknowledge the painstaking care with which the original draft of this paper was read by Richard A. Easterlin. His comments and suggestions have contributed much to the final form the paper has taken, although he must be absolved of any responsibility for the shortcomings that remain.

virtue of this duality is an internal check upon the final estimates. However, there are further gains as well. It is the inconsistency between movements in the trackage and equipment index and Ulmer's investment series that calls attention to the inadequacies in our existing dollar measures of investment. This is an instance where the check is not internal, but relates to other previous estimates. A second advantage of the physical approach to the capital stock—spelled out in greater detail below—is its avoidance of some particularly nasty deflation problems. Cumulative railroad investment can be interpreted unequivocally as the reproduction cost of the current-year stock with base-year prices and technology, but current-year quality. Ordinary price indexes, built up of quotations on an ever-changing unit, when applied to current expenditures, do not produce the same straightforward result.

The second part of the paper goes on from these estimates to consider the record of productivity change implied by them. It both corrects Kendrick's post-1869 measures of productivity in the railroad sector, and extends those measures considerably backward in time. Then, in an effort to explain the observed pattern of increasing productivity, we take up the contribution of certain key innovations of the latter nineteenth century: steel rails, automatic couplers, power brakes, improved rolling stock. Of more general interest in this section is the use of a simple technique for integrating the up-to-now distinct approaches from the side of sectoral productivity measures, and from the side of specific technological developments.

Quantitative Measures of Railroad Development, 1840–1910: Output

Table 1 sets out estimates of passenger- and ton-miles, and the rates that make possible their combination into a single output index. This index is of the link-relative form, one familiar to users and producers of statistics covering relatively long spans of years.[1] The virtue of the method is the comparability it affords between adjacent observations when structural change is quite rapid—in this instance, the relative importance of freight and passenger service, and their rates. Laspeyres or Paasche indexes freeze the price structure or the output mix at that of a given year, with the consequence that they are very sensitive to the choice of the base.[2]

[1] See, for example, John W. Kendrick, *Productivity Trends in the United States*, Princeton for NBER, 1961, p. 55, for a discussion of the link-relative approach and its use in that study.

[2] Although the Paasche index is currently weighted in the sense that prices of the current year are used to evaluate output of the current and the base year, there are different results when alternative bases, i.e., output mixes, are used.

TABLE 1

PHYSICAL OUTPUT ESTIMATES, 1839-1910

Fiscal Year[a]	Passenger-Miles (billions) (1)	Ton-Miles (billions) (2)	Passenger Rate (cents) (3)	Freight Rate (cents) (4)	Output Index (1910=100) (5)
1839	0.09	0.03	5.0	7.5	0.04
1849	0.47	0.35	2.9	4.05	0.31
1859	1.9	2.6	2.44	2.58	1.74
1870	4.1	11.7	2.8	2.18	6.03
1880	5.7	32.3	2.51	1.29	13.78
1890	12.1[b]	80.0[b]	2.20	.92	32.79
1900	16.2	144.0	2.00	.73	54.79
1910	32.5	255.0	1.94	.75	100.00

Source

Cols. 1-4
1839-59: Albert Fishlow, *American Railroads and the Transformation of the Ante Bellum Economy* (in press), Cambridge, Mass., Appendix A.
1870: Interpolated on sample series between 1859 and 1880; see text.
1880: U.S. Bureau of the Census, *Tenth Census, 1880*, Vol. IV, *Agencies of Transportation*, p. 11.
1890: U.S. Bureau of the Census, *Eleventh Census, 1890*, Vol. XI, *Agencies of Transportation*, p. 593.
1900, 1910: Harold Barger, *Output in the Transportation Industries*, Princeton for NBER, 1951, Appendix B.
Col. 5: The formula for the link-relative index is:

$$\frac{O_t}{O_{1910}} = \frac{P_t\left(\frac{p_t+p_{t+1}}{2}\right) + T_t\left(\frac{f_t+f_{t+1}}{2}\right)}{P_{t+1}\left(\frac{p_t+p_{t+1}}{2}\right) + T_{t+1}\left(\frac{f_t+f_{t+1}}{2}\right)} \cdot \frac{P_{t+1}\left(\frac{p_{t+1}+p_{t+2}}{2}\right) + T_{t+1}\left(\frac{f_{t+1}+f_{t+2}}{2}\right)}{P_{t+2}\left(\frac{p_{t+1}+p_{t+2}}{2}\right) + T_{t+2}\left(\frac{f_{t+1}+f_{t+2}}{2}\right)} \cdot$$

$$\cdots \cdot \frac{P_{1900}\left(\frac{p_{1900}+p_{1910}}{2}\right) + T_{1900}\left(\frac{f_{1900}+f_{1910}}{2}\right)}{P_{1910}\left(\frac{p_{1900}+p_{1910}}{2}\right) + T_{1910}\left(\frac{f_{1900}+f_{1910}}{2}\right)}$$

, where P are passenger-miles from col. 1, p the passenger rates of col. 3, T the ton-miles of col. 2, and f the freight rates of col. 4.

[a] Fiscal year for 1880-1910 ends June 30. For previous years the typical practice of individual roads was to report either on September 30 or at year's end, which would make the covered year correspond to a fiscal year ending in the autumn (cf. Edwin Frickey, *Production in the United States, 1860-1914*, Cambridge, Mass., 1947, p. 115).

[b] Coverage was extended by means of the ratios of mileage of roads reporting tonnage and passengers to mileage of roads reporting ton-miles and passenger-miles. It was assumed that the average haul per ton was half the national average for roads not reporting ton-mileage, while passenger trips were assumed to be of equivalent length. This adjustment is preferable to one based upon mileage coverage of receipts relative to output since the latter leads to average rates much smaller than those reported elsewhere. The mileage disparity between receipts and output is probably as large as it is because of incorrect tabulating of subsidiary roads rather than of actual differences in coverage.

Thus an 1839 Laspeyres (or a 1910 Paasche) yields an 1839 output of .03 (1910 = 100), whereas the 1910 Laspeyres (or 1839 Paasche) gives a result almost three times as large. The link-relative index presented functions like a weighted average of these two extremes, with weights changing over the course of the period.

As the source notes to Table 1 indicate, 1880 marks the start of official tabulation of passenger- and ton-miles. Rapidly thereafter, there was a proliferation of sources so that 1890 affords the option of three alternative estimates from *Poor's Manual*, the report of the Interstate Commerce Commission, and the Census volume, *Agencies of Transportation*. The latter was used because its coverage is more complete than that of the other two. By 1900 the ICC's official status and continuous collection of data led to the cessation of further Census inquiry, and it is these data, as adjusted by Barger, that are entered here.

For roughly half the period, therefore, no official tallies exist.[3] One survey was taken by the Secretary of the Treasury in 1856, it is true, but that effort is marred by considerable error.[4] The apparent total of 3.4 billion ton-miles for fiscal 1856 must be scaled down by about 1 billion ton-miles, the output incorrectly reported for a single New York railroad of insignificant size and a financial failure to boot. That adjustment is insufficient. For the residual is now too small because many railroads failed to return ton-miles, even when responding to other parts of the questionnaire. Stated passenger-miles, although free of gross errors, are unreliable for the same reason.

Since passenger- and ton-miles were rarely reported by individual railroads before the Civil War, and tabulated by less than a handful of states, it is necessary to proceed from receipts. The reduction to physical quantities is accomplished by dividing through by the average charge for transport services. This further information on rates is required in any event to combine passenger- and ton-miles into a single output measure.

Elsewhere I have described in some detail the derivation of the receipts and rate information for the antebellum period;[5] only a brief summary

[3] There is also a nonofficial, continuous series on ton-miles going back to 1852 developed by Carl Snyder in his *Business Cycles and Business Measurements*, New York, 1927. This has had a reasonably wide circulation as a consequence of its appearance in Joseph A. Schumpeter's *Business Cycles*, New York, 1939. Snyder extrapolated back from the 1880's on a sample series of ton-miles for the large systems. At its maximum the coverage is substantial, but for the antebellum period it is limited to four railroads at most and to only a single road for part of the 1850's. The series is sometimes useful for cyclical analysis, but the absolute levels are not always reliable, as one would anticipate from the method of derivation.

[4] It is published in the *Report on Finances for 1855-56*, S. Ex. Doc. 2, 34th Cong., 2d Sess.

[5] In Appendix A of Albert Fishlow, *American Railroads and the Transformation of the Ante Bellum Economy* (forthcoming), Cambridge, Mass.

noting the relative reliability of these estimates is presented here. The receipts data are of good quality. Less than 10 per cent of the 1839 total could not be obtained directly, and a still smaller 2 per cent in 1849 and 1859 had to be estimated. The allocation of receipts to their passenger and freight origins is slightly more inexact, but that is still of small consequence. The caliber of the passenger and freight rates is subject to more concern, especially because small absolute errors in these are transformed into large percentage deviations in the output aggregates. For 1839, a small sample of roads and the considered judgment of Franz Anton von Gerstner, a visiting Belgian engineer who chronicled early American railroad development in a very thorough fashion, must suffice.[6] By 1849 information on rates had become more abundant and a survey of charges by almost all operating railroads in 1848 forms the basis of the estimates.[7] So detailed is the source that regional, and even state, disaggregation is possible, and this contributes to greater confidence in the final output entry for that year. The 1859 rates rest upon an elaboration of the 1855-56 Treasury survey mentioned above. Exact 1855-56 rates were calculated from simultaneous output and receipts information for roads carrying 60 per cent of aggregate ton- and passenger-miles. Only the extension of these rates to 1859 affords a cause for concern, and this not a serious one. The time span is so short, and rates so stable—as sample data of individual roads and the state reports of New York and Massachusetts testify—that the potential error is minimal. Freight rates were extrapolated on a small sample of roads, chosen for their geographic representativeness, and passenger rates left unaltered except for a slight reduction in the initially above-average charges of southern railroads. All in all, the resultant 1859 output estimates are the best of the antebellum period and subject to relatively small deviations.

This elaborate procedure for the pre-Civil War years could be avoided for the 1870 estimate because it was straddled by firm benchmarks in 1880 and 1859 and a suitable interpolating series was available. By that time the five trunk lines reporting ton-miles at all three dates not only represented one-fourth of the national aggregate, but a stable proportion as well.[8] Hence interpolation is a quite satisfactory procedure here. And

[6] Gerstner visited almost every railroad in the country in the late 1830's and collected a storehouse of information, published (much of it posthumously) as a series of articles in 1839-41 in the *American Railroad Journal* and *Journal of the Franklin Institute* and as a two-volume book in German, *Die Innern Communicationen der Vereinigten Staaten von Nord Amerika*, Vienna, 1842-43. There is no question of his knowledgeability.

[7] See *American Railroad Journal*, Vol. XXI, 1848, pp. 467 ff.

[8] The sample consists of the Boston and Albany, the New York Central, the New York and Erie, the Pennsylvania, and the Pittsburgh, Fort Wayne, and Chicago railroads (see *Wholesale Prices, Wages and Transportation*, S. Rept. 1394, 52d Cong., 2d Sess., 1, pp. 618-620).

although there is no equivalent set of observations for passenger-miles, the results of extrapolating on Frickey's two sample series—one forward from 1859, the other backward from 1880—are so close that large errors are doubtful.[9] The average of the respective projections of 4.2 and 4.0 billion passenger-miles is used in Table 1. The same technique applied to ton-miles yields an estimate almost identical to that obtained by interpolation, a finding which supports the extrapolation approach and also the ton-mile estimate itself.

Logarithmic interpolation was also used for 1870 rates. Two alternative series produced quite similar estimates for the freight rate: the average New York State rate derived from the state reports for the three years, and the average rate of a sample of large railroads roughly comparable to those used in the derivation of 1870 ton-miles.[10] The absolute level was further corroborated by comparison with the average freight charge reported for thirteen trunk lines in 1870.[11] Information on passenger rates is less abundant and more confusing. The passenger-mile rate on all New York railroads shows an increase from 1859 to 1870—but only a moderate one—and a national rate derived from it stands substantially below the rate reported on the thirteen trunk lines. Since the 1880 rate of the trunk lines is comparable with the known 1880 national average, there is a disparity in 1870 between these two methods. With the trend of the New York State rate also at variance with the movement in the national rate from 1859 to 1880, the interpolated estimate is further suspect. In its stead, an average rate somewhat lower than that prevailing on the thirteen trunk lines, but higher than that indicated by the New York results, was selected.[12]

The greater uncertainties of the 1870 rate estimates have less effect on the accuracy of the final output index than might appear at first blush.

---

[9] Edwin Frickey, *Production in the United States, 1860–1914*, Cambridge, Mass., 1947, pp. 87 ff. John W. Kendrick also uses Frickey's sample to extrapolate 1880 passenger and ton-miles to an earlier date, 1869. But he does not make an adjustment for the increasing coverage of Frickey's sample over time, and hence underestimates both passenger- and ton-miles (Kendrick, *Productivity Trends*, pp. 509–510).

[10] New York Board of Railroad Commissioners, *Annual Report on the New York Railroads*, 1859, 1870, and 1880 (title varies); S. Rept. 1394, 1, pp. 615–617. The sample this time included eight railroads rather than five as before.

[11] The information on the thirteen lines was gathered by *Poor's Manual of Railroads*, and presented in various issues; see, for example, the volume for 1900, p. xlix. The average of these trunk railroads is some .12 cents less than the 2.11 actually used, a deviation whose direction is correct and whose magnitude seems to be of the right order.

[12] This procedure obtains support from a pattern of passenger-rate increase in Massachusetts during the same interval that is closer to the national trend estimated in Table 1.

In the first instance, the gross receipts conjointly implied by the output and rate estimates appear quite consistent with other information in this sphere. Thus the $388.4 million total for 1870 (including allowance for additional revenues from mail, express, etc.) compares with Poor's estimates for 1867 and 1871 of $334.0 and $403.3 million.[13] Furthermore, extrapolation of 1859 and 1880 national receipts upon Frickey's two sample receipts series yields totals of $386 and $360 million, which at worst fall within about a 5 per cent margin of our implied figure. In the second instance, since it is the relative weight of passenger charges to freight charges averaged over two decadal observations that determines the level of the output index, small changes in a single year are of small matter. As a case in point, if the passenger rate implied by the New York State series were in fact used, the index for 1859 and 1870 would be smaller by scarcely more than 1 per cent than the entries in Table 1.

In general, with the availability of official information after 1880 and the low levels of the index before that date, the advance it portrays—while subject to the usual disclaimers of perfect representation—is probably one of our more adequate long-term measures of industrial growth.

Quantitative Measures of Railroad Development, 1840–1910: Capital

The same cannot be said for existing capital estimates. For no part of the period in question are there official statistics of gross investment in current dollars, let alone the desired magnitude of the net value of road and equipment in constant dollars. Capital expenditure was first recorded by the ICC in 1912 and, although book value was collected from 1890 on, this is subject to the vagaries of financial manipulation as well as being limited by its original cost form. This lack led Melville Ulmer to estimate annual gross and net investment in constant and current dollars, as well as the capital stock, as far back as the beginning of 1870.[14] Close examination suggests serious defects in these series that make them unacceptable for use here, or in many other contexts. Since the data have already gained a wide audience, it is important to set forth in some detail the grounds for such a judgment.

The most telling objection is the inconsistency of Ulmer's constant-dollar estimates with physical series corresponding to various parts of the

[13] These figures are reprinted from their original source, *Poor's Manual*, in *Historical Statistics of the United States, Colonial Times to 1957*, Bureau of the Census, Washington, 1960, p. 428, Ser. Q-38.

[14] Melville J. Ulmer, *Capital in Transportation, Communications, and Public Utilities: Its Formation and Financing*, Princeton for NBER, 1960, Appendixes A and C.

capital stock. Thus while Ulmer's value of the capital stock in 1929 dollars increases by 167 per cent over the period 1870–1910, mileage grows by some 400 per cent (trackage even faster), and equipment goes up by a larger factor.[15] Within certain decades—the 1900's are a prominent example—the divergence is more striking. To be sure, Ulmer's series is net whereas trackage and equipment are gross (undepreciated), and this explains part of the difference. But if Ulmer's stock estimates are reconstructed gross on the most favorable terms by adding the flow of constant-dollar gross investment to the original net value of the capital stock for 1870, the growth rate is increased only to 321 per cent and continues to be significantly below that of the physical series. Since we would expect the "true" gross investment series to capture the increasingly important expenditures on stations, realignment, electrification, etc., not included in the mileage and equipment series, the shortfall would be still greater if the two series were of comparable scope.

To appreciate the defects in Ulmer's estimates that contribute to this inconsistency necessitates a brief description of his method. Ulmer starts from current-dollar additions to capital account reported by railroads in certain sample states and sample years. National investment for the sample years is equal to the sample total scaled up by the ratio of national book value of road and equipment to book value of the reporting roads; interpolation on a trackage series fills in the missing years. The total investment in current dollars during the interval 1870–1915 is then checked by comparing this sum with the change in original cost of road and equipment over the period. The two magnitudes are almost identical, a finding which Ulmer takes as confirmation both of a low level of retirements—cumulative gross investment should include expenditures for equipment replacement whereas original cost will not be affected—and of the correctness of his investment flows. The resultant annual series is converted to 1929 dollars by applying a weighted price index of input components, namely, labor, building materials, metals, and implements. A depreciation rate applicable to current-dollar investment is derived from unpublished ICC calculations and some skillful manipulation. The capital stock estimates in constant dollars follow from an adjusted ICC 1936 benchmark; values for later years are the 1936 value plus cumulative net constant-dollar investment; values for earlier years, the 1936 value less cumulative net investment.

[15] For mileage changes and the terminal stock of trackage and equipment, see *Historical Statistics*, 1960, pp. 427 ff. Trackage and equipment estimates have been made for 1870 and are presented as part of this paper.

Despite its frequent ingenuity and a welcome reliance upon the rich resources of railroad reports, Ulmer's technique is seriously marred by inadequate execution. Specifically, the cumulative gross investment flow in current dollars is both understated and subject to important distortions in its annual distribution. A dubious price deflator compounds these difficulties, and thereby contributes to the further underestimate of the growth of the capital stock from 1870 to 1910.

Current-dollar gross investment is not large enough, principally because Ulmer misinterprets nineteenth century railroad accounting categories. Until the ICC regulations of 1907 instituted depreciation accounting, the practice of replacement accounting was almost universal.[16] With the latter, equipment replacements—that is, retirements—were charged to current operating expenses at the time expenditures were made, and no entry ever appeared in the capital account. Hence Ulmer's exclusive focus on the capital account excludes one component of gross investment.

Were retirements as modest as the $239 million Ulmer finds during the period 1870–1914, this objection, however valid, would not be vital. The point is that this estimate of retirements and the simultaneous check upon the magnitude of cumulative gross investment are both far off the mark. The basic identity involved, which can be written $\sum_{0}^{\tau}$ gross investment$_t$ ≡ book value of road and equipment$_\tau$ minus book value of road and equipment$_0$ minus $\sum_{0}^{\tau}$ write-ups$_t$ plus $\sum_{0}^{\tau}$ retirements$_t$, is obviously not at fault. But the estimate of 1870 book value, $3.4 billion, is. This figure is the result of linear interpolation between 1860 and 1876 book values, a procedure justifiable only if approximately equal absolute increments to the capital account occurred in each of the intervening years. In fact, the average annual increment in mileage (a reasonable proxy for investment despite changing price levels) was about 1,500 miles during the period 1860–69, against more than 4,000 miles from 1870 to 1876. We expect, therefore, that the true book value at the beginning of 1870 should be smaller than $3.4 billion. And so it is. The *American Railroad Journal* tabulation of construction accounts of individual roads yields a more

---

[16] For rapid confirmation of this point one need only examine the very state reports Ulmer used in making up his sample investment series. In Massachusetts in 1873, for example, all roads clearly charged such replacement purchases of new equipment to current account. Also see the reports of individual companies like the Baltimore and Ohio, Lake Shore and Michigan Southern, and others, where clear distinction is made between replacements and additions to the capital stock in reckoning changes in construction account.

appropriate $2.2 billion.[17] Substitution of this value in the above formula suggests a deficiency of cumulative current-dollar gross investment of well over $1 billion, and thus over 10 per cent of Ulmer's sum.

Indeed, if we apply this method rigorously, and substitute an independent estimate of retirements derived from a model to be described presently, it suggests possible additional understatement from sources other than exclusion of retirements. Thus the correct difference in book value from 1870 to 1910 is $12.4 billion which, with the addition of estimated retirements of $1 billion, brings gross investment to $13.4 billion less write-ups over the period. The ICC estimate of original cost in 1915, when its series began, is only 8.8 per cent less than the corresponding book value, and if this is accepted as a measure of write-up included in the latter, the cumulative investment is reduced to $12.3 billion or $1.6 billion more than Ulmer's series during the same interval. On the other hand, it may well be that the ICC original cost estimates, composed as they were largely of reproduction costs in 1914 prices, understate the write-up inherent in the book value statement.[18] Certainly *Poor's Manual*, a competent contemporary authority, suggests a higher margin, as does the qualitative evidence of the era.[19] A more ample allowance still supports my basic contention that retirements are excluded from Ulmer's estimates, and that they represented a sizable proportion of gross investment.

Not only is the total flow from 1870 to 1910 too small, but also its distribution through time is irregular. Ulmer does nothing to ensure that the level of investment in his various sample years is consistent with the level of the expenditure index used to interpolate changes between such years. That is, if in a given year x miles of trackage is built and sample investment is y, in a subsequent year trackage can well be $2x$ and sample

---

[17] *American Railroad Journal*, Vol. XLIII, 1870, p. 2. Had Ulmer even consulted *Historical Statistics* (1949, p. 201, Ser. K-18 and K-19), the excess of his estimate would have become apparent.

[18] The original cost series can be found in W. H. S. Stevens, *Analysis of Steam Railway Dividends*, Interstate Commerce Commission, Washington, 1943, Table H. Ulmer reprints it in his Table C-12.

Stevens (*ibid.*, p. 43, n. 3) describes the method of its derivation. There can be little doubt that much of the original cost at that early date was estimated by taking reproduction costs in 1910–14 prices. In the valuation cases before the Commission at the time, actual original costs were a rarity. Moreover, an original cost figure even from company records easily might embody the implicit exaggeration associated with payment to contractors in securities accepted at par but actually selling at less than the market price.

[19] *Poor's Manual* for 1900 (pp. liv ff.) asserts that some 40 per cent of the capital stock outstanding in that year was fictitious. Even if all bonds were sold at par, total cost would be overstated by about 20 per cent, and correspondingly more dependent upon the actual discount on bonds. The implication of the ICC estimate that there was no write-up by 1914 is doubtful at best.

investment $\frac{1}{2}y$, after price changes are taken into account. So, although the expenditure index in the early 1900's stands at levels as great as those of the early 1870's, investment comes to just half as much. Since the ratio of equipment to trackage increased rapidly in the later decades, along with other capital expenditures, the difference is not due to the limitations of the index, which is biased in exactly the opposite direction. The failure to gear the value estimates properly to the underlying physical series is the fundamental flaw. It leads not only to cumulative and decadal distortion but also to annual misrepresentation. In no fewer than six pairs of years do changes in Ulmer's expenditures and expenditure index diverge in *direction*: 1876-77, 1888-89, 1891-92, 1895-96, 1905-06, and 1908-09. A particularly striking case is 1891-92. The 1891 sample yields expenditures of $234 million, that of 1892, $251 million, but trackage opened in the latter year is less than half that in the former. Poor methods of interpolation and of sampling are responsible for these incongruities. Allowance for the lead of outlays over track emplacement and incorporation of equipment changes in the index would improve the annual distribution. Stratification of the sample roads into those in the process of extension and those building *de novo* would tie the levels of the value estimates more closely to the interpolating index than simple escalation by the ratio of assets does.

As mentioned earlier, the conversion of these imperfect current-dollar investment flows to real terms makes matters still worse. As is often done for construction investment, a price index of inputs is used to deflate current expenditures. But since labor bulks large as an input, the part of increased wages that reflects productivity change in the building trades results in substantial exaggeration of the increased price of final output. Labor productivity had an upward trend over the period, and thus the measured change in the real capital stock is biased downward. Adding to this defect, which is admittedly difficult to circumvent, is the selection of a bad index. In particular, the price of metal inputs is an unweighted conglomeration of prices of door knobs, butcher knives, and files, among others, but not of steel rails.[20] At the end of the century, rails were perhaps the largest single metallic input in value, and subject to a sharp downward trend in price. From 1880 to 1889, their price plummeted from $67.52 a ton to $29.25, a decline of almost 60 per cent, but the Ulmer index declines only 25 per cent. In addition, the weights of the index allot less

---

[20] This is the Aldrich Committee unweighted index of prices of metal implements (excluding pocket knives) contained in S. Rept. 1394. The composition would be irrelevant as long as the series moved approximately like a rail-price series. The point made here, however, is that it does not.

importance to metal inputs than is appropriate and more to other materials with a lesser rate of price decline.[21] The combination of these two factors is sufficient to transform a correctly weighted aggregate price reduction perhaps as great as 30 per cent into price stability during the decade. As a consequence, real capital formation at the beginning of the 1880's is overstated relative to that at the end, and it is toward the end that the large spurt of current-dollar outlay occurred. The use of a skilled-wage series rather than one for laborers and of a materials index comparable in its misrepresentation to the metals index distorts the real series as well, but the effects are not as obvious.

This rejection of Ulmer's post-1870 estimates made it impossible to link my own pre-Civil War investment series to his in 1870, as had been originally intended. Complete correction of his series was not feasible because of the incomplete detail of the published materials; nor in any event would it represent the most efficient method of deriving continuous and consistent capital-stock estimates of the sort required here. For this limited objective, an approach from the physical rather than the value side seems especially promising. Few industries possess such extensive records of physical components of the capital stock, namely, trackage and equipment, which retain reasonable homogeneity over time and which encompass such a large proportion of total investment. On the other hand, the resultant estimates are only a partial replacement for Ulmer's comprehensive coverage of the sector. The task of reworking his investment series, both current and constant, to correct for the error in trend and the annual aberrations still awaits attention.

The construction of capital-stock estimates from physical components is not without its difficulties. Despite the notable extent to which capital expenditures consisted of additions to trackage and rolling stock, over time there was some substitution of other outlays, for stations, realignment of way, electrification, and so forth. But these introduce only a small bias, if any, during the period of our interest, a period of initial extension of the system. As late as 1914–18, equipment alone accounted for more than 40 per cent of total investment, and such direct outlays associated with construction of road—like grading, track laying, rails, and ties, etc.—make up more than half the residual. In total, therefore, the activities to which the index is most obviously sensitive represent almost three-fourths of capital expenditure, and if an allowance were made for the "usual" amount of station and shop construction associated with

---

[21] Ulmer's weights of 4.8 for other materials—4 for labor and 1.2 for metals—are derived from maintenance outlays of class 1 roads in 1925 and 1935 and are singularly inappropriate to initial construction of road in the late nineteenth century.

mileage extension, an even higher proportion. Very much earlier a similar ratio was not very different—85 per cent in the 1850's—and the maintenance of such a continuing high level so late leaves little scope for trend distortion in a combined trackage and equipment index.[22] Accordingly, in the absence of firm indications of the temporal incidence and magnitude of changing coverage, no adjustment is made here.

Another and more serious problem arises from the changing nature of the physical units over time. On the one hand, there is technological change. An index of equipment designed to measure changes in this segment of the capital stock cannot treat the primitive 10-ton locomotives of the 1830's the same as the later behemoths; nor can a trackage index fail to differentiate between mileage laid with strap iron and later with steel rails. On the other hand, there were geographic shifts in construction that were associated with altered real costs: a mile on the prairies of the Middle West was not as expensive as one scaling the Rocky Mountains.

The principle involved is clear. Each physical unit, whether of track or equipment, should be weighted by its cost of reproduction in the prices and technology of a given year before being compared. In this way, one gets a consistent measure of the capital stock over time, where real investment is measured, as is conventional, by its cost, not its capacity.[23] In the subsequent calculations we apply this rule, as far as the data permit, to deal both with technological change and the aggregation of different types of track—main, second, yard, etc.—and equipment at a moment of time.

No geographic weights are used, however. The shift in the regional concentration of construction is dramatic—almost one-third of total 1890 mileage was located in states west of the Missouri compared with one-tenth in 1870—but its significance depends upon the interregional variance of costs of road construction, and these are surprisingly limited. Total cost per mile of track in the eight regions distinguished in *Poor's Manual* ranges from $35,000 to $73,000, with six of the observations clustered from $35,000 to $45,000.[24] The differences here are exaggerated

---

[22] ICC, *Statistics of Railways in the United States for 1918*, p. 87; for comparable distribution in the 1850's (after subtraction for excess interest and discount charged to construction account), see my *American Railroads*, Table B-4.

[23] It is well to point out two characteristics of this technique. First, the choice of a technology base is effectively limited to later years, since the type of construction before that date could be duplicated, whereas the same condition does not hold for early years. Second, there may well be different results corresponding to different technology bases, since the range of products that must be reproduced may be relatively cheaper to duplicate at different later dates.

[24] *Poor's Manual for 1891*, pp. ii, iii, and xviii. The slightly different Census regions would not alter the terms of the comparison.

TABLE 2

NET TRACK IN EQUIVALENT MAIN-TRACK MILES OF 1909

End of Year[a]	Miles of Main Track (1)	Miles of Other Track (2)	Total Track in Equivalent Miles, Specified Date (3)	Rail Weight, Gross (tons per mile) (4)	Index of Resource Content (1909=100) (5)	Total Track in Equivalent Main-Track Miles of 1909 (6)	Cumulative Depreciation Since 1828 in 1909 Equivalent Main-Track Miles (7)	Net (Surviving) Track in Equivalent Main-Track Miles of 1909 (8)
1838	2,633	132	2,699	41	78.6	2,121	135	1,986
1848	6,279	936	6,747	80	89.9	6,065	607	5,458
1858	27,621	4,598	29,920	85	91.3	27,317	2,440	24,877
1869	46,844	9,369	51,528	85	91.3	47,045	6,512	40,533
1879	86,566	18,200	95,666	90	92.8	88,778	13,872	74,906
1889	161,276	40,812	181,682	100	95.6	173,688	27,739	145,949
1899	190,046	64,418	222,255	107.5	97.8	217,574	47,936	169,429
1909	238,116	108,916	292,574	115	100.0	292,574	73,677	218,897

Source

Main track

1838: Gerstner total of 3,172 miles at end of 1839 shifted to 1838 by relative change between two dates in 1880 Census mileage-added series (reprinted in *Historical Statistics*, 1960, Ser. Q-43). For Gerstner mileage total, see *Journal of Franklin Institute*, N.S. XXVI, 1840, pp. 89-102, 227-230, and 301-307; also my *American Railroads*, Appendix B.
1848, 1858: 1860 Census figures for 1850 and 1860 extrapolated as above. *Preliminary Report of the Eighth Census*, Washington, 1862, p. 238.
1869-89: *Poor's Manual* for 1900.
1899-1909: Average of ICC *Statistics of Railways* totals for adjacent fiscal years.

Other trackage

1838: 0.05 times mileage. (For justification of this and other antebellum figures, see my *American Railroads*, Appendix B.)
1848: 0.10 times change in mileage between 1838 and 1848 plus 600 miles of double track in 1850 interpolated to 1848 on 1880 Census mileage series.

NOTES TO TABLE 2 (concluded)

1858: 0.10 times change in mileage between 1848 and 1858 plus 1,700 miles of 1860 double track interpolated to 1858, as above.
1869: Average of 1858 and 1879 ratios of other track to mileage times 1869 mileage.
1879-1909: Same as main track, 1899-09.
Col. 3: Col. 1 + .5 col. 2.
Col. 4: See text.
Col. 5: Col. 4, converted to an index with 1909 base, weighted one-third. Thus the 1838 value, 78.6, is equal to $\frac{41}{115}$ + 200.0/300.0.
Col. 6: Col. 3 x col. 5.
Col. 7: Cumulative depreciation$_t$ = cumulative depreciation$_{t-10}$ + 10 x .01 x (col. 3_{t-10}) + 5 x .01 x (col. 3_{t-10}).
Col. 8: Col. 6 - col. 7.

[a]These years have been selected to conform as nearly as possible to the midpoints of the years for which output has been estimated. From 1880 on, the calendar-year basis of the stocks and the fiscal-year basis of the output flows result in perfect agreement. Beforehand, because of a later fiscal year, the stock may be centered slightly before the midpoint.

owing to variations in rolling stock per mile; roads in the Middle Atlantic group, with the highest cost, operated less than one-sixth of the mileage but one-third of the equipment. Total cost divided only by trackage thus distorts the extent of regional differences in construction of road. Nonetheless, even these imperfect weights applied to the mileage increase from 1870 to 1890 produce a rate of change of only 215 per cent, as opposed to 200 per cent for the unadjusted totals. Accordingly, the price paid for the neglect of this complication is small.

Table 2 presents the calculations translating the initial mileage series into an index of real capital invested in road and structures according to the principle of equivalent costs. First, the mileage of main-line and other track have been aggregated into the single, uniform trackage series in column 3 by assigning weights of 1 and ½, respectively. This approximate relative cost of other track has been derived from both financial and engineering considerations. We start from the general observation that, despite large changes in relative prices over the period, direct superstructure costs do not deviate much from 30 per cent of the total expense of construction. To this outlay for track must be added the investment in the permanent way, grading, masonry, and so on. For double, third, and fourth track, this did not escalate proportionally, but by a much smaller factor. Two engineers, arguing from technical considerations, claim the incremental investment in this other construction to be less than 50 per cent of that required for first track. A weight of one-half for double and other main-line track is therefore suggested.[25] The same value is used for sidings, although the possible substitution of lighter rails could reduce the proportion somewhat; however, one railroad commission valuation in the opening years of the twentieth century indicates a relative cost of .55, so the one-half weight cannot be too far off the mark.[26] Yard track is assigned the same weight also. Although newly constructed from the start, because the selection of the site was determined by advantageous physical circumstances—and hence less excavation, grading, etc.—and because multiple tracks were always involved, the per unit cost of the trackage inevitably was less than that for first main track. In the most unfavorable circumstances, where only two tracks were built upon a substructure as expensive as the average for main line, the per mile relative cost for yard trackage would be .75. For parallel trackage many times greater, built under favorable conditions, the previous one-half weight therefore retains its applicability. This is fortunate, since the distinction among the various types of other track is never very clear in the mileage statistics until very late in the period.

The second step in the process is to convert the equivalent main track miles of the various dates to equivalent mileage of a specific date. This requires temporal relative cost weights to supplement the previous cross-sectional ones. That is, we must determine what it would have cost to construct the mileage at the specified dates, with 1909 prices and technology, but as it actually was; this cost relative to 1909 costs can then be used to derive a new mileage series homogeneous through time.

The most apparent way in which mileage differed over time is the increase in rail weights. A network laid in the 1830's principally with strap iron of 25 gross tons to the mile gave way to one in the first decade of the twentieth century with steel rails weighing something like 115 gross tons to the mile. At a minimum, therefore, the rail component of total investment in road must be reduced in earlier years relative to 1909 if temporal uniformity is to prevail. At a maximum, all investment could

[25] See W. M. Camp, *Notes on Track*, Chicago, 1904, p. 623; and Arthur Wellington, *The Economic Theory of the Location of Railways*, 6th ed., New York, 1887, p. 765. Wellington frames his judgment directly in terms of the increment to total cost associated with the additional grading and masonry. His 10 to 15 per cent increment to total cost is roughly equivalent to Camp's evaluation of a 50 per cent increase in substructure outlays, since grading and masonry made up about 40 per cent of the total.

The addition of the 10 to 20 per cent required for the substructure of a double track to direct superstructure outlays of about 30 per cent leads to the total incremental cost of one-half used here. (Wellington's 10 to 15 per cent range is probably too low because it excludes some other variable costs like interest, station facilities, etc., and that is the reason for a one-half weight rather than something smaller.)

[26] Cited in F. Lavis, *Railway Estimates*, New York, 1917, pp. 35–36.

be scaled in this manner if the increase in rail weights represented parallel improvements in the substructure, bridges, shops, and so on. We follow an intermediate course here, and allocate to the variation in rail weights shown in column 4 an influence of one-third the approximate relative importance of the entire superstructure in order to derive the index of resource content in column 5. The rationale for such a weight is the constancy of quality of much of railroad investment over time. Excavation and grading, masonry work, and the like were probably as well done in the antebellum period as at the end of the nineteenth century, quite independently of the continuous improvement in superstructure. All expenditure on the superstructure is considered, because not only did rail weights increase, but the number of ties per mile, the depth of ballast, the type of rail chairs, and other costs of the superstructure also changed as the rails did. Two cross-section observations are relevant here. The first is the evidence that the investment in the substructure at a moment in time is determined by the circumstances of terrain, etc., and not by rail weight. The second is the almost proportional movement of rail weights and costs per mile for the entire superstructure (of which direct purchases of rails constitute less than half); as rail weights increase from 50 to 60 to 80 to 100 pounds per yard, estimated costs climb from $7,600 to $8,800 to $10,600 to $12,300 a mile.[27] Although neither piece of evidence relates exactly to the problem at hand—and could not since our difficulties lie with reconstructing the unobserved—they do afford some support for our decision to weight the increased size of rails only by the superstructure component of road investment and not to apply it either more or less universally. That costs perhaps did not increase quite as rapidly as rail weight did provides a small additional, and warranted, allowance for the better bridges and other structures along the line.

Since the index in column 5 is central to the final results, it may be well to emphasize its insensitivity to errors in the average rail weights specified in column 4. These last are only approximations, based upon scattered contemporary reports on the subject, except for the terminal observations. The 1838 weight is firmly established from Gerstner's researches into the type and weight of rail on almost all railroads then in operation. Similarly the 1909 observation is sufficiently close to the first ICC report on the subject in 1920 so that its order of magnitude is well founded. The

[27] *Ibid.*, pp. 34, 240. Different relative prices of superstructure inputs will influence the results, of course. At this date, the first decade of the twentieth century, rails were relatively cheap. The more expensive rail costs are, the more nearly proportional is the movement. Over much of the period, therefore, the relationship was presumably better than that described here.

gradations from decade to decade are limited in importance because of the minor weight attached to the series;[28] from 1848 to 1909, while rail weight goes up almost by 50 per cent, the index goes up by 9 per cent. The crucial assumption, therefore, is the one-third weight to be assigned to the rail index rather than the precision of that index itself.

The series of 1909 equivalent main track miles must be further adjusted to reflect depreciation if it is to measure the capital stock at various dates. Wear of rails and ties can be excluded because their lives are short, and their replacement is implicit in the trackage figures. (Of course, to the extent that renewals are not exactly uniform, the proportion of track requiring replacement may differ slightly in each of the terminal years of Table 2, and may thereby distort short-term comparisons.) For the inevitable obsolescence of the more permanent components of the road, such as earthworks, culverts, etc., and for shorter-lived structures inadequately maintained, a 1 per cent charge per annum is levied. This corresponds to the ICC finding of 0.86 per cent in 1917 and 0.82 per cent in 1949, and is set slightly higher to reflect the small increase in longevity of the permanent way over time.[29] In addition, because the depreciation allowance is deliberately applied to the mileage series unadjusted for resource content (column 3), the effective rate is tapered downward from 1.28 per cent in the 1830's to 1 per cent in the early twentieth century. Subtracting the cumulative depreciation so calculated in column 7 from the gross series of column 6 yields the final net series in column 8.

Tables 3 to 5 present a transformation of equipment to 1909 equivalent freight cars through application of the same real cost principle. The procedure in this instance is slightly more complicated because the shorter life of equipment necessitates the derivation of production statistics rather than simple manipulation of the stocks as before. The stock of locomotives, passenger cars, and freight cars is already reduced to a quasi-net basis owing to prior retirements, but it also contains an element of depreciation which can be determined only by exact knowledge of its vintage. Hence the need for production statistics.

[28] The allocation to intervening years is intended to convey a consensus of comment on the trend to heavier rails. For the antebellum years, see my *American Railroads*. Subsequently, cf. *Railway Gazette*, Vol. IV, 1872, p. 107; XII, 1880, p. 334; Wellington, *Economic Theory*, p. 119; Lavis, *Railway Estimates*, p. 242.

Note that there is no distinction here between iron and steel rails since in 1909 their prices had converged to the point where no significant difference between the two existed. Since 1909 technology is the basis of the reduction, it is conditions at the latter date that are relevant.

[29] These are cited by Ulmer, *Capital in Transportation*, p. 225. For evidence of pre-Civil War rates of similar (but slightly higher) magnitude, see my *American Railroads*, Appendix B.

Table 3 presents their determination from prior information on the stocks and the retirement rates. The stock at any date is equal to the stock a decade earlier plus intervening purchases, less intervening retirements. Once the retirement rates are specified, retirements are determined, and purchases then follow directly. The equipment lives used in Table 3 are twenty years for both locomotives and freight cars, and twenty-five years for passenger cars (with the exception that twenty years has been applied to the production of the 1830's). These are a fair consensus of informed engineering opinion, and they incorporate a longer service life over time, since the average mileage run each year showed an upward tendency.[30] Fortunately, the validity of the assumptions can be checked from 1879 on by comparing the implied production estimates against actual production statistics. The results are quite favorable as the figures below reveal:[31]

	1880–89	1890–99	1900–09
Locomotives			
Predicted	18,957	18,326	40,893
Produced	17,200	17,060	44,234
Passenger cars			
Predicted	16,771	13,022	25,015
Produced	12,857	11,393	25,113
Freight cars			
Predicted	686,701	643,819	1,473,837
Produced	532,400	723,900	1,455,300

The large overstatement of the 1880's is as much due to the inadequate coverage of the production series as to the inaccuracy of the retirement model. The production series excludes rather extensive production by railroad repair shops. As late as 1900, railroads themselves produced an addition of 20–25 per cent to commercial output of cars and the trend, if any, would make the earlier proportion even larger. Self-manufacture of locomotives was less widespread. The inclusion of exports in the production tabulation only partially offsets the understatement, since these amounted but to 5 per cent of output.[32] For the same reason, the closer correspondence after the 1880's exaggerates the adequacy of fit.

[30] For retirement ages, see Wellington, *Economic Theory*, p. 419; J. L. Ringwalt, *Development of Transportation Systems in the United States*, Philadelphia, 1888, p. 320; *Railroad Gazette*, Vol. XVIII, 1886, pp. 201, 869; Emory Edwards, *Modern American Locomotive Engines*, Philadelphia, 1883, p. 119.

[31] These production statistics come from *Historical Statistics*, 1960, p. 416, Ser. P-213-215.

[32] *Twelfth Census, 1900*, Vol. X, *Special Reports on Selected Industries*, pp. 273 and 279, gives production of cars in railroad repair shops. For the importance of exports, see William Howard Shaw, *Value of Commodity Output Since 1869*, New York, NBER, 1947, pp. 56–57.

TABLE 3

DERIVATION OF EQUIPMENT PRODUCTION ESTIMATES

End of Year	Locomotives	Passenger Cars[a]	Freight Cars[b]	Locomotives	Passenger Cars[a]	Freight Cars[b]
		STOCK				
1838	375	417	1,250			
1848	1,168	1,238	14,415			
1858	5,158	4,703	80,402			
1869[c]	10,000	10,000	185,000			
1879	17,084	16,528	480,190			
1889	30,566	28,524	1,051,141			
1999	37,183	34,282	1,330,520			
1909	59,119	46,422	2,117,656			
	RETIREMENTS			PURCHASES		
1829–38				375	417	1,250
1839–48				793	821	13,165
1849–58	375	417	1,250	4,365	3,882	67,237
1859–69	950	533	16,000	5,792	5,830	120,598
1870–79	4,525	2,450	69,250	11,609	8,978	364,440
1880–89	5,475	4,775	115,750	18,957	16,771	686,701
1890–99	11,609	7,264	364,440	18,226	13,022	643,819
1900–09	18,957	12,875	686,701	40,893	25,015	1,473,837

Source

Stock

1838–58: See my *American Railroads*, Appendix D, for estimates of equipment stocks at end of 1839, 1849, and 1859. These were extended to the stated years (after allowance for replacements in the instance of 1858) by an annual series of locomotive production to be found in *ibid.*, Appendix B, Table B-8.

1869: These are approximated to the nearest thousand and are derived by interpolating the aggregate ratio of each type of equipment to earnings on the sample ratios derived from twenty-five railroads reporting at the three dates, 1859, 1869, 1880. The sample was representative at the terminal dates of 1859 and 1880; that is, the sample ratios corresponded to the national ratios, so interpolation is a satisfactory procedure. The sample data came from the share lists of the *American Railroad Journal* in 1860 and 1870, and the 1880 Census. The aggregate fiscal 1870 receipts needed to convert the ratios to absolute stocks are estimated from the physical outputs and are discussed above.

The final estimates are consistent with the contemporary report of 10,000 locomotives and 214,000 cars in the early 1870's (*Railway Gazette*, Vol. III, 1873, p. 287).

1879–1909: Same as trackage; see Table 2.

Retirements: $P_{t-9 \to t} = P_{t-29 \to t-20}$ for locomotives and freight cars; $= P_{t-34 \to t-25}$ for passenger cars, after 1829–38. A linear distribution of production was assumed to obtain $P_{t-34 \to t-25}$. (Since there is one eleven-year interval from 1859 to 1869, the formula had to be adjusted for the estimated output in the overlapping year; this is the reason for the lack of exact correspondence between retirements and production entries two decades previously for some of the figures.)

Purchases: Equals change in stocks plus retirements.

[a] Includes baggage and mail cars in passenger service. For 1838–58, the number of cars is given in terms of eight-wheel equivalents. Thereafter, most cars took this form, with the exception of some twelve-wheel cars.

[b] For 1838–58, noncoal cars in eight-wheel equivalents. Thereafter, most box cars were of this variety. Coal cars are reported as units, and no attempt has been made to render the gradual change from four-wheel to eight-wheel cars.

[c] For 1869, since an earnings estimate was essential to the method, the equipment stock corresponds more closely to the end of the output year rather than its midpoint.

But only in the instance of freight car output in the 1890's is the divergence —including an allowance for understatement in the production series— troublesome. Even here, the apparent understatement of retirements only lends authority to our expressed view of the seriousness of their neglect by Ulmer. From 1870 through 1909, we estimate retirements at 40,666 locomotives, 27,364 passenger cars, and 1,236,141 freight cars. At prices of $8,000 per locomotive, $2,500 per passenger car, and $400 per freight car—prices that are among the lowest of the period—retirements are seen to amount to $900 million, or far more than the Ulmer results yield.[33]

Just as adjustment for changes in the weight of rails increased the comparability of the trackage series over time, so the weight of equipment provides initial guidelines for a similar transformation to a consistent real cost standard. The average weight of engines rose from something like 10 tons in the 1830's to exactly 72 tons in 1909 as measured by an ICC count. The increase in weight seems to have been fairly regular, and modest, until the 1870's. Then the introduction of the Mogul and Consolidation types to replace the standard American 4-4-0 led to more rapid change. Accompanying this acceleration, and interdependent with it, was the development of more capacious freight cars. Not only did the use of steel in trucks and underframes make possible an increase in size with far less than proportional investment, but also newer designs using conventional materials came forth as soon as driving power was available.[34] Therefore the virtual doubling of capacity between the 1870's and 1880's was accomplished with little increase in dead weight (and resource cost).

[33] For prices twice as large as these in the 1870's, see *Railway Gazette*, Vol. III, 1873, p. 34, and Vol. I, 1871, p. 516. These prices are representative of the 1880's (cf. Wellington, *Economic Theory*, pp. 163, 411, and 565). In 1889 the average price of locomotives was $8,200, and in 1899, $9,500; in the latter year, passenger cars were sold at $7,500 and freight cars at $530. By 1909 the three types of equipment were priced at $12,065, $8,638, and $843. All these prices, except that for locomotives in 1909, come directly from Census reports. The 1909 locomotive price was calculated by multiplying the 1899 price by the ratio of the prices of locomotives manufactured in railroad repair shops at the two dates, i.e., $P_{1909} = P_{1899} \times p^1{}_{1909}/p^1{}_{1899}$, where p^1 is the price of locomotives built in repair shops and is given in the Census. This technique is necessary because the average locomotive built by railroads themselves was undoubtedly larger than average since such railroads had heavier traffic than average.

[34] Credit for the rapid increase in the ratio of payload to dead weight before 1900 must be sought elsewhere than in the use of steel. As late as June 30, 1915, out of some 55,000 passenger cars only 10,884 were steel and 5,197 had steel underframes. Freight cars took part in this materials revolution to a greater degree: of more than 2,300,000 freight cars, 515,000 were steel (almost all for the transportation of coal) and another 681,000 had steel underframes (ICC, *Statistics of Railways in the United States for 1915*, pp. 22–23).

In light of this record, it is difficult to attribute the early decline in the dead-weight ratio to the substitution of steel. That transformation, with its significant productivity effects, discussed later, is one that will repay much more careful study.

TABLE 4

REAL COST OF EQUIPMENT

	Locomotive Weight (tons) (1)	Index of Real Cost, Locomotives (1909=100) (2)	Passenger-Car Weight (tons) (3)	Index of Real Cost, Passenger Cars (1909=100) (4)	Freight-Car Weight (tons) (5)	Index of Real Costs, Freight Cars (1909=100) (6)
1830-38	10	42.6	14	48.7	8.0	56.6
1839-48	20	51.8	14	48.7	8.0	56.6
1849-58	25	56.5	14	48.7	8.0	56.6
1858-69	30	61.1	16	52.8	10.0	65.7
1870-79	35	65.7	20	61.0	11.0	70.3
1880-89	45	75.0	27	75.4	12.5	77.1
1890-99	58	87.0	35	91.8	15.7	91.8
1900-09	78	105.6	45	112.3	18.3	103.7
1909	72	100.0	39	100.0	17.5	100.0

Source

Col. 1, 1830-1858: *Railway and Locomotive Historical Society Bulletin*, No. 35, 1934, pp. 10-37.

Col. 1, 1858-1909: *Ibid.*; Wellington, *Economic Theory*, pp. 407, 499, 544. *Twelfth Census, 1900*, Vol. X, *Manufacturing*, p. 245. ICC, *Statistics of Railways* for 1902 and 1909, giving average weights at those dates.

Col. 2: $2/3 \left(\dfrac{\text{weight}_t}{\text{weight}_{1900-09}} \right) + \dfrac{1}{3} 100.0$.

Col. 3: J. L. Ringwalt, *Development of Transportation Systems in the United States*, Philadelphia, 1888, p. 337; Wellington, *Economic Theory*, p. 524; Edwin Pratt, *American Railways*, London, 1903, p. 89; Lavis, *Railway Estimates*, New York, 1917, p. 424.

Col. 4: $4/5 \left(\dfrac{\text{weight}_t}{\text{weight}_{1900-09}} \right) + \dfrac{1}{5} 100.0$.

Col. 5: Ringwalt, *Development of Transportation*, p. 336; Wellington, *Economic Theory*, pp. 114, 135; *National Car Builder*, Vol. VIII, 1877, p. 9; Pratt, *American Railways*, p. 51; Master Car Builder's Association, *Car Builder's Dictionary for 1906*, New York, 1906, pp. 166 ff. *Twelfth Census, 1880*, Vol. X, *Special Reports on Manufactures*, p. 264. ICC, *Statistics of Railways* for 1902 and 1909, giving average capacities at those dates. For the fares associated with those capacities, see *Car Builder's Dictionary*.

Col. 6: $4/5 \left(\dfrac{\text{weight}_t}{\text{weight}_{1900-09}} \right) + \dfrac{1}{5} 100.0$.

From the 1880's to the first part of the twentieth century another doubling in capacity occurred, again with less than equivalent increases in weight. Precise information is lacking on the development of passenger cars, but it seems a substantial upgrading of passenger comfort occurred during the latter part of the nineteenth century. Giant palace and sleeping cars, as well as the enlargement of the more prosaic coaches, contributed to perhaps a tripling in the average size of units between the antebellum decades and 1910. Moreover, increased relative prices of passenger cars versus freight cars testify to this greater change in real costs in the former.

Table 4 summarizes this history. Associated with the weight changes are real cost changes but, as before, not in exact proportion. Fortunately, there is more to go on here than the indirect methods used to convert variations in rail weight to variations in unit resource consumption. Actual prices of locomotives per unit weight at a moment of time afford a measure of the relationship we are after. What data have been found tend to indicate a decline in cost per ton as size increases, that is, a change in real cost substantially smaller than the simple increase in weight. Thus, in 1876, as locomotives increased in weight from 20 to 40 tons, the price went up from $7,000 to $9,200. In another instance ten years later, Baldwin locomotives varied from a 24-ton American to a 59-ton Consolidation, with selling prices rising from only $5,750 to $9,750. The marginal cost was only $110 per ton against an average cost of almost $250 per ton for the smaller unit. Later, in 1906, the same circumstances prevailed: a 40-ton freight engine was quoted at $8,000; an 80-ton engine, at $12,800; and a 140-ton engine, at $20,160. Over a somewhat narrower range, from 46 to 60 tons, a final observation dating from the late 1880's reveals constant unit costs.[35] All four bodies of information diverge from the findings of Dorothy S. Brady which assert increasing unit prices over the size range.[36]

Despite this unsatisfactory current state of knowledge about price-size relationships for locomotives, I have accepted the evidence pointing to a declining cost pattern. Materials inputs made up only slightly more than half the expense of locomotive production, so doubling of size would yield increased costs of only 50 per cent on this account. For such a specialized industry as locomotive production in which engineering salaries and wages of skilled workmen loom large, and with machine tools capable

---

[35] Wellington, *Economic Theory*, pp. 411 and 565; Lavis, *Railway Estimates*, p. 418; *The American Railway*, New York, 1889, p. 126.

[36] See her "Relative Prices in the Nineteenth Century," *Journal of Economic History*, June 1964, especially Table 2. The evidence presented here also seems to contradict her generalization of price-size relationships to either increasing or U-shaped forms.

TABLE 5

STOCK OF EQUIPMENT IN 1909 EQUIVALENTS

End of Year	Gross Stock, Locomotives (1)	Net Stock, Locomotives (2)	Gross Stock, Passenger Cars (3)	Net Stock, Passenger Cars (4)	Gross Stock, Freight Cars (5)	Net Stock, Freight Cars (6)
1838	160	114	203	152	708	531
1848	571	348	603	371	8,159	5,765
1858	2,877	1,952	2,290	1,672	45,507	30,405
1869[a]	5,924	3,102	5,108	3,130	115,684	65,645
1879[a]	10,994	6,562	9,391	5,617	332,689	211,273
1889[a]	21,867	12,570	19,585	12,453	786,087	461,135
1899	30,097	15,447	27,336	14,895	1,120,912	575,631
1909[b]	59,062	36,352	46,367	27,888	2,119,865	1,294,033

Source

Gross stock of all types of equipment:
Cols. 1, 3, 6: (production$_t$ times index of real cost$_t$) minus (retirements$_t$ times index of real cost$_t$), where the inputs are taken from Table 3 and 4.

Cols. 2, 6: .75 production$_{t-9 \text{---} t}$ plus .25 production$_{t-19 \text{---} t-10}$.

Col. 4: .8 production$_{t-9 \text{---} t}$, plus .4 production$_{t-19 \text{---} t-10}$ plus .1 (.5) production$_{t-29 \text{---} t-20}$.

[a] There is some small modification in these years of the formulae for the derivation of the net stock due to the eleven-year interval from 1858 to 1869, and hence a slightly different average age of equipment.

[b] The 1909 entries deviate from the comparable gross stocks of Table 3 only because of rounding.

TABLE 6

REAL NET CAPITAL STOCK, 1838-1909
(million 1909 dollars)

End of Year	Equipment (1)	Track (2)	Total[a] (3)	Index (1909=100) (4)	Index, Variant II (5)
1838	2.9	79.8	82.7	0.8	0.8
1848	11.4	219.4	230.8	2.2	2.3
1858	59.2	1,000.1	1,059.2	10.1	10.6
1869	111.7	1,629.4	1,741.1	16.6	17.3
1879	286.1	3,011.2	3,297.4	31.5	32.5
1889	606.8	5,867.2	6,474.0	61.9	63.7
1899	749.6	6,811.0	7,560.6	72.3	74.2
1909	1,658.2	8,799.7	10,457.9	100.0	100.0

Source

Col. 1: Net stock of locomotives times $11,000; net stock of passenger cars times $8,000; net stock of freight cars times $800.
Col. 2: Net surviving mileage times $40,200.
Col. 3: Col. 1 plus col. 2.
Col. 4: Total divided by 10,457.9.
Col. 5: Index of col. 1 times .1 plus index of col. 2 times .9.

[a] May not add to total because of rounding.

of turning out a variety of shapes and sizes, the variable costs of other inputs well might decline per unit, and it is difficult to imagine their increase. In the subsequent calculations, therefore, I have allowed all materials inputs to increase proportionally with the increase of weight, but only about three-tenths the other costs. This is equivalent to assuming that two-thirds of total cost is influenced by the increase in weight, and roughly accords with the weight-price data cited earlier.[37] Column 2 of Table 4 presents the index of resource content for locomotives based upon this weighting scheme. The ratio of materials purchases to value of final product is not the same for the manufacture of passenger and freight cars, and so this relationship between weight and resource content will not hold there. But the same general method can be used, in which all materials inputs and one-third of other inputs vary with weight, inasmuch as scattered data affirm declining unit costs for cars as well.[38] Since materials were more important in car manufacture, the more nearly proportional indexes of columns 4 and 6 result.[39]

To estimate the net stock of equipment in consistent equivalents first requires conversion of all production flows to the common 1909 denominator; that is why the indexes of real cost have been put in decadal terms rather than in those of a single date. This done, the gross stock at each date is easily reconstructed in 1909 equivalents from the information in Table 3. To these are applied depreciation rates of 5 per cent for locomotives and freight cars, and 4 per cent for passenger cars, following the useful lives of the various types of equipment developed earlier. The results are presented in Table 5.

Two additional pieces of information are required to transform the materials already developed into a measure of the net capital stock. The first is a set of weights to combine the various types of rolling stock; the second, the weights to aggregate equipment and mileage units. For both, 1909 prices are used in order to maintain a consistent technology and price base, but as we shall show, the results are not especially sensitive to this specification. Let us begin, however, with the capital data in 1909 dollars and technology presented in Table 6.

[37] In both 1890 and 1900, materials outlays absorbed slightly more than one-half the total costs of locomotive construction, with labor inputs about 40 per cent (*Twelfth Census, 1900*, Vol. X, p. 243).

[38] See Lavis, *Railway Estimates*, pp. 423 and 428–429.

[39] Materials were about two-thirds of the value of cars in 1889, 1899, and 1909. So variable costs of .67 + .3(.33) yields variable costs of .77 (or four-fifths) the change in weight. The larger fraction (rather than three-fourths) was used, since the very large range encompassed here would tend to cause convergence to proportionality. For the materials ratios, see the *Twelfth Census, 1900, Special Reports on Manufactures*, Vol. X, p. 267, and *Thirteenth Census, 1910*, Vol. VIII, p. 474.

The equipment stock in Table 5 has been valued at prices of $11,000 per locomotive, $8,000 per passenger vehicle, and $800 per freight car. These are derived from corresponding Census valuations of $12,045, $8,638, and $843 for the year 1909-10, after adjustment for the greater size of equipment produced in that year than the average for the stock as a whole. The prices are slightly higher than direct application of the earlier formulas would indicate in order to allow for delivery and set-up costs met by the railroads and included in capital outlays.[40]

This equipment total for 1909 is an input in the determination of the 1909 price for construction of an average mile of main track.[41] This latter is not observable since in a year of limited construction, concentrated geographically, the cost per mile of road actually built is not a good proxy for the national average reproduction cost. What has been done, therefore, is to subtract this equipment estimate of $1.7 billion from Ulmer's 1909 total net capital stock estimate and thereby obtain a corresponding aggregate for net road investment alone. This is convertible into a per mile statistic by division by the earlier series of net 1909 equivalent main trackage. The resultant unit price of $40,200 is applied to the quantities in all previous years.

Following so soon after my severe criticism of Ulmer's pre-1909 figures, this ready acceptance of his 1909 net capital stock estimate may give the impression of a double standard. There is none. Ulmer's investment flows beginning with 1912 are derived from the ICC reports, not the sampling method under attack earlier. Likewise, his post-1914 price index is also entirely of ICC origin. Finally, his capital stock estimates throughout are developed by adding his net investment figures to an ICC benchmark of January 1, 1937. This means that, as long as the constant dollar investment flows between a given date and January 1, 1937, are

[40] To determine the prices for the average equipment stock as of 1909, the prices of the average increment to the stock in 1909 were divided by the relevant indexes of real cost. The average weight of the additions to locomotives and freight cars was calculated by using the ICC information on purchases and retirements, and the weights of the stock in 1908 and 1909. These gave rise to indexes of 114.8 and 109.1, respectively; the index of real cost in Table 4 was used for passenger cars in the absence of exact information. These divided into prices of $12,271 for locomotives, $8,313 for passenger cars, and $843 for freight cars yield 1909 prices applicable to the 1909 stock of $10,689, $7,402, and $772. (Note that the 1909 prices used here differ slightly from the Census values of footnote 32 because they are adjusted for manufacture in railroad repair shops.) These prices in turn were rounded to the values cited in the text.

[41] The exclusion of cars in company service (information on which is not available continuously) leads to a very small and insignificant bias here because it overstates the investment in road relative to equipment and these two do not grow in parallel fashion. How limited this distortion is may be seen from the later comparison between columns 4 and 5 of Table 6, which are the results of applying two different sets of weights for road and equipment.

accurate, so is the capital stock estimate. The date 1909 satisfies this condition; earlier dates do not because the flows are subject to the flaws detailed before. Beyond this, there is further corroboration from various attempts by state railroad commissions around 1909 to estimate current reproduction costs less depreciation of railroads lying within their borders. In the instance of Nebraska, the 1909 valuation, less cost of right-of-way and station grounds, was $34,000 per mile. Since the book value of railroads in that state was some 30 per cent below that for the country as a whole, the implied national valuation of $47,000 per mile of roadway corresponds quite well with the $44,000 implied by Ulmer's net capital stock estimate.[42] Accordingly, the 1909 Ulmer total can be equated with mine, a procedure which has the added virtue of allowing a smooth linkage with Ulmer's subsequent, and reliable, capital stock totals.[43]

In light of the successive layers of assumption which underlie these final estimates of the net capital stock, it is essential that there be some independent checks upon their validity. The cause of the most easily allayed doubt is the influence of alternative relative prices upon the movements of the capital stock series. Column 5 of Table 6 presents an alternative index of total capital based upon a share of equipment in total investment of 10 per cent, and a 90 per cent share of road.[44] The 1909 prices actually used imply constant shares of 16 per cent and 84 per cent, which, although representative of the situation at the terminal date, overstate the role of equipment during much of the period—hence the relevance of the weighting scheme of Variant II. Note how little difference the substitution makes. A similar test on the equipment total was performed by substituting relative prices for locomotives, passenger cars, and freight cars that were more typical of nineteenth century conditions than

---

[42] *Fourth Annual Report*, Nebraska State Railway Commission, 1910-11, pp. 328, 330, and 498. The book value of road and equipment was $42,000 per mile for Nebraska railroads versus $58,000 for all United States railroads, as calculated from ICC statistics for 1909.

[43] His constant-dollar estimates can be converted to 1909 dollars by multiplying by .568, the ratio of the value of his price index in 1909 to that in 1929. Since the ICC index is based upon 1910–14 weights, the earlier constant dollars are probably more appropriate in any event.

[44] This ratio is probably more indicative of the situation during most of the nineteenth century. See my *American Railroads*, Table B-4, for the antebellum period, and Ulmer, *Capital in Transportation*, p. 225, citing the ICC, for the latter part of the period.

The use of shares as weights for the index is identical with the use of prices for the original quantities:

$$\frac{R_t}{R_0} \cdot \frac{P_{R_0} R_0}{P_{R_0} R_0 + P_{E_0} E_0} + \frac{E_t}{E_0} \cdot \frac{P_{E_0} E_0}{P_{R_0} R_0 + P_{E_0} E_0} = \frac{P_{R_0} R_t + P_{E_0} E_t}{P_{R_0} R_0 + P_{E_0} E_0}$$

where R is the quantity of road, P_R its price, E the quantity of equipment, and P_E its price.

those in 1909 were; the principal difference is the increase in the relative price of passenger cars by the latter date. Although not shown here, the two alternatives move in virtually identical fashion. The further force of these results is the independent validity of the rates of growth of the capital stock estimates even if the absolute values based on Ulmer's 1909 total are in error.

Beyond this reassurance of relative inconsistency, there are some measures of absolute reliability as well. Table 7 brings together the relevant information. The first test compares the antebellum flow and stock estimates with a set derived by deflating current dollar expenditures developed from railroad accounts.[45] The almost perfect agreement in the decades of the 1830's and 1840's is heartening. At their worst in the 1850's, moreover, gross capital formation obtained from the present model is within 20 per cent of the other series, with the corresponding terminal stocks still closer. In terms of deviation from a long-term trend, the error is negligible, in sharp contrast to the break in 1869 when the Ulmer series begins.

A second check involves extension of the method to the decade beyond 1909 when Ulmer's data assume the authority of an official basis. The last line in Table 7 presents the results. My capital stock projection of $11,277 million in 1909 dollars falls short of the Ulmer $12,053 million because of a corresponding underestimate of gross investment.[46] That

---

[45] The conversion from prices of 1860 to those of 1909 was accomplished by substituting the corresponding 1909 prices of rails, daily wages of unskilled labor, materials, and locomotives for the 1860 ones. The 1909 locomotive price first was converted to a 1909 price for an 1860-size locomotive. Actual 1909 and 1860 prices are: rails, $29.40 and $49.98 per ton; labor, $1.38 and $1.04 per diem; locomotives, $6,710 and $9,250. The relative materials price of 1.493 was calculated from the Warren and Pearson and BLS index of wholesale prices of building materials. Sources for the 1909 prices are: rails, *Historical Statistics*, 1960, p. 123, Ser. E-108, with an allowance of 5 per cent for delivery; labor, ICC *Statistics of Railways for 1909*, daily wages of "other trackmen"; locomotives, see above; materials, *Historical Statistics*, 1960, pp. 115, 117, Ser. E-8 and E-21.

The use of a daily-wage relative avoids overstating the increase in costs between the two dates, since some of the productivity advance was absorbed in a shorter working day. This, and a large weight for materials, helps explain the very different index found by Ulmer (based after 1890 on hourly wage rates). His 1909-1860 relative is 1.75 and would place the pre-Civil War estimates in sharp contradiction. (But the rate of change within the antebellum period would still agree; at issue only would be the change from 1858 to 1869, and even with Ulmer's cost index, his 1869 estimate is out of line.)

[46] As further evidence of the satisfactory fit of the retirements model in this period, there are the ratios of predicted to actual equipment purchases during the decade 1910–19: locomotives, 1.056; passenger cars, .953; freight cars, .863. Information on actual purchases and retirements during the decade may be found in ICC *Statistics of Railways for 1919*, p. 15.

TABLE 7

ALTERNATIVE INVESTMENT AND STOCK ESTIMATES
(million 1909 dollars)

	Gross Capital Formation		Capital Consumption Allowance		Net Capital Stock (terminal dates)	
	Variant I (1)	Variant II (2)	Variant I (3)	Variant II (4)	Variant I (5)	Variant II (6)
1828–38	89.3	88.8	6.6	4.6	82.7	84.2
1839–48	172.2	172.5	24.1	30.0	230.8	226.7
1849–58	926.9	761.2	98.5	114.5	1,059.2	873.4
1859–69	919.9		238.0		1,741.1	3,911.0
1870–79	2,010.4	2,458.0	454.1	843.0	3,297.4	5,526.0
1880–89	4,094.6	3,478.0	918.0	1,271.0	6,474.0	7,733.0
1890–99	2,498.6	2,659.0	1,412.0	1,767.0	7,560.6	8,625.0
1900–09	4,945.8	3,967.0	2,048.5	2,133.0	10,457.9	10,459.0
1910–19	3,554.1	4,414.0	2,784.5	2,820.0	11,227.5	12,053.0

Source

Col. 1: Purchases of equipment in 1909 equivalents and change in undepreciated 1909 equivalent main trackage times 1909 prices.
Cols. 2, 6, 1828–58: Gross investment in 1860 dollars from Table B-10 (of my *American Railroads*) times 108.23.
Cols. 2, 6, 1870–1919: Gross investment in 1929 dollars from Ulmer, *Capital in Transportation*, times 56.8.
Cols. 3, 4: Gross capital formation minus changes in net capital stock.
Col. 5: Table 6, col. 3.

the procedure should yield such favorable results in that decade with the smallest increase in mileage of road since the 1860's, and consequently a much larger expenditure upon betterments than additions, is confirmation of its merit. The shortfall is significant not only because it represents the anticipated deviation, but also because it reinforces the criticism earlier made that Ulmer's gross investment estimates before 1910 are too small. It would require a sharp reversal indeed, and within a single decade, in the coverage of the road and equipment index to explain the exactly opposite divergence between Ulmer's investment flow of 1900–09 and my own.

A final test consists of the very magnitude of that divergence over the longer span 1870–1909. Ulmer's total gross investment is about $1 billion (1909) smaller than mine. Earlier I have contended that the principal cause of understatement in Ulmer's investment estimates was his neglect of retirements. Accordingly, total retirements in 1909 equivalents valued in 1909 prices should approximate this sum. In fact, they do, aggregating $1.16 billion. Thus the flows in column 1 are at least reasonably correct.

The same judgment applies to the stocks. Although the above difference in gross investment explains but half the difference in 1869 stock estimates, with the remainder due to Ulmer's much higher capital consumption allowances, the latter are also easily shown to be in error by the appropriate amount. What is responsible for Ulmer's excess is not higher depreciation rates, but rather the understatement of flows in conjunction with the erroneous 1869 original cost estimate to which attention was drawn earlier. In the decade 1910–19 when the depreciation base is identical, the two estimates of capital consumption virtually coincide, and draw progressively apart only as one moves back in time. That is because Ulmer calculates his allowances as a percentage of an exaggerated total; his 1869 original cost is more than one-third too large, and his estimates for the intervening years to 1909 are likewise too great because they are interpolated backwards by the inadequate gross flows. If we reduce his capital consumption allowances by the ratio of the two different estimates of the capital stock at the beginning of the decade, i.e., substitute a smaller depreciation base, his total is almost at parity with mine.

The positive fruits of this extended discussion of railroad capital formation in the nineteenth and early twentieth centuries are a new series of constant-dollar gross and net investment, by decades, and a corresponding set of net stocks. At the same time there is a negative product. No longer can we safely rely upon the pre-1910 Ulmer estimates. Thus there remains the task of providing new annual series, both current and constant dollar, to fill the gap. Reworking the sample results obtained by Ulmer from railroad accounts to include retirements represents one approach, to which must be tied revision of the method of annualization to make the year-to-year changes consistent with the levels. For trend analysis, however, the capital stock estimates presented here should suffice in the interim.

Quantitative Measures of Railroad Development, 1840-1910: Labor

The estimation of labor inputs poses less difficulty than the reconstruction of the capital stock series. First, employment is a much more straightforward magnitude that was measured reasonably accurately at the time. Thus from 1880 on, there are counts of employment for the whole industry, and, even from 1850 on, national tallies of an occupational group termed "railroad men." Secondly, employment is a good measure of labor inputs because the standard number of hours per week remained constant over the entire period under investigation. Not until 1917 did the length of the workweek decline from sixty to forty-eight hours.[47] Finally, close relationships between labor inputs and measures of output like receipts, ton-miles,

THE RAILROAD SECTOR, 1840-1910

TABLE 8

LABOR INPUTS, 1840-1910

End of Fiscal Year[a]	Employment (thousands)
1839	5
1849	18
1859	85
1870	230
1880	416
1890	750
1900	1,018
1910	1,699

Source

1839-59: *American Railroads*, Appendix C.
1870: .592 (the sample regression coefficient) times $388.4 million, the estimated total receipts in that year derived in developing Table 1.
1880: *Tenth Census, 1880*, Vol. IV, *Agencies of Transportation*, adjusted to reflect corrected Atchison, Topeka, and Santa Fe report presented in the note (*ibid.*, p. 13).
1890-1910: Same as Table 1.

[a] In principle, the employment figures should be annual averages to represent accurately labor inputs corresponding to the flow of output. But until the appearance of the ICC data averaging counts, these are estimates of end-of-year totals.

etc., afford the possibility of estimating industry-wide employment in the earlier part of the period when it was not directly reported.

Such an approach is preferable to working with the Census of Occupation totals. These are badly understated because the category "railroad men" excludes a considerable number of maintenance and shop workers listed under other titles. In 1880 when industrial and occupational classifications are presented together for the first time, the ratio of the occupation to the industry total is .56, and increases to .64 by 1910.[48] An upward trend in coverage extends back still further, but is not sufficiently stable to permit the effective use of the occupation statistics. According to our independently derived industry estimates, the ratio is as low as .27 in 1850, rising thereafter to .42 in 1860, and .67 in 1870.

These pre-1880 industry estimates, arrayed in Table 8 along with the

---

[47] See the Aldrich Report, S. Rep. 1394, pp. 1365 ff., and Leo Wolman, *Hours of Work in American Industry*, NBER Bulletin 71, New York, 1938. As a consequence, Kendrick's man-hour and employment series show an almost exactly parallel change over the period.

[48] These are computed in Daniel Carson, "Changes in the Industrial Composition of Manpower since the Civil War," *Studies in Income and Wealth, 11*, New York, NBER, 1949, p. 127.

later and more reliable Census and ICC reports, have been developed by extending cross-section relationships between employment and receipts fit to sample observations in each of the years except 1839. For 1839, because of the limited information available, application of the 1849 parameter was the only feasible procedure.

We first evaluate the strengths and weaknesses of this method, and then compare its results with alternative estimates. A minimum criterion that there should be a significant relationship between employment and receipts for the sample observations is certainly satisfied by this approach. In 1849, with thirty-six reporting railroads in Massachusetts and New York, the share of the variance in employment explained by a single pooled regression is 89 per cent. In 1859, with a total of sixty-three observations, and three different regressions for each of the three states represented, R^2 varies from .82 to .99. Finally in 1870, a single pooled function relating employment to receipts on thirty-six railroads in Ohio and Massachusetts yields an R^2 of .98.[49]

This internal consistency of the relationships is not conclusive evidence in favor of the technique, however. An equally important requirement is that the sample relationships derived from these two or three states be appropriate to all regions of the country. If all railroads in Massachusetts in 1849, say, conform closely to the practice of having one employee per $1,500 of receipts, but southern and western railroads have two for

---

[49] Receipts are in thousand dollars, employment in units. Thus the coefficients are to be read as 64 men per $100,000 of receipts, 71 men, etc. The regressions are homogeneous in each case, that is, without a constant term, to facilitate aggregation. If $E = a + bR$, $\Sigma E = Na + b\Sigma R$, whereas if $E = bR$, $\Sigma E = b\Sigma R$ without the necessity of specifying the number of railroads in the country. In all instances the constant a was determined not to be significantly different from zero before the homogeneous form was fitted.

Only in 1859 were the regressions so distinct that an analysis of covariance test rejected the hypothesis of a single population. The reason is the extremely large weight of the New York observations in the total and the very close fit achieved in that state. As is pointed out, the coefficients are sufficiently close that the failure of a single regression to hold does not invalidate either the method or the results.

For 1870 two adjustments were made. The Massachusetts State report covered only ten months of operation, requiring multiplication of earnings by 6/5 to convert them to an annual basis. Three Ohio railroads were excluded from the sample, the Central Ohio, the Sandusky, Mansfield and Newark, and the Marietta and Cincinnati, because they deviated obviously and radically from their counterparts in the direction of an excess of employment. This is almost certainly due to part-time employment, for in the next year, all three railroads exhibited decreases in employment with increases in receipts.

Exact references to antebellum state reports are to be found in Appendix C of my *American Railroads;* for 1870 they are Massachusetts Board of Railroad Commissioners, *Second Annual Report,* 1870, and Ohio Commissioner of Railroads and Telegraphs, *Annual Report,* 1870 and 1871.

the same volume of earnings, the Massachusetts statistic will do rather poorly as a projection device. No direct demonstration of the satisfaction of this condition can be made since it is the lack of complete information that necessitates estimation in the first place. Nonetheless a blend of a priori analysis and indirect checks supports the approach.

The selection of receipts as the independent variable for the regressions carries the promise of greater regional stability than alternative choices like physical output or operating expenses. By rewriting the equation $E = bR$ in the following form, $E = \left(\dfrac{O}{R} \cdot \dfrac{W}{O} \cdot \dfrac{1}{w}\right) R$, where E is employment, O operating expenses, W the total wage bill, and w the annual wage per worker, it is easy to see why. The same parameter b may obtain in all regions even when there are differences in labor productivity (w), or profit margins (O/R), or income shares of capital and labor (W/O) as long as they all move in such a way as to produce a constant or near constant product. Since greater labor productivity will yield larger shares in income when substitution possibilities are limited (as in railroad technology), there are reasons to expect a positive correlation between w and W/O; at the empirical level, it also seems to be true that high profit margins coincide with relatively low wages, as in the South, where lack of railroad competition not only maintained rates but also kept wages low. Supporting this line of argument is the considerable stability of b over time, while all three components in its determination varied widely. Its value is .64 in 1849, .71 in 1859, .59 in 1870.[50] The ability to pool observations, except in 1859, also confirms geographic stability, although admittedly within a narrow compass; note too that in 1859 the range of variation in b is limited to the difference between .63 in Ohio and .73 in New York. Finally, the sample states represent the regions with the greatest employment so that errors in extrapolation to others are less crucial. Thus in 1859 the Middle Atlantic and New England regions plus Ohio account for 60 per cent of national railroad receipts. Suppose that the coefficient used to determine employment in other regions is as much as 20 per cent too small, total employment is in error by only 8 per cent. In earlier years the sample coverage is greater, and it is probably not much smaller in 1870.

Three less conjectural checks buttress the case. First, there is the 1880 distribution of receipts and employment. Although the very extensive difference between the lowest and highest ratios of employment to receipts —.48 in the Far West and .88 in the South Atlantic states—seems to belie

[50] For simplicity here, we have pooled the 1859 Massachusetts, Ohio, and New York observations, despite the circumstances elaborated in footnote 49.

the logic developed above, initial appearances are somewhat misleading. For the three regions with 80 per cent of total employment, the simple ratios range only from .67 to .76.[51] Moreover, because our 1870 sample states are drawn from two of these regions, a second check that consists of comparison of 1880 total employment projections using sample Ohio and Massachusetts relationships with the known total also comes out well. An 1880 Ohio sample as small as the 1870 one yields an almost exact estimate of the true 1880 figure, falling short by less than 1,500 from the observed 416,300. The Massachusetts coefficient in 1880, however, is smaller relative to the Ohio one—.63 vs. .71—and so leads to an underestimate of about 12 per cent. This is still a creditable performance; and since the disparity in coefficients did not prevail in 1870. it does not necessarily imply an equivalently large error earlier.

The last test draws upon new employment and receipts information in a southern state (Virginia) in 1859 and contrasts the calculated coefficient there with the statistic estimated from Massachusetts, New York, and Ohio data. A close relationship holds in Virginia between employment and receipts, but with a larger coefficient (.82) than the .63 to .73 recorded in the others.[52] A maximum error of about 25 per cent is involved if one uses the Ohio value to predict Virginia employment, and less than 15 per cent for the pooled .71. If this result holds for the South generally, the implied total 1859 error is well within the 8 per cent limit discussed above. But this is the worst possible representation of the comparison. If a single Virginia observation is eliminated from the sample, the coefficient for that state is altered to a much more consistent .70, and virtually without reduction in R^2. To the extent that this single road is indeed atypical, there apparently need be no concern at all for the application of the sample regressions to other regions, at least in the early years.

For these reasons, therefore, we can expect the pre-1880 employment estimates in Table 8 to be accurate measures of railroad employment.[53] Stanley Lebergott's estimates in this volume derived by an entirely different technique—extrapolation of 1880 regional employment-mileage ratios based on sample ratios—confirm these expectations for the three antebellum years.[54] His 7,000, 21,000, and 82,000 totals are quite comparable to our 5,000, 18,000, and 85,000, and leave little to choose between.

---

[51] This information is drawn from the *Tenth Census, 1880*, Vol. IV, pp. 25 and 257.

[52] The source is the *Annual Report* of the Virginia Board of Public Works for 1858/59. There are ten observations and the R^2 of .82 is significant at the 1 per cent level.

[53] This applies even to the 1839 estimate obtained by using the 1849 parameter without change. The relative constancy of the coefficient over time is the reason for confidence in this procedure.

[54] See pp. 117–204.

But his 1870 estimate of 159,000, even allowing for a slight difference in timing in our measures, is so far short of our 230,000 to compel comment. Lebergott's shortfall can be traced to a single region, the Middle Atlantic; his data indicate a rise of more than 85 per cent in employers per mile between 1870 and 1880 in that region, while the ratio increases in the rest of the country only by 13 per cent. This inconsistency suggests that the 1870 Middle Atlantic value may be too low, but not necessarily: the rate of increase in the same regional ratio between 1860 and 1870 (if my 1870 estimated employment is correct) considerably exceeds that of others due to much greater increases in output per mile of line. To document this allegation further, therefore, it has been necessary to examine the employee-mileage ratio in Ohio both in 1868 and in 1880; at the former date, it stands at 6.10, well above Lebergott's regional value of 3.95, and actually declines to 6.07 in 1880.[55] Both the earlier absolute level as well as the direction of change are inconsistent with Lebergott's analysis, and this for a state with 20 per cent of the mileage in the area. The other states would have to have a much smaller number of employees per mile (3.5) for the results to be consistent, and this would imply a difference between Ohio and other Middle Atlantic railroads much greater than that observed in either earlier or later years.

These technical details are sufficient reason to cast extreme doubt on Lebergott's 1870 alternative. Another, more general argument is the inconsistency of the lower employment estimate with the pattern of output growth. Ton-miles per mile of road increased from about 100,000 in 1859 to more than 270,000 in 1870 and to 370,000 in 1880.[56] Yet Lebergott's employment per mile climbs more rapidly from 1870 to 1880 than from 1860 to 1870. Unless one can explain the almost unparalleled productivity growth of the decade 1860–70 and the negative movement in the next ten years, it is best not to rely on the 1870 Lebergott magnitude.

The only other possible comparison also relates to the disputed 1870 figure. But in this instance, our 230,000 is substantially *below* Daniel Carson's 270,000, which was obtained by scaling up the Census of Occupations count by the undercoverage observed in later years.[57] It is tempting to invoke the virtue of being at neither extreme as further justification for our 1870 value; but a little reflection should establish

---

[55] The Ohio data are to be found in the *Annual Reports of the Commissioner of Railroads and Telegraphs* for 1867/68 and 1879/80. One reason why Lebergott—who also used Ohio in his tabulations—may have gone awry is that employment is listed for the Ohio portion of the line only, not the total. If total mileage is then used in the denominator, the number of employees per mile is incorrectly depressed.

[56] According to the estimates presented in Tables 1 and 2.

[57] See Carson's paper in *Studies in Income and Wealth, 11.*

that a middle-of-the-road position has no intrinsic merit in statistics, whatever its advantages in politics. Rather, the ground for preferring our 1870 estimate to Carson's is its basis in contemporary sample data, while Carson's depends upon extrapolation of later nonbehavioral relationships. There would seem to be little reason why the undercount of the Census of Occupations in 1870 (a rather questionable Census in many respects) should be the same as in 1880, whereas there is good and obvious reason for employment to be related to receipts, and also for a cross-section extrapolation to yield a close approximation to the correct national total.

The 1880 and subsequent entries require no comment beyond that already presented in Table 8 itself.

Quantitative Measures of Railroad Development, 1840-1910: Fuel

In principle, an analysis of productivity changes ought to be concerned with inputs of materials as well as of capital and labor. For reasons of convenience, however, materials are usually neglected in favor of working directly with value added. This simplification overstates productivity change in the typical case and also obscures one important source of increased output per unit of labor and capital.[58] Although this criticism

---

[58] When technological change is positive, a value-added index (geometric or arithmetic) will exaggerate the rate of change for a given industry. See Evsey D. Domar, "On the Measurement of Technological Change," *Economic Journal*, December 1961, p. 716, and "On Total Productivity and All That," *Journal of Political Economy*, December 1962, p. 603.

But frequently *gross* output and *net* inputs are used. This formulation actually may eliminate the value-added bias, depending upon changes in the relative importance of materials. I will illustrate with the arithmetic, or total factor productivity, index, although the argument can also be extended to the geometric index. Let Y be value of real output in base-period prices; L, K, and R physical inputs of labor, capital, and raw materials; and w, i, and r their respective prices. Then there are three alternative indexes corresponding to (1) all inputs, (2) value added, and (3) gross output, net inputs:

$$C_1 = \frac{Y}{w_0 L + i_0 K + r_0 R}$$

$$C_2 = \frac{Y - r_0 R}{w_0 L + i_0 K}$$

$$C_3 = \frac{Y}{w_0 L + i_0 K} \cdot \frac{w_0 L_0 + i_0 K_0}{Y}$$

$C_2 > C_1$ when $C_1 > 1$, but $C_3 = C_1$ when $\dfrac{w_0 L_0 + i_0 K_0}{V} = \dfrac{w_0 L + i_0 K}{w_0 L + i_0 K + r_0 R}$,

that is, when the share of materials in total inputs valued at base-year prices is constant. If the share declines, C_3 will understate C_1, and if it increases, $C_3 > C_1$, but for certain ranges $C_1 < C_3 < C_2$, making C_3 a better measure than the value-added index.

applies with less force to the railroad sector, because of its high value-added ratio, nevertheless, I have undertaken to include fuel inputs explicitly among the series. Although not exhaustive of the complete range of materials necessary to railway operation, fuel does account for the largest part of current material inputs.[59]

The inclusion of fuel inputs has the further virtue of highlighting a significant nineteenth century technological change. Before the Civil War, despite some preliminary experimentation that began to yield tangible progress by the 1850's, American locomotives were fired almost exclusively by wood. Within a score of years thereafter, a transformation so rapid had occurred that twenty times more coal than wood was being consumed annually, and more than a fourth of bituminous coal output was regularly absorbed by the railway sector. The underlying mechanism is almost a text book illustration of substitution in response to changing relative prices. To begin with, eastern railroads with large coal deposits along their lines, and hence both low coal prices and elastic supply, invested in research necessary to eliminate the troublesome technical problems that had limited the development of coal-burning locomotives. Once successful, the eastern railroads penalized by high wood prices and the western railroads favored by low coal prices led the parade to mineral fuel. Note that it was not so much the initially high price of cordwood that motivated the first research as the prospect of much higher future cordwood prices due to rapidly growing demand and inelastic supply. But once the breakthrough had come, the New England railroads that paid high prices found it profitable to switch. The last holdouts, naturally enough, were southern railroads whose scant demands and favorable environment meant continued low prices for wood.[60]

Unfortunately, the pace of this transition, as well as the quantities of fuel consumed at various dates, is largely unrecorded at the national level. A single observation on the 1880 volume of cordwood consumption must suffice until ICC tabulations of coal purchases begin in 1917. All we can glean from railroad records are accurate reports of total fuel expenditures (and not before 1880).

[59] The other large material inputs are those of rails and ties to compensate for depreciation of capital. To ignore these is to measure outputs gross but inputs net, or, in terms of footnote 58, to use C_3 as our index. Since depreciation actually declined due to technological changes, C_3 will then understate the observed increase in productivity, but the magnitude of the effect is limited. In 1880 the neglected nonlabor costs and purchases of all additional services amount to about 20 per cent of output, in 1910 to 18 per cent.

[60] For further discussion of the shift before the Civil War, see my *American Railroads*, Chapter III. Appendix D of the *1880 Census of Transportation* confirms the concentration of wood consumption on southern railroads.

TABLE 9

MATERIALS INPUTS, 1840-1910

Fiscal Year	Expenditures on Fuel (mill. dollars) (1)	Locomotive Mileage (mill. miles) (2)	Cordwood Consumption (mill. cords) (3)	Coal Consumption (mill. net tons) (4)
1839	.4	5.4	.12	--
1849	1.5	15.9	.35	--
1859	8.1	81.0	2.0	.3
1870	25.2	233.0	3.3	4.3
1880	32.8	465.3	2.0	13.4
1890	64.0	919.7	--	36.8[a]
1900	90.6	1,170.6	--	58.8[a]
1910	217.8	1,714.4	--	128.1[a]

Source

Col. 1, 1839-1859: Estimated operating expenses times 10 per cent in 1839 and 1849 and 12 per cent in 1859. See my *American Railroads*, Appendix A and Chapter 3. The proportion of expenses accounted for by fuel was estimated by a sampling technique for 1859.

Col. 1, 1870: Estimated operating expenses times 10 per cent. Expenses were derived from receipts by applying 1871 operating ratio reported in *Poor's Manual*. The fuel account proportion for this period is given in Wellington, *Economic Theory*, pp. 176, 180.

Col. 1, 1880, 1890: *Tenth Census, 1880*, Vol. IV, and *Eleventh Census, 1890*, Vol. XI.

Col. 1, 1900, 1910: ICC, *Statistics of Railways* for 1900, 1910.

Col. 2, 1839, 1849: Terminal locomotive stock[b] multiplied by 12,000 miles. The 1849 average service is determined by extrapolating the 1859 average on the New York average train-miles per locomotive in both years. (The New York average train-miles per locomotive was the same as the national average locomotive mileage in 1859.)

Col. 2, 1859: Terminal locomotive stock[b] times 15,000 miles, the average service of 1,571 locomotives on twenty-five railroads in 1858. See *American Railroad Journal*, Vol. XXXI, 1858, p. 297.

Col. 2, 1870: .6 times $388.4 million; the coefficient was determined from a random cross-section sample of fifteen roads in 1870 with $R^2 = .82$, and the earnings total is the one discussed in connection with Table 1.

Col. 2, 1880, 1890: *Tenth Census, 1880*, Vol. IV, and *Eleventh Census, 1890*, Vol. XI. Total train mileage in these years includes switching and non-revenue service and is therefore equivalent to locomotive mileage.

Col. 2, 1900: ICC, *Statistics of Railways*. Revenue train mileage adjusted to locomotive mileage by interpolating between the ratios of locomotive mileage to train mileage in 1890 and 1910; the interpolating series is the ratio of other track to total mileage, on the grounds that switching mileage is dependent thereon.

Col. 2, 1910: *Statistics of Railways*.

Cols. 3, 4: See text.

[a]Contains some wood and oil consumption, to a maximum extent of 5 per cent in fuel equivalents.

[b]These terminal stocks differ slightly from the 1838, 1848, and 1858 totals of Table 3. The 1839, 1849, and 1859 values are 450, 1325, and 5400, respectively.

There are two alternative approaches to take full advantage of what few data are available. For the period beginning with 1880 when solid information on expenditures exists, coal was used almost exclusively. This means that division of the financial aggregate by an average unit coal price will yield the relevant physical volume. And, fortunately, for the one year in which wood loomed largest, 1880, we can separate out expenditures for it with the help of the Census tabulation of cordwood consumed by railroads in that year. Earlier, our most reliable knowledge is of the number of locomotives and the approximate consumption of wood and coal per mile run. It is not too difficult to convert the locomotive stock to an annual flow of service performed based on sample data from railroad reports. Nor is it perilous in the antebellum period to specify the division between mileage logged by wood and coal burners since, even in 1859, the latter accounted for a meager proportion of the total. This method does encounter difficulty with the 1870 estimate due to the uncertain division between wood and coal in that year. With the further use of state reports, and some variation upon the technique just described, we treat that problem in a fashion described below.

Neither of these two methods is mutually exclusive, and although each has reason to be favored in particular periods, the other is also employed as a check. Accordingly Table 9 presents in column 1 a complete series of fuel outlays to serve as the basis of the expenditure approach, and in column 2 a full complement of locomotive mileages. Generally, the figures for 1880 on are official, or comparable in quality thereto. Earlier, they are the product of indirect calculations. For example, the fuel outlays are estimated as a proportion of operating expenses, which are themselves obtained from an array of contemporary sources in the same fashion as the total receipts discussed in connection with Table 1. Similarly, locomotive mileage in 1870 is tabulated from earnings, and at earlier dates from sample accounts of annual services.[61]

It is easiest to explain the derivation of the cordwood and coal consumption totals of columns 3 and 4 in chronological order. The antebellum estimates do not pose great problems despite the lack of any official benchmarks. Abundant contemporary evidence confirms an average run of 30 to 40 miles per cord in the later 1850's, a distance that must have been greater in earlier years as a result of shorter and less frequent freight

---

[61] Although 1870 locomotive mileage was determined by a regression technique utilizing earnings as the independent variable, other data on annual mileage per locomotive also support the aggregate reached. The Pennsylvania average in 1870 is 23,000 miles per engine, as calculated from Wellington, *Economic Theory*, p. 141 (after allowance for switching mileage). Since the Pennsylvania experience was close to the national average both in 1858 and 1880, this is relevant evidence.

trains and greater concentration upon passenger accommodation. Fuel requirements for both 1839 and 1849 are, therefore, pegged at the lower level of 45 miles per cord. This may be immediately divided into the stated locomotive mileage to obtain cordwood consumption for 1839 and 1849 since coal was, for all practical purposes, not used. For 1859, the intermediate step is necessary of allocating the locomotive mileage between wood and coal burners due to the first appearance of the latter in any number. Contemporary discussions suggest that at most 500 coal-burning locomotives were in use in 1859, and that their total mileage could not have exceeded 10 million miles with aggregate consumption of 300,000 tons. The residual 71 million miles gives rise to the cordwood entry.[62]

These inputs, in combination with the corresponding financial magnitudes, yield prices of $3.33, $4.29, and $3.60 per cord of wood. Towne and Rasmussen's prices for comparable years are far lower at $0.84, $1.10, and $1.60 and imply consumption levels at least twice as high as our estimates.[63] I am not inclined to give this disparity much weight, however. Towne and Rasmussen's prices are based upon extrapolation of a single 1880 observation on a single series of prices received by Vermont farmers. In 1880 this average price is itself 20 per cent below a comparable Census price of wood purchased by railroads, and earlier the divergence seems to be greater. At the end of the 1850's, all indications point to a price of over $2.00 per cord for even the most favorably situated railroads, and the average price for Massachusetts railroads as a whole was $4.46.[64] The implied prices, therefore, seem quite reasonable and consistent with recorded prices, far more so than Towne and Rasmussen's which embody the dubious assumption that the trends of national scarcity and demand replicated the Vermont experience.

Cordwood and coal requirements in 1870 are more difficult to gauge. The distribution between the fuels at this time is quite uncertain although there are clear indications both of the absolute superiority of coal and a rapid, continuing tendency in that direction. In Ohio in 1868, wood burning was still more than twice as prevalent as coal consumption; by 1870 equality prevailed; and by 1872 the proportions had been almost reversed. Massachusetts railroads in 1871, too, were using more than twice as much coal, and Illinois roads, due to the accessible bituminous

---

[62] My *American Railroads*, Chapter III, gives detailed sources and elaborates the procedure summarized here.

[63] *Trends in the American Economy in the Nineteenth Century*, Studies in Income and Wealth 24, Princeton for NBER, 1960, p. 311.

[64] The average national price in 1880 reported by the *Tenth Census* (Vol. IX, p. 479) is $2.21; the corresponding railroad price is $2.60. Also in 1876 the *Railroad Gazette* (Vol. III, 1872, p. 114) indicates prices of $3.50 a cord; Towne and Rasmussen, only $2.44.

deposits in that state, even more.[65] If we accept a 2-to-1 ratio in favor of coal in the East in 1870 (as suggested by the Massachusetts experience), parity in the West (as indicated by the Ohio transition), and zero coal consumption in the South, we can reach a national ratio by weighting each region by its relative importance in traffic operations. Such a procedure points to a 30 per cent disparity in favor of coal consumption in 1870.[66]

To obtain aggregates of cords of wood and tons of coal, it is necessary to add information on fuel requirements per locomotive-mile. Here varied evidence suggests the equivalent of some 60 pounds of coal a mile. Ohio locomotives in 1861 were apparently using an even greater amount, 69 pounds, and Illinois railroads are found to have used over 80 pounds in 1872, but these seem to be exaggerated in the light of later observations in these states and simultaneous Massachusetts use of less than 50 pounds and consumption of 64 pounds on the Pennsylvania Railroad.[67] The small Massachusetts requirements are natural in view of a passenger predominance in that state, and the significantly smaller fuel requirements associated with passenger trains—half as much—whereas it is doubtful that the national average exceeded the experience of the heavily freight-oriented Pennsylvania Railroad.

At an input of 60 pounds a mile, aggregate coal-equivalent consumption in 1870 reached a level of 7 million tons. With a national ratio of coal to cordwood consumption of 1.3 tons to 1 cord, this is translated into estimates of 4.3 million tons of coal and 3.3 million cords of wood.[68] The financial ledgers are quite consistent with this allocation. At prices

[65] These accounts of consumption are derived from the relevant state reports, except for the 1870 Ohio observation which is reprinted in *Railway Gazette*, Vol. II, 1872, p. 377. For Massachusetts, the average expenditures on wood and coal were converted to physical units by dividing by prices of $5.50 per cord and $8.00 a ton, prices determined from prices paid by individual railroads within the state.

[66] That is, with ratios of coal to wood of 2, 1, and 0 and weights of .434, .432, and .132, we derive the national relationship of $x = 2(.434) + 1(.432) + 0(.132) = 1.3$. The weights are 1871 regional gross earnings relative to the national total as reported in *Poor's Manual* for 1872-73, p. xxviii. (The Pacific region is included in the West.)

[67] These data are derived from the state reports mentioned earlier. The observation relating to the Pennsylvania Railroad is found in Wellington, *Economic Theory*, p. 141. One reason for the overstatement both in Illinois and Ohio is the tendency of railroads to err in returning locomotive mileage; gross errors such as substitution of car mileage are easy to detect and eliminate, but more subtle distortions cannot be rectified.

For the purposes of a coal equivalent consumption total, cordwood is converted to coal at the rate of 1 cord of wood = .8 ton of coal. This is the same factor used in Sam H. Schurr, Bruce C. Netschert et al., *Energy in the American Economy*, Baltimore, 1960, p. 499.

[68] Algebraically, this converts to the solution of two simultaneous equations. Let x be million tons of coal consumed, and y million cords of wood. Then

$$x + .8y = \frac{60 \times 233}{2000} = 7.0$$

and

$$x = 1.3y.$$

of $4.00 a cord for wood, and $3.00 a ton for coal, drawn from quotations in the railroad literature, total expenditure is $26.1 million or within $.9 million of the entry obtained from railroad operating expenses. As further evidence, one may adduce the vintage of the 1870 locomotive stock. Almost 60 per cent of the locomotives extent at the beginning of that year were built during the decade 1859–69. If most were coal burners, as they probably were, the postulated distribution of coal and wood consumption in 1870 is to be expected.[69] Whatever its shortcomings, such a picture is undoubtedly closer to reality than the exaggerated claim of 6 million cords of wood consumed by railroads at the end of the 1860's.[70]

The publication of accurate fuel expenditures beginning with 1880 signals a corresponding change in estimating technique. Now the task becomes one of determining accurate coal prices paid by railroads. In preference to simple extrapolation of the prices actually paid by railroads—which first became available in 1917—a somewhat more complicated procedure has been substituted. To the minehead price, we add a variable transport charge equal to 100 miles of transportation at .8 the average ton-mile rate in the given year. This formulation has been determined from the data of the period 1917–25.[71] How well it does as far back as 1880 depends upon changes in the average haul for coal and the structure of rates. Since railroads on the average were closer to coal deposits in 1880 than thereafter, I have altered the specification to assume only 80 miles of transport in that year. Some notion of the maximum error is given by the largest difference between predicted and actual level of fuel purchases during the 1917–25 period; it occurred in 1917 when it amounted

---

[69] Systematic tabulation of the locomotive rosters published in the *Bulletin of the Railway and Locomotive Historical Society* would be necessary to establish firmly the contention that most new locomotives were coal burners. But this is the impression I have gained from scanning the records. In addition, conversions increased the ranks of the coal business. Thus the Michigan Central, which increased its locomotive stock only by two between 1859 and 1869, and retired a few additional units, therefore lagged in conversion to coal, whereas the Burlington, whose locomotive stock doubled, was entirely dependent upon coal (*ibid.*, No. 19, pp. 24–25, and No. 91, 127).

[70] Cited by Schurr et al., *Energy in the American Economy*, p. 52.

[71] Let Q be tons of fuel consumed, p the price of bituminous per ton at the mine, f average freight rates per ton-mile, and X fuel expenditures. I have solved the equation $\sum_{1917}^{1925} Q_i(p_i + mf_i) = \sum_{1917}^{1925} X_i$ for m, the number of miles coal was transported at the average rate. This turns out to be 83.8. Since rates for coal lay below the average rate, I have interpreted this as 100 miles of transportation at .8 the average rate, an equivalent formulation.

Extrapolation can be expressed as $\dfrac{X}{Q} = \dfrac{X_0}{Q_0} \cdot \dfrac{p}{p_0}$

Q and X (1917–25) came from the ICC, *Statistics of Railways*; p and f from *Historical Statistics*, 1960, Series M-91 and Q-86, respectively. The same p and f series are used for the 1880–1910 estimates, except that 1880 freight rates are given in Table 1.

to 19 per cent; the average absolute deviation is a much lower 5.4 per cent, however. This is a good deal better than that obtained by extrapolation of the 1925 price over the same period, when it underestimates aggregate consumption by more than a tenth and is too small by 28 per cent in 1917. The reason for the bias is the upward tendency in rates during 1917–25. Since the opposite tendency in rates is observed between 1917 and the 1880–1910 period, simple extrapolation necessarily leads to substantial overestimates of consumption if applied earlier.

An equally error-prone method is that adopted by H. S. Fleming in order to derive his estimates of bituminous coal consumption by railroads, a series which the volume *Energy in the American Economy* seems to sanction.[72] What he has done is to multiply ton- and passenger-miles by fuel factors for each. These are roughly constant, and hence make no allowance for the increasing efficiency of longer trains and less deadweight, and in addition are set at levels that far exceed what the later data of the ICC affirm. Not surprisingly then, Fleming's estimates are pegged higher than those in Table 9 for 1890 and diverge increasingly, to the point where the 1900 upper bound set by price extrapolation actually falls below Fleming's figure. Indeed, because he also assumes that railroads not only used coal exclusively, but also used only bituminous coal, his estimated consumption of bituminous in 1905 is almost equal to the ICC 1919 bituminous total, despite a doubling of output between the two dates.

Although the disparity with Fleming's series can easily be dismissed, it nonetheless is useful to check the estimates for 1880 and thereafter against independent information on coal consumption per locomotive-mile. The coal equivalent consumption per mile implied by columns 2 and 4 of Table 9 rises from 65 pounds in 1880 to 150 pounds in 1910, a trend determined by the shift to heavier trains. The early levels are consistent with Wellington's 1885 fuel requirements of 25–50 pounds for passenger trains and 75–125 pounds for freight trains, as well as with the 1870 value of 60 pounds used earlier. The later observations accord with the ICC measured consumption of 166 pounds in 1917 and an informed 1906 estimate of "not less than 90 million tons," which when translated into requirements per mile comes to 120 pounds.[73] Finally, extrapolation to earlier dates of the 1917 national consumption per mile by the use of state reports from Kansas, Iowa, and Illinois, encompassing as much as 20 per cent of locomotive mileage, lends additional confirmation to the

---

[72] See the approving citation on p. 73. Fleming's estimates are found in *A Report to the Bituminous Coal Trade Association on the Present and Future of the Bituminous Coal Trade*, New York, 1908.

[73] Wellington, *Economic Theory*, pp. 132 ff; ICC, *Statistics of Railways for 1917*, pp. 32, 61; W. F. M. Goss, "The Utilization of Fuel in Locomotive Practice," *U.S. Geological Survey Bulletin* No. 402, 1909, p. 17.

estimates. The average of the three projections in 1910 is 3 per cent smaller than the entry in Table 9, within 1 per cent in 1900, and at worst in 1890—when extrapolation becomes a more dubious tool and only two states are reporting—8 per cent smaller. If anything, the suggestion is a slight underestimate, but it is so small as to require no revision.

Output, Input, and Productivity Change

The pattern described by the first eight decades of railway operation, as we have just derived it, is summarized in Table 10 and Chart 1. The

TABLE 10

SUMMARY OF RAILROAD OUTPUT, INPUT, AND PRODUCTIVITY
(1910 = 100)

Fiscal Years	Output (1)	Persons Engaged (2)	Capital			Fuel (6)	Total Input (7)	Total Factor Productivity (8)
			Road (3)	Equip. (4)	Total (5)			
1839	.04	.3	.9	.2	.8	.07	.5	8.7
1849	.31	1.1	2.5	.7	2.2	.2	1.4	22.1
1859	1.7	5.0	11.4	3.6	10.1	1.5	6.6	26.4
1870	6.0	13.5	18.5	6.7	16.6	5.4	13.9	43.4
1880	13.8	24.5	34.2	17.3	31.5	11.7	25.9	53.2
1890	32.8	44.1	66.7	36.6	61.9	28.7	49.3	66.5
1900	54.8	59.9	77.4	45.2	72.3	45.9	63.2	86.7
1910	100.0	100.0	100.0	100.0	100.0	100.0	100.0	100.0

Source

Cols. 1–6: Tables 1–9.
Col. 7: Weights for employment, capital, and fuel are .52, .38, and .10, which are the proportional 1910 shares of the factors as determined from ICC, *Statistics of Railways for the United States for 1910*, pp. 45, 57, 74, and 76.
Col. 8: Output divided by total input.

growth of output is no different from that exhibited by many industries in their initial phases. An extraordinary advance from the small pre-Civil War base yielded to a still impressive, but much lower, stable rate of 7.3 per cent per annum from 1870 through 1910. This record far exceeded that of such aggregates as national income or total commodity production. Over the entire interval 1839–1910, railroad services grew at an annual rate of 11.6 per cent, with income and commodity output proceeding at a pace only one-third as rapid. Indeed, no single major sector grew as rapidly. With an 1870 benchmark, these same observations obtain, albeit with a somewhat narrowed margin of superiority. Nonetheless, every single decade saw the railroads as a pace setter.[74] There is little reason,

[74] The income and commodity output estimates are the earlier cited ones of Robert E. Gallman. The sector breakdown of commodity output includes agriculture, manufactures, mining, and construction; and services can be approximated as the difference between income and the comomdity flow.

CHART 1

Output, Input, and Productivity Change, 1839–1961

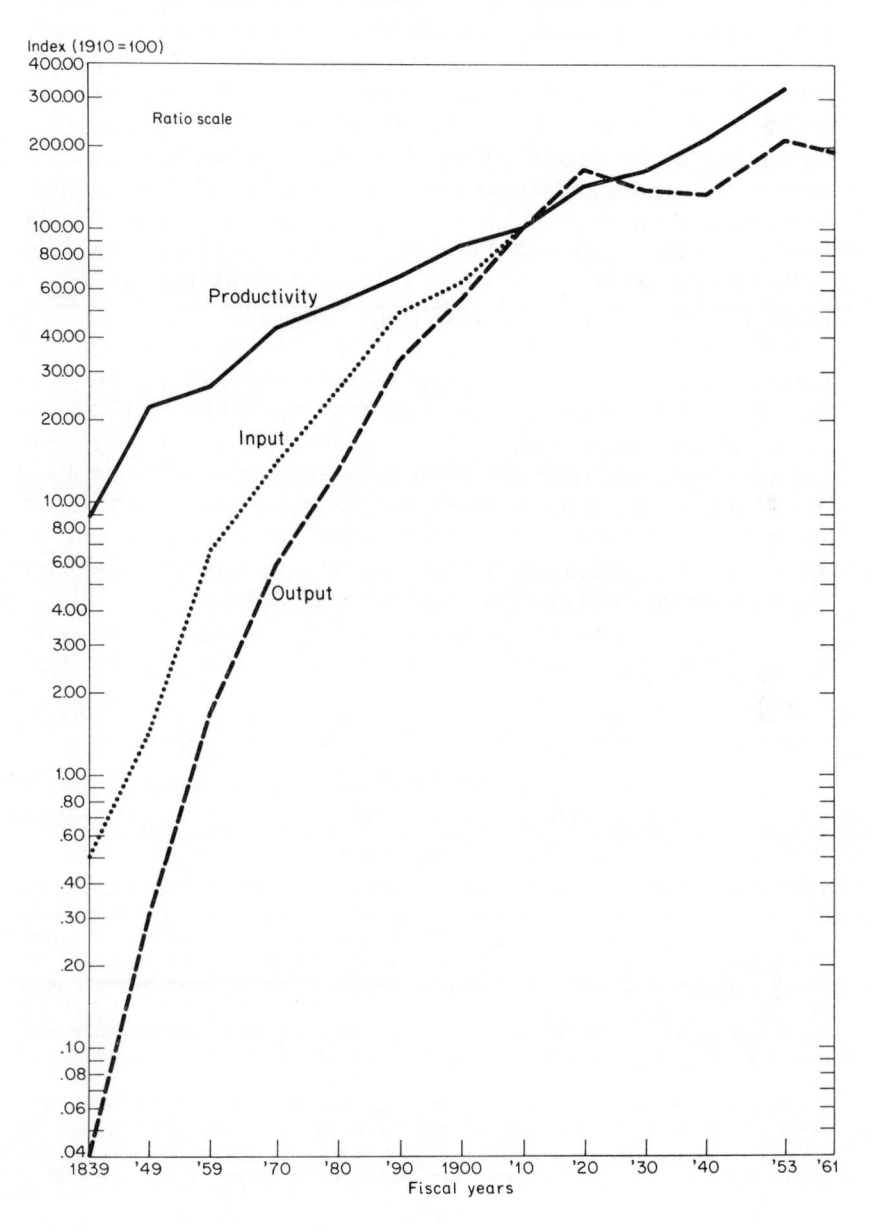

then, to wonder at the prominent position railroads won in the hearts and minds of contemporaries.

After World War I, the forces of retardation increased in strength. Chart 1 tells a tale of continuing deceleration, broken only by episodes of wartime prosperity. Indeed, the final stage of absolute decline is already here or very close upon us. The reason is not hard to find. Output growth before World War I derived considerable impetus from a substantial geographic extension of the system; it suffices to recall that the 1910 mileage was almost eight times the 1860 level; as a consequence, from 1859 to 1910, geographic extension bears the largest share of the explanation of output growth. Mileage in use increased just about twice as rapidly as traffic density, and if we credit the lengthening of average haul to the influence of extension as well, the role of geographic widening in output growth is 2.2 times that of intensified demand.[75]

Because the system reached its peak in 1916, this dominant source of previous expansion gave out by World War I. At the same time railroads were faced with mounting competition from new forms of transport, revitalized waterways, and, especially, motor vehicles. Finally, unit transport requirements may well have declined as the industrial composition of national income in the twentieth century altered. From virtually all sides, therefore, railroads have been subject to retarding tendencies, against which diversified regional development with a concomitant longer average haul has stood as a lone and inadequate defense.

This pattern of output changes was duplicated by input trends, with one crucial exception. Both the stock of capital and employment grew

[75] Railroad output can be decomposed into three factors, tons (and passengers) originating per mile, number of miles of road, and average distance carried:

$$O = T/M \times M \times d.$$

Therefore, $\quad O_{1910}/O_{1859} = \dfrac{T/M_{1910}}{T/M_{1859}} \times \dfrac{M_{1910}}{M_{1859}} \times \dfrac{d_{1910}}{d_{1859}} = 4.51 \times 8.62 \times 1.16 = 45.1.$

The relative size of the ratios of the components is the basis for the assessment of the importance of each. Note that the output index of Table 1 gives a slightly larger ratio of output between the two dates because 1910 weights have been used for this comparison; this probably leads to understatement of the change in average haul, and the role of extension since it was calculated as a residual. Offsetting this is the use of total tonnage and passengers carried rather than those originating. Consolidation probably meant that tonnage originating was a higher proportion of total tonnage at the later date, and so T/M_{1910} divided by T/M_{1859} may underestimate the change in intensity of demand.

Tonnage and passengers carried are taken from Frickey, *Production in U.S.*, p. 100; mileage from Table 2; and ton-miles and passenger-miles from Table 1; average distance was calculated as the residual. 1910 average revenues per passenger and per ton were used to weight those quantities and 1910 rates for passenger- and ton-miles.

much less rapidly, and it is this divergence which gives rise to the impressive productivity advance recorded by the railroad sector. As with output, the accomplishment in this sphere outstrips that of the economy as a whole. Between 1839 and 1910, railway output per employee expanded 2.8 per cent annually; Gallman's aggregate product estimates divided by Lebergott's labor force totals yields a rate of 1.3 per cent, or less than half as great.[76] A productivity index including capital services enhances the margin in favor of railroads. In that sector, unlike the economy as a whole, labor productivity increased in the face of a *declining* capital-output ratio, and a total factor productivity comparison therefore pits a higher railroad rate of growth of 3.5 per cent against an aggregate rate indeterminately lower than 1.3 per cent.

This disparity is reflected, although inexactly, in a dramatic relative decline in the price of railroad services. In terms of the bundle of commodities making up Warren and Pearson's wholesale price index, it cost little more than one-tenth as much to purchase a ton-mile of transportation in 1910 as in 1839, and about two-fifths as much for a far more comfortable passenger-mile. Even excluding the very large decline in rates (and increase in productivity) between 1839 and 1849, real freight rates fell more than 80 per cent from their 1849 level, and real passenger charges 50 per cent.[77]

The continuation of the data beyond 1910 in Chart 1 reveals a striking difference between the trends of output and productivity that had so much in common earlier. Kendrick's data affirm a continuing upward movement in productivity from 1909 through 1953 at the creditable rate of 2.7 per cent each year, while output, as we have seen, manages hardly to expand after 1920. This pace, although below the 3.5 per cent rate maintained through the nineteenth century, actually exceeds the 1870–1910 rate of 2.2 per cent. Not only is there scant evidence of retardation, therefore, but there are possible signs of accelerated advance. One distinction between the earlier and later period is relevant, however. After 1919 the gains in railway productivity relative to those achieved throughout the

---

[76] These are found in this volume, and in Stanley Lebergott, *Manpower in Economic Growth*, New York, 1964, p. 512. Gallman's 1840 and 1909 outputs were divided by Lebergott's labor force estimates for the same years.

[77] The BLS all-commodity price index stands at 101 in 1909-10, and the Warren-Pearson continuation of the same at 112 in 1839 (1910–14 = 100). Thus the decline of rates in current prices only slightly overstates its real descent. The indexes are found in *Historical Statistics*, 1960, pp. 115–117.

The failure of the real price decline to equal precisely the ratio of sectoral to aggregate productivity is due to inappropriate index number construction and to a more rapid decline in the price of inputs in the railway sector than in the whole economy. (See last few pages of this paper, where we show that the dual of Kendrick's productivity index is a price index of inputs and output.)

economy drop off sharply. To this extent, the productivity series mirrors the same decline in importance of the sector as do the output statistics.

With these trends in mind, we may now turn to the final task before us—explanation of that impressive nineteenth century productivity advance. The fundamental factors operating then (and later) may be grouped into two major categories: improved quality of inputs and economies of scale. The latter rubric is used to shelter a number of different positive effects all associated with size, and it will be taken up first.

To begin with, it is useful to make allowance for an effect that is peculiar to railroads and other related industries. For capital-intensive and capital durable sectors faced with indivisibilities, the size of the capital stock is not a good proxy for the annual flow of services it delivers. At their inception, firms will typically be burdened with higher capital-output ratios than current demand seems to dictate, due both to technical considerations and to positive expectations. Only over time will capital services attain a stable relationship with the magnitude of the stock. As can be seen in Table 11, the railway industry during its period of geographic extension is a prime example of what has just been described. From a capital-output ratio of over 70 in 1839, it descends to less than 4 in 1910. The pace of descent varies over time. Periods of rapid extension of the rail network like the 1850's and the 1880's exhibit much smaller declines than adjacent decades when increased output is accommodated by existing road and equipment. Conversely, a period of curtailed construction, coupled with greater demand such as that of the Civil War, saw dramatic declines. After 1910, as could be expected, the capital-product ratio moves within relatively narrow limits.

Because the capital stock has been used as an input, part of the measured productivity gain of railroads recorded in Table 10 derives from this phenomenon of increasing utilization. Two very crude adjustments assist in isolating the effect of this special factor. On the one hand, we may assume that the true production relationship involved capital inputs exactly proportional to the employment of labor; that is, over time each worker used no less capital, as the capital stock and employment data of Table 10 tell us, but actually the same amount. On the other, we may suppose that the capital-labor ratio actually increased—as it did for the economy as a whole—and postulate a constant ratio of capital services to output. In the former case, total factor productivity converges to the lesser growth of labor productivity, while in the latter it is further reduced because the implied increase in capital per worker partially explains the greater output.

Table 11 computes each of these alternatives. Series I for total factor productivity results when the capital-output ratio is held constant at its 1910 level, while series II reflects a 1910 capital-labor relationship. The rate of growth of series I is 2.3 per cent annually, while that of II is a higher 2.9 per cent. Contrasted with the initial 3.5 per cent, it seems that secular variation in capital utilization may explain between one-sixth and one-third of the recorded rate of productivity gain. A constant capital-output ratio undoubtedly removes too much of the gain. Such capital saving innovations as increased motive power per unit of weight, the higher ratio of load to dead weight in freight cars, and the greater durability

TABLE 11

PRODUCTIVITY EFFECT OF INCREASING UTILIZATION

Fiscal Years	Capital-Output Ratio (1910 dollars) (1)	Capital-Labor Ratio (2)	Total Factor Productivity (3)	Total Factor Productivity Variant I (1910=100) (4)	Total Factor Productivity Variant II (5)
1839	73.5	16.6	8.7	20.0	13.3
1849	26.4	12.8	22.1	44.3	31.0
1859	22.2	12.5	26.4	50.0	37.0
1870	10.3	7.6	43.4	61.2	47.2
1880	8.5	7.9	53.2	71.9	59.5
1890	7.0	8.6	66.5	85.6	77.0
1900	4.9	7.4	86.7	96.8	93.7
1910	3.7	6.2	100.0	100.0	100.0

Source

Col. 1: ICC fiscal 1910 receipts of $2,812 million extended back upon output index of Table 1, and divided by capital estimates of Table 7.
Col. 2: Capital estimates of Table 7 divided by employment estimates of Table 8.
Col. 3: Table 10.
Col. 4: Output divided by new input index calculated by weighting employment index of Table 10 by .52, materials of Table 10 index by .10, and new capital index (equated to output index) by .38.
Col. 5: Derived like col. 4, except new capital index is equated to employment index.

of rails probably superimposed an exogenous downward trend. Disaggregation to the firm level could help in extracting the appropriate utilization correction. But note that even after a maximum allowance for this factor, in the shape of series I, the railway sector continues to exceed the performance of the economy as a whole.

In explaining this still quite large residual, we may invoke more conventional economies of scale. First, there are the familiar increasing returns at the level of the firm production functions, that is, nonproportionalities associated with higher levels of labor and capital inputs (correctly measured). Another possibility is the traditional Marshallian external economies of specialization enjoyed as the industry approaches an optimal size; separate supplying firms may absorb tasks formerly carried out in the industry less efficiently, for example. A third mechanism ties firm size to productivity increases via technological progress. Edwin Mansfield's important research has suggested that large firms typically have been the most aggressive innovators and also the most rapid followers (the latter point is less well established).[78] There is a corresponding industry analogue in technological and production effects, as has been demonstrated by Jacob Schmookler's results relating inventive activity to the size and growth of industries.[79]

Historically, all four of these linkages probably operated at various times in the nineteenth century evolution of the railway sector. Firm size certainly increased quite significantly. Between 1860 and 1880, as output rose more than sevenfold, the number of firms increased by a modest 50 per cent. As a result, the average output of an 1880 firm was five times that of an 1860 firm. From 1880 to 1910, further intensification of demand and, more significant, financial consolidation led to another quadrupling.[80] Any nonproportionalities inherent in railroad production conditions—and their size is a subject for further research—thus had abundant scope for exploitation. So did the technological effects, but here there is positive evidence of their existence. Of five important innovations (as measured by frequent citation in discussions of railroad technological change)—the use of the telegraph to control train movements (1851), the substitution of steel rails for iron (1862), the development of block signaling (1863), the adoption of air brakes (1869), and the employment of automatic couplers (1873)—four can be traced unequivocally to major roads, and one is uncertain.[81] A larger sample of innovations that

---

[78] See Mansfield's paper, "Innovation and Technical Change in the Railroad Industry," in *Transportation Economics*, Special Conference 17, New York, NBER, 1965.

[79] Jacob Schmookler, "Economic Sources of Inventive Activity," *Journal of Economic History*, March 1962, pp. 1–20, and other sources cited there.

[80] I estimate the number of railroads in 1860 at between 400 and 500, based upon the partial count made by the Treasury Department in 1856. The 1880 Census enumerates 631 corporations, and ICC *Statistics of Railways* gives 926, 1224, and 1306 for 1890, 1900, and 1910, respectively.

[81] I am indebted to Jacob Schmookler for making available a chronology of inventions in the railroad industry. This facilitated the identification of specific innovators. The Erie Railroad introduced telegraphic communication; the Pennsylvania, steel rails, after Edgar Thomson had viewed them abroad, and also the automatic coupler; the Camden and Amboy pioneered with block signaling, again an original English innovation. There is some mystery surrounding the air brake. An *American Railroad Journal* reprint credits the Boston and Providence and Eastern with first use, but other indications point to the Baltimore and Ohio as the earliest.

The following sources (in the order of the innovations) were accepted: *Railway Age*, Vol. 141, 1956, p. 285; *Twentieth Annual Report of the Pennsylvania Railroad*, 1867, p. 25; *Railway Gazette*, Vol. XXXII, 1900, p. 506; Stewart H. Holbrook, *The Story of American Railroads*, New York, 1947, p. 290; *Railway Age*, Vol. 141, 1956, p. 259; *American Railroad Journal*, Vol. XLVI, 1873, p. 776.

include less dramatic but perhaps equally significant changes probably would show as much concentration on large lines increasingly staffed with skilled personnel.[82]

Expansion of demand naturally contributed to increased firm size. It also led to external economies and induced technical progress on its own. Thus, as railroads grew in importance, production of rolling stock was more and more consigned to specialty firms that had evolved from a previous general machinery orientation. Similarly, a class of rail mills emerged from what had been manufacturers of the whole gamut of rolled products. Other examples, including all the special commodities used in railway operation, could be cited. As to the potential feedback to technology associated with increased levels of output, no one who has even casually studied the industry can fail to be impressed by the volume of literature that disseminated and critically evaluated new ideas. Formal industry-wide associations and committees also blossomed. Standardization was one result. In the instance of rails, there were 119 patterns of twenty-seven different weights produced in 1881; within a few years after an 1893 report by a special committee of the American Society of Civil Engineers, three-fourths of the rails were of the standard ASCE sections.[83] If efficient flow of information is essential to technological progress, the conditions in the large railway industry of the late nineteenth century tended toward the optimal.

The exact contribution and timing of these effects require much additional study. What may well be the most promising approach is a limited assault upon specifics. For example, how much more would it have cost to produce car wheels directly in railroad repair shops rather than in specialized firms? At what size industry were maximum economies reaped? Or, how much more slowly or rapidly would an industry of smaller or larger firms have introduced steel rails than was actually done? 1880 Ohio cross-section data show a significant positive correlation between firm size and adoption of steel rails, with unitary elasticity of response.[84] Given the very much greater increase in adoption of steel rails than

---

[82] Greater technical sophistication also occasionally has its costs. Charles B. Dudley, Ph.D., chemist for the Pennsylvania Railroad, and hence commanding wide respect, persuasively advocated low-carbon, or soft, steel rails in the 1870's. The error was not corrected until ten years later when breakage soared under the ever heavier motive power. William H. Sellew, *Steel Rails*, New York, 1912, p. 13; *Railway Gazette*, Vol. XXII, 1890, pp. 702 ff.

[83] See the discussion of the development of rails in the United States by Robert W. Hunt in *Railway Gazette*, Vol. XXXII, 1900, pp. 505–507 and 522–523.

[84] The data for this regression come from the 1880 Census. $R^2 = .56$ with twenty-six observations so the relationship is significant. Size is measured here by total receipts, not assets, it should be noted.

indicated by changes in size, this hardly casts size in a crucial role. The subject awaits fuller treatment, however. The typical analysis of industry production functions cannot supply answers to such questions; and most of the answers it can provide actually confound the individual mechanisms distinguished here.

It is precisely in this modest spirit that I wish to go on to examine the impact of quality changes in capital—or technological progress—upon recorded productivity advance.[85] The object is to measure the absolute and relative importance of four significant innovations introduced in the latter half of the nineteenth century, namely, steel rails, increased equipment capacity, air brakes, and automatic couplers. Typically their contribution is assumed to be both large and uniform: "The ... increases in efficiency brought by ... a host of other technical advances permitted lower railroad rates. The use of steel rails, the adoption of standard gauge, the utilization of faster and more powerful locomotives pulling longer and heavier trains, the introduction of standard time, better brakes, and improved couplers all helped to create a truly national rail network." [86] Yet preliminary study suggests that these innovations varied substantially in significance and also only partially explain the total productivity gains achieved at the end of the nineteenth century.

As evidence of the extent of variation among the innovations, there is first the differential rapidity of their diffusion. Neither air brakes nor couplers were greeted with much enthusiasm. Although first used in 1869 and 1873, respectively, and despite the designation of a standard form of coupler by 1887 and the definitive proof of the Westinghouse brake in the third Burlington trial in the same year, it finally required national legislation to secure their adoption. In 1890, although almost all passenger locomotives and cars had been fitted with brakes, not many more than half of the freight locomotives and less than one-tenth of the cars used the appliance. Couplers had attained still lesser acceptance: only 3 per cent of locomotives and 10 per cent of cars (but almost all passenger cars) were so equipped. The low marginal rate of adoption at that time quite justifies the conclusion of the statistician of the Interstate Commerce Commission that the railroad "claim that the adoption of uniform safety

---

[85] We do not examine quality changes of labor in this paper. This is not because they did not apply. On the contrary, the railroads were quite interested in industrial education. The Baltimore and Ohio Railroad even set up a special comprehensive training program for employees in 1885 (W. T. Barnard, *Report on Technical Education*, Baltimore, 1887). To deal adequately with the subject here would overburden the already sorely taxed patience of the reader. I hope to publish some findings on this subject elsewhere in the near future.

[86] John F. Stover, *American Railroads*, Chicago, 1961, pp. 143–144.

devices is progressing with satisfactory rapidity is not supported by the facts." [87] Congress acted in 1893, requiring both appliances to be installed on equipment used in interstate commerce before January 1, 1898, a deadline later extended to August 1900. Compliance followed, and the beginning of the twentieth century saw air brakes and automatic couplers as standard equipment, although it took almost another decade for them to become universal.

This diffidence can be contrasted with the reception afforded steel rails. Although steel rails commanded a substantial premium in price over the iron variety, by 1880, less than twenty years after their first use on the Pennsylvania Railroad, almost 30 per cent of the nation's trackage was so equipped. By 1890, 80 per cent of track mileage consisted of steel rails.[88] In the single decade 1871 to 1880, the investment in steel rails—measured as the price differential multiplied by consumption—totaled more than $80 million. Not only was such a sum commensurate with the extra absolute cost of the safety appliances installed twenty years later, but it was much larger relative to the smaller railroad assets of the time.

The assimilation of successive improved designs of locomotives and freight cars likewise proceeded without external pressures. Its cumulative character and the lack of a single impressive innovation should not obscure its rapidity. Within the space of some forty years—from 1870 to 1910—freight-car capacity more than trebled. The remarkable feature of the transition was its apparent small cost; capacity increased with only a very modest increase in dead weight, the ratio changing from 1:1 to 2:1. Over the same interval, locomotive force more than doubled as powerful engine types, such as the Mogul, the Consolidation, etc., replaced the familiar and faithful American 4-4-0.

Either railway decision-makers were irrational in their apparent hesitancy to adopt air brakes and automatic couplers, or their economic value relative to steel rails and improved equipment was markedly smaller. Both qualitative evidence and some crude calculations support the latter inference. In the first place, the properties of the innovations were well known, which rules out ignorance as a factor in the delay. Nor can undue weight be given to the claim that lack of uniformity of design

---

[87] ICC, *Statistics of Railways for 1891*, p. 45. The railroad opposition to legislation was contradictory. One group argued that the new techniques were unproven; another that railroads were moving ahead as rapidly as possible. Cf. U.S. Senate Interstate Commerce Committee, *Hearings in Relation to Safety Couplers and Power Brakes in Freight Cars*, 51st Congress, 1st Sess., 1890.

[88] *Poor's Manual for 1891*, p. xxi. The 1880 Census records a slightly higher percentage in that year, namely 35 per cent, but its result may be based upon an unrepresentative sample.

compelled caution on the part of individual railroads in outfitting themselves. Many systems were large enough that interchange of equipment was not an overriding concern. Moreover, the ultimate Congressional legislation specified no single type, but allowed the choice to emerge from within the industry. Finally, there is the unmistakable emphasis on safety considerations in the debate about brakes and couplers. Proponents before Congress spoke of savings in terms of human lives, not operating expenses. On at least two occasions when the question was directly put, there were no assurances of reduced personnel requirements attendant upon the innovations; otherwise, the various railroad labor organizations might have been less whole-hearted in their support.[89] Twenty years later an air-brake engineer commented upon the same tendency of railroad men to emphasize the safety features of the appliances.[90] Perhaps they were justified: in Europe, where adoption was not compulsory, diffusion was notably slow; and in the United States, prior to compulsion, only the western lines with steep grades felt that the potential economies of train brakes outweighed their cost.[91]

Calculation of the cost savings realized in 1910 from these four innovations seems to support the minor economic contribution of the air brake and automatic coupler. The principal advantage afforded by these devices was increased speed. Although longer trains were also claimed to depend upon them, there seems to be little direct relationship. Automatic couplers were not notably stronger, and it was possible without great hazard to extend the number of cars in the absence of air brakes simply by adding more trainmen. That a trend toward heavier freight trains was under way well before these appliances were installed is proof of the virtual independence of these developments.[92] Greater speed in itself is not an unmixed blessing, however. Unless engine capacity is not being fully utilized, higher speeds can be attained only by the sacrifice of load. What the air brake and coupler really did, therefore, was to allow a greater element of choice in train operation, permitting higher speed when it was more desirable than larger loads. But exactly because its

[89] U.S. Senate ICC, *Hearings*, 1890, p. 53, and *ibid.*, 1892, p. 55.
[90] W. V. Turner and S. W. Dudley, *Development of Air Brakes for Railroads*, Pittsburgh, 1909, p. 6.
[91] For the concentration of brake installation upon the four major western railroads—the Union Pacific, the Southern Pacific, the Atchison, Topeka, and Santa Fe, and the Northern Pacific—see ICC, *Annual Report for 1889*, Appendix 10.
[92] It is, of course, probable that the great increase in train size and load after 1910 may have required prior advance in the design of couplers and brakes. A longer time horizon, therefore, might alter the relative economic significance of the innovations. (Note, however, that the absolute importance of other changes also increased at the same time.)

influence manifested itself in this literally marginal fashion, the size of the economies was limited.

The two positive features of greater speed are reduced capital costs, through increased utilization of the rolling stock, and smaller train wages, through fewer hours of travel time. On the other side are increased repairs to equipment and roadway at higher speeds of operation, as well as larger fuel expenses. Quantitatively, the economies are circumscribed because so much rolling stock time in transit is spent off the road—in loading, redirection in yards, repair, etc.—and because train wages are relatively small. If later experience is any guide, and scattered late nineteenth century evidence affirms that it is, a 100 per cent increase in speed could have made possible a less than 10 per cent reduction in the number of freight cars.[93] Likewise, with train wages accounting for some 12 per cent of operating expenses, the same radical change in speed could have led to a saving in total outlay of only 6 per cent. Against these must be reckoned the diseconomies of speed, moreover. Fuel outlays were widely regarded as increasing more than proportionally with speed, and the fuel account is almost .10 per cent of expenses. Rail replacement and equipment repair also may have been adversely affected. Indeed the principal discussions of speed in the technical literature relate to its disadvantages rather than its virtues.[94]

To ascertain the absolute sum saved in 1910 by the more nearly optimal speed made possible by the air brake and automatic coupler, we have to specify how far from the actual circumstances a hand brake and link-and-pin-coupler regime would have been. Evidence on this point suggests a very small divergence. Freight and passenger train speeds in 1910 hardly differ from those recorded twenty years before.[95] But because trains were much heavier at the later date, let us suppose that attained speed, both freight and passenger, would have been one-third smaller

---

[93] A report by the Federal Coordinator of Transportation in 1935 reveals that freight-car time in trains was responsible for only 16 per cent of total time (cited in James C. Nelson, *Railroad Transportation and Public Policy*, Washington, 1959, p. 266). A discussion of car utilization in *Railroad Gazette* (Vol. XLII, 1907, p. 200) affirms the same low proportion at that earlier date.

[94] Cf. the extended comments running through *Railroad Gazette*, Vol. XXXII, 1900. The most favorable report was an increase in speed of 30 per cent with only a 20–25 per cent *increase* in costs (*ibid.*, p. 215).

[95] The Pennsylvania State Railroad Report for 1888 lists average freight-train speeds, including stops, as ranging from 12 to 18 miles per hour. The New York 1885 report records similar responses. The average speed in 1920 was 10.3 miles per hour (Julius H. Parmalee, *The Modern Railway*, New York, 1940, p. 210). Since the latter is based on actual performance, it is not surprising to find it somewhat smaller than the earlier cited speeds. We can safely infer rough equality at the two dates. The same approximate parity holds for speed of passenger trains.

without the benefit of the safety appliances. On this generous basis, the rolling stock required would have been about 3 per cent larger than it was; since equipment was roughly equal to 20 per cent of the total capital stock, and returns to capital were $821.9 million, the added expense would have come to $4.9 million. Increased train wages of $113.7 million augment this, from which we deduct fuel savings of $42 million, to reach a final tally of $77 million.[96] In sum, railroad rates would have been about 3 per cent greater in 1910 to compensate for the inefficiencies of the older technology.

An additional contribution of speed is a reduction in transit time and hence smaller inventories. Also there are the direct safety gains to be dealt with. Neither factor influences the previous total greatly. Delivery time is reduced only 3 per cent because freight-car train time is such a small proportion of elapsed time. The average reduction in working capital involved is an insignificant $40 million, even if the railroads are assumed to have transported the entire value of gross national product.[97] At an interest rate of 6 per cent, this is a negligible gain of $2.4 million. As far as safety is concerned, the 1910 losses were a higher proportion of operating expenses than in either 1880 or 1890. It is conceivable that the absence of safety appliances could have forced the percentage up even higher, but it is equally possible that a reduced pace of operations, such as we have necessarily hypothesized, may have reduced it to its previous level. In any event, the 1910 total loss is only $50 million, so an increment on either side is safely neglected.

Granting the crudeness of these calculations, it is still difficult to see how the state of affairs envisaged by a railroad spokesman in 1912 could have prevailed: "It may safely be said . . . that if we were dependent,

[96] Increased train wages are one-half of actual 1910 train wages of $227.4 million, or $113.7 million (since the hypothetical hours worked are 50 per cent greater than the actual ones). Reduced fuel outlays are less than one-third of actual expenses of $189.0 million, however, to compensate for fixed consumption independent of running speed. It is assumed that two-thirds of consumption is related to speed, and hence the figure in the text is 2/9 × $189.0 million (since speed and associated fuel outlays are assumed to vary proportionally).

[97] GNP in 1910 was approximately $30 billion. Even if the value of railroad freight was this large, this is equivalent to a daily transport of about $90 million. The average length of trip was fifteen days, tying up $1350 million of capital. A speed one-third greater could have reduced delivery time to a little more than fourteen and a half days, or reduced required working capital by $40 million.

Later indications (1936) are that the aggregate value of goods transported by rail is something like twelve times the freight payment. This would make the actual value of freight in 1910 less than $25 billion, and reduce the saving even more.

See L. F. Loree, *Railroad Freight Transportation*, New York, 1926, pp. 264 ff., for an analysis of 1910 freight-car trip time, and Parmalee, *Modern Railway*, p. 236, for the 1936 ICC valuation of railroad freight.

for one day, on the brakes of 20 years ago, the business of the country would be practically paralyzed, and the loss of both life and property very great." [98] The fact of technological advance is not synonomous with its indispensability.

In contrast to these results stand the estimates of the saving attributable to a steel-railed 1910 track in place of one of iron. Steel rails influenced railway operations in two respects: first, and most obviously, they wore much longer; second, and perhaps more important, they bore much heavier loads without breakage. Both technical advantages were of economic significance.

The durability of iron rails in the 1870's was variously placed between 4 and 15 million tons of traffic, although the Lehigh Valley reported tonnages as small as 1.5 and 2.3 million on two different stretches of road.[99] The total traffic in 1910, including the weight of rolling stock, amounted to some 830 billion ton-miles; the average total tonnage passing over each mile is estimated by dividing this sum by the number of miles of track in the system—351,767. The resultant figure, 2.4 million tons, divided into a modal durability of iron rails of 8 million tons, say, means an average life under 1910 conditions of 3.3 years. To put it another way, 30 per cent of the track mileage in 1910, or 105,530 miles, would have required replacement in that year. Expense for the 12.1 million tons of rails, allowing for the scrap of the worn-out rails, then would have come to almost $200 million.[100] Actual 1910 outlays for rails were less than $17 million. Steel rails, therefore, wore more than ten times as well as iron would have. Such a ratio agrees with independent engineering assessments placing the life of steel rails at from 100 to 250 million tons, against the earlier cited range of 4 to 15 million tons for iron rails.[101]

From this aspect alone, steel rails were almost three times as important as the air brake and automatic coupler. To the differential materials

[98] *Journal of the Franklin Institute*, Vol. 173, 1912, p. 35. Appropriately enough, the author was also arguing against the virtue of scientific management à la Taylor.

[99] See *Transactions of the American Society of Civil Engineers*, Vol. III, 1870, p. 104; C. P. Sandberg, "The Manufacture and Wear of Rails," in *Minutes of Proceedings of the Institution of Civil Engineers*, Vol. 27, 1867/68, p. 325; *Railroad Gazette*, Vol. II, 1870, p. 243; *ibid.*, Vol. III, 1871, p. 510.

[100] The average weight of rails per mile in 1910 as cited in Table 2 is 115 gross tons. The wholesale price of rails in that year, as in other years after the formation of U.S. Steel, was $28. With the addition of delivery charges to make the retail price close to $30 and a $14 or 50 per cent scrap allowance, the net charge becomes equal to $16 per ton.

[101] Camp, *Notes on Track*, pp. 97–98. The variety of experimental evidence cited there points to an average of 140 million tons; the implied 1910 average is just about 100 million tons, which is of the same general magnitude.

cost, moreover, must be added the increased labor inputs associated with more frequent replacement. Although accurate information on the labor cost of rail renewal is difficult to obtain, various technical sources point to a coefficient of between 50 and 100 man-days per mile of replacement. At 1910 daily wage rates of $1.50 for trackmen, the aggregate cost of relaying 105,000 miles of track is $11.8 million, and the differential $10.8 million.[102]

The second-order effect of greater strength considerably extends the economic consequences of greater durability. Nonhomogeneous iron rails were limited in the locomotive weight they could bear. Under heavy loads they were subject to lamination and crushing. Steel sections, on the other hand, wore evenly. Consequently it is impossible to maintain that 1910 engines could have been accommodated by iron rails: the maximum engine weight capable of being borne by eighty-pound track would have been closer to fifty tons, and quite below the seventy-ton average locomotives actually upon the scene.[103] Smaller engines, in turn, would have meant more frequent trains. The simplest case is one in which we assume that total tonnage, load factors, direction of traffic, etc., were unchanged. Then required freight-train mileage—passenger trains would have remained almost unaffected—would have been 40 per cent greater, i.e., the ratio of the seventy-ton average weight in 1910 to the fifty-ton maximum (and average) that would have prevailed. Under these conditions, and accepting Wellington's computations indicating a .5 ratio of marginal cost per train-mile to average cost, the incremental outlay in 1910 comes to $279 million.[104]

---

[102] The practical literature gives a rather varied set of figures that extends from less than 30 man-days per mile of replacement to more than 100. The 75 man-days actually used is not likely to introduce a very great absolute error. For discussions of labor requirements, see *ibid.*, pp. 563–565; E. E. Tratman, *Railway Track and Track Work*, New York, 1901, pp. 335–340.

[103] *Description of Sandberg's Standard Rail Sections*, London, 1872, Table 1; *Transactions of the American Society of Civil Engineers*, Vol. III, 1870, p. 113; Sellew, *Steel Rails*, p. 13.

[104] Wellington, *Economic Theory*, p. 571. An allowance for interest on the increased stock of required locomotives is included in the determination of the ratio.

Freight expenses are not broken down separately in 1910, and they have been estimated using the 1916 proportion, when the ratio between ton- and passenger-miles is virtually identical to that in 1910. The imputed freight expenses on this basis are almost $1.4 billion, of a total of $1.8 billion.

There are two offsetting biases in this simplified calculation that should be mentioned. More frequent, but smaller, trains might have permitted more efficient loading and reduced the proportion of empty to full cars per train; on the other hand, the comparison of the 1910 *average* weight of locomotives with the hypothetical maximum makes no allowance for the fact that the 1910 hypothetical average would have been somewhat smaller, and so increased the divergence under the two alternative technologies.

In all, therefore, steel rails directly and indirectly saved something like $479 million in 1910. How much do the increase in freight-car capacity and the autonomous increase in locomotive tractive power from 1870 to 1910—which grew by more than the difference between iron and steel rails can explain—add to this total? With the same kind of simplification just resorted to, we can approximate the answers. The total increase of average locomotive tractive power between 1870 and 1910 was more than 100 per cent. The calculation of savings due to steel rails has subsumed less than half of this change. At least double the actual number of 1910 trains would have been required under conditions of 1870 motive power. Beyond the $279 million reckoned above, therefore, there is another $420 million in outlay that improved locomotive design obviated.

Had the powerful twentieth century engines been developed without that simultaneous remarkable advance in freight-car construction, much more of the increased power would have been dissipated in the non-productive task of hauling dead weight. A higher ratio of dead weight requires either more or heavier trains to deliver the same payload, both involving additional expense. If 1910 tonnage had to be moved in 1870 freight cars, it would have required about 3.3 of them to equal one 1910 car, and at twice the weight. With identical load factors under both technologies, the same loads would have been carried in four trains of identical weight (but with 3.3 times as many cars) as were actually transported in three. Making the simplifying, and savings minimizing, assumption that trains of so many more cars but equal weight could be operated at the same cost, we still reach the substantial additional expense of $329 million.[105] These important savings credited here to modern equipment reflect the observed secular association of increased loads per train and railroad productivity growth.

To recapitulate, our computations indicate that the incremental expenses incident upon meeting 1910 railroad demands with an 1870 technology of iron rails, light engines, low-capacity freight cars, hand brakes, and manual couplers would have amounted to about $1.3 billion. This is not likely to be much of an exaggeration. The estimate does not suffer from

[105] Operating expenses for the one-third increase in train mileage, using the same .5 relationship between marginal and average train-mile costs as before, total $233 million; interest in catpital outlays for the tripled number of smaller freight cars comes to a further $96 million since each smaller freight car would have cost about two-thirds of the actual 1910 price. The aggregate capital outlay required is, therefore, equal to twice the valuation of the 1910 freight-car stock, or $3.2 billion. Interest on the differential $1.6 billion, at 6 per cent, is $96 million.

Once more, there is a possibility that smaller cars, more efficiently loaded, could have compensated. But the explicit American choice of large, capacious cars, with an attendant lower load factor seems to suggest that such economies were limited.

the usual substantial upward bias associated with requiring an older technology to meet later demands, because in this instance exactly the same bill of goods was not imposed. Trains were not required to go as fast as they did in 1910; only the direct costs of a slower pace were reckoned. Equally heavy trains with a tremendous toll upon iron rail wear were eschewed in favor of the feasible alternative of lighter ones. Thus, one element of substitution enters the calculations, although not the response of demand for railroad output to its inevitably higher price. At every stage, moreover, assumptions that mildly understated savings were introduced.

This $1.3 billion is an impressive total against the backdrop of 1910 railway operations. To compensate, revenues in that year would have had to be more than 40 per cent greater. The question remains, however, as to the importance of this saving in the continual cost reduction—i.e., productivity advance—that railroads had been enjoying. These innovations, after all, were not the product of a single decade, or even two. Their effects were transmitted over some forty years, and by 1910 they had fully worked themselves out. It is misleading to limit the comparison to a single year; the entire range of years from 1870 to 1910 must be included.

In embarking upon such a task, it is useful first to reinterpret our previous total factor productivity measurements and to show how they relate to the foregoing analysis of technological change. An index of total factor productivity, TFP, is the ratio of an index of output to a weighted index of inputs, where the weights are the base-period income shares of the factors:

$$TFP_{t/0} = \frac{\frac{X_t}{X_0}}{\Sigma a_j \frac{I_{jt}}{I_{j0}}} = \frac{\frac{X_t}{X_0}}{\frac{W_0 L_0}{P_0 X_0} \cdot \frac{L_t}{L_0} + \frac{i_0 K_0}{P_0 X_0} \cdot \frac{K_t}{K_0} + \frac{r_0 R_0}{P_0 X_0} \cdot \frac{R_t}{R_0}}. \quad (1)$$

X represents output; L, K, and R stand for inputs of labor, capital, and raw materials; P is the unit price of output, and w, i, and r are factor prices; the subscripts 0 and t refer to the base period and the period of comparison, respectively.

Now, eq. (1) can be rewritten simply as the ratio of the value of output in base-period prices to the value of inputs in base-period prices:

$$TFP_{t/0} = \frac{P_0 X_t}{W_0 L_t + i_0 K_t + r_0 R}. \quad (2)$$

It follows from (2) that Kendrick's measure of technological change is nothing more than the ratio of an index of input prices to one of output prices. When output prices fall relative to input prices, technological change has occurred, and vice versa. For eq. (2) divided by $P_t X_t$ numerator and denominator (remembering that $P_t X_t = W_t L_t + i_t K_t + r_t R_t$) is the same as

$$TFP_{t/0} = \frac{\dfrac{P_0 X_t}{P_t X_t}}{\dfrac{W_0 L_t + i_0 K_t + r_0 R_t}{W_t L_t + i_t K_t + r_t R_t}} = \frac{\dfrac{W_t L_t + i_t K_t + r_t R_t}{W_0 L_t + i_0 K_t + r_0 R_t}}{\dfrac{P_t X_t}{P_0 X_t}} \qquad (3)$$

when the final numerator and denominator are seen to be indexes of input and output prices, respectively.[106]

This result recasts productivity analysis in a value frame of reference similar to one that has been used here to evaluate the specific innovations. Explicitly, we have computed C_0^t, the cost of producing output in the base year (1910) using the technology of another year t (1870). This can be expressed as

$$C_0^t = P_0^t X_0 = X_0 \left(w_0 \frac{L_t}{X_t} + i_0 \frac{K_t}{X_t} + r_0 \frac{R_t}{X_t} \right); \qquad (4)$$

that is, the hypothetical cost is equal to the volume of 1910 output, X_0, multiplied by 1870 unit input requirements, $\dfrac{L_t}{X_t}$, etc., valued in 1910 prices, w_0, etc. Dividing both sides by P_t and X_0, we get

$$\frac{P_0^t}{P_t} = \frac{w_t L_t + i_0 K_t + r_0 R_t}{w_t L_t + i_t K_t + r_t R_t} \qquad (5)$$

whence, from eq. (3), we reach the desired equation

$$TFP_{t/0} = \frac{P_0}{P_0^t}. \qquad (6)$$

The index of total factor productivity is, therefore, nothing more than the ratio of the actual price of output in the base year to the price that would

---

[106] I am grateful to Dorothy Brady for suggesting this price index variation to my original proof. It is useful since it points up the omnipresent duality between price and quantity results in a competitive equilibrium framework. The simplicity of this dual relationship for the Kendrick arithmetic index is a virtue that has gone unnoticed. It helps to offset the valid criticism of its unconvincing production implications. Because both geometric and arithmetic indexes give similar results over reasonable ranges, the practical advantages of the Kendrick index are enhanced by the ease of its alternative value formulation.

have prevailed in the absence of change in the technological coefficients. $P_0 = P_0^t$ if the matrix of technical coefficients, L/X, K/X, and R/X, remains constant; $P_0 > P_0^t$ if the technology at t is superior, and $P_0 < P_0^t$ if technology at t is inferior. In the case at hand, because t precedes the base period in time, $TFP_{t/0}$ less than one, as it is, constitutes progress between the two dates.

It is now possible to measure the significance of the selected innovations relative to all influences operative on productivity advance. The P_0^t that emerges from the previous cost calculations is partial, encompassing only the effects of the four innovations. Consequently the ratio of this limited $P_0^{\prime t}$ to the total P_0^t is an index of their explanatory power. There are three alternative P_0^t's to choose among, the highest one corresponding to the original productivity calculations and the other lower two reflecting the alternative capacity utilization adjustments made earlier. The respective ratios $\dfrac{P_0^{\prime t} - P_0}{P_0^t - P_0}$ are .36, .42, and .75.[107] Roughly, therefore, the technical changes dealt with account for half the productivity gains registered in the railway sector between 1870 and 1910.[108]

From the standpoint of the limited variety of changes taken up, this accomplishment is no mean achievement. From another aspect, however, it is a rather small contribution compared with what might have been anticipated. The two innovations of greater engine power and freight-car capacity include a host of lesser changes that are reflected in the final result: increased steam pressure, substitution of steel trucks, etc. Thus, the cost reductions made possible by these two advances sum up much of the residual technical change that occurred; it is not surprising that they make up almost two-thirds of the $1.3 billion estimate. The fact that, despite this, only half the productivity change from 1870 to 1910 can be accounted for shows how thick is the veil of ignorance surrounding the causes of the rapid increases that advanced economies have experienced in output per unit of input.

---

[107] What is termed the total P_0^t is, of course, the actual 1910 price divided by the productivity index. It is easiest to work in 1910 relatives throughout, however, which then makes P_0^t/P_0 simply equal to the inverse of the productivity index.

Note that the ratio used in the text is preferable to the alternative $\dfrac{P_0^{\prime t}}{P_0^t}$ because, in the absence of economic impact, $P_0^{\prime t} = P_0$; such a ratio would then be greater than zero regardless of the lack of change.

[108] It will be remembered that the capacity utilization adjustment obtained by holding the capital-output ratio constant at its 1910 level exaggerated the explanatory value of this factor. Hence the .75 ratio credits too much to technology in a relative sense, because the denominator is too small.

In part, this doubt will be resolved by extending the analysis to factors not taken up here, such as the increased educational level of railway employees and the value of an experienced labor force. In part, too, it must be resolved by a finer analysis of technical progress than has been performed here. The underlying engineering relationships employed, such as that relating the wear of rails to tonnage passing over them, can be made more sophisticated and exact. Testing these against actual experience is also necessary. Likewise, the cost relationships taken on the authority of Wellington can find more accurate substitutes in statistical cost functions fitted to the data. A third direction of advance is the use of smaller time intervals than the forty-year period employed here. The relative importance of innovations within shorter spans may be quite different. It is possible that between 1870 and 1890 the substitution of steel rails made up much more of the productivity gain than over the longer haul since it was concentrated in those earlier years.

Despite these admitted deficiencies of the present calculations, they do point to certain conclusions that seem likely to hold up under closer scrutiny. One of these is the limited economic significance of air brakes and automatic couplers in the array of late nineteenth century improvements. Another is the important role of the larger engines, a trend that is not attributable to a sudden major breakthrough, but rather to a cumulation of knowledge of design. The apparent sudden increase in the ratio of load to dead weight in freight cars in the late 1870's is more of a discontinuity. Yet virtually nothing has been done to determine the origins and causes of this major shift. Finally, these quantitative results confirm one major contention on which historians have been nearly unanimous: the importance of steel rails. Yet even in this instance, an amendment is necessary. What made such rails so crucial was less their wearing properties than their strength, and the opportunity thus presented for heavier trains.

Concluding Comments

Economists have come more and more to appreciate the role increased efficiency plays in the present growth of income per capita. The American economy in the nineteenth century was probably no exception. Economic historians must evaluate and interpret that experience from the same vantage point. In this paper we have made a tentative start in this direction in the railway sector. Two objectives have been pursued: first, extension and correction of the underlying data to provide a reasonably firm long-term record of inputs and outputs in the industry; second, explanation of the notable productivity growth that emerges. The latter endeavor also illustrated the wider possibilities of placing the analysis of specific innovations within an aggregate context, as well as yielding important substantive results.

Obviously, there is much left to do. Within the railway sector, the tasks range from more detailed evaluation of the factors influencing productivity advance to application of such information to such important historical questions as regional rate differences, discrimination between short and long hauls, and excessively high railway profits. All of the latter topics need to be restudied with more reliance upon underlying and changing conditions of production of railway services. Railroads, however important, are only one activity among many. Other industry studies must be carried out too. These conference proceedings include three such ventures into the extractive industries, albeit with rather limited attention to productivity per se. For manufacturing, there is nothing, however. Recalcitrance of the data helps to explain this, although one suspects that this is an area of research where important gains can be and are yet to be made.

Index

Abbott, Edith, 185n
Abramovitz, Moses, 3n, 21–24, 206, 243n, 287, 526
Adams, H. C., 138n
Agriculture:
  capital coefficients, 129
  effect on economic growth, 126–131
  employment, 117, 121–123, 127–131, 150–162
    farmers, 152–156, 161–162
    female, 153
    free farm laborers, 152–156
    "laborers not specified," 156–161
    present estimates, 159–161
    previous estimates, 157–159
    slaves, 150–152
  role in economic growth, 126–131, 209–210
  United Kingdom, 205–206
  value of output, 28, 56
Aldrich Committee, 593n
Aldrich report, 93, 198n, 199, 613
Allen, Zechariah, 182
American Appraisal Co., 482, 518
American Petroleum Institute, 355n
American Society of Civil Engineers, 633
Amusements, value of, 59
Andreano, Ralph L., xii, 349–404
Animal products, 79, 83
  current price series, 39
  output estimates, 39
  value of output, 30–31, 56
Arnold, Ralph, 355n, 356n, 358n, 390n
Assessors and assessments, 243–245, 252–259
Associated Oil Co., 394
Atkinson, Edward, 161

Bagnall, William R., 183n
Bakeries, 74–75
Balance of payments, 39
Baltimore and Ohio Railroad, 634n
Bancroft, Gertrude, 148n
Barger, Harold, 31–32, 36–37, 85, 132, 402n, 427n, 586
Barnard, W. T., 634n
Barnett, Harold J., 437–439
Bartlett, Richard A., 320n
Barton, Glen T., 580–582

Batchelder, Samuel, 182, 184n
Bateman, Fred, Jr., 523n
Bauxbaum, Bertold, 479n, 487
Belknap, Jeremy, 164n
Benchmark years, 39, 41–55
  price indexes, 41
Bigelow, Erastus, 239n
Bigelow, John P., 165n
Bishop, J. Leander, 101, 112
Bissell, George H., 383
Blank, David, 260, 270, 271–276, 284, 286
Blodget, Samuel, 137–138
Bogue, Allan, 130n
Bolles, Albert S., 101, 112
Borenstein, Israel, 307n
Bornstein, Morris, 7
Bowman, Isaiah, 359n
Brackenridge, H. M., 174n
Brady, Dorothy S., ix–xiv, 3n, 39–41, 62–63, 85, 91–115, 284, 605, 643n
  price index, 32, 36, 56
Brainerd, Carol, 130n, 158–161, 181
Briggs, Isaac, 184n
Brookings Institution, 523n
Brown, David, 500
Brown, Joseph R., 500
Brown and Sharpe Co., 497, 499–519
  business records, 489–490, 497
Buildings:
  cost index, 256, 289
  costs, 255
  Ohio statistics, 243–290; see also Ohio building statistics
  ratio of appraised to market values, 252–259
Bull, Marcus, 439
Bullard, E. P., 500–501
Bullard Co., Bridgeport, Conn., 499, 500–510, 518
Bullock, Mary G., 523n
Burns, A. F., 502, 511n
Business cycles, 20–23, 205–210, 511
  comparison of U.S. and U.K., 207–208
  long-swings, 20–24, 207–208, 225–226
  textile industry, 222–226
  volume-of-trade index, 222, 225

California, gold discovery, 300, 320–321
Camp, W. M., 598n, 639n

INDEX

Canals, construction of, 60–61, 176
Capital formation, 319
  claims against foreigners, 14, 24
  comparison of U.S. and U.K., 207
  composition of, 14–17
  construction, 16–17, 24
  domestic investment rate, 14, 24
  domestic savings rate, 14
  farm improvements, 15–16, 24
  gross domestic (GDCF), 10–11
  gross national (GNCF), 10–11
  long swings, 207–208
  manufactured producer durables, 17, 24
  product data and, 12–13
  share in GNP (1834–1908), 10–14, 23–24
Capital stock:
  as measure of railroad productivity, 589–618
  railroad development and (1840–1910), 583–584, 589–612
Carey, Harry, 171–172
Carey, Matthew, 138
Carll, J. F., 358n, 364n, 365n
Carson, Daniel, 86, 132, 133, 157–160, 175n, 177, 189, 199, 613n, 617–618
Carter, James, 124
Casey, M. Claire, 133, 147, 149n, 161, 163n, 200n
Cattle and sheep industry, 313n
Censuses, *see* United States, Bureau of the Census
Cereals, *see* Grain production
Charcoal production, 420–421
Chawner, L. J., 260, 271, 272
Children, employment of, 146–148
Churches, construction of, 41, 57, 63
Cincinnati, Ohio, metalworking machinery, 489–490, 499, 506, 510–511, 518
Claims against foreigners, 14, 24, 61
  GNP estimates (1869–1909), 39
Clark, Colin, 130
Clark, Victor S., 101, 112, 183n
Clay, Henry, 186
Clinch, D. J., 479n
Clocks, prices, 95
Clothing industry, 43, 75–76
  employment, 139, 142
Coal mining industry (1839–1918), 173, 293, 405–439
  consumption and use, 405, 408, 417–423, 437–439
    charcoal, 420–421
    coke, 421
    foreign trade, 417–418, 437
    iron and steel industry, 418, 420–421, 437
    locomotive fuel, 421–422
    per capita, 418, 438
    sources of information, 417, 437–439
  economic history, 405–408
  employment, 405, 423–430
    adjustment of basic data, 423–425
    output per man day, 427–429, 436
    ratio of output to, 427–430
    by regions, 425, 437
  fuel oil competition, 388
  Illinois, 415, 425, 427
  industrialization and, 405, 418
  Kentucky, 407, 415, 425, 427
  literature and studies, 405
  markets, 406–407
  Maryland, 407
  mechanization and, 405, 418, 430
  Ohio, 407, 415, 425, 427
  output, 405–417, 439
    adjusted statistics, 413
    anthracite and bituminous, 411–412, 417, 421–422, 427–430, 437
    Eavenson's series, 405–406, 409, 411, 436–437, 439
    1830 to World War I, 410–413
    ratio of employment to, 427–430
    by regions, 415–417
    sources of information, 408–409
    value of, 413–415
  Pennsylvania, 407, 415, 417, 425, 427, 430
  position in U.S. economy, 408
  prices, 419–420
  problem of excess capacity, 436
  shift from wood to coal, 405, 418–420, 437–438, 619–624
  statistical data, 405–408, 436–439
  transportation costs, 436–437
  urbanization and, 418, 437
  Virginia, 407, 415, 417, 425, 427
Cobbett, William, 101, 112
Coke, production, 421
Cole, Arthur H., 222–223, 225, 227n
Cole, W. A., 4, 7, 205
Commodity flow, 77–85
  Barger's estimates, 31–32
  Census data, 28–29
  composition of, 17–20
  consistency tests of data, 79–85
  1844, 1854, 55
  estimating procedures, 25
  Kuznets' estimates, 25–26, 29–31
  major benchmark years, 41–55

manufactured, in producer prices, 41–55
manufacturing (1834–59), 41–62
  value of output, 41–55
  output estimates, 41–42
  perishables, 17–20, 24
  producers, 61
  rents, 57–59
  services, 17–20, 24
  technique, 90
  trade markup, 36
  value added by home manufacturing, 20, 74–76
  value of output, 41–55, 80
    since 1869, 25–26, 28–31
Construction, 37
  activity determined by assessment statistics, 288
  business cycles and, 208–209
  canals, 60–61
  component of capital formation, 16–17, 208
  employment, 121–124, 133
    1840–1900, 175–177
  estimating procedures, 60–62, 71, 85
  GNP estimates (1869–1909), 37–41
  long swings, 208–209
  Ohio building statistics, 243–290; see also Ohio building statistics
  output growth, 37–39
  railroads, 40, 60–66
  rate of increase, 21–23
  ratio of appraised to market values, 252–259
  value of materials, 175, 177
Consumers, commodities flowing to, see Commodity flow
Cootner, Paul H., xiii*n*
Copper, 293, 294, 312, 314, 316, 319, 321
  employment, 302, 325–326, 337
  output, 296, 200–201, 325–326, 337
  per mine, 317–319
Corn, 64
  intraregional shifts, 535–540
  labor requirements, 556–562, 566–570
  land yields, 528
  output growth, 541, 582
  productivity growth, 523
  See also Grain production
Cotton textile industry, employment, 133, 139, 140*n*, 182–186
Coxe, Tench, 138*n*, 183
Cranmer, H. Jerome, 60*n*, 64, 85
Crew-Levick, 392, 396, 397–398
Cudahy Oil and Refining, 397–398

Custom production, 29, 42*n*
Cycles, *see* Business cycles

Dales, John, 52
D'Amico, Mary, 243*n*
Daum, Arnold R., 349*n*, 355*n*, 359*n*, 362*n*, 366*n*, 368*n*, 369*n*, 370, 371, 374*n*, 379*n*, 383*n*, 386*n*, 387*n*, 388*n*
David, Paul A., 3*n*, 281–288
Davidson, Paul, 358*n*
Davis, Lance E., xii, xiv, 213–242
Davis, Samuel, 164*n*
Deane, Phyllis, 4, 7, 123*n*, 205
De Bow, James, 131*n*, 151, 168, 484
Developing nations, 130
Diffusion processes, xiii
Distribution:
  estimates of value added by, 62
  geographic, xiii–xiv
  trade markups, 56–57
Domar, Evsey, D., 618*n*
Domestic investment rates, 14
Domestic savings rates, 14
Domestic service, 136, 139, 142, 145
  employment, 59, 60, 133, 203–204
  United Kingdom, 205
Drake, Daniel, 171*n*
Duane, William, 138
DuBoff, Richard, 461
Dudley, Charles B., 633*n*
Dudley, S. W., 636*n*
Durables, 24
  manufactured producer, 17
  price index, 39
  semidurables and, 24
  value of output, 44
Durand, John, 132
Dwelling units:
  estimates, 57–59, 61
  ratio of population to, 58
  statistical records, 243–290; *see also* Ohio building statistics
  stock of, 21, 41, 63

Easterlin, Richard A., 3*n*, 76–90, 130*n*, 133, 150*n*, 154*n*, 158, 160, 181, 207*n*, 443*n*, 583*n*
Eaton, Lucien, 315
Eavenson, Howard, N., 405, 409, 411, 413*n*, 436–437, 439
Eckle, E. C., 315*n*
Economic growth:
  role of agriculture, 126–131
  See also Regional economic development

Economic History Association, ix, xiii
Economic Research Service, 582
Education, 59, 86
  economic effects of, 124–126
  quality and quantity of, 125–126
Edwards, Alba, 133, 134n, 140n, 147n, 148, 149, 153n, 156–160, 162n, 175n, 188n, 203n
Edwards, Emory, 601n
Electric power, 443–444
  capacity, 444
  importance of, 451, 455
  industrial distribution, 447
  regional distribution, 447
Eliasberg, Vera F., xii, 405–439
Ely, Northcut, 359n
Employment (1800–1960), 117–210
  agriculture, 117, 121, 122, 123, 126–131
  coal mining industry, 405, 423–430
  construction, 121–124
  estimates, 132–204
    agriculture, 150–161
    construction, 175–177
    cotton textile industry, 182–186
    CPS series, 132
    derivation and explanation of, 132–133
    domestic service, 203–204
    gainful workers, 1800–1960, 134–148
    iron manufacturing, 186–188
    manufacturing industries, 177–182
    methods of, 134–304
    mining, 173–175, 301–307
    navigation, 162–173
    railroads, 191–200
    teachers, 200–203
    trade, 188–190
    USDA series, 132
  fishing and navigation, 121–123, 162–173
  historical background, 117–133
  labor force and, 117–210
  manufacturing, 177–182
  metal mining industries, 301–307
    compared with all employment, 305–307
    regional differences, 305–307
    by regions and industries, 302–303
    statistical data, 301–302
  mineral industries, 303–305, 425–427
  railroads, 121, 122, 123, 583, 612–618
  total U.S., 426–427
    compared with U.K., 117–133
  trade, 121, 122, 123, 124
  United Kingdom, 117–133, 205–210

Energy:
  coal, 405–408; *see also* Coal mining industry
  consumption, 418, 437–438
  work done by work animals, 408
Eppley, Martha Ann, 479n
Estimating procedures:
  construction of constant price estimates, 58–60
  errors, 43
  gross national product, 3
  income and wealth, 88–89
  reliability, 89
  tests of, 3, 43
Evelyth, J. F., 383
Exports and imports, 17, 77
  data on, 52–53
  estimates, 54
  international trade adjustments, 45, 55
  mineral mining, 319–320, 347
  prices, 53
  value of, 45, 52–55

Fabricant, Solomon, 90n, 132, 133, 139n, 140n, 149, 156n, 177, 188, 461
Factory system, 139
  employment, 177–182
Family size, 58
Farms and farmers, 152–156, 161–162
  improvements, 15–16, 24, 71–74
  laborers, 152–156, 161–162
    estimates, 134, 136–138, 152–156, 161–162
    free, 152–156
    slaves, 150–152
    white, 136
  value of output, 31, 55–56
  See also Agriculture
Faulkner, H., 229n
Fearon, Henry Bradshaw, 101, 113
Females, employment, 139, 140, 142–145
Fenichel, Allen H., xiv, 443–478
Final product flow, *see* Gross national product, final product flow
Firewood, 39, 62–63
  output estimates, 39
  price index, 32, 39, 56
  value of output, 28, 31–33, 56
Fisher, Allen, 130
Fishing, employment, 121–123, 133, 134, 137, 163–164
Fishlow, Albert, xii, xiv, 3n, 37–39, 60, 63, 64, 83, 84n, 85, 89, 133, 198–200, 583–646
Fleming, H. S., 625

INDEX

Flour, prices, 93, 95, 97–98
Flow of commodities, *see* Commodity flow
Fogel, Robert, 3*n*
Folger Report, 358*n*, 364*n*
Food:
  prices, 92, 95–96
  value of output, 31–32, 57
Ford Foundation, 3*n*, 243*n*, 497*n*, 523*n*
Foreign trade:
  employment, 166–167, 169
  estimates, 62
  metalworking machinery, 484
  petroleum products, 382–383
  See also Exports and imports
Forest land, clearing, 16
Forestry, 159, 160*n*
Foster, Ray R., 284
France, gross national product, 4–7, 23
Frasch, Herman, 375, 388
Free colored population, 135, 137, 145
Freedley, Edwin T., 485–487
Freeman, James, 165*n*
Frickey, Edwin, 90, 198*n*, 588–589
Friedman, Milton, 29, 481*n*
Frost, Miles F., 523*n*
Fruits, value of output, 55–56
Fuel consumption, as measure of railroad productivity, 583, 618–626
Fur industry, 313*n*

Gallatin report, 178–179, 183
Gallman, Jane, 3*n*
Gallman, Robert E., xi, 3–76, 77, 83–86, 88, 90, 91, 127, 181, 205, 206, 461, 502, 626*n*, 629
Garcia, Marz, 3*n*
Gasoline, output, 385–386, 397
Geier, Frederick V., 488*n*
Gerstner, Franz Anton von, 587, 599
Gibb, George S., 229, 230*n*, 239*n*
Gilbert, Milton, 4–7
Gold, 293, 312, 313–314, 319, 320
  output and employment, 296, 299–302, 327–328, 338–339
  output per mine, 317–318
Goldsmith, Raymond W., 59, 63, 270
Goss, W. F. M., 625
Gottlieb, Manuel, xii, 3*n*, 58, 61, 63, 64, 206, 243–290
Government, 59, 86
Grain production (1840–1910), xii, xiv, 530–582
  alternative growth paths, 545–546
  capital inputs, 524*n*
  definitions and assumptions, 525–528
  effects of agricultural chemistry and plant biology, 525, 546, 581
  factors, measuring importance of, 555
  influences on, 523–524
  labor inputs, 526–527, 531–533, 541
  labor requirements, 523, 525, 555–575
    corn, 556–562, 566–570
    harvest, 562–565, 566–570
    period 1, 556–570
    period 2, 566–575
    postharvest, 570, 573–575
    preharvest, 556–562
    regional abbreviations and definitions, 555, 556
    wheat and oats, 562–565, 570, 573–575
  land yields, 525, 527, 531–533, 541, 549–555
  mechanization, effects of, 524–525, 534, 543–544, 581
  Northeast, 541–546
  periods of the study, 526, 556–575
  principal findings, 532–539
  productivity growth, 523–582
  regional shifts, 535–540, 546–549, 556
  regional weights, 532–533, 535
  source bibliography, 575–580
  South, 541–546
  statistical techniques, 529–530
  technological changes and, 523, 524, 545–546
  value of output, 30
  westward expansion and, 523, 545–546
Gray, G. A., Co., 499, 506, 510–511
Great Britain:
  gross national product, 4–7, 23
  metalworking machinery, 485
Grebler, Leo, 270*n*, 271, 272–276, 286
Greene, Thomas, 165
Gross domestic capital formation (GDCF), 10–11
Gross national capital formation (GNCF), 10–11
Gross national product (GNP), 1834–1909, 3–90
  composition, 10–20, 25–36, 41–56
  concepts, 3
  cycle averages, 23
  estimates for 1869–1909:
    changes in inventories, 39
    claims against foreigners, 39, 41
    construction, 37–39, 40–41
    Kuznets' estimates, 25, 29, 31
    manufactured producer durables, 37

INDEX

Gross national product (*Continued*)
  services flowing to consumers, 37, 85–88
    Shaw's estimates, 25–36
    trade markups, 36
  final product flow, 62–63
    construction, 60–61
    current prices, 25–39, 41–62
    1860 prices, 39–41, 62–71
    manufactured commodities, 41–55
    trade markups, 56–57
    unmanufactured commodities, 55–56
  flow of commodities, 77–85
  fluctuations in rates of growth, 20–23
  France, 4–7, 23
  Great Britain, 4–7, 23
  gross new construction, 37–39, 60–61
  interpolators and extrapolators, 64–71
  levels and rates of growth, 3–10, 33
    national product per capita, 3–10
    real GNP, 7–10
    long-term changes, 8–10
  services flowing to consumers, 37, 57–60
  value added by home manufacturing, 20, 74–76
  value of improvement to farm land, 71–74
Grossack, Irvin M., 479n
Grosse, Anne, 231–232
Groves, H. M., 274
Gulf Oil, 394, 397–398

Hall, James, 202n
Hardware, prices, 94
Hardwicke, R. E., 359n
Heating equipment, price index, 100–101
Henry, J. T., 358n
Herfindahl, Orris C., xii, 293–348
Hibbard, Benjamin H., 73n
Higgins Oil and Fuel, 397–398
Hill, Joseph, 160
Hill, Robert, 365n
Holbrook, Stewart H., 632n
Home manufacturing, 43n, 90
  value added by, 20, 74–76
Hoover, Ethel D., 40, 85, 92
Horsepower, 443
Horses and mules, 76
Houses, *see* Dwelling units
Hoyt, Homer, 283n
Hu, Stephen, 3n
Hubbard, G., 498–499
Hughes, J. R. T., 221n
Hunt, Robert W., 633n
Hutcheson, Harold, 138n
Hutton, F. R., 480n

Illinois, coal mining, 415, 425, 427
Immigration, 130, 156, 240
  effect on economy, 209–210
Imports, 240; *see also* Exports and imports
Indexes, use of, xii
Indian Oil and Refining, 397–398
Industries:
  Census data, 451
    adjustments, 462–468
    revisions (1899), 466–468
  distribution of primary-power capacity by, 447–451, 455
  nonmanufacturing, value of output, 28
  two-digit classification, 451, 461–462
Innovations:
  measuring significance of, 642–645
  railroads, 632–634
Input-output tables, xiii-xiv
Internal combustion engines, 443, 444
International trade adjustments, 45, 83
Interpolators and extrapolators, 64–71
Inter-University Project for American Economic History, 497n, 523n
Inventories, 240
  changes in, 12–13
  GNP estimates (1869–1909), 39
Investments, domestic rates, 14
Iron and steel industry, 127–128
  coal consumption, 418, 420–421
  employment, 133, 173, 186–188
  output, 128–129
Iron ore, 293, 319, 321
  employment, 302
  output (1839–1909), 296, 299–301, 312, 316
  output and employment, 324–325, 336–337
  output per mine, 317
Irrigation, value of, 72
Irving, J. Washington, 101, 115

Johnson, Arthur M., 403–404
Julihn, C. E., 315

Kaplan, David L., 133, 147, 149n, 161, 163n, 200n
Katz, Leo, 555
Kemnitzer, William J., 355n, 356n, 358n, 390n
Kendrick, John W., 8n, 90n, 132, 133, 177, 186, 189, 584, 588, 629, 643n
Kentucky, coal mining, 407, 415, 425, 427
Klein, Judith L. V., 523n
Klose, Gilbert C., 349n, 355n, 362n, 368n, 371n, 374n, 386n

INDEX 653

Knowles, James, 132
Kravis, Irving, B., 4–7
Kuznets, Simon, 3n, 6n, 8n, 13, 25–26, 31, 36, 37–39, 41, 56, 76n, 85–86, 88, 90, 127, 133, 147n, 148n, 177, 185n, 205, 206, 207–208, 526

Labor force (1800–1960), 117–210
  comparison of U.S. and U.K., 117–133
  domestic service employees, 59–60, 203–204
  education, 125–126
  historical background, 117–133
  manufacturing, 177–182
  railroads, 191–200
  teachers, 200–203
  wage rates, 40
  See also Employment
Land:
  clearing, 129, 206
  growth in value, 274
  improvements, 90, 288
  tax assessments, 243–245
Landsberg, Hans H., 132, 405n, 439n
Lanman, James H., 190n
Laspeyres price index, 97–99, 240, 584
Lavis, F., 598n, 600n, 607n
Layer, Robert, 218n, 227n, 229
Lead, 293, 294, 316, 321
  employment, 302
  output and employment, 296, 300, 326, 338
  output per mine, 317
Lebergott, Stanley, xi, 33, 60, 85, 86, 89, 117–210, 616–617, 629
Lee, Everett, 130n, 158, 160, 181
Leiper, B. M., 101, 115
Leontief, Wassily, 129, 205, 232n
Lerner, Joseph, 405n, 439n
Levine, Jules, 479n
Light, Diane, 523n
Long, Clarence, 132, 133, 139, 140n, 147, 149n
Loree, L. F., 638n
Lumber series, 71

MacDougall, Duncan M., xii, xiv, 484, 488n, 490, 491, 497–519
McGann, Paul W., 347–348, 436–437
McGouldrick, Paul F., 239–242
McGreevey, William P., 3n
Machine tools (1891–1910), xii, xiv, 497–519
  automotive demand and, 513
  Brown and Sharpe data, 497, 499–514

Census data, 485, 498
cyclical behavior, 511–512
definition, 479n, 498
foreign trade, 504, 506, 512, 519
importance of, 501, 518
markets, 506–511, 512
milling machines, 497, 500
output, 129, 497–519
Philadelphia companies, 498
price index, 482, 491
production, 501–506, 511–512
profits, 513–514
sources of data, 499, 512–513
technological changes, 491
types of, 518–519
Machines, capacity of, 443
Machlup, Fritz, 124n
Macroresearch, 88
Maeshiro, Asa, 243n
Males, employment estimates, 134, 135, 138–139, 140–142, 144, 147
Manpower-population ratio, 157–158
Mansfield, Edwin, 631–632
Manufactured producer durables, 39, 60
  component of capital formation, 17
  GNP estimates, 39
Manufacturing:
  average hourly earnings, 272
  employment, 133, 177–182
    cotton textiles, 182–186
    hand trades, 189
    iron, 186–188
    proprietors and salaried workers, 181–182
    wage earners, 177–181
  home, 74–76
  outputs and inputs, 79
  power in (1838–1919), 443–478
  value of output, 31
Marburg, Theodore, F., 57, 85, 90, 190n
Marryat, Capt. Frederick, 171–172
Marshall, J. H., 359n
Martin, Edward W., 58–59, 88
Martin, Max, 523n
Maryland, coal mining, 407
Massachusetts:
  Bureau of Statistics of Labor, 91
  industrial censuses, 93–94, 96–97, 100
  report on wages and prices (1752–1860), 93, 96–97
Mathews, R. C. O., 218n
Measurements:
  estimating missing observations, xii
  observations and, x–xiii
  significance of, ix–xi
  source material, x–xiii

Meat packing industry, 75, 79, 83
  value of output, 31
Medical services, 40, 57, 59
Menezes, Carmen, xii, 349–404
Merchant marine, *see* Navigation
Merrill, Josiah, 386
Metal mining industries, 293–348
  discovery-exhaustion sequence, 313–315
  effect of transportation on, 300–301, 321–322, 347–348
  effect on regional development, 293–294, 347
  employment, 301–307, 347, 425–427
    compared with all employment, 305–307
    compared with all mineral employment, 303–305
    by region and industry, 302–303
    statistical data, 301–302
  hydraulic mining, 316, 318
  large-scale production methods, 315–316, 348
  open-pit mining, 316
  output, 295–301, 324–330
  output and employment (1839–1909), 293–348
    capital goods part, 307, 310
    change in rate of growth, 293, 295–299
    factors affecting, 293
    by mineral product, 294
    number of mines and, 317–319
    by region, 293, 296–301, 336–339
    in the West, 296–300
  quality of deposits, 312–313
  ratio of employment to output, 307–319
    behavior, 310–312
    factors affecting, 312–319
    long-term changes, 307
    rate of change, 311–312
  regional development and, 319–346, 348
  search and development, 320–321
  statistical data, 293–295, 301–302, 347
  technological changes, 315–316
  transportation and, 300–301, 313–315, 321–322, 347–348
  world output, 294
Metalworking machinery:
  access to business records, 495
  automotive demands, 491
  capital goods, 490
  Census data, 480, 481, 485–487
  definition, 479*n*
  estimation difficulties, 480
  foreign trade, 484
  importance of, 479
  industry, 484
  interpolation of time series, 481
  made for own use, 491
  price changes, 480, 482, 490
  production, 479–495, 518
    rate of increase, 490
    real output and in current dollars, 481–484
    by regions, 482, 484
  sources of information, 480, 484–485, 491
  see also Machine Tools
Meyers, N. L., 359*n*
Microresearch, 88
Miller, Ann, 130*n*, 133, 148*n*, 158–161, 181
Milligin, Jacob, 136*n*
Milliman, Jerome, 293*n*
Mineral mining industries, *see* Metal mining industries
Mining industries, employment estimates, 159–160, 173–175, 187
Mitchell, B. R., 123*n*
Mitchell, W. C., 511*n*
Moody, Paul, 230
Morison, Samuel Eliot, 165*n*
Morrison, Chaplain W., 523*n*
Mott, Edward H., 193*n*

National accounts, xiii
National Bureau of Economic Research:
  business cycles, 222, 225
  reference cycle chronology, 511
National income:
  estimates, 205
  France, 4
  Great Britain, 4
  rate of growth, 626
National Machine Tool Builders' Association, 481, 498, 499
National Petroleum, 394
National product, 3–10
  See also Gross national product
National Refining Co., 393, 397–398
Navigation:
  employment, 134, 137, 162–172
  fishing and, 163–164
  foreign trade, 166–167
  sources of data, 162–163, 168–172
  whaling industry, 164–165
Navin, Thomas, 239*n*
Nelson, James C., 637*n*
Netschert, Bruce C., 405, 439*n*, 623*n*, 624*n*
New companies, 239–242

INDEX 655

New England:
　machine tool industry, 488, 497 ff.
　power capacity, 445, 447, 451
　railroad employment, 191–192
　textile industry, xii, xiv, 213–242
　　See also Textile industry
Nimmo, Joseph, 172
Nonfarm industries, employment, 135–136
North, Douglass C., 53–54, 61, 64, 85
Nurses, employment, 143

Oats:
　intraregional shifts, 535–540
　labor requirements, 527
　　harvest, 562–565
　　postharvest, 570, 573–575
　output:
　　growth, 541, 582
　　per acre, 527
　　per man-hour, 580, 582
　　See also Grain production
Ohio building statistics (1837–1914), 243–290
　adjusted time series, 276–278
　adjustment and deflation techniques, 243, 245, 259–260, 289
　appraisal standards, 252–259
　building costs, 255, 256, 275, 289
　building cycles, 262
　churches and schools, 244, 247–249, 250
　collected by tax assessors, 243–244
　completions, 259–260
　data evaluation, 260–278
　deficient returns, 245–247
　erosion slippage, 278, 282–284
　estimates of new building, 281
　extrapolation to 1837, 249–252
　growth in land values, 274
　industrial and commercial types, 244, 246, 278
　market values, 289
　new building, 244
　permit data, 245, 259–260, 271, 273–274, 284
　ratio of appraised to market values, 252–256, 282–284
　regression model, 265, 285–286, 290
　residential building, 243–245, 274
　taxable new building, 251
　tax-exempt construction, 244, 247–249
　use of statistics, 287–288
　validity and reliability, 245
　values per unit, 257–276, 281
　　farm residential, 260–268

　　indexes, 259
　　nonfarm residential, 268–276
　　residential buildings, 257–258, 266
　　standard of purchasing power, 252–259
　　total new building, 257–258, 281
Oil Investor's Journal, 265n
Olenin, Alic, 188n
Orton, Edward, 365n
Ouseley, William G., 174n
Output:
　current price series, 39
　of final products, 213–290
　　building in Ohio (1837–1914), 243–290
　　textile industry, 213–242
　goods flowing to consumers, 17–20
　nonmanufacturing industries, 28
　railroads, 583, 584–589
　rate of growth, 626
　　long-swings, 205–210
　　See also Production

Paasche price index, 97–98, 240, 584
Palfrey, John G., 164n
Paragon Oil Co., 393
Parker, William N., xii, xiv, 3n, 83, 85, 89, 129, 210, 523–582
Parmalee, Julius H., 637n, 638n
Parsons, A. B., 315n
Peabody, Robert R., 170n
Pearce, Congressman, 170
Pearson, F. A., 64, 180, 610n, 629
Peckham, S. F., 358n
Pennsylvania:
　coal mining, 407, 415, 417, 425, 427, 430
　price lists, 93, 96–97
　railroad employment, 193–194
Pennsylvania Railroad, 623, 635
Pennzoil, 392
Perishables, 17–20, 64, 71
　price index, 39–40
　rate of growth, 20–23
　value of output, 44
Perloff, H. S., 426n
Petroleum industry, 294, 349–404
　Appalachia, 363–364, 390, 392–393
　California, 365, 387–388, 390, 392, 396–398
　changes in structure, 390–403
　crude oil production, 350–354, 397
　　drilling costs, 363–366
　　dry wells, 355–359
　　inputs, 362–363
　　institutional factors affecting, 359–361

Petroleum industry, (*Continued*)
   new wells drilled, 355–357
   output, 355–359
   sectoral developments, 355–380
   technology, 361–362
   transportation and storage, 366–373
   exploration, 355–360
     "flush fields," 360–361
     "rule of capture," 360
   fuel oil, 374, 387–389, 397
   gasoline, 385–386, 397
   Gulf Coast, 362, 365, 388, 390, 392, 393–394
   Illinois, 390, 392, 396
   illuminating oil, 382, 383–385, 397
   kerosene, 382, 383–384
   Lima-Indiana, 365, 375, 390, 393
   lubricating oil, 386–387, 397
   major producing areas, 390
   methodological note, 401–403
   mid-continent, 365, 390, 394
   naphtha, 385–386, 397
   physical outputs, 350–355
   position in American economy, 350–355
   refined products:
     changes in relative value, 389–390
     demand for, 380–390
     factors affecting, 383–389
     foreign distribution, 382
     production, 350–354
     transportation and storage, 369–373
   refining, 373–380
     capacity, 397–398
     capital and labor, 377–379
     construction costs, 379–380
     costs, 378–379
     distillation and treating, 373–375
     inputs, 377–379
     lubricants, 374
     output record, 375–376
     technology, 375
   Russia, 369, 382, 386
   size distribution of firms, 396–397
   Standard Oil complex, 350
   structure of firms (1911), 390–403
   transportation and storage, 366–373
     bulk handling, 369–370, 372
     costs, 372–373
     facilities, 366–367, 397
     gathering lines, 367
     inputs, 370–372
     pipelines, 367–368, 371
     tank cars, 367, 371, 373
     tankers, 368–369, 371–373
   value added, 350–355
   waxes, 397
Pipelines, 367–368, 371
Pitkin, Timothy, 163n
Poor, H. V., 192n
Poulson, Barry, 3n
Power in manufacturing (1838–1919), 443–478
   capacity, 443–44
     by geographic division, 444–447
     horsepower, 443
     by industry groups, 444, 447–451
     measuring, 443
     rates of growth, 444
     steam power, 444
   Census data, 455–461
     adjustments, 462–468
   distribution by type of, 451–455
   electric, 444
     importance, 451, 455
     industrial distribution, 447
     regional distribution, 447
   industrial distribution, 444, 447–451
   internal combustion engines, 444
   manufacturing, xiv
   percentage distribution of total primary power by type, 444, 451–455
   primary power, 443
     capacity, 443–444
   prime movers, 443
   regional distribution, 444–447
   scope, sources, and methods, 445–465
     classification of industries, 461–462
     coverage, 455–461
   steam power, 444
     distribution in primary-power capacity, 451–455
     importance, 451, 455
     industrial distribution, 449–450
     regional distribution, 446–447
   steam turbines, 444, 468
   total primary-power capacity, 444–449
     by geographic division, 444–445
     industrial distribution, 447–449
   water power, 444, 449
     importance, 444, 451, 455
     regional distribution, 445, 447
Pratt and Whitney Co., 500, 504
Price deflators, 91–115
   bibliography, 112–115
   chain indexes, quality changes and new goods, 98–100
   derivation of indexes for classes of commodities, 95–96
   geographic price differences and population changes, 96–98

INDEX

nature of price data, 91–92
outline of items included in price indexes, 102–105
price relatives, 92–95
sources and notes, 100–111
Prices:
  benchmark estimates, 41
  Brady indexes, 39–41
  constant price estimates, 36, 58–60, 63
  current price series, 36, 39
    GNP final product flow, 25–62
  deflators, *see* Price deflators
  distribution of, 92–94
  effect of population changes, 96–98
  geographic differences, 96–98
  indexes, 610, 629n
    bias, 97–99
    chain, 98–100
    commodity, 39
    deflating output estimates, 97
    derivation for classes of commodities, 95–96
    1860 base, 39
    firewood index, 32, 62
    items included in, 102–105
    Laspeyres, 97–98, 99
    new qualities, 98–99
    Paasche, 97–98, 240, 584
    retail prices, 36
    wholesale, 36
  metalworking machinery, 482
  nature of data, 91–92
  new goods, 98–100
  quality changes and, 98–100
  sources of information, 100–111
  weighted averages, 95–96
Primack, Marlene, 523n, 536n
Primack, Martin, 3n, 71–73, 85, 129
Prime movers, definition, 443
Production:
  data and capital formation, 12–13
  estimates of flows of materials, 71
  for "own use," xiv
Productivity:
  changes in, xiii
  grain production (1840–1910), 523–582
  railroad sector (1840–1910), 583–646
  textile industry, 227–232, 240
Professional services, 59
Property, ratio of appraised to market value, 252–259
Providence Tool Co., 500
Pure Oil Co., 392, 394, 396

Quincy, Josiah, 169

Radford, William A., 275
Railroads (1840–1910), 583–646
  air brakes, 632, 634, 637, 645
  Arkansas, 196
  automatic couplers, 632, 634, 637, 645
  block signaling, 632
  capital-output ratio, 630–631
  capital stock, 583–584, 589–612
    checks on estimates, 609–610
    in constant and current dollars, 589–593, 612
    defects in Ulmer's estimates, 589–594, 608–609, 611–612
    estimates, 589–612
  competition, 628
  construction series, 60
  decline of, 628
  economies of scale, 630, 631, 633
  effect on mineral mining, 313–315
  1840, 197–198
  employment, 121–123, 133, 191–200, 583, 612–618
    alternative estimates, 614
    data, 612–614, 616–617
    length of workweek, 612
    productivity, 593
    receipts and, 612, 614–616
    training, 645
    wages, 637–638
  firm size, 631–632
  fuel consumption, 583, 618–626, 637
    effect of technological change, 619
    expenditures, 624, 637
    per locomotive-mile, 623, 625
    wood and coal, 418–420, 437–438, 619–624
  hauling dead weight, 641, 645
  increase in train size and load, 636–637
  innovations, 632–634
    cost savings, 636
    measuring significance of, 642–645
  inputs and outputs, 630, 645
  Iowa, 195–196
  locomotives, 595, 601, 603–605, 619, 621n, 635, 645
  Louisiana, 196
  measures of development, 584–626
    capital stock, 589–612
    fuel consumption, 618–626
    labor, 612–618
    output, 584–589
  Minnesota, 195–196
  Missouri, 195–196
  Mountains and Pacific, 197

Railroads (1840–1910) (*Continued*)
  output, 583, 584–589
    estimates of passenger-miles and ton-miles, 584–587
    factors, 628n
    rate of growth, 626, 629
    receipts and rate information, 586–588
  passenger and freight cars, 601, 635, 641
  price index, 608–610
  productivity, 583–646
    changes, 584, 645–646
    effect of air brakes, 584, 632, 634, 637, 645
    effect of automatic couplers, 584, 632, 634, 637, 645
    effect of improved rolling stock, 584, 633
    effect of innovations, 584, 642–645
    effect of steel rails, 584, 632, 633, 635, 639–641, 645
    evaluation of factors, 631–646
    index, 629
    output, input, and, 626–645
    total factor productivity index, 642–644
  profits, 646
  rate differences, 646
  safety features, 636
  standardization, 633
  statistical methods, 583–584
  steel rails, 593, 600n, 619n, 632, 633, 635, 639–641, 645
  tank cars, 367
  technological change, 633–642
  trackage and equipment, 594–608
  train speed, 637–638
Randall, S. S., 202n
Rasmussen, Wayne D., 83–84, 90, 179, 622
Real estate, 257, 266, 268–276
Records:
  state and local, 281
  study of business, 214–215, 216n, 233
Reed, Congressman, 170–172
Rees, A., 272
Regional economic development:
  coal mining, 415
  effect of mineral mining on, 319–346
  export base, 319–320, 347
  metal mining industries and, 319–346
Reid, Margaret G., 273–274, 286
Rein, Selma, 293n
Rents, 21, 40, 57–59, 63
Residential dwellings, 257, 266, 268–276

Richey, H. G., 125n
Riem, J., 274n
Ringwalt, J. L., 601n
Ripley, Samuel, 183n
Robertson, Ross, xii, xiv, 479–495, 518
Rockefeller Foundation, 243n
Roe, Joseph Wickham, 479n, 487n, 497, 499, 500, 510
Rogin, Leo, 570n
Rosenberg, Nathan, xiii$n$, xiv, 221n, 491
Rossiter, W. R., 160
Rostow, W. W., 502
Rural population, estimates, 134–135
Ruschenberger, William S. W., 171n, 172
Russel, R., 229n
Russia, petroleum industry, 369, 382, 386

Salt mining, 175
Sampson, Paul, 243n
Sandberg, C. P., 639n
Savings, domestic rates, 14
Schilling, Don C., 523n
Schmookler, Jacob, 632n
Schoolcraft, Henry, 174n
Schools, construction of, 247–250, 276
Schumpeter, Joseph A., 210, 583, 586n
Schurr, Sam H., 293n, 405, 427n, 439n, 623n, 624n
Scott, Franklin D., 170n
Seaman, Ezra, C., 4–7, 33, 44, 57–59, 61, 75, 76, 85, 90, 253
Sellers, William, Company, 495n, 498
Sellew, William H., 633n, 640n
Semidurables, 71
  price index, 39
  value of output, 44
Servants and laundresses, 145
Services, 17–20, 85–88
  components, 21, 86
  1834-59, 57–60
  estimates of value, 62
  flow of material inputs, 87
  fluctuations in rate of growth, 20–23
  GNP estimates (1869–1909), 37
  nonrent, 59–60
  price index of, 17–20, 40
  value of output, 28, 62
Sewing machines, 497, 512, 519
Seybert, Adam, 163, 164n, 170n, 171
Shannon, C. W., 365n
Sharpe, Lucien, 500
Shaw, William Howard, 25–26, 28–29, 36, 37n, 39–40, 41–43, 45, 52, 54, 55, 79, 80, 85, 90n, 175, 177, 185n, 601n
Shen, T. Y., 232

INDEX 659

Shipping, employment, 166, 172–173
  See also Navigation
Shotwell, Larry, 3n
Silliman, Benjamin, 383
Silver, 293, 312, 313–314
  employment, 302
  output, 296, 300–301, 317–318
  output and employment, 328–330, 339
Simon, Matthew, 39
Simpson, Herbert D., 283n
Slaughtering and packing, 75, 79, 83
  value of output, 31
Slave population, 58, 134–135, 137, 140n, 141, 143, 144, 146, 150–152
Sloan, Samuel, 101, 114
Smith, Walter B., 222–223, 225, 227n
Snyder, Carl, 586n
Southern states:
  agriculture, 130–131
  power capacity, 445, 447
Spence, Mary Lee, 523n
Spencer, Milton H., 490n
Spencer, Vivian E., 427n
Standard Oil group, 350, 368, 371–372, 379, 390, 392–398
Standardization, 633
Steam engines, 443
  Census data, 462–468
Steam power, 444, 451–455
  capacity, 444
  importance of, 451, 455
  industrial distribution, 449, 450
  regional distribution, 446
Steam turbines, 443, 468
  Census data, 468
Stearns, Onslow, 192n
Steel industry:
  coal consumption, 418, 420–421
  production, 421
Stettler, H. Louis, III, xii, xiv, 213–242
Stevens, W. H. S., 592n
Stover, John F., 634
Strassman, W. Paul, 479n, 518–519
Sun Oil Company, 393, 397–398

Talbert, Lewis, 188n
Teachers, 60, 124–126, 133, 143, 145, 200–203
Technological change:
  effect on mineral mining, 315–316
  measurement of, 618n, 619
  railroads (1840–1910), 583–646
  value-added ratio, 619
Temin, Peter, xiiin
Texas Company, 379, 384, 397–398

Textile industry, 26–27, 75–76, 127–128, 213–242
  analysis of fluctuations, 221–227
    long swings, 225–226
    specific cycles, 222–225
  capital and labor productivity, 230, 240
  Census data, 213–214
  employment, 122, 123, 230, 240
  estimates, 214–221
    adjustment to regional levels, 219–221
    output measure, 218, 240
    regional approach, 214
    uniform accounting period, 216–218
  firms, 215
    Massachusetts-type, 215, 220, 227
    Rhode Island-type, 215–216, 220
  formation of new companies, 239–242
  New England (1825–60), xii, xiv, 213–242
  productivity change, 227–232, 240
    "best-practice techniques," 231–232
    due to technological advances, 229–230, 232
    output per spindle, 215n, 230, 231n, 233
  use of business records, 214–215, 216n, 233
  value of output, 29, 30
Thomas, Brinley, 205–210
Thomson, Edgar, 632n
Tidewater Oil Co., 392, 396–398
Tidewater Pipe Co., 368
Time series, 205, 206
  interpolation of, 481
  long-swings, 205–208
Tobacco products, 57
Tool, Kent, 479n
Tools, prices, 94
Total factor productivity index, 642–644
Towne, Marvin W., 83–84, 90, 179, 622
Trade markups, 36, 56–57, 85
Trades, employment, 133, 135, 188–190
Transportation, 79
  effect on mineral mining, 300–301, 313–315, 321–322
Tratman, E. E., 640n
Trout, L. E., 365n
Tryon, F. G., 315n
Turner, W. V., 636n

Ulmer, Melville, J., 37–38, 40, 85, 589–594, 600n, 608–609, 611–612
Underdeveloped nations, 130–131
Union Oil of California, 397–398
Union Petroleum, 392, 396–398

United Kingdom:
  domestic service, 205
  employment, 117–133
  long-swings in capital formation, 207
  role of agriculture, 205–206
United States:
  Bureau of the Census:
    adjustments, 140, 147, 149, 154
    benchmark years, 41–55
    coal mining, 408–409, 413
    employment statistics (1800–1960), 25–26, 80n, 132, 133, 135, 139–140, 142, 144, 146, 147, 149, 154–156, 162, 168, 169, 178
    industry data, 41–43
    metal mining industries, 295
    power in manufacturing, 455–468
    product data, 42
  Bureau of Labor Statistics, 132–133, 610, 629n
  Bureau of Wildlife and Fisheries, 133
  Department of Agriculture, 29–30, 31, 33, 132
  Forest Service, 32
  Geological Survey, 295
    coal mining, 408–409, 413, 425
    petroleum data, 358n
  National Archives, 140n, 154
  Office of Education, 133
  Treasury Department, 30, 52–53, 64
    data on navigation employment, 162, 170–173
    foreign trade data, 43
Urban population:
  estimates, 134–135
  females, 136
  white males, 135–136
Uren, Lester C., 358n

Virginia, coal mining, 407, 415, 417, 425, 427

Wage earners, 177–181
Wage rates, 40, 72
Walker, Census Commissioner, 25–26
Ware, Caroline, 215n
Warren, G. F., 64, 180, 610n, 629
Water power, 444, 451, 455
  capacity, 444
  industrial distribution, 449
  regional distribution, 445, 447
Water wheels, 443
Waters-Pierce Co., 393

Weeks, Joseph D., 92–93, 95–96, 100, 112
Wellington, Arthur, 598n, 600n, 601n, 603n, 621n, 623n, 625, 640n, 645
Wells, David A., 122n, 171
Western states:
  agriculture, 131
  effect of mineral mining on, 319–346
  power capacity, 445, 447
  surveys of, 320
Whaling, employment data, 164–165
Wheat, 64
  intraregional shifts, 535–540
  labor input per acre, 527
  labor requirements:
    harvest, 562–565
    postharvest, 570, 573–575
  output:
    growth, 541, 582
    per acre, 527
    per man-hour, 580, 582
  productivity growth, 523
  See also Grain production
Whelpton, P. K., 133, 139, 148, 149–150, 156, 159
Whipple, John, 182
White families, 134–135
Whitney, J. D., 294
Wickens, David L., 271, 275, 276, 284
Williams, R. M., 275n
Williamson, Harold F., xii, 349–404
Wing, George, 479n, 489–490, 497n, 499n, 518
Winnick, Louis, 270n, 271–276, 286
Wisconsin Urban Program, 243n
Wolman, Leo, 613n
Wood:
  consumption, 418, 420–422, 437–438
  shift to coal, 405, 418–420, 437–438, 619–624
Woodbury, Robert S., 479n
Woolens and worsteds, 43
  See also Textile industry
Wright, Carroll, 185n
Wrigley, H. E., 358n, 364n

Young, Edward, 195n

Zarley, Arvid M., 497n
Zinc, 293, 294, 316, 321
  employment, 302
  output and employment, 296, 301, 326–327, 338
  output per mine, 317